ENVIRONMENTAL PROBLEMS and HUMAN BEHAVIOR

SECOND EDITION

Gerald T. Gardner • Paul C. Stern

Pearson
Custom
Publishing

Printed in the United States of America

13   16

Please visit our web site at www.pearsoncustom.com

ISBN 0–536–68633–5

BA 995649

PEARSON CUSTOM PUBLISHING
75 Arlington Street, Suite 300, Boston, MA 02116
A Pearson Education Company

Contents

Environmental problems—the greenhouse effect and global climate change, depletion of the Earth's protective ozone layer, the large-scale clearing of tropical rain forests and the resulting loss of genetic diversity, regional air pollution, water pollution, and others—are the result of human behavior. Everyone recognizes this fact, but few use the science of human behavior to understand the roots of the problems and create strategies for solving them. This book helps establish important new links between environmental science and behavioral science. It develops a framework for addressing key questions about human behaviors that harm the environment, summarizes knowledge from psychology and related fields about these behaviors, and uses that knowledge to point the way to realistic solutions.

The key behavioral questions include these: Do the roots of environmental degradation lie in a system of ethics, values, and beliefs that gives the environment second-class status compared to human desires? Are there inherent human behavioral tendencies—for example, innate selfishness—that contribute to environmental degradation? How are individual behaviors that harm the environment controlled by the situations people find themselves in? How can we intervene to change these behaviors? Is education an effective way to encourage proenvironmental behavior? Which changes in individual behavior are likely to have the greatest proenvironmental impact? Are global and regional ecosystems too complex for government policymakers and others to understand and manage? Are members of the public and policymakers likely to significantly misperceive the risks posed by looming and unprecedented environmental problems?

This book develops a framework for addressing these questions, drawing on behavioral theory, real-world case studies, field experiments, and other evidence. Because its central focus is individual behavior, it draws most heavily on concepts from social, cognitive, and behavioral psychology. However, it puts behavior in the context of the economic, institutional,

and policy forces that shape it and emphasizes arenas where individual action makes a real difference to the natural environment. The result is an interdisciplinary treatment, rooted in behavioral science but addressing practical issues of environmental policy.

The book is written at a level suitable for undergraduate students in psychology, social science, and environmental studies and science. Because of the multidisciplinary audience, we assume little specialized knowledge in any one discipline. Chapters begin with vignettes that illustrate the major problems or issues of the chapter; we use these concrete examples to set the stage for discussions of theory and evidence in the rest of each chapter. We introduce key concepts from various fields as needed, either in boxes or woven into the text. Rather than exhaustively reviewing the literature on each topic, we selected examples that illustrate major principles and present them in enough detail to allow the reader to imagine the place, time, and social context and consider their general applicability.

This book fills an important gap in the four environmental psychology texts now in print, addressing issues of great interest to students that are given only cursory treatment in these texts. The four current texts devote no more than a chapter to environmental problems and human behavior, but focus, instead, on the impacts of physical environments, such as buildings, offices, and classrooms, on behavior. This book advances the environmental-problems branch of environmental psychology by presenting a broad, new conceptual framework, reviewing the limited quantities of directly relevant psychological research, integrating key insights from psychological research on nonenvironmental subject matters, and incorporating relevant material from other disciplines. Professors who teach environmental psychology courses will find this book useful as a primary or supplementary text. We hope the book will also encourage behavioral and social science professors who do not now teach courses on environmental problems to develop new courses in the field. Potential new adopters include professors who teach courses in social ecology, applied psychology, applied social psychology, social problems, and others.

This book also fills an important gap in environmental studies and environmental science curricula. Entry-level texts in these fields are broad and multidisciplinary, and usually cover the economic, political, and ethical aspects of environmental problems. For example, Miller's best-selling text, *Living in the Environment,* devotes one chapter to each of these topics. However, extant texts cover little, if any, of the behavioral science material covered in this new book. Because there is growing awareness among environmental scientists that the behavioral components of environmental problems are vitally important but heretofore neglected, many professors of biology and other sciences who teach courses in environmental studies and science will find the book attractive as a supplementary text. The following quote from a publication of the National Academy of Sciences (Silver and DeFries, 1990) gives evidence of this new awareness:

> *No longer is it sufficient to explore only the physical dynamics of the earth system. This effort, daunting in itself, may be dwarfed by the effort to decipher the confounding behavior of* Homo sapiens, *the planet's most powerful inhabitant. . . . For the first time, scientists from disciplines ranging from geochemistry to ecology are realizing that human action is the critical element in their studies. So potent is the human impact on the earth system that knowledge of physical processes ruling terrestrial or atmospheric change will be incomplete until scientists better understand the human dimensions of that change.*

We believe that our book takes a forward step toward understanding the human dimension, and we hope it will encourage its readers to take further steps in that direction.

PREFACE TO THE 2ND EDITION

This edition updates certain key material in the first edition. All information on the state of the environment has been updated including global population growth, global warming, ozone-layer erosion, and human interference with the nitrogen cycle. Similarly, all information on U.S. public concern about the environment has been updated including the socio-demographic correlates of concern and variations in the level of concern over recent years. There is an updated discussion of environmental attitudes, values, beliefs, and worldviews and also of the *tragedy-of-the-commons* process. The analysis of the U.S. energy system has been updated. Smaller changes appear throughout the text.

The authors are grateful for the support and help of Ann, Sue, Sadie, and Bob, and that of other family members, friends, and colleagues. We are also grateful to Tom Dietz for valuable assistance with our discussion of environmental values, attitudes, and action, and to Riley Dunlap for vital help with public opinion data and data on environmental values and worldviews. We also thank Barbara Essenmacher and Christy Rybak of Pearson Publishing for unstinting help and support. Finally, we would like to thank readers of the first edition who gave us enthusiastic feedback and encouragement.

ACKNOWLEDGMENTS

Many fine colleagues, students, and friends made substantial contributions to this book. We deeply appreciate the unstinting help and support of Peter Amann, who read and critiqued the entire manuscript as well as the original Prospectus for the book. Also, Phyllis Morris, Orin Gelderloos, Olga Klekner, and Gloria Trudell read and critiqued the entire manuscript or large portions of it and offered invaluable advice and encouragement. Camille Ward, Don Ward, Cynthia Tchilinguirian, John Perkins, Willett Kempton, Ray DeYoung, students in Gerald Gardner's environmental psychology classes at the University of Michigan–Dearborn, and the students in Neil Weinstein's environmental psychology class at Rutgers University provided generous and helpful feedback on blocks of chapters. Stephen Kaplan, Paul Slovic, Bill Wolf, Richard Erickson, Barry Bogin, Michael Akiyama, Bonnie McCay, and Vikas Mittal carefully reviewed chapters. We also thank John Presley, Noriko Kimachi, C. Stern, K. King, and colleagues in the Behavioral Science Department at U. of M.–Dearborn, who were extremely supportive. Ann Epstein and Becca Epstein provided very special help and encouragement. Shelley Perlove's help and support were also vital. Our families and friends deserve credit for putting up with a long and time-consuming writing process. Without the generous help, encouragement, and enthusiasm of all of these people, this book would not exist.

We also acknowledge the debt we owe to our participation several years ago in Yale University's Institution for Social and Policy Studies interdisciplinary Program on Energy and Behavior. Our research, writing, and general outlook profited enormously from interactions with our colleagues in the program, including Leroy Gould, Stan Black, Leonard Doob, Charles Lindblom, Charles Walker, Guy Orcutt, Jan Stolwijk, Joseph Morone, Edward Woodhouse, Don DeLuca, Bill Freudenberg, and Bill Burch. We are similarly indebted to the National Research Council, and particularly the members of the Committee on Behavior and Social Aspects of Energy Consumption and Production, and the Committee on the Human Dimensions of Global Change, for educating us to think about environmental problems in a more balanced way than our training as psychologists would probably have allowed. We are also indebted to our research colleagues, particularly Tom Dietz, Linda Kalof, and Gregory Guagnano, for their contributions to our thinking.

PART I

INTRODUCTION

CHAPTER 1

THE EARTH'S ENVIRONMENTAL PROBLEMS AND THE ROLE OF HUMAN BEHAVIOR

I. Chapter Prologue

II. Introduction

    Box 1-1: Human Activities and the Global Nitrogen Cycle
    A. Focus of This Book: The Behavioral Dimensions
    B. Format and Organization of This Book
    C. More on the Role of Individual Behavior

III. More on Global Environmental Problems

    A. Global Warming
    B. Ozone Layer Depletion
    C. Rapid Loss of Tropical Forests and Loss of Genetic Diversity
    D. Destruction of Ecological Capital
    E. Driving Forces That Underlie Environmental Problems
    Box 1-2: More on Momentum in Growth Processes

IV. Reasons for Guarded Optimism

CHAPTER PROLOGUE

For the first time in history, in July of 1969, a manned spacecraft left Earth and landed on another celestial body—the moon. Millions of Americans watched their televisions raptly and saw astronaut Neil Armstrong climb down the ladder of the *Apollo* lunar module and take the first human step on the lunar surface. The photos taken by the astronauts of the moon were fascinating. But perhaps more striking were the photos they took from the moon of the Earth. Our Earth was breathtakingly beautiful. The

Earth as Seen from the Moon
(Courtesy of NASA)

colors of its surface were vivid and distinct: Forested areas were dark green; deserts were red and tan; the poles, snow-capped mountains, and cloud cover were white; but the main color was the deep blue of the Earth's oceans.

From these and other photos taken from space it was clear as never before that our planet was a very small, beautiful island in a vast sea of stars. Our planet was our only home. It was alone in an otherwise lifeless solar system, or perhaps even lifeless universe. We came to understand more vividly than ever before our dependence on the Earth's natural systems and the critical importance of protecting them.

INTRODUCTION

Even before the *Apollo* lunar landing, public concern about the environment had been growing. The modern U.S. environmental movement began in the mid-1960s, drawing inspiration from Rachel Carson's groundbreaking book on pesticides, *Silent Spring* (1962) and the writings of Aldo Leopold (1949) and others. Public concern about environmental problems, measured in public opinion surveys, surged dramatically in the late 1960s and early 1970s (Dunlap, 1991). This sharp increase had several causes: Continuing reports by scientists about new pollution problems, extensive media coverage of these reports (e.g., in-depth coverage on national TV news shows), media publication of the space-based photos of Earth, and also the effects of the first Earth Day in 1970. Earth Day 1970 was the largest educational event on the environment that had ever taken place. In hundreds of American schools, colleges, and universities, thousands of people learned more about air pollution, water pollution, the threats of pesticides, and other problems associated with human misuse of the environment (DeBell, 1970). As a result of all these influences, public concern about the environment was so great that legislation to create the Environmental Protection Agency passed the U.S. Senate in 1970 without a single dissenting vote.

The American public's new environmental consciousness was not, as it turned out, a passing fad. Although opinion polls revealed some lessening of concern during the mid- and late 1970s, the polls showed a steady increase during the 1980s and into the 1990s. As a result of this strong public consciousness and concern, many efforts to combat local and regional environmental problems were mounted, and many were successful. Thus, levels of several air and water pollutants decreased in U.S. cities. However, in the thirty or so years since Earth Day 1970, new, largely unanticipated, and more ominous environmental problems have emerged: For the first time in history, human activity is beginning to have a major negative effect on global, rather than just local and regional, environmental systems. Human actions are now disrupting global natural systems upon which all life depend and that have operated for millions of years without human interference.

This global effect derives from the large size of the Earth's human population, now more than 6.1 billion, the characteristics of modern technology, and the intensity with which humans use it. For example, the combustion of vast quantities of fossil fuels to heat homes, run factories, and power cars and trucks has released enough carbon dioxide into the atmosphere to increase the strength of the "greenhouse effect," a natural process that traps solar heat. The result is an increase in the earth's average temperature—global warming. Most atmospheric scientists believe that average global temperature has already increased and that further increases are inevitable, though they are unsure of the

magnitude and timing. However, even small increases— only a few degrees Fahrenheit—could alter rainfall patterns and disrupt agriculture in many parts of the world, increase sea level enough to inundate many of the world's coastal cities, and change global storm patterns, possibly causing serious damage along coastlines and elsewhere (Silver and DeFries, 1990).

Human activity, also, is eroding the layer of ozone in the stratosphere which shields the Earth from solar ultraviolet radiation. This erosion results from the release of freon and other chemical gases from an ever increasing number of air conditioners and refrigerators, and from industrial processes. The damage to the ozone layer has already caused an increase in skin cancers, and it may impair human immune system functioning. It may also interfere with the growth of important food crops (Miller, 1994).

Humans are clearing huge tracts of tropical rain forest, destroying the habitats of plants and animals, and causing the extinction of species at record rates. Because of human activity, global fisheries are declining, underground drinking-water aquifers are being depleted, and agricultural topsoil is being lost to erosion (Brown, 1995). Human population continues to increase exponentially. Even the "medium" United Nations projection, which assumes an early end to exponential population growth, holds that the Earth's population will grow to about 9.3 billion by the year 2050 (U.N. Population Fund, 2001).

In a later section of this chapter (titled "More on global environmental problems") we say more about global warming, the ozone depletion problem, and the extinction of species caused by the clearing of tropical forests. We also present an analysis of the main forces and influences that underlie global environmental problems, namely: the size of the human population, the affluence of that population, and the environmental impacts per unit of affluence. Finally, in Box 1-1, we describe a newly discovered potential environmental threat—disruption of the global nitrogen cycle. Like the damage to the ozone layer that we discussed above, this problem was unanticipated and has taken scientists by surprise. It is reasonable to wonder if there are yet other unanticipated and potentially catastrophic human-produced environmental changes that may occur in the future. Humanity is conducting a

Box 1-1

Human Activities and the Global Nitrogen Cycle

The Earth's environment is, despite advances in science, still far more complex than our understanding of it. This fact was illustrated when the hole in the ozone layer was discovered in the atmosphere over Antarctica in the 1980s. Scientists, for whom ozone depletion had been only a theoretical possibility, then began working hard to understand the phenomenon, its causes, and its likely future course. Given our limited understanding, and the unprecedented scale of human activity on the earth, we should not be surprised to live to hear about more such environmental surprises.

One surprise of potentially catastrophic proportions, first brought to light in 1994, concerns human interference with the global nitrogen cycle. Nitrogen is essential to life on earth, as it is an ingredient in all proteins. It is a highly abundant element, but according to scientific understanding as of the 1990s, only a small fraction—less than one tenth of one percent—of the earth's nitrogen is in the inorganic forms believed to be required by the great majority of living organisms. The rest consists of the nitrogen molecules that make up almost 80 percent of air and are dissolved in the ocean. Before the advent of human technology, living organisms got their needed nitrogen in three ways: from the action of a few species of nitrogen-fixing bacteria (some in nodules on the roots of plants in the bean family [legumes]), from algae that directly convert nitrogen molecules into soluble ammonium ions, and by lightning, which converts some atmospheric nitrogen into soluble nitrates, which fall in rain. The global nitrogen cycle was closed by some other bacterial species, which convert nitrogen compounds back into nitrogen molecules. In preindustrial times, this cycle was believed to have involved no more than 140 million metric tons (MMT) of nitrogen per year (Kinzig and Socolow, 1994; Ayres, Schlesinger, and Socolow, 1994; Vitousek et al., 1997).

Human activity now fixes (alters elemental nitrogen into compounds that organisms can use) perhaps as much as 210MMT of nitrogen per year—about eighty in the synthetic production of nitrogen fertilizers for crops, forty by planting leguminous crops and thereby increasing the total amount of nitrogen fixed by bacteria in crop roots, twenty by burning fossil fuels and releasing nitrogen from them, forty by burning wood and crops, twenty by clearing land, and ten by draining wetlands, all of which activities release nitrogen that had been stored in organic matter. All these kinds of human activity are expanding rapidly. Nitrogen fertilizer use, for example, has increased almost tenfold since 1960. It is expected to continue to grow because of the demands of a growing population for food: Most of the additional food will have to be produced by using existing land more intensively, which means increased fertilization. In all, the rate of human nitrogen fixation is expected to increase rapidly over the next fifty years if present trends continue, so that it may be several times the preindustrial rate.

What are the environmental implications of doubling the prehuman rate of nitrogen fixation, or of increasing it further by a factor of up to nine? A number of very disturbing possibilities appear to follow from current scientific knowledge. Nitrogen compounds will flow from crop fields into other terrestrial and aquatic ecosystems and rapidly increase plant growth there, with unknown effects on the ecological balance of these systems. Nitrogen fertilization increases the acidity of soils, making some nutrients, such as calcium and magnesium, rapidly available for plants. What plants do not use quickly leaches out of the soil and is unavailable for future crops. Some excess nitrogen enters the atmosphere as nitrous oxide, which contributes to the greenhouse effect and to ozone depletion. And there are potential impacts on human health from nitrates in water supplies lies and from nitrogen oxides in air, which are chemical precursors to smog. Some scientists expressed great confidence in predictions of such consequences in the late 1990s (Vitousek et al., 1997). But there are remaining uncertainties that may lead to serious rethinking and re-estimation of the consequences of human disturbances of the nitrogen cycle. A report released early in 2002 (Perakis and Hedin, 2002) suggested that preindustrial forests used nitrogen from different sources from those used by most postindustrial plants. This finding suggests that some forests have been less dependent on inorganic nitrogen than crop plants are and that human activity may already have changed forest ecology fundamentally from its preindustrial state. The implications for the future are still being debated. A comment from 1994 remains apt: "the changes being created by humans today may be too rapid for ecosystem adaptation to outpace ecosystem damage on time scales relevant to humans" (Kinzig and Socolow, 1994, pp. 30–31). In sum, human interference with the global nitrogen cycle is a major environmental surprise. It looks to be an unpleasant one, but it is still too early to be certain.

crude, giant "experiment" on its natural environment and cannot afford to fail. We have only one Earth.

Focus of This Book: The Behavioral Dimensions

Many books have been written about global environmental problems like global warming, ozone layer damage, and the loss of genetic diversity, as well as other important environmental problems that occur at local and regional levels. However, this book's coverage and orientation are unusual. Rather than examine the chemistry, biology, meteorology, physics, ecology, or economics of environmental problems, as do other books, this book explores the behavioral dimensions of these problems. The book repeatedly asks: What does our knowledge of human behavior, both individual and collective, tell us about the causes of environmental problems and about ways to lessen or solve these problems? Our goal in the book, more specifically, is to survey the most important theoretical and empirical contributions that psychology and the allied behavioral and social sciences can make to the understanding and solution of environmental problems.

The behavioral components of environmental problems have, in the past, been largely ignored by natural and physical scientists, engineers, and government policymakers. Only recently has the importance of these components been widely recognized. As Silver and DeFries (1990) note: "For the first time, scientists from disciplines ranging from geochemistry to ecology [who are doing research on environmental problems] are realizing that human action is the critical element in their studies [p. 47]." After all, environmental problems are ultimately traceable to human activities that consume natural resources and generate pollution. These activities include people's purchase and use of certain goods and services (e.g., autos, air conditioners, furnaces, air travel) and the manufacturing and commercial operations that produce and distribute these goods and services. To quote Harvard University's William Clark, "Ultimately, it is certain patterns of human behavior that lead to environmental degradation, and other patterns that result in sustainable development (quoted in Silver and DeFries, 1990, p. 48)."

However, though many scientists now recognize human activities as the cause of serious environmental threats, very few scientists study those human activities. Environmental scientists do not often ask the critical questions: What explains the human actions that are disrupting the environment? How can those actions be changed? And when they do ask such questions, they often draw conclusions based on their intuitive understanding, as if understanding human behavior does not require the same careful methods of study needed to understand ecosystems or climate—experimentation, mathematical modeling, and the other systematic tools of science. As a result, much of what we hear about the human causes of environmental problems and many of the proposals for solving them are based only on plausible but untested, and often misleading or mistaken, presumptions.

Myths about Human Behavior and Environmental Problems. The work we cover in this book strongly suggests that many intuitively compelling assumptions about human behavior, the role of human behavior in global and regional environmental problems, and the implications for solutions, are misleading, mistaken, or incomplete. In Table 1-1, we list 10 widely held beliefs about human behavior and environmental problems, for example, that strong financial incentives are sufficient to encourage proenvironmental behavior or that seeing frightening images of environmental disasters will lead people to take these problems more seriously. Each one of the entries in Table 1-1 is plausible, but as you will see throughout this book, on closer examination each turns out to be seriously misleading. This book examines these beliefs in the light of the evidence to reveal what is right and wrong about each and to develop a better, scientifically based analysis of the causes of environmentally destructive behavior and of the possibilities for changing it.

Format and Organization of This Book

While the 10 myths in Table 1-1 provide a useful and dramatic overview of this book, the book is not organized around them. Instead, it is based on a systematic and, we hope, coherent analysis of the behavioral

TABLE 1-1 Myths about Human Behavior and Environmental Problems

1. Radical proenvironmental shifts in Western values, ethics, and religions will solve regional and global environmental problems. (Chapter 3)
2. Educating people—changing their attitudes and providing them with information—is an effective way to change their behavior in a proenvironmental direction. (Chapter 4)
3. Environmental degradation is the result of innate human egoism or selfishness. (Chapter 5)
4. Strong financial incentives (e.g., subsidizing 90 percent of the cost of energy conservation upgrades for private homes) are sufficient and effective ways to encourage proenvironmental behaviors. (Chapter 5)
5. Local, community-based resource management programs are not effective in modern, urbanized countries. (Chapters 6 and 7)
6. People almost always misperceive the likelihood of major hazards, including possible future environmental disasters. (Chapters 8 and 9)
7. Showing people vivid images of environmental disaster (e.g., pictures showing how past civilizations collapsed due to environmental mismanagement) is an effective way to get people to take global environmental threats more seriously. (Chapters 8 and 9)
8. The most effective thing individuals can do to save energy and lessen pollution is to use existing equipment less intensively; for example, to set their home thermostats lower in the winter, drive their autos more conservatively, avoid unnecessary trips, and so on. (Chapter 10)
9. "People start pollution; people can stop it." (We discuss this myth in a section below and in Chapter 10.)
10. Scientific breakthroughs, human ingenuity, and the methods of problem solving now used by large governments are sufficient to understand global environmental systems and to avert environmental disasters. (Chapters 11 and 12)

dimensions of global and regional environmental problems. We begin the analysis with an in-depth look (Part II—Chapters 2–7) at the most widely cited general model of the role of individual behavior in environmental problems: Garrett Hardin's (1968) "tragedy of the commons" (TOC) conception. In TOCs, the consumption of a natural resource by many self-interested individuals who have unrestricted access to the resource inevitably leads to the resource's destruction. TOCs may be seen as underlying all of the resource-depletion problems—and also the pollution and population problems—the world now faces. In Chapter 2 we critique Hardin's approach and then describe four methods that have been used in various cultures for centuries to encourage prosocial individual behavior and that might be applied to avert TOC situations—religious and moral controls on behavior, efforts to educate and change attitudes, government laws and incentives, and certain small-group/community management arrangements—and briefly outline their application to today's environmental problems. We examine each method in detail

in its own chapter (Chapters 3–6). We review historic examples of each method, the method's overall record of success and failure, and key factors that appear to influence its effectiveness.

In Part III (Chapters 8 and 9) we take a more bird's-eye view and examine some very broad and general human behavioral predispositions that might aid or impede the four behavior change approaches reviewed in Part II. In Chapter 8, we explore the possibility that humans have genetic behavioral tendencies that were shaped during the Stone Age but that are maladaptive under the space-age conditions of modern life, and that could impede efforts to encourage proenvironmental behavior. In Chapter 9, we explore psychological processes that may cause humans to significantly underestimate the magnitude of some environmental hazards, overestimate the magnitude of some others, and, as a result, fail to take appropriate action even when encouraged to do so via the methods of Part II (Chapters 2–7).

In the last part of the book, Part IV (Chapters 10–12), we examine environmentally relevant human behavior in the broad context of the ecological, physical, political, and economic systems in which the behavior takes place. In Chapter 10, we outline analytic methods for targeting those individual and household behavior changes that have the greatest potential to conserve resources and/or lessen environmental pollution. We use these methods to analyze U.S. energy problems, U.S. litter/solid waste problems, and the greenhouse effect and global climate change.

In Chapters 11 and 12, we examine some possible discrepancies between the characteristics of the human mind and human institutions, on the one hand, and, on the other, the nature of the highly complex systems with which humans interact. We begin, in Chapter 11, with sociologist Charles Perrow's (1984) controversial claim that certain complex technological systems that can damage the environment (e.g., nuclear power plants) are inherently prone to serious and unforeseeable accidents. We then evaluate engineer Jay Forrester's (e.g., 1971, 1987) claim that humans cannot successfully understand and manage large cities, countries, and regional and global ecological systems because these systems are too complicated for the unaided human mind to comprehend, and also because these systems often work in the reverse of the way human intuition would predict. We go on to discuss social mechanisms and institutions that humans have used throughout history to manage systems too complicated for the human mind to comprehend. In Chapter 12, we conclude that these mechanisms and institutions can no longer be counted on to work, because the environmental impacts of human activity now involve long time delays, sharply exponential growth processes, and the possibly catastrophic and irreversible consequences of policy errors. We conclude by presenting some strategies—including learning from computer simulations of complex systems and strategies for slowing the rate of human impact on environmental systems—that might help *Homo sapiens sapiens*[1] live in harmony with our environment into the twenty-first century and beyond.

More on the Role of Individual Behavior

As we emphasized above, all of today's regional and global environmental problems are traceable to

human actions. However, though human actions play the central role, that role is *not* confined to the actions of individuals. Thus myth 9 in Table 1-1, "People start pollution, people can stop it"—a well-known slogan from an antilittering campaign sponsored by the container industry several years ago—is seriously misleading because it implicitly equates human actions with individual behavior. It blames individuals for pollution and suggests that by changing their daily behavior they can protect the environment.

In fact, most pollution is caused by organizational behavior, not individual behavior. As we show in Chapter 10, the majority of energy use, releases of water and air pollutants, and many other environmentally destructive activities are traceable to the acts of corporations and governments, not individuals and households. It follows that changing the resource-using and polluting behaviors of individuals, even if perfectly effective methods could be found, would not eliminate most pollution. We do not conclude that there is little individuals can do to protect the environment, or that it is a waste of time to try to change individual behavior. Many proenvironmental individual behaviors *do* significantly lessen pollution and conserve energy. In addition, individuals can take effective collective action—usually in the political system—so as to encourage proenvironmental changes in government policies and in the behavior of corporations. Further, social and organizational/industrial psychologists have research findings and theories that can be of some use in understanding "large actors" (corporations and governments) and encouraging proenvironmental changes in these actors.

But, psychologists—whose focus of study is the behavior of individuals and small groups—can shed light on only some of the causes and possible solutions of environmental problems. Environmental problems are multifaceted and multidimensional. Their full understanding requires insights from a broad range of disciplines, as we noted earlier in this chapter, including chemistry, biology, physics, ecology, meteorology, engineering, political science, history, law, business, and economics, as well as from psychology and the other behavioral sciences. Thus, this book deals with only some important, and heretofore neglected, facets of environmental problems.

MORE ON GLOBAL ENVIRONMENTAL PROBLEMS

Before we begin our formal discussion of the behavioral dimensions of environmental problems (Chapter 2), we complete our review of the nature and severity of the three global problems we described earlier in the chapter. These problems are ones judged by science advisers to the U.S. Environmental Protection Agency as most serious, based on the potential magnitude of environmental damage they might cause and the irreversibility of that damage.

Global Warming

Some readers may know that the five hottest years on record occurred in the 1990s (Miller, 2002). These warm years are probably harbingers of things to come. As we mentioned above, most atmospheric scientists believe that a human-produced change in global climate, of which an increase in global temperature is one indicator, has already begun, and that further changes are inevitable. These scientists can't predict exactly the severity of these changes or their pattern over time.

Global warming results from an increase in the strength of the greenhouse effect. The greenhouse effect is a natural process that works roughly this way: Certain gases in the atmosphere act like the glass windows of a greenhouse, allowing sunlight through but trapping the resulting heat and reflecting it toward the ground. Carbon dioxide is a main greenhouse gas, and its atmospheric concentration has increased significantly. Approximately 70 percent of the global carbon dioxide emissions that result from human activity are caused by the combustion of fossil fuels in factories, power plants, motor vehicles, commercial establishments, and homes. The remaining 30 percent result from the clearing and burning of tropical forests (Miller, 1994). Over time, as human population has grown and as industrial and forest clearing activity have in creased, atmospheric levels of carbon dioxide have increased as well. Carbon dioxide levels in 1988, for example, were 25 percent higher than those in the mid-1800s. Other gaseous chemicals besides carbon dioxide, also produced by human activity, contribute to the greenhouse effect.

These include nitrous oxide, methane, and the chlorofluorocarbons or CFCs; (the CFCs are doubly damaging, since they also damage the atmosphere's ozone layer—see the section below).

The reason atmospheric scientists can't predict the future of global warming with certainty is that the underlying climatic processes are exceedingly complex and not completely understood, and precise computer models of the global climate system have not yet been developed. (There is even the possibility that increased air pollution, or increases in airborne water vapor caused by initial global warming, might cause an overall global *cooling* and a new ice age (Miller, 1994). Despite this inability to make detailed predictions, however, a majority of atmospheric scientists agree with the following general statements (Silver and DeFries, 1990; Miller, 2002):

— Between 1860 and 2000, mean global temperature increased by 0.6–0.7 degrees Celsius (1.1–1.3 degrees Fahrenheit). Most of this increase occurred after 1946.

— The increase in global temperature in the last 50 years is likely the result of human activity.

— Global temperature will continue to increase in the future. Between 2000 and 2100, the mean temperature will likely increase by 1.4–5.8 degrees Celsius (2.5–10.4 degrees Fahrenheit).

The temperature increases predicted in the paragraph above may seem quite small, but they are actually enormous compared to past changes of global climate caused by natural processes. Thus, a 4-degree Celsius (7-degree Fahrenheit) increase would create an average global temperature higher than that at any other time in the last 40 million years (Silver and DeFries, 1990)!

The concern about global warming is not mainly about average global temperature changes, however. Relatively small changes in average temperature can cause surprisingly large changes in global wind, rainfall, and temperature distribution patterns. Many scientists believe, for example, that the increases of global temperature predicted above could cause significant decreases in rainfall and agricultural productivity in the U.S.; one study predicts a 50 percent reduction of rainfall in grain belt states and a major drop in food production there.

Small increases in global temperature could also increase sea level significantly, due to the resulting expansion of ocean water and the partial melting of polar ice. For example, a 3-degree Celsius (5-degree Fahrenheit) increase in average global temperature could raise sea level by 0.2 to 1.4 meters (0.6 to 4.6 feet) (Miller, 1994). Even modest increases of sea level such as these would inundate large tracts of agricultural lowlands and deltas in China, India, Bangladesh, and other developing countries. The resulting drop of agricultural production would probably cause serious food shortages. Sea level increases might also disrupt coastal wetlands around the world, and lower populations of important fish and shellfish species (Miller, 1990). Finally, many of the world's low-lying cities would face massive flooding problems, especially during storms, including New Orleans, New York, Galveston, Shanghai, and Cairo (Miller, 1994).

We remind the readers of this book that these climatic changes could affect *their* lives, not just the lives of their children and grandchildren. Finally, we note that, though atmospheric scientists can't predict future climate changes with precision, most are convinced that the greenhouse effect poses a serious enough potential threat that action should be taken now to slow the release of carbon dioxide and the other greenhouse gasses (Silver and DeFries, 1990; French, 1995; Miller, 2002).

Ozone Layer Depletion

Naturally occurring ozone gas (each molecule of which consists of three atoms of oxygen) is found in the atmosphere at altitudes between from 12 to 24 kilometers (7.5 to 15 miles) above the Earth's surface. This "ozone layer" prevents 99 percent of the biologically harmful ultraviolet radiation from the sun from reaching the Earth's surface. The ozone layer has already been damaged, as we noted earlier in the chapter, by gaseous chemicals released into the atmosphere from industrial products and processes. These chemicals mainly include chlorofluorocarbons (commonly called freons), which have been widely used—as coolants in refrigeration and air-conditioning equipment, to manufacture Styrofoam insulation and fast-food containers, as propellants in aerosol cans, and to clean electronic microchips. The harmful chemicals also include halons, which are bromine compounds used in fire extinguishers. Releases of chlorofluorocarbons and halons (like those of carbon dioxide and other gases) have increased over time as human population and industrial activity have increased.

Scientists first detected significant damage to the ozone layer over Antarctica in 1985 (some damage first appeared in 1979) (Gardiner et al., 1985). The average concentration of ozone above the Antarctic continent dropped for several months by approximately 50 percent. This decrease, sometimes referred to as a hole, in the ozone layer reappeared in 1986 and has reappeared every subsequent year. More disturbingly, the overall global concentration of ozone has dropped by approximately 2 percent in the period between 1969 and 1986. Levels above North America have dropped by 4 to 5 percent between 1978 and 1990 (Miller, 1992). Some scientists predict that global ozone levels could drop by 10 to 30 percent as chlorofluorocarbons and other chemicals *already released* in the atmosphere reach the ozone layer (Miller, 1994). Although the decreases in ozone initially took the scientific community by surprise, as we mentioned earlier in the chapter, research conducted by large, international teams of scientists has definitively confirmed these decreases as well as their human cause.

The erosion of the ozone layer and the resulting increase in ultraviolet radiation reaching the Earth's surface will have a significant negative impact on human health; indeed, it is already having an impact. Specifically, a 10 percent thinning of global ozone layer could: increase by approximately 300,000 the annual number of cases in the United States of basal-cell and squamous-cell skin cancers, increase by 14,000 the annual number of cases of melanoma skin cancer (a form of cancer that is frequently fatal), sharply increase the incidence of cataracts, and possibly weaken human immune systems, lowering resistance to infectious disease (Miller, 1994). In addition to these health effects, a 10 percent erosion of ozone could reduce yields of agricultural crops, including wheat, corn, and soybeans, and might also disrupt aquatic ecosystems and aquatic food chains by interfering with the growth of phytoplankton, algae, and the larvae of important fish and shellfish species.

To add insult to injury, the chlorofluorocarbon and halon gases that damage the ozone layer also contribute to the greenhouse effect and global warming. Keep in mind, too, that it can take as long as one hundred years for some of these chemicals to break down naturally in the atmosphere. Fortunately, significant progress has been made in arriving at international agreements—such as the 1987 Montreal accord, and two more-stringent U.N. accords, the latest in March 1994 (French, 1995)—that will decrease and ultimately phase out the use of these chemicals, though the damage already done will continue to affect life on this planet for decades.

Rapid Loss of Tropical Forests and Loss of Genetic Diversity

Although tropical forests cover only a small fraction of the Earth's land surface, they contain approximately half, possibly even three-quarters, of all the Earth's plant and animal species (Silver and DeFries, 1990; Miller, 1994). However, large portions of these forests are being cleared annually to provide lumber, fuelwood, and land for growing crops and raising cattle. When tropical forests are cleared, the natural habitats of many native plant and animal species are damaged, and some of these species are permanently lost. The clearing of tropical forests also contributes to the global greenhouse effect, as we mentioned above, and has several major negative impacts on regional and local ecosystems, which we will discuss below. Losses of tropical forests, further, threaten the cultures of thousands of indigenous peoples.

Each year, an area of tropical forest roughly two and one-half times the size of Ireland is cut down (Miller, 1994). In the last few decades, approximately half of all the inland tropical forest area existing on the Earth before the middle of the nineteenth century has been cleared (Wolf, 1988). Continuing habitat destruction of this magnitude could, by the beginning of the next century, cause the irreversible loss of as many as one million plant and animal species (Miller, 1990).

There are several reasons the extinction of species in tropical forests should greatly concern us. First, this loss destroys vital chunks of the overall global gene pool, which is ". . . one of our planet's most important and irreplaceable resources (Wilson, quoted in Silver and DeFries, 1990, p. 125)." Scientists who develop improved strains of agriculturally important plant and animal species must call on this gene pool constantly. The genetic traits of wild, or nondomesticated, organisms alive today are the products of millions of years of biological evolution, and enable these organisms to live successfully in the particular environments they now occupy. Scientists regularly crossbreed wild varieties that have particular desirable traits, such as heat and drought resistance, pest and disease resistance, and high yields, with domestic agricultural species. Since so many of our most important agricultural plants—including corn, sugar, oranges, and potatoes—came originally from tropical regions, future efforts to breed better strains of these plants will be seriously impaired by the loss of genetic material in tropical forests.

Second, as many as one-quarter of the prescription and nonprescription drugs used in this country—including aspirin—were derived originally from plants that grow only in tropical forests (Miller, 1990, 1994). For example, two drugs that are highly effective against Hodgkin's disease and other cancers of the blood come from a periwinkle species that grows in Madagascar. The National Cancer Institute estimates that tropical rain forests are now the source of 70 percent of the most promising new anticancer drugs (Miller, 1994). Since scientists have only identified and studied a small fraction of all the plant species that live in tropical forests, they can only speculate on how many potentially medicinally useful species have already been lost forever.

In addition to its effects on genetic diversity, the destruction of tropical forests has negative impacts on global, regional, and local ecosystems. The clearing and burning of tropical forests produces about 30 percent of all the carbon dioxide released into the atmosphere by human activity, as we mentioned above, and is therefore a major contributor to the greenhouse effect and global warming (Miller, 1994). At regional and local levels, intact tropical forests absorb rainwater during rainy seasons, re-evaporate this water during drier seasons, and thus actually help create new rainfall. When tropical forests are cleared, rainfall levels drop, and the rainfall that does occur erodes the thin forest soil and also causes flooding downstream. Some scientists blame deforestation for causing the enormous 1988 floods in Bangladesh that left 25 million people, out of a total population of 110 million, homeless (Silver and DeFries, 1990).

Finally, as we discuss in Chapter 3, many writers, theologians, and others argue that from a moral or ethical point of view, the elimination of any plant or animal species is undesirable, regardless of the species' usefulness for meeting human needs.

Destruction of Ecological Capital

All three of the environmental problems discussed in the paragraphs above—ozone depletion, global warming, and the wide-scale clearing of tropical forests—are examples of a phenomenon sometimes called the "destruction of ecological capital" (Ehrlich and Ehrlich, 1990; Miller, 1994). Ecological capital refers not to money or economic wealth but to the Earth's natural resources and the Earth's "environmental services" that people alive today have inherited and upon which our survival depends. These resources and services include vast, but not limitless, quantities of freshwater, topsoil, forest land, and grassland; various atmospheric and climatic conditions and processes necessary for life; the ability to absorb human wastes; and on and on. They are called "capital" because, like economic capital, they automatically generate "dividends" in the form of a steady stream of services and because they can "depreciate" or become depleted if they are used too much and are inadequately maintained.

The Earth's resources and environmental services, if not overwhelmed by human demands, could supply its human occupants with the wherewithal for a reasonably comfortable life for long periods into the future, as they have for so many of us until now. If, however, we humans severely waste, damage, or otherwise mismanage these resources and services, we erode the basis of our own well-being and that of future generations.

Unfortunately, the destruction of ecological capital is now more the norm than the exception. As we already noted, the three global environmental problems discussed above are examples of this destruction. Another example is the depletion of underground freshwater reservoirs (aquifers) faster than they can be naturally replenished. In the U.S. Great Plains states, water levels in the underground Ogallala aquifer are dropping at a rate of four to six feet a year as water is pumped out for agricultural and other uses. The water in this aquifer was originally formed during the last ice age and is naturally replenished by rainwater at the rate of only one-half inch per year (Ehrlich and Ehrlich, 1990).

As yet another example, improper agricultural practices in countries throughout the world now cause the erosion by wind and water of as many as 25 billion tons of topsoil each year. This is equal to the topsoil now covering all the wheat fields in Australia (Brown, 1989). Overall, it is claimed that the world now loses 7 percent of its agricultural topsoil every ten years (Miller, 1994). The natural processes that replenish or form new topsoil typically require hundreds or thousands of years to produce just a few inches of it (Miller, 1994).

A particularly tragic pattern of soil erosion, aquifer depletion, and other forms of the destruction of ecological capital is found in many of the world's poorest nations, including many in Africa, the Indian subcontinent, and elsewhere (Brown, 1987; Postel, 1994). Population growth, unwise government policies, and, in some cases, war in these countries are beginning to overwhelm the capacity of local croplands, forests, grasslands, and aquifers to provide environmental services. In increasingly desperate efforts to get food, fuel, and shelter, the population has begun to consume its resource bases by overgrazing grasslands, denuding forests, eroding cropland, and pumping aquifers dry. This destruction of ecological capital causes agricultural yields to shrink and per capita food production to drop (tens of thousands of infants now die of malnutrition in these and other third world countries every day [Brown, 1990]). Many of these countries then switch from being food exporters to food importers, their monetary debts to other nations become enormous, and living conditions in them deteriorate further, in a self-reinforcing downward spiral.

Driving Forces That Underlie Environmental Problems

The I = P-A-T Model. Global and regional environmental problems are the result of an extremely complex set of interacting causes. A way to begin thinking more clearly about these causes is with a simple model put forth by Ehrlich and Holdren (1971) (see also Commoner, 1971), one that we briefly mentioned earlier. The model identifies three main types of causes: The size of the human population (P), the level of material affluence of that population (A), and the impact on resources and the environment per unit of affluence (T). T is determined by the characteristics of the technologies that produce the affluence. T is also a function of various factors that influence how technologies are developed and used, and that differ from country to country, including: a country's geography, its climate, cultural beliefs, government policies, social institutions, and others. For example, the size and sparse settlement of Canada lends itself to more travel and therefore more energy use than the geography of the Netherlands despite similar levels of affluence. The relationship of the three main causes—P, A, and T—is expressed in an equation[2] that represents the total impact of human activity on the environment, (I), as the product of the three causes: $I = P \times A \times T$.

In different countries, the absolute and relative sizes of P, A, and T differ greatly. Wealthy, or "developed" countries (the United States, Western European countries, Japan, Canada, Australia, and others) contain only 22 percent of the world's population (P), but as a result of their affluence (A), their technologies and other factors (T), they consume a disproportionately large share of the Earth's resources (approximately 88 percent of the natural resources consumed each year, including 73 percent of the energy resources) and generate most of the pollution and waste, as we discuss in Chapter 12 (Miller, 1994). These countries, a minority of the world's population, are the major responsible parties for global environmental problems.

In contrast, poor, or "developing," countries have a larger total population than the developed nations, but their level of affluence is much lower, as is their impact on the global environment. For example, India, a single country, contains approximately 16 percent of the world's population but uses only 3 percent of the natural resources and produces only 3 percent of the pollution and solid waste (Miller, 1994). Even so, efforts to feed and shelter India's rapidly growing population, provide for economic development, and pay international debts are destroying ecological capital in that country: Agricultural soil is being degraded and lost by overuse, huge areas of forests have been destroyed, and underground aquifers are being depleted (Dwivedi and Tiwari, 1987; Postel, 1989; Postel, 1990). Hence India faces a number of localized environmental crises that may be more devastating than the local environmental problems of the developed world.

Among countries that are roughly equal in affluence, there are also large differences in environmental impact per person. These differences reflect T, which, as we noted above, refers to several different things. Consider, for example, the reasons the United States uses about twice as much energy per capita as Japan despite roughly equal average incomes. One reason is the geography of the two countries, which makes travel distances greater in the United States and mass transit less practicable. Another is government policies that have kept energy prices low in the United States, causing increased use of all forms of energy. Behavioral patterns also differ: Japanese live in smaller homes and keep them much colder in winter than American homes, often sitting as a family unit around a space heater built into a living room table.

One important feature of the "P-A-T" analysis is that it makes clear that the major causes of environmental degradation can be traded off. Thus, increasing population in a country will not increase total impact on resources and environment if technology or social innovations (T) can proportionately decrease the environmental impact per person. Similarly, national economies can grow and affluence increase without increasing demands on resources/environment if what people produce and buy is more environmentally benign. Modern science and technology have, so far, generally kept up with population and affluence increases in wealthy countries by finding substitutes for scarce resources, ways to increase agricultural yields per acre, and ways to decrease the environmental impacts of technology. As an example of the latter, in the late 1970s and early 1980s sophisticated auto emissions control technologies cut the pollutants emitted by new cars in the U.S. by approximately 70 percent.

Exponential Growth. Population (P) and economic activity (P × A), however, have a property that makes it difficult in the long run for technology, resource substitution, and other factors (T) to keep the total human impacts on the environment within safe limits: Their growth tends to be exponential. Exponential growth starts off slowly but accelerates constantly, rounds a bend, and becomes explosive. It is unlikely that global population (P) and total economic activity (P × A) can increase indefinitely without causing some kind of environmental catastrophe, because other factors (T) would have to improve indefinitely to compensate.

Because exponential growth is not easy to understand intuitively, we examine it in more detail here. It is easy to confuse exponential growth with a simpler and more familiar type of growth, linear growth. In linear growth, a quantity increases by a fixed amount every fixed period of time. A good example is adding $10 a week to a glass jar or piggy bank. In exponential (sometimes called "geometric") growth, a quantity *doubles* every fixed period of time. The growth of

money in a compound-interest bank account is a good example. Thus, an account yielding 7 percent interest annually doubles in size every ten years. The pattern of growth follows a *J*-shaped curve, rather than a straight line.

As we noted above, population growth and economic growth both tend to be exponential. Here is a simple and prosaic example showing how exponential population growth works:

Assume that the first two human beings on Earth, "Sam" and "Sue," appear in the year 500. Assume, also, that they have access to modern medicine, public health measures, and ample food and that they and their descendants all live exactly eighty years. Assume that when they reach the age of twenty, they have four children, two boys and two girls. Finally, assume that the females in every subsequent generation (always half the children) all have four children at age twenty. Table 1-2 provides an analysis of the "world's" population over time. Each line of the table represents a generation: The first line is for Sam and Sue, the second for their children, and so on. Note how the 8 on the third line under the year 540 is derived: Sam and Sue's four children (two of them female) have, given the above assumption of four children per female, a total of eight children.

The bottom line of the table shows the total population of the world. Note that after an initial start-up period of several decades, population regularly *doubles every twenty years.* This doubling every fixed time is, again, the defining property of exponential growth. (Of course, the doubling time value depends on both life expectancy and the number of children per woman.) In this example, the

TABLE 1-2 An Example of Exponential Population Growth

YEAR	500	520	540	560	580	600	620	640
S & S	2	2	2	2	—	—	—	—
S & S's children		4	4	4	4	—	—	—
S & S's grandchildren			8	8	8	8	—	—
S & S's great-grandch.				16	16	16	16	—
Et cetera					32	32	32	32
						64	64	64
							128	128
								256
Total World Population:	2	6	14	30 ↔	60 ↔	120 ↔	240 ↔	480

population of the world reaches 1,000 in about 660, 1 million two centuries later, a billion in another century and a half, and so on. Figure 1-1 illustrates part of the accelerating growth.

There's an additional and important point. The Sam and Sue example assumes the existence in the initial year of modern medicine, public health measures, and so forth, and a resulting life expectancy of eighty years. However, those advances are actually relatively recent ones in human history, and they evolved over a period of several hundred years. Human life expectancy increased and death rates dropped during this period and, as a result, global doubling time fell. Thus, for this period in history, global population growth was superexponential; that is, growth was exponential but at successively higher rates, i.e., with successively shorter doubling times. The doubling time of the global population was approximately 1,000 years around 1500, 200 years in the 1800s, and 35 years in the mid 1900s. Figure 1-2 illustrates actual global population growth over time.

Currently, average global life expectancies and death rates are relatively stable (though in the poorest countries, life expectancies are still lower and death rates higher than in developed countries). Further,

global birth rates have declined, especially in wealthy countries. The current global rate of population growth is 1.4 percent, less than the all-time high of 2.06 percent in 1970 (Miller, 2002; Postel, 1994). As a result, global population now has a doubling time of fifty-two years (Miller, 2002). Although a major improvement from 1970, it is still a rate of exponential growth that many ecologists, demographers, and biologists think is still too great. We discuss these issues further in Chapter 12.

Having described exponential population growth, we turn now to a simple parallel example illustrating exponential economic growth:

Suppose a small country invests $2,000 and earns a constant rate of 3.5 percent per year, after accounting for inflation. Further suppose that the country reinvests all the earnings from this modest investment in order to save for future generations. In twenty years, the money would have roughly doubled, to about $4,000, just as the population of the world doubled in the previous example. In two centuries, this country's future generations would have an account worth about $2 million, and in another two centuries, the account would be worth over $2 billion and would be earning money at over $70 million per year.

A final example that may better convey the explosive nature of exponential growth is adapted from Miller (1990) and from what Donella Meadows and her colleagues describe as an old Persian legend (Meadows et al., 1972, p. 29):

[A clever citizen who, seeking favor from a king, presented the king with] a beautiful chessboard . . . and requested that the king give him in return 1 grain of [wheat] for the first square on the board, 2 grains for the second square, 4 grains for the third, and so forth. The king readily agreed and ordered [wheat] to be brought from his stores. The fourth square of the chessboard required 8 grains, the tenth square took 512 grains, the fifteenth required 16,384 grains, and the twenty-first square . . . [required] more than a million grains of [wheat]. By the fortieth square a million million [wheat] grains had to be brought from the storerooms. The king's entire [wheat] supply was exhausted long before he reached the sixty-fourth square. [Indeed, this last square would have required 2^{64}, probably more than all the wheat ever harvested in the world (Miller, 1990, p. 4).]

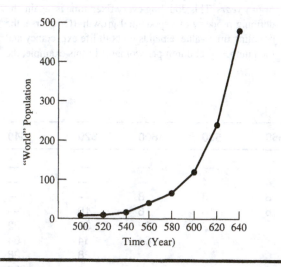

FIGURE 1-1 "World" Population over Time in "Sam and Sue" Example

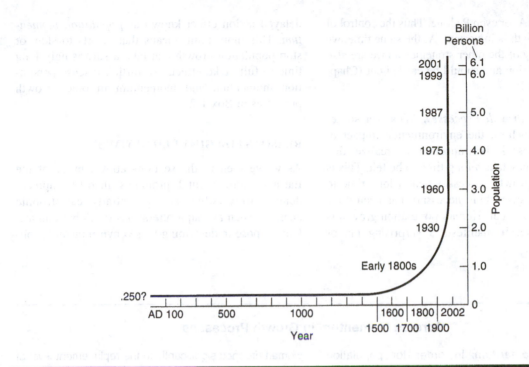

FIGURE 1-2 Actual World Population over Time.

Exponential increase is deceptive because it generates immense numbers quickly.

Again, because of the explosive character of exponential growth, the environmental effects of human activity may seem small for a long time, and then rapidly and surprisingly shoot up to intolerable levels. One of the properties of exponential growth is that each additional unit of environmental impact takes less time than the last. So it follows that if the value of I—the total impact of human activity on resources and the environment—is to be kept under control, the sooner it is done the easier it is likely to be.

Different Ways to Prevent Environmental Catastrophes. As we briefly noted above, the $I = P \times A \times T$ equation suggests that there are different ways to prevent environmental catastrophes. If the value of $P \times A$ continues to increase exponentially, the value of I can be held constant *if T also improves at the same*

rate. Remember that T is not only technology. Economies can grow without harming the environment if they produce different things that have less environmental impact per dollar. For example, a dollar's worth of computer software is more environmentally benign than a dollar's worth of petroleum. Economies can change their production processes to reduce environmental demands. For example, fiber optic telephone lines are an environmental improvement over copper because they decrease environmental damage from mining and smelting. Governments can create incentives to shift spending and production in more environmentally benign directions. There are many other possibilities for improving T, and this book examines a number of them. It is also possible that people can give up the goal of ever-increasing material affluence in favor of other values and activities. We discuss the potential for such changes in values in Chapter 3. However, as we noted earlier, it is unlikely that changes in T can continue to keep up with exponential growth in P and A, or even P alone. Thus the control of exponential growth is important. At the same time, we will argue, many of the other strategies above are also parts of the solution and should be carried out (Chapters 7, 10, and 12).

Momentum in Growth Processes. Whatever strategies are used to limit the environmental impact of human activities, it is important to realize that improvements may take a long time to be felt. This is because many growth processes take a long time to stop, even after everything necessary has been done to bring them to a halt. Human population growth is a good example. It manifests a surprising lag or delayed action effect known as *population momentum*. This momentum means that efforts to slow or stop population growth can take a surprisingly long time to fully take effect. We further discuss population momentum and momentum in other growth processes in Box 1-2.

REASONS FOR GUARDED OPTIMISM

As we've seen in the sections above, many of the Earth's environmental problems involve unprecedented, irreversible, and potentially catastrophic changes, such as major alterations of global climate. Our purpose in describing these environmental prob-

---

Box 1-2 _____

More on Momentum in Growth Processes

Population Momentum. In order for population growth to stop (that is, to reach zero population growth, or ZPG), the birthrate must be at the replacement level, that is, an average woman must just replace herself by having slightly over two children. (We refer only to women, rather than families or couples, because family structures change over time, but only women give birth. Slightly more than two children are required for a woman to replace herself because about half are males, and some children do not live to adulthood and to reproduce.) Because of population momentum, however, if, starting today and forever into the future, all women in the world had only the replacement number of children, it would take about sixty years before ZPG would be achieved, and population would continue to grow significantly in those sixty intervening years.

The population momentum phenomenon is similar to the long delay that occurs between the time a huge ship's engines are stopped and the time that the ship actually stops moving forward in the water. The mechanism that produces population momentum is easy to grasp and is best conveyed with a concrete example and some graphs. Population trends in the United States since the 1960s provide an ideal example:

In the late 1960s, the average U.S. woman had more than three children. However, in the early 1970s, for a variety of reasons, the number of desired children per woman dropped significantly to the replacement level of about two, the number needed to ultimately achieve zero population growth (ZPG). The figures below illustrate the process that would finally achieve ZPG about sixty years after the new average of about two children takes effect.

Figure 1B2-1 shows a simplified version of the age structure of the U.S. population in 1970, just before the new lower birthrate became predominant. Each vertical bar in the graph represents the number of people in the labeled age group.

Note that every bar is shorter than the one to its immediate left. This is not because people die. Indeed, this example assumes that no one dies until the age of eighty. Instead the reason is the average of three children per woman that had prevailed for decades: Women in the 20–40 year, or childbearing, age group averaged more than two children, and they therefore did more than replace themselves.

Beginning in the early 1970s, American norms on childbearing changed. The new social norm affected only women who had not yet borne their children, that is, the youngest group in the graph above, the bar highlighted with a "*." When these women had children, they stopped at two. As a result, by 1990, the left two bars were of about the same height, as shown in Figure 1B2-2. (The 60–80 age group from the 1970 graph has died and has disappeared from the 1990 graph.)

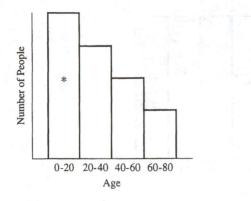

FIGURE 1B2-1 Population Structure in 1970

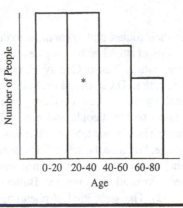

FIGURE 1B2-2 Population Structure in 1990

Going forward in time another twenty years to the year 2010, the age structure of the American population would correspond to that shown in Figure 1B2-3.

Finally in the year 2030, ZPG would be achieved, as shown in Figure 1B2-4. The United States would be in a true population steady-state condition. From 2030 onward, for every two people who die, another two people are born.

The important feature of this process is the delay of about sixty years between the change in social norms that achieves the replacement level of childbearing and the actual achievement of ZPG. During the intervening period of about sixty years, the population continues to grow—in the United States example, increasing by approximately 55 million people, over

and above any increases due to immigration. The increase is shown in the shaded area in Figure 1B2-5.

Other Cases of Momentum or Lag. Population is not the only growth phenomenon that has the property of momentum or lag. Another important example is growth in the built environment, and particularly in urban and suburban infrastructure. New automobiles are likely to be operated for ten years or more, so that any polluting or fuel-inefficient aspects of the technology in those vehicles will accumulate its effects for that long. New buildings usually last fifty years or more. The phenomenon of momentum may be clearest with projects like roads and sewer lines which, once built, encourage

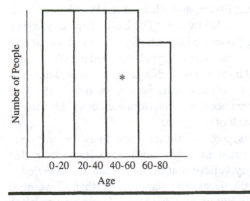

FIGURE 1B2-3 Population Structure in 2010

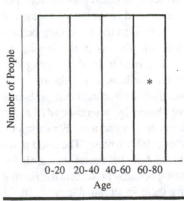

FIGURE 1B2-4 Population Structure in 2030

BOX 1-2   Continued

people to build along their routes, and to continue to do so for some time. A study of suburban housing development and population growth in Fairfax County, Virginia (now a suburb of Washington, D.C.), traced the building boom of the 1970s and beyond to a sewer-building project of a decade or more before. People had not built much in Fairfax because a high water table made it difficult to dispose of sewage, but once the sewer lines were put in, houses and shops sprang up until the entire area became suburbanized (Population Reference Bulletin Population Report, 1972). The same kind of phenomenon is responsible for much of the deforestation in the Brazilian Amazon. Until the 1960s, it was difficult to cut trees in the Amazon because the forest was impenetrable. The Brazilian government's highway building program changed all that, and the effect of the highways on the Amazon forest is still being felt almost thirty years later.

The phenomena of momentum give reason to be especially concerned about growth of the human population and also about other kinds of growth that

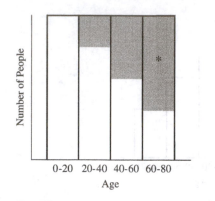

Figure 1B2-5 Population Structure in 2030

have long-term effects. Both exponential growth and momentum underscore the importance of not delaying global action to stabilize population or to stop other environmentally threatening kinds of growth. Once they reach an obvious danger point, it may be too late to bring them under control. We discuss these issues further in Chapter 12.

---

lems is not to overwhelm or depress you. Instead, our desire is to alert you to the severity of the problems and to motivate you to join us in taking a close look at their behavioral dimensions. Though global environmental problems are serious and unprecedented, they are, we are convinced, solvable. The scientific, technological, political, and economic measures needed to solve them, as well as the necessary behavioral measures that we will discuss in this book, are available now. It's just a matter of developing the will, commitment, and know-how to use them.

Our conviction that global environmental problems are solvable is widely shared by scientists who agree with us that the problems are very serious. For example, ecologist G. Tyler Miller (1990) writes: "The good news is that we [already] know how to sustain the Earth for human beings and other species by protecting the Earth's resources and by using them in sustainable ways [p. 3] (also Miller, 1994)." Silver and DeFries (1990), in a book approved and published by the U.S. National Academy

of Sciences, suggest that: ". . . [E]ncouraging signs abound." Robert Repetto (1986), an economist at the World Resources Institute, writes: "[If we] . . . mobilize now to achieve the global possible . . . , the future can be bright. We have sufficient knowledge, skill, and resources—if we use them [p. 137]." And Sandra Postel, Christopher Flavin, and others at the Worldwatch Institute (1991, p. 188) write: "The basic elements involved in getting [to an environmentally sound and sustainable future] . . . are no mystery; all the needed technologies, tools, and instruments of change exist. The real hurdle is deciding to commit ourselves to a new path. That commitment needs to come from each of us individually. And from all of us together."

We disagree with these scientists in only one respect, but it provides our reason for writing this book. They believe that humanity has the knowledge to make the necessary changes, and that all we need is the will. But two kinds of knowledge are needed to make these changes, and we have much more of one than the other. One is knowledge about the natural systems in the environment—how they work, how they respond to human intervention, and how much of this intervention they can withstand. Many books have been written on these natural-scientific subjects. The other is knowledge about the human activities that alter the natural environment—which human actors and activities are responsible, what causes their actions, and how to change them. This field has also produced valuable knowledge, but it is less widely available, understood, and appreciated. This book synthesizes some of that knowledge in order that readers may put it to use in solving environmental problems.

NOTES

1. *Homo sapiens sapiens* is the full Latin genus-species name of modern humans.

2. The equation is highly simplified in two important ways. First, T is really a combination of very many factors rather than a single thing. Because it is not a single thing, it cannot be measured directly, and it is misleading to interpret it simply as "technology." For example, some kinds of new technology increase the environmental impacts of human activities and some kinds decrease it. The same is true of institutions, policies, beliefs, and the other elements of T. So, it is hard to know exactly what "an increase in T" means, if it is represented by a number. The second simplification is that the equation suggests that P, A, and T are independently acting causes, when in fact each affects the others in many ways that are not yet well understood. Still, the equation is a simple way of conveying several important insights.

PART II

ENVIRONMENTAL PROBLEMS AS "TRAGEDIES OF THE COMMONS"—BEHAVIORAL SOLUTION STRATEGIES

CHAPTER 2

ENVIRONMENTAL PROBLEMS AS TRAGEDIES OF THE COMMONS

I. **Chapter Prologue**

II. **Introduction**

III. **Intellectual Roots of the Tragedy-of-the-Commons Concept and Different Solutions to Tragedy-of-the-Commons Situations**

    A. Four Solution Approaches
    B. Government Laws, Regulations, and Incentives
    C. Education
    D. Small-Group/Community Management
    E. Moral, Religious, and/or Ethical Appeals
    F. Which of the Four Solution Approaches to Use? Champions of the Different Approaches
    Box 2-1: How Tragedy Was Averted on the English Agricultural Commons

IV. **The Four Solution Approaches: Strategy for the Next Five Chapters**

V. **Chapter Epilogue**

CHAPTER PROLOGUE

This chapter begins with two true stories. The stories, though simple and brief, illustrate an important process, one that plays a major role in almost all environmental problems.

One. On a sultry evening in late August of 1982, a group of graduate students and professors, including the two authors of this book, gathered in a backyard in a small Connecticut town to celebrate what was left of the summer. One memorable feature of the evening was the food: Alaskan king crab legs. The legs, each over a foot long (the name *king* really applies—a whole king crab measures several feet across), were steamed and served with melted butter. The host and hostess had discovered king crab legs several years earlier when they had lived briefly in Alaska. The two continued to buy these expensive delicacies at local fish stores after they moved back to Connecticut. None of the guests had eaten king crabs legs before, but few forgot the experience. Most of us agreed that the legs were the best-tasting seafood we had ever eaten.

Three years later, in the spring of 1985, one of the authors and his wife went into a fish store in Michigan. Preparing for a dinner for an old family friend from out of town, they decided to splurge and asked for Alaskan king crab legs. They were told that Alaskan king crab legs were no longer available. "You mean you're out of them?" they asked. "No, we can't get them anymore—no one can, and we may never be able to get them again," replied the clerk. "Alaskan king crabs have been overfished to the point of near extinction."

The author and his wife were astonished. They had tasted king crab legs only once, and might never taste them again. The Alaskan king crab was a unique natural resource; the species had been around for tens of thousands of years, but humans had managed to almost, or totally, destroy it in the course of just a few years!

The author later came upon the following account of the collapse of the king crab population in a book chapter written by L. Brown (1985):

Found principally in the seas off Alaska and the nearby Aleutian Islands, the king crab became an international delicacy. . . . The seventies witnessed rapid growth in the harvest of this species, [and l]andings of this widely sought after seafood peaked in 1980 at 86,000 tons. [However,] by 1984 the catch had fallen to 7,000 tons, less than a tenth the earlier level. Surveys [also] showed a precipitous drop in the number of fertile females. . . . Shocked by these findings, the Alaskan [state government] . . . declared an emergency closure of the fishery. The unanticipated collapse of the Alaskan king crab population . . . raised questions about whether the species [would] . . . ever recover (Brown, 1985, pp. 78–79).

Two. John Sweeney's alarm clock went off at 4 A.M., but he was already awake. A fisherman his whole working life, Sweeney was used to the rhythm of his job. By 4:45 A.M. he was down at the dock and readied his boat, the *Volsunga IV,* for its daily trip out to sea. His crew members, Eddie and Brian, arrived a few minutes later. All three had been born and raised in Alaska, and despite several decades on the job, they still enjoyed working outdoors and being on the water.

For the last eight years, the three had fished mainly for king crabs, and they received high prices for their catch. However, John's overhead was also high; his outlays for fuel, insurance, maintenance, and boat loan payments were considerable, not to mention the thousands of dollars he had invested in equipment specially designed to catch king crabs. So John's take-home pay, and Brian's and Eddie's pay, were modest.

Although the three had brought in large catches of king crab in several previous years, their catch last year, 1983, had been much smaller. This year, so far, it was dismal, and John was having trouble meeting expenses. The three noticed many more large, well-equipped crab-fishing boats on the water out of Alaskan ports, as well as more Japanese boats in adjacent waters beyond the U.S. territorial limit. It finally became clear to them that the king crab population—which in the past they had somehow assumed was inexhaustible—was indeed finite and was being decimated by overfishing.

The three discussed the possibility of limiting their own catches to help conserve the crab population, but they were already too financially burdened to do so. They also realized that any unilateral efforts on their part to decrease the size of their catches would be futile: Their own impact on the depletion of the king crab population was relatively small, and any crabs they refrained from catching would simply be taken by other fishers.

As it turned out, the year's fishing season came to an abrupt end. The state of Alaska closed down king crab fishing because of the collapse of the crab population. Even if the state had not taken this action, John would soon have stopped fishing for king crabs anyway. His catches were simply too small to allow him to make a living wage.[1]

INTRODUCTION

The sequence of events and forces that led to the collapse of the king crab population, described in the stories above, is an example of what is usually called the *tragedy-of-the-commons* process. This name was coined by Garrett Hardin, a biologist at the University of California at Santa Barbara. Hardin's article on the subject, published in the journal *Science* in 1968, is probably the most widely quoted, cited, and reprinted article ever published on the social and behavioral dynamics of environmental problems. Now, many people might judge the loss of a specialty-food item like Alaskan king crab legs as hardly a catastrophe. However, the tragedy-of-the-commons process, Hardin argues convincingly, plays a major role in causing most—if not all—the resource, pollution, and population problems the world now faces. These problems include air and water pollution, global warming, acid rain, ozone depletion, deforestation, loss of biodiversity, the depletion of key energy resources, and many others. Given the importance of the tragedy-of-the-commons process, we devote a sizable portion of this book—Part II (Chapters 2–7)—to an in-depth examination of it.

In this chapter, we provide a general introduction to our multichapter coverage: We start with an analysis of the tragedy-of-the-commons process as Hardin (1968) conceived of it. We then review the history of tragedy-of-the-commons situations and find that they are not new phenomena first appearing in the twentieth century; they are, instead, variants of a very old problem, one that all human societies have faced for thousands of years. Further, we find that there have been, over the centuries, only a few methods for dealing with the problem that have been used successfully, and that might be used to stop or prevent modern tragedy-of-the-commons situations. We briefly describe each of these methods and their application to contemporary environmental problems. We argue that, though Hardin (1968) overlooked or rejected some of the methods, all of the methods have value and should be used. We conclude by outlining the organization of our more detailed coverage of the methods in Chapters 3–7.

Let's begin with a brief definition of the tragedy-of-the-commons process as Hardin (1968) conceived of it. In tragedy-of-the-commons situations, behavior that makes sense from the individual point of view, when repeated by enough individuals, ultimately proves disastrous to society. More specifically, the consumption of a natural resource by each of many individuals who have unrestricted access to the resource inevitably leads to the resource's destruction—a disaster for all. The underlying logic is simple: Each individual, Hardin assumes, is self-interested, that is, behaves mainly so as to advance his or her own interests (we examine this assumption more carefully later in the chapter). Each individual gains, financially or otherwise, by consuming the natural resource. Each, furthermore, sees little harm in doing so since the resource is so huge in size and their impact on it is so small. Each, therefore, is driven to consume as much of the resource as rapidly as possible. Now, this unrestricted consumption causes no problem when the Earth's human population is small and the resource-intensity of human activity (as we discussed in Chapter 1) is low. However, when human population and resource-intensity both reach high levels, as they have in recent decades, vital natural resources can be rapidly destroyed.

The collapse of the Alaskan king crab population, described at the beginning of the chapter, clearly fits Hardin's tragedy-of-the-commons framework. Let's step through the logic: Each crab fisher, presumably, wants to maximize his own financial earnings. Each crab he catches can be sold at market and is worth a certain amount of money. On the other hand, each crab he catches decreases the regional and global population of king crabs. It thus decreases the number available for his, and everyone else's, fishing in the future. However, the fisher knows that the total, overall crab population is large and that the dent he personally makes in it by fishing is small. The fisher also knows that any crabs he forgoes will be harvested by others. Each fisher, therefore, is driven by the logic of the situation to catch as many crabs as possible and as quickly as possible.

This pursuit of individual self-interest causes no problems as long as the total rate at which fishers catch crabs—determined by the number of fishers, times the number of crabs each catches—does not exceed the rate at which the crab population can naturally replenish itself. However, catch size and number of fishers both increase sharply over time as global human population, affluence levels, and the efficiency of fishing methods increase. Eventually, the number of fishers and catch sizes together exceed the rate of natural replenishment. When this happens, unrestricted access

Harbor Crowded with Commercial Fishing Boats
(Photo copyright 1994, Corel Corporation.)

to the resource leads inevitably to its total destruction—a disaster for all. Note, in this connection, that a population collapse of a commercial marine species can be difficult, or even impossible, to reverse: The species' ecological niche (its characteristic role in the aquatic ecosystem, that is, its particular set of relationships with other animals and plants) can become occupied by other, noncommercially useful species, which then cannot be dislodged (Anon, 1983; Repetto, 1986). Note also that overfishing is already a major global problem: Besides king crabs and other species found in Alaskan waters, 13 out of 19 commercially important fish species found in northwest Atlantic waters, for example, had been overfished to the point of decline or collapse by the early 1980s (Brown, 1985).

Fish populations are not the only natural resources that can be destroyed in tragedy-of-the-commons situations. There are many others. Indeed, any natural resource—either renewable or nonrenewable—to which people have more or less unrestricted access can be rapidly depleted through the tragedy-of-the-commons process. Political scientists and other social scientists usually call resources of this type *common-pool resources*. Common-pool resources found in oceans (besides fish) include valuable plant species and minerals on the sea floor. Similar resources are found in freshwater bodies. Yet other common-pool resources lie in publicly owned forests, wilderness areas, rangelands, and watersheds (which are a source of drinking water). Even underground petroleum reservoirs that are accessible to many different drillers are common-pool resources. Again, these resources all can be rapidly destroyed via the tragedy-of-the-commons process.

Like a fishery or a rangeland, the cleanliness of the air, the waste-processing capacity of a river, and the atmosphere's ability to regulate the earth's temperature are also common-pool resources that can be degraded by overuse. Here, however, the direction of human action is reversed: Rather than actions that take away parts of a resource pool or population, as with overfishing, at issue are actions that put things such as sewage, chemical wastes, or garbage into a resource, thereby degrading it. For example, the greenhouse effect and the erosion of the Earth's protective ozone layer are caused by human activities that release chemicals into the earth's atmosphere. For the greenhouse effect it's mainly the carbon dioxide produced by the burning of fossil fuels in cars, furnaces, and factories; for ozone-layer destruction it's mainly the chlorofluorocarbons emitted by air-conditioning and refrigeration equipment. Local and regional air pollution, acid rain, water pollution, and solid waste problems all have similar causes. Thus, the tragedy-of-the-commons process underlies most problems of environmental pollution, just as it underlies the depletion of most natural resources.

We have already noted that unrestricted access to natural resources can trigger the tragedy-of-the-commons process only when human numbers, and the resource- and pollution-intensity of human activity, have reached certain high levels. Hardin (1968), however, claims another connection between the tragedy-of-the-commons process and human population growth: He argues that population growth *per se* results from a tragedy-of-the-commons dynamic of its own (this is, in fact, the main point Hardin makes in his classic 1968 article). Specifically, couples are driven to have children because they find children rewarding, and they perceive little environmental cost or harm in having them; this logic repeated over millions of couples causes, according to Hardin, population growth levels that are too high to be environmentally sustainable in the long run. Consider, in greater detail, how this tragedy-of-the-commons process operates, for example, in poor countries. Couples in these countries want children because—among other reasons—grown children are the only source of support people have when they grow old and sick (there are usually no government social security programs); further, couples must have many children if enough are to survive to their old age (there is a high child mortality rate in poor countries because of inadequate diets and poor medical care). On the other side of the ledger, the pollution and the depletion of local and regional resources caused by any one couple's children are, relatively speaking, very small; the damage is usually not even visible to the couple. Given all of the above factors that couples face, it is not surprising that birthrates are high in poor countries. Unless governments act to improve social security, health care, and other programs in these countries (and related changes occur in rich countries),[2] populations will eventually reach levels that are higher than local, regional, and global ecosystems will be able to sustain. To repeat, Hardin argues that exponential population growth—a main driving force behind tragedies of the commons—is itself caused by its own tragedy-of-the-commons dynamic.

Finally, it is important to highlight one other key characteristic of tragedy-of-the-commons situations, a characteristic emphasized by political scientist William Ophuls (1977) and illustrated in the crab-fishing stories above: Tragedies of the commons usually involve ordinary people doing ordinary things, rather than villainous or greedy people doing especially nasty things.[3,4] The king crab fishers are not evil people who want to harm others deliberately (nor are the people in poor countries who have many children). Furthermore, even if each fisher understands that his actions are contributing to the ongoing destruction of the resource, each is powerless to stop the process via unilateral individual action.

Tragedies of the commons can therefore be seen as tragedies in the classical Greek sense: The participants are locked into a process that leads inexorably to calamity. To convey this quality, Hardin (1968) quotes Whitehead's (1948) description of *tragedy* as involving "the solemnity of the remorseless working of things . . . the inevitableness of destiny . . . the futility of escape (quoted in Levine, 1986, p. 93)."

INTELLECTUAL ROOTS OF THE TRAGEDY-OF-THE-COMMONS CONCEPT AND DIFFERENT SOLUTIONS TO TRAGEDY-OF-THE-COMMONS SITUATIONS

We have argued, so far, that the tragedy-of-the-commons process tends to be triggered when many individuals have unrestricted access to a resource, and that this process underlies most, if not all, environmental problems. Further, the process can not be stopped by unilateral individual action. However, despite the apparent "remorseless workings" of the tragedy-of-the-commons process, it is *not*, as it turns out, inevitable or unstoppable. Going an important step further, it appears that tragedy-of-the-commons situations are not even fundamentally new phenom-

ena in the history of our species. Ophuls (1973, 1977) argues persuasively that tragedy-of-the-commons situations are actually a sub-case of a basic problem or dilemma that all human groups or civilizations have faced for the last 40,000 years. This problem, furthermore, is one that political philosophers and others have been working on for many hundreds of years. And it is a problem for which there are several solutions. The basic problem is this: How can behavior of individual group members that in some way threatens the welfare of the group be discouraged, and behavior that is prosocial, or beneficial to the welfare of the group, be encouraged? Note the parallel with tragedy-of-the-commons situations, in which the specific issue is: How can the consumption of a resource by individual members of society be restrained so as to avoid the destruction of the resource—a disaster for all—and behavior that conserves the resource be encouraged?

Ophuls (1973, 1977) points out that Plato's book, *The Republic*, written over 2,000 years ago and considered to be the "fountainhead of all Western thinking about politics," is concerned with the basic, age-old form of this problem. Ophuls (1973) writes:

> . . . *Plato [who lived in Athens from 427–347 B.C.] tried to design a "ship of state" in which the most skilled navigator would be spontaneously accepted as captain by the crew members, each of whom would perform the shipboard duties for which he was best qualified rather than striving to lay hands on the ship's tiller himself. The political philosophy of the Republic . . . emerged directly from Plato's own historical experience: he lived through the tragedy of the Peloponnesian War [a war between city-states Athens and Sparta in 431–404 B.C.] and [he] attributed the Athenian defeat to the sordid struggle [by Athenian politicians] for individual political advantage at a time of common peril (p. 216).*

As we noted above, the problem of coordinating individual behavior for the common good is one that all human groups have faced during the entire 40,000-year period that *Homo sapiens sapiens* have been around. So, for example, in Stone Age times, humans lived in small groups of 12 to 250 people (Bogin, 1994) and hunted and gathered their food. These people survived, in part, because they cooperated with each other and divided up the various activities necessary for survival. For example, the successful stalking and killing of large game animals (with only Stone Age weapons) required the cooperative action of several hunters, each with a specialized role to play. A single hunter couldn't do the job alone. This and other instances of cooperation and division of labor required that relevant individual behaviors be coordinated and limited for the good of the entire group.

In more modern times, as human populations have grown in size and as human technology has become vastly more sophisticated, there is an even greater interdependence between people, and an even greater need to properly coordinate individual behavior for the common good. For example, a nation's complicated highway, rail, and air transportation systems, and systems to distribute electricity and other utilities require considerable coordination and limitation of individual behavior.

Four Solution Approaches

Having established that the problem of coordinating individual behavior for the common good is an eternal one, Ophuls (1973, 1977) goes an important step further: He points out that, over the centuries, there have been *only a few basic methods* for promoting pro-social individual behavior that political philosophers have written about and that societies have used. And these basic methods are the same ones that contemporary scholars and writers have proposed for encouraging individual behaviors that avert regional and global tragedies of the commons—though these scholars and writers don't usually mention the historical origins of the solutions they propose. Specifically, Ophuls identifies the following *four basic solution types*, or ways to encourage individual behavior for the common good: 1) the use of government laws, regulations, and incentives to encourage prosocial behavior; 2) programs of education, which attempt to encourage prosocial behavior by giving people information and trying to change their attitudes; 3) the encouragement of prosocial behavior via certain informal (nongovernmental) social processes that operate in small social groups and communities; and 4) the use of moral, religious, and/or ethical appeals to encourage prosocial behavior. We discuss each solu-

tion type in more detail below, using 1990s examples that involve both environmental and nonenvironmental subject matter.

Government Laws, Regulations, and Incentives

In this solution type, government laws, regulations, and incentives are used to encourage specific pro-social individual behaviors and/or discourage antisocial behaviors. To use a contemporary, nonenvironmental example, state and local laws regulate how people operate motor vehicles, including maximum speeds, signals to stop at, places to park, and so on. As a second example, national laws require that each citizen pay the government an annual income tax. The laws and regulations in these two examples are enforced by the threat of fine and/or imprisonment. In contrast, other government laws and regulations use rewards—monetary or nonmonetary—to encourage people to perform certain behaviors that are in the public interest. To use an environmental example, U.S. tax regulations, in place in the late 1970s but not currently, awarded sizable income tax deductions to taxpayers who installed energy conservation equipment (e.g., additional attic insulation or high-efficiency furnaces) and/or solar energy devices in their homes.

Note that a defining characteristic of the government laws/regulations/incentives approach is that the laws, regulations, and incentives encourage people to behave in the public interest by making it in each individual's personal best interest—monetary or otherwise—to do so. This same defining characteristic is found in an important, and environmentally relevant, variant of the laws/regulations/incentives approach known as *privatization*. As an example, imagine that the State of Alaska divided the coastal ocean into small segments marked with buoys, and awarded each segment to a different king crab fisher as his private property, that is, for his exclusive use for some period of time. (Assume for this example that crab "migration" from one segment to another is negligible.) Under this arrangement, it would be in each fisher's personal best interest to harvest crabs carefully so as to avoid the collapse of the crab population in his segment. Failure to do so would be catastrophic to the fisher, as he would have no other place to go to catch crabs and thereby earn a living. Again, this arrangement attempts to motivate environmentally sound behavior by making such behavior in each person's own best interest.

Although the laws/regulations/incentives method encourages individuals to behave in the public interest by making it in each individual's personal self-interest to do so, the other three basic solution approaches try to encourage prosocial individual behavior in a fundamentally different way: These methods assume that under the right conditions, people will want to behave in a public-spirited fashion, whether or not such behavior is in their own narrowly defined personal interest. As we discuss more fully below, Hardin (1968) rejects the use of two of these three methods by themselves and neglects to mention the third. He, instead, advocates the first approach—laws/regulations/incentives. Hardin's judgment appears, in part, to be based on his assumption, noted above, that humans are intrinsically egoistic, that is, behave mainly in ways that advance their own interests. But before we can discuss any of this further, we must finish describing the three other basic solution approaches.

Education

In the second approach, people are educated about a societal problem in schools, via the mass media, and/or by other agents. The program of education typically has two main thrusts: First it describes the nature and severity of the problem in an effort to change people's attitudes toward it, that is, to convince them that the problem is serious and important enough to warrant their immediate personal action. Second, it outlines the specific actions individual citizens can take to help solve the problem.

High school Driver's Education classes provide a good, nonenvironmental example of the education approach. These classes present information about how an auto operates, how accidents occur, and the best ways to drive safely. These classes also try to change students' attitudes toward safe driving. By pointing out that 40,000 lives are lost on the nation's highways each year, that 10,000 lives could be saved if everybody simply buckled his or her seat belt, and

so on, these classes hope students will regard using seat belts and other safe-driving practices as urgently necessary things for all citizens to do, and that the students will act accordingly. (Keep in mind for this example that safe driving is in society's, as well as the individual's, best interests: Auto accidents mean lost worker productivity, medical expenses that must be spread over all the holders of insurance policies, and other costs to society.)

Note, finally, that educational programs sometimes use a well-known public figure to help convince people to take the recommended actions. For example, a popular movie star appears in a TV ad and urges people to "buckle up for safety" or to "just say no to drugs" or to practice safe sex.

The application of the education solution-approach to environmental problems is straightforward: Programs in schools, on TV, in newspapers, books, magazines, and other media would describe the nature and severity of environmental problems in an effort to convince the public that action on their part is essential; the programs would also describe the specific actions individuals should take to help solve those problems.

Small-Group/Community Management

In the third solution approach, a small group or community of people informally develop and mutually enforce their own rules of individual behavior to solve some group problem, without the involvement of any government authority. To use an environmental example, a small group of crab fishers who fish in the same harbor would get together and agree to certain rules that would help conserve the crab population. These rules might include specific weeks of the year to fish crabs, the maximum number that can be caught per fisher, the minimum allowable size for a crab, the exclusive use of the area by group members, and so on. The fishers in this group would tend to follow the rules, in part because of mutual observation and social pressure. However—and this is important—they would also follow the rules out of mutual respect and concern for one another and out of a sense of obligation to the group or duty as a group member. All the above processes would work most effectively if the fishers knew each other well and formed a cohesive group or community. As we note in the text below and in Box 2-1, such small-group arrangements were extremely common in Europe and elsewhere hundreds of years ago. These arrangements, or certain aspects of them, may be adaptable to contemporary times to encourage proenvironmental individual behavior.

Moral, Religious, and/or Ethical Appeals

In this fourth and final solution approach, religious, moral, and ethical teachings and principles are used to encourage proper individual behaviors. For example, the ten commandments of the Judeo-Christian religious tradition proscribe certain behaviors (e.g. murder and adultery) and prescribe others (e.g. honoring one's parents). However, religious/moral/ethical pro/prescriptions go beyond the mere lists of dos and don'ts, that are usually conveyed in education (the second approach, above); they also have major intuitive, emotional, and spiritual components, and they are usually conveyed and enforced in religious ceremonies and rituals.

As a possible application of this approach, Western countries could adopt a set of proenvironmental religious/moral beliefs and practices. Several writers and scholars have suggested that these be derived from Native American religions and moral codes, which, they argue, helped these societies live in harmony with their natural environments. But a few of these writers and scholars make an additional claim: They argue that not only do the current Judeo-Christian religious beliefs of Western countries lack a proenvironmental orientation, but these beliefs are actually strongly *anti*environmental; indeed the beliefs are a major cause of all contemporary Western environmental problems. We devote the next chapter to an in-depth discussion of these issues.

Which of the Four Solution Approaches to Use? Champions of the Different Approaches

Given the four basic solution approaches that could be used to encourage prosocial individual behavior, the obvious next question is: Which one(s) should be used? In other words, which one(s) are most effective and desirable? Over the centuries, different scholars

and writers have answered this question differently. For example, according to Ophuls (1973, 1977), English political philosopher Thomas Hobbes (1651) mainly advocated something like the first approach above—government laws, regulations, and incentives; in contrast, the French political philosopher J. J. Rousseau (1762) championed elements of the third approach—small-group/community management. Of course, neither Hobbes nor Rousseau wrote about solving global environmental problems, as such problems did not exist at the time. Instead each wrote about solving the more general problem, noted above, of how to encourage pro-social individual behaviors and discourage individual behaviors that might in some way threaten the common good.

But most important for our purposes in this book is that Garrett Hardin (1968), author of the highly influential tragedy-of-the-commons article we've discussed at length, takes a basically Hobbesian position. He argues, as noted briefly above, that only government laws, regulations, and incentives (including privatization) will ensure widespread proenvironmental behavior on the part of the general public. Hardin appears to take this position, in part, because of his assumption that humans are innately egoistic, that is, inclined to act only in ways that advance their own interests. He thus concludes that the most effective solution approach is the one that channels human egoism into proenvironmental behavior, that is, makes proenvironmental behavior in each individual's best interest—in other words, government laws/regulations/incentives. Hardin rejects the use of education programs and religious/moral/ethical appeals by themselves, apparently because he considers them too weak to overcome human egoism.[5]

Significantly, Hardin (1968) fails to mention the third, or small-group/community management, approach altogether. Recall that this approach assumes, in part, that people in small groups and communities will cooperate to advance their common interests out of mutual respect and concern for one another or a sense of obligation or duty, without consideration of their narrowly defined self-interest. Perhaps Hardin overlooks this approach because he assumes that egoism (the pursuit of self-interest) is an inevitable characteristic of human behavior. Alternatively, he may not be aware of the prevalence of

Box 2-1

How Tragedy Was Averted on the English Agricultural Commons

In his classic article on the tragedy-of-the-commons process, Garrett Hardin (1968) carefully develops the illustrative example of a commonly held pasture used by herdsmen to graze cattle (Hardin does not use the crab-fishing example presented earlier in this chapter). Hardin writes: "Picture a pasture open to all. It is to be expected that each herdsman will try to keep as many cattle as possible on the commons, [etc., etc., etc.] (p. 1244)." Hardin continues this example to its ultimate tragic conclusion: the complete defoliation of the pasture and the ruination of all the herders.

Hardin, further, credits English mathematician William Forster Lloyd (1833) with having written the first account of the tragedy-of-the-commons process (though in a later article Hardin [1980] notes that tragedy-of-the-commons–like situations were described in earlier scholarly works dating back to Aristotle). Lloyd (1833), in turn, used the very same illustration of a commonly held pasture for grazing cattle, presumably referring to such arrangements in England. He wrote: "Why are the cattle on a common so puny and stunted? Why is the common itself so bare-worn and cropped so differently from the adjoining [private] enclosures (quoted in Cox, 1985, p. 51)."

Several other scholars, however, argue that Lloyd's example is, from a historical point of view, extremely misleading. Based on a review of historical evidence, political scientist Susan Cox (1985), community psychologist Bruce Levine (1986), economist David Feeny (1990), and several other scholars (e.g., see Feeny et al., p. 10) conclude that commons-type grazing arrangements were widely found in medieval and postmedieval England, as well as over much of Europe, in some cases dating back as much as 1,000 years. However, and most important, these arrangements did not, in fact, operate in the manner that Lloyd (and Hardin) describe, and were not usually subject to "tragedy." Instead, two key features made these arrangements remarkably successful in avoiding a collapse of the shared resource: First, access to a commons was strictly limited; commons were not open to all individuals who wished to use them (e.g., Feeny et al., 1990). Second, those individuals with access to a given commons were limited in the intensity of their use; for example, in one simple and widely used practice known as *stinting*, each user was allowed to graze

only a limited number of animals each year (e.g., Levine, 1986).

According to Levine (1986), agricultural village communities with shared pasturages of this type first appeared in England in the 1100s and became universal in the 1400s. These villages were often located on land owned by lords and were usually inhabited by 200 to 500 tenants. Each tenant was assigned a private, nonshared plot of land on which to raise his/her crops, but areas for grazing livestock were held and used in common by all village residents.

There is historical evidence that agricultural villages with common pasturages of this type existed in many other places besides England and Europe, including Africa, Asia, India, and Central and South America (Levine, 1986). (Indeed, anthropologist J. T. McCabe [1990] points out that there are several million nomadic pastoralists in Africa who now support themselves by grazing livestock in common pasturages.) In the great majority of the above cases, tragedies of the commons were (or have been) successfully averted by, as mentioned above, restricting access to the commons to certain individuals and by restricting the intensity of use by those individuals.

Further, these restrictions on access and intensity of use were accomplished by means of informal social mechanisms that evolved over time within the individual communities (Cox, 1985; Levine, 1986; Feeny et al., 1990). The residents of small agricultural villages knew each other well, understood that their survival hinged on cooperation, felt responsible for each other, and even cared deeply about one another (Levine, 1986). The residents fully understood the capacity limits of their communal pasturages and developed their own mutually agreed upon and mutually enforced systems (for example, stinting) to regulate use, without the need for a government authority to determine and impose these regulations. Note how different this type of arrangement is from the one assumed by Lloyd (1833) and Hardin (1968) in their examples of pasturages shared by herders who are motivated exclusively by self-interest and who needed regulatory intervention by the state to prevent ruin.

We devote a later chapter (Chapter 6) to a more detailed look at this type of community-managed commons arrangement, and provide both successful contemporary examples of it and an in-depth discussion of its inherent problems and limitations.

But there is one final topic we should discuss now: If community-managed commons arrangements in England and elsewhere were as successful as the social scientists cited above claim they were, why did Lloyd (1833) refer to despoiled English commons-type pasturages in his article? A likely answer is provided by Cox (1985) and Levine (1986). They point out that Lloyd wrote his paper during a period in history—the late 1700s to late 1800s—in which the centuries-old community-managed commons system was falling apart for a variety of reasons. These included ". . . land 'reforms' chiefly designed to increase the holding of a few landowners [such as the so-called 'enclosure movement' that broke apart villages and drove tenants from communal land], improved agricultural techniques [which favored

large-scale farming operations], and the effects of the industrial revolution (Cox, 1985, p. 49)." (Note, in this connection, that McCabe, 1990, suggests that somewhat similar changes account for the failure of some community-controlled commons systems in contemporary Africa.)

Thus, Lloyd's (1833) account and Hardin's (1968) example mislead, as they do not recognize the long historical period of ". . . hundreds of years—and perhaps thousands. . .—[during which] land was managed successfully by communities . . . , not a 'tragedy of the commons' but rather a triumph (Cox, 1985, p. 60)." Cox goes on to suggest that "Our re-examination of [community-managed agricultural] commons requires . . . [us] to search for the ideas and practices which led to successful commoning for centuries and to try to find lessons and applications for our own times (p. 61)."

---

this approach in history: As noted above, historians claim the small-group approach was used successfully in Western and other countries for hundreds of years. In fact, these historians argue, the approach was used to prevent the very example of the tragedy of the commons that Hardin discusses at length in his article—one that can occur when a bounded pasture is shared by many herders to graze their cattle. We further examine Hardin's omission of the small-group/community approach in Box 2-1.

THE FOUR SOLUTION APPROACHES: STRATEGY FOR THE NEXT FIVE CHAPTERS

Clearly, there's much more to discuss concerning whether Hardin's conclusion—that only government laws, regulations and incentives will prevent or stop the tragedy-of-the-commons process—follows from his assumptions concerning human nature. We also need to discuss whether those assumptions are correct. Indeed, our brief description above of the four general solution approaches and their historical origins barely scratches the surface of this important topic.

For this reason, we devote each of the next four chapters (Chapters 3–6) to a careful, in-depth, behavioral-scientific examination of each of the four solution approaches. In each chapter we will present

detailed historical and contemporary examples of the use of the corresponding approach, evaluate the approach's level of effectiveness, and thoroughly explore its limitations, advantages, and disadvantages. The general conclusion we come to in these chapters—and the one we hope to convince you of—is that *none of the four solution approaches is, by itself, likely to work effectively;* no one approach will successfully prevent tragedies of the commons, and the resulting problems of resource depletion, pollution, and exponential population growth. Instead, we argue that only a coordinated effort involving all—or at least most—of the four solution types will succeed. In this spirit, we devote Chapter 7, the final chapter of Part II, to a discussion of several successful environmental behavior-change programs that involve multiple, integrated solution strategies.

We conclude this chapter with a brief Epilogue that describes an unexpected ending to the two stories at the beginning of the chapter.

CHAPTER EPILOGUE

This chapter began with two brief stories about the collapse of the Alaskan king crab population caused by overfishing in the early 1980s. As we noted above, the population collapse of an aquatic species can be difficult or even impossible to reverse, since the species' ecological niche can become occupied by other species that cannot be dislodged. As it turned out, however, fate took a positive turn; our stories have a surprise happy ending: The Alaskan king crab population recovered. Alaskan king crab legs can, at the time of this writing, be found in many supermarkets and specialty stores (king crab legs are not to be confused with "krabs legs," which are made from ordinary fish that has been cut, processed, colored, and flavored to resemble the real thing). Indeed, the authors hope to repeat their backyard picnic (described in the first story) with Alaskan king crab legs as the featured food.

Here's a brief account of what happened: The Alaskan Department of Fish and Game did some careful research after it issued its total prohibition on king crab fishing in 1983–84. It concluded that the king crab population, though decimated, was still sufficiently intact to permit, with careful management, an eventual recovery.

In the ensuing years the department permitted king crab fishing but strictly limited the total catch. Each year the department carefully audited population levels and determined the number of crabs that could be harvested without jeopardizing the long-term recovery of the species. The department then adjusted the maximum total harvest for that year accordingly. It enforced that maximum by, quite simply, limiting the total number of hours and days in the year's legal fishing season.

As a result of these annually adjusted regulations, the availability of king crab legs in food stores varied from year to year between 1984 and the present, though availability generally increased over the period. Some years, the quantity available was small (at times making king crabs difficult to find) and retail prices were high. In other years, king crabs were more plentiful and less expensive (Dee, 1991).

In any case, the species was brought back from the brink of extinction, and a tragedy of the commons that had almost run to completion was successfully reversed. Unfortunately, eleventh-hour rescue efforts of this type with other species, both aquatic and terrestrial, have not always succeeded.

In the next five chapters we take an in-depth look at different methods—including, but not limited to, the government laws and regulations approach that was used in the Alaskan king crab case—that may prevent and stop tragedies of the commons.

NOTES

1. While the names in the story above are fictional, the account is based on: Brown, 1985, pp. 77–79; Anon., 1983; Gay, 1991, pp. 16–18.

2. Note in this context that although birthrates are lower in rich countries than in poor countries, each citizen in rich countries has far greater negative environmental impact than does each in poor countries.

3. Hardin's (1968) view of tragedy-of-the-commons situations is slightly different. He assumes that humans pursue economic self-interest in a rather calculated and relentless manner (p. 1244). Yet, as Ophuls points out, tragedies of the commons can equally well occur if sufficient numbers of people pursue run-of-the-mill existences, rather than economic self-aggrandizement.

4. We remind the reader of the limited role of individual behavior relative to organizational actions (actions of corporations and governments) in causing environmental problems, as we discussed in Chapter 1 in the section titled "More on the Role of Individual Behavior in Environmental Problems."

5. Hardin (1968) rejects the use of religious/moral/ethical appeals for the purpose of decreasing the birth rate and slowing population growth on additional grounds: He argues that those individuals who respond to religious/moral/ethical suasion will have fewer children than those who are not responsive to this approach, and the former will be outbred by the latter to the point of becoming extinct. As he puts it: "Conscience is self-eliminating" in the long run (p. 1246).

CHAPTER 3

RELIGIOUS AND MORAL APPROACHES: CHANGING VALUES, BELIEFS, AND WORLDVIEWS

CHAPTER PROLOGUE

In 1854, Chief Seattle, leader of the Suquamish tribe in the northwest United States, gave a speech in reply to President Franklin Pierce's offer to buy a large tract of Indian-occupied land and to provide a reservation for Seattle's people.

While there is no written record of Seattle's words, one person present, Dr. Henry Smith, took brief notes (Seed et al., 1988). Based on Smith's notes, Ted Perry, a film scriptwriter, attempted to recreate Seattle's speech for a 1970 movie. Perry's re-created speech became somewhat of a classic and was frequently reprinted in books and articles on the environment. Though Perry's words must be regarded as fiction and while they may over-roman-ticize Native American religious beliefs, they do convey the reverence with which many Native American tribes regarded (and still regard) their environments (Viola, 1992, quoted in Jones, Jr. and Sawhill, 1992). We quote some of Perry's words below:

> *How can you buy [the land,] . . . or [we] sell the sky. . . ? This idea is strange to us. . . . If we do not own the freshness of the air and the sparkle of the water, how can you buy them? . . .*
>
> *Every part of the earth is sacred to my people. Every shining pine needle, every sandy shore, every mist in the dark woods, every clearing, and humming insect is holy in the memory and experience of my people. The sap*

Reflections in a Lake
(Copyright 1994 Corel Corporation.)

which courses through the trees carries the memories of the red man.

The white man's dead forget the country of their birth when they go to walk among the stars. Our dead never forget this beautiful earth, for it is the mother of the red man. We are part of the earth and it is part of us. The perfumed flowers are our sisters; the deer, the horse, the great eagle, these are our brothers. The rocky crests, the grasses in the meadows, the body heat of a pony, and man—all belong to the same family. . . .

The shining water that moves in the streams and the rivers is not just water but the blood of our ancestors. If we sell you land, you must remember that it is sacred, and you must teach your children that it is sacred and that each ghostly reflection in the clear water of the lakes tells of events and memories in the life of my people. The water's murmur is the voice of my father's father.

The rivers are our brothers, they quench our thirst. The rivers carry our canoes, and feed our children. If we sell you. . .[the] land, you must remember, and teach your children, that the rivers are our brothers, and yours, and you must henceforth give the rivers the kindness you would give any brother. . . .

The air is precious to the red man, for all things share the same breath: the beast, the tree, the man, they all share the same breath. The white man does not seem to notice the air he breathes. Like a man dying for many days, he is numb to the stench. But if we sell you [the] . . . land, you must remember that the air is precious to us, that the air shares its spirit with all the life it supports. . . .

[If we sell you the land,] . . . the white man must treat the beasts of this land as his brothers. I have [heard about]. . . .a thousand rotting buffalo on . . . [a] prairie, left by the white man who shot them from a passing train. . . . [How can] the smoking iron horse . . . be more important than the buffalo that [red men] kill [only enough of] to stay alive? . . . What is man without beasts? If all the beasts were gone, men would die from a great loneliness of spirit. . . . [W]hatever happens to the beasts soon happens to the man. . . .

This we know—the earth does not belong to man, man belongs to the earth. . . . All things are connected like the blood which unites one family. Whatever befalls the earth befalls the sons of the earth. . . . Man does not weave the web of life, he is merely a strand in it. Whatever he does to the web, he does to himself. (Quoted from Seed, J. et al., Thinking like a mountain, pp. 68–73. 1988. New Society Publishers. Reprinted with permission.)

INTRODUCTION

This beautiful and moving speech describes a proenvironmental religion that resembles the religions of some Native American tribes. These tribes and religions predate the industrial revolution and the westward wave of European invaders/settlers across the American continent by centuries, maybe even millennia. It is therefore ironic that a number of contemporary scholars, theologians, and writers claim that unless modern Westerners adopt a religion like the one in the speech, we will fail to solve the environmental problems that threaten human survival (e.g., Naess, 1989; Devall and Sessions, 1985; and Ehrenfeld, 1978). In other words, these people argue that we must set aside our current religious teachings and practices in favor of the proenvironmental ones in the speech. Nothing short of radical change of this type will ensure that Westerners—both individually and collectively—behave in proenvironmental ways. In this chapter, we closely examine this argument and explore the impacts of religious and moral teachings and practices on the origins, and possible solutions, of global and regional environmental problems.

Though the Chief Seattle "speech" is brief, it illustrates the main components of most religions and religious systems. In the case of Seattle's religion, all of these components act to encourage proenvironmental individual and collective behavior. (The different components overlap and intertwine, so it is somewhat artificial to discuss them separately.) First, a religion upholds certain *basic values*, that is, things, qualities, and principles it considers important and worthwhile. Thus, portions of the speech urge a reverence and respect for nonhuman forms of life and for natural processes. Some specific examples: ". . . [E]very part of the earth is sacred. . . . Every shining pine needle, . . . sandy shore, . . . humming insect . . . is holy. . . . The air is precious." The speech also portrays nonhuman forms of life as having as much importance and worth as human life: "Man does not weave the web of life, he is merely a strand in it."

Second, religions include basic *beliefs* and *worldviews* (collections of beliefs about the world and an overall perspective from which an individual and culture view it). Thus, the speech stresses the interrelatedness of all forms of life and the human dependence on nonhuman forms. For example: "The flowers are our sisters; the deer, the horse, the great eagle, these are our brothers. The rocky crests, the grasses . . . , and man—all belong to the same family. . . ." And: "All things are connected. . . . Whatever befalls the earth befalls the sons of the earth." Further, the speech portrays the Earth as the creator of life and views humans as strongly connected with it. For example: ". . . [T]his beautiful earth is the mother of the red man. We are part of the earth and it is part of us. . . ."

Third, religions include a system of *ethics or morals,* that is, they specifically encourage *specific individual behaviors* and enjoin other behaviors. Some ethics and morals are implicit in basic values and beliefs. Other ethics/morals are mentioned explicitly. Thus the speech urges: ". . . [Y]ou must give [the rivers] the kindness you would give . . . [a] brother." And: "[Red men] kill [only enough buffalo] to stay alive. . . ."

Fourth, religions generally include *ceremonies, rituals,* and other practices that convey and reinforce their values, beliefs, and behavioral injunctions. Ceremonies and rituals are not mentioned in the brief speech above, but they play a major role in Native American religions, the Western Judeo-Christian religious tradition, and most others.

Fifth, religions have *spiritual elements,* that is, elements involving deities or other supernatural forces. Spiritual elements arouse our feelings and emotions and they appeal to our intuitive sides. Thus, the speech asserts that the "the air shares its spirit with all the life it supports. . . ." and the speech refers to "ghostly reflections in the clear water of the lakes [that] tell . . . of events and memories in the life of my people." Note also how moving, beautiful, and inspiring the speech is overall. The speech is much more than a simple intellectual or rational description of the different forms of life in nature and how these forms interact.

Although a number of scholars, theologians, and writers argue that no resolution of environmental problems is possible without a major shift in religion and morals in the direction of the speech, there are a few who make a *more radical* claim. These schol-ars/writers/theologians argue that not only do current Western religions lack a proenvironmental orientation, but these religions are actually *antienvironmental;* indeed these religious beliefs, values, and practices are a major and active cause of all contemporary Western environmental problems. Probably the best-known of the writers who make this claim is historian Lynn White, Jr. In a provocative, often-quoted, though no longer widely accepted, article in *Science* magazine (1967), White argued that the Judeo-Christian religious tradition is a root cause of all Western environmental problems. White focused especially on the *Genesis* portion of the Bible. Consider the following Genesis quote:

> And God said, Let us make man in Our image, after Our likeness, and let them have dominion over the fish of the sea, and over the fowl of the air, and over the cattle, and over the earth, and over every creeping thing that creepeth upon the earth.
> So God created man in His own image, in the image of God created He him; male and female created He them.
> And God blessed them, and God said unto them, Be fruitful, and multiply, . . . and subdue [the earth]; and have dominion over the fish of the sea, and over the fowl of the earth, and over every living thing that moveth upon the earth. (Genesis I:16–28, King James version).

Note how different this well-known biblical passage is from the speech attributed to Chief Seattle. The speech depicts humans not as a special species but as one strand in an overall "web of life." Humans are seen as dependent upon other forms of life and are urged to protect them. Humans are an integral part of nature, rather than being separate from, and superior to, the other forms of life. In contrast, the biblical passage depicts humans as a unique and exalted species, since only humans were created in God's image. The passage also exhorts humans to control and "subdue" the other forms of life. Thus the passage makes a major distinction between humans and the rest of nature, with humans in a primary and dominant role, and other forms of life in a secondary and subservient role. Finally, the last paragraph ("be fruitful, and multiply") appears to encourage unlimited growth in human numbers.

The biblical view, White claimed, permeates Western culture, creating a general disregard for nonhuman forms of life and natural processes, a feeling of human invulnerability, and a push toward limitless growth. "Especially in its Western form," White argued, "Christianity is the most anthropocentric [i.e., human-centered] religion the world has seen. . . . [In the biblical story of creation,] God created light and darkness, the heavenly bodies, the earth and all its plants, animals, birds, and fishes. Finally God . . . created Adam and . . . Eve. . . . Man named all the animals, thus establishing his dominance over them. God planned all of this explicitly for man's benefit and rule; no item in the physical creation had any purpose save to serve man's purposes [p. 1205]."

White's (1967) basic position was shared by several other scholars and writers, including historian Arnold Toynbee (1973), biologist Paul Ehrlich (1978), and regional planner Ian McHarg (1971). However, White's (1967) thesis is no longer very widely accepted. Some have criticized both White's interpretations of the Judeo-Christian religious tradition and his analysis of Western culture and history. A number of scholars and theologians argue that—though Western values, beliefs, and morals *are* anthropocentric and *do* legitimate the exploitation of nature for human ends (and *are*, therefore, root causes of environmental problems)—the Judeo-Christian religious tradition is *not* the main source of these values, beliefs, and morals. Some of these scholars and theologians claim that the "multiply and subdue the earth" portions of Genesis are taken out of context and misinterpreted (e.g., Shaiko, 1987). Some point out, further, that many other portions of the Old and New Testaments emphasize the concept of "stewardship," or care, of nature (e.g., Shaiko, 1987; Naess, 1989; Gelderloos, 1992; and Whitney, 1993). These scholars and theologians thus argue that the Judeo-Christian tradition is more correctly seen as a major source of *pro*environmental values and beliefs, rather than of antienvironmental values and beliefs. We discuss the work of these "ecotheologians" in more detail later in the chapter.

Other scholars/writers blame the human-centeredness of Western culture and the belief in the legitimacy of exploiting nature for human ends not on our Judeo-Christian heritage but on elements of ancient Greek philosophy that are, in part, bases of modern scientific thought (Callicott, 1983), or on the view of nature as mechanical and inert that emerged in Western countries at the beginning of the scientific revolution in the 1600s (Shiva, 1989; Whitney, 1993), or on the development of capitalism in Western countries beginning in the late 1700s (Whitney, 1993).

Finally, yet other scholars/writers, including Ophuls (1977) and Brown (1981), argue that excessive levels of materialism and consumerism in Western countries are main causes of environmental problems, and these scholars/writers urge radical decreases in these values. To quote Brown (1981):

> *None of the political philosophies today embraces the values essential to a sustainable society. Indeed, as scientist B. Murray noted in an address to theologians, "Capitalism and Marxism have one thing very much in common: they both presume man's fundamental needs are material." Murray believes that for this reason both fall short. Whether capitalist or socialist, materialism is neither sustainable nor satisfying over the long term [p. 350].*

To quote Ophuls (1977):

> *. . . [T]he sickness of the earth reflects the sickness in the soul of modern industrial man, whose whole life is given over to gain, to the disease of endless getting and spending that can never satisfy deeper aspirations and must eventually end in cultural, spiritual, and physical death [p. 232].*

To summarize: Many scholars and writers argue that Western morals, beliefs, values, and/or religious teachings and practices play a major role in causing global and regional environmental problems, and that these morals, beliefs, and so forth must be shifted sharply in a proenvironmental direction if the problems are to be solved.

DO VALUES, MORALS, BELIEFS, AND RELIGIOUS TEACHINGS AND PRACTICES AFFECT HOW INDIVIDUALS AND CULTURES TREAT THEIR ENVIRONMENT?

As our discussion above indicates, the religions/values/morals approach is popular with scholars and theologians as a framework for understanding and potentially solving major environmental problems. The approach also has considerable intuitive appeal. It is manifestly true that Western culture is anthropocentric and materialistic, regardless of the historical sources of these orientations, and it makes sense that such a culture would treat its environment roughly.

However, much of the evidence we will review in the rest of this chapter suggests that the religions/ morals/values framework has only limited power in explaining why major environmental problems occur, and also suggests that major changes in societal religious beliefs, values, and morals are not likely to be effective by themselves in solving environmental problems. We remind the reader, however, of our emphasis in this book on *multiple* strategies for understanding and solving environmental problems rather than the total reliance on any single strategy. That a good deal of the evidence to be reviewed below is negative does not detract from the importance of this solution strategy as *one ingredient* in a multidimensional strategy for understanding and solving environmental problems.

We begin, in the next section, by reviewing the results of survey-research studies that fail to show a consistent relationship between individual Americans' religious beliefs and practices and their levels of environmental concern. We move on, in the section after it, to review the results of cross-cultural studies that fail to find a relationship between the degree of proenvironmentalism in a culture's religion, values, and so on, and its success in averting major environmental problems. We will see in later sections of the chapter, however, that values, beliefs, and morals can influence individuals' actions toward the environment, within limits.

Survey-Research in a Single Country (the U.S.)

One way to test the idea that Western religious beliefs and practices play a major negative role in environmental problems is to see if, within a single country, differences in people's religious affiliations and beliefs are associated with differences in their concern or behavior regarding environmental problems. Some psychologists, sociologists, and anthro-

pologists have done studies of this type in the United States. Unfortunately, these studies have reached contradictory conclusions. A few early survey-research (public opinion) studies by Eckberg and Blocker (1989), and also Hand and Van Liere (1984) and Shaiko (1987), concluded that Christians and Jews report less concern about the environment and environmental problems than do non-Christians/non-Jews; further, those individuals who most strongly believe in the Bible and its literal truth report less concern about the environment and environmental problems than those whose biblical beliefs are weaker. These results lend support to White's (1967) argument. However, in a more recent study, Greeley (1993) found, in survey-research data collected in 1988, no relationship between people's religion or their belief in the literal truth of the Bible and the level of concern they report about the environment. On the other hand, the authors of this book analyzed similar survey data collected in 1993 that appear to contradict Greeley's findings and support those of Eckberg and Blocker. Finally, Kempton, Boster, and Hartley (1995) provide evidence, though based on a small, nonrepresentative sample of Americans, that religious belief, defined generally, is *positively* correlated with expressions of environmental concern. In the next sections we explore these inconsistent findings.

Eckberg and Blocker (1989) interviewed 300 residents of the Tulsa, Oklahoma, metropolitan area via telephone. They asked each respondent several questions about religious beliefs and affiliation. One question asked whether the respondent was Christian or Jewish, or a member of some other religion (or none at all); (a large majority of the respondents were, as it turned out, Christian or Jewish). Another question asked the respondent to choose from the following three statements the one statement that most accurately described his/her beliefs about the truth of the Bible: "The Bible is the actual word of God and it should be taken literally. . . ," or "The Bible is the inspired word of God, but it was written by men and contains some human errors.," or "The Bible is an ancient book of history and legends [written by humans]. . . ."

Eckberg and Blocker also asked each respondent twelve questions about environmental problems. Eight addressed the respondent's level of concern about environmental problems in general: Four questions dealt with whether this country should accept pollution for the good of the economy (e.g., "Pollution control measures have created unfair burdens on industry"); four questions addressed whether we should protect the environment despite the economic costs of doing so (e.g., "We should maintain our efforts to control pollution, even if this slows down the economy and increases unemployment"). Response alternatives for these questions ran from "strongly disagree" through "strongly agree." The remaining questions concerned environmental problems in the Tulsa area: Two questions concerned air and water pollution, and two concerned waste disposal. Response alternatives for these questions ran from "not very serious" through "very serious."

Eckberg and Blocker began their data analysis by calculating the correlations between respondents' answers to the religious questions and answers to the questions about environmental concern. The left-hand data column (headed *Zero-order correlations*) of Table 3-1 displays the results. To explain how to interpret the column entries, consider a specific example: The table shows a small negative correlation (−.25) between respondents' degree of biblical literalism and their level of environmental concern expressed in answers about "accepting pollution for the good of the economy." In other words, respondents who expressed *a high level* of belief in biblical literalism were slightly less likely to express proenvironmental sentiments (i.e., were *more* willing to accept pollution for the sake of the economy) than were respondents who expressed a low level of biblical literalism. Looking at the left-hand column entries overall, we see small, but in many cases statistically significant, negative correlations between being Judeo-Christian and level of environmental concern as measured via the four different types of concern questions, and also between strength of belief in the literal truth of the Bible and level of environmental concern: Judeo-Christian respondents tended to show less environmental concern than non-Judeo-Christians, and those respondents with the strongest beliefs about the literal truth of the Bible tended to show the weakest levels environmental concern.

However, there is a major problem in interpreting the results in the left column of Table 3-1: A correlation between two variables (e.g., the inverse relationship between strength of belief in the literal truth of the Bible and level of environmental concern) does not necessarily mean that one variable (literal belief in the Bible) *causes* the other (lack of environmental concern). Instead, some third variable may be responsible for the correlation. For example, the people in the Eckberg and Blocker study who believed in the literal truth of the Bible may have been, on average, somewhat *older* than those who did not believe in the literal truth, and age may be the real cause of their lack of environmental concern. Alternatively, believers may have been *less educated* than nonbelievers, or believers may have belonged to *more strict, fundamentalistic denominations;* in turn, education level or denomination may have been the real cause of lack of environmental concern.

In anticipation of these problems of interpretation, Eckberg and Blocker (1989) asked each respondent several additional questions about his/her age, income, education, specific religious denomination, and other relevant variables. Eckberg and Blocker then calculated partial correlations (using what are known as regression analyses). The *right*-most data column of Table 3-1 displays the results of these analyses. The entries are the correlations between being Judeo-Christian and belief in the literal truth of the Bible and environmental concern, *independent of* the possible influences of age of respondent, income, level of education, gender, conservatism of religious denomination, and the importance of religion in respondents' lives. Looking at the pattern of the partial correlations in the right column of the table, note that belief in the literal truth of the Bible was negatively correlated (to a modest but statistically significant extent) with all four of the measures of environmental concern; also being Judeo-Christian was negatively correlated with the "protect the environment" questions on environmental concern. These results, then, do lend support to Lynn White's (1967) hypothesis

TABLE 3-1 Selected Results of the Eckberg and Blocker (1989) Study

CORRELATIONS BETWEEN RESPONDENTS' BELIEF/NONBELIEF IN THE LITERAL TRUTH OF THE BIBLE, AND THEIR JUDEO-CHRISTIANITY/NON-JUDEO CHRISTIANITY ON FOUR MEASURES OF ENVIRONMENTAL CONCERN

	ZERO-ORDER CORRELATIONS	PARTIAL CORRELATIONS
Use of the environment for the economy (4 questions)		
Judeo-Christian vs. non J-C	−.14*	—
Belief in literal truth of Bible	−.25***	−.17**
Protect the environment (4 questions)		
Judeo-Christian vs. non J-C	−.21***	−.14*
Belief in literal truth of Bible	−.16**	−.13*
Concern about Tulsa air and water (2 questions)		
Judeo-Christian vs. non J-C	−.13*	—
Belief in literal truth of Bible	−.09	−.14*
Concern about Tulsa waste disposal (2 questions)		
Judeo-Christian vs. non J-C	−.10	—
Belief in literal truth of Bible	−.10	−.18**

Source: Adapted from Eckberg, D., and Blocker, J., Varieties of religious involvement. *Journal for the Scientific Study of Religion,* Volume 28, p. 514. Copyright 1989. Society for the Scientific Study of Religion. Used with permission.
Legend: *significant at the .05 level of significance; **significant at the .01 level of significance; ***significant at the .001 level of significance; —This correlation is only minimal.

that Judeo-Christian biblical teachings are responsible for environmental degradation in Western nations: Within Eckberg and Blocker's sample of 300 Americans from Oklahoma, those individuals who most strongly believed in the Bible and its literal truth reported less concern about the environment and environmental problems than those whose biblical beliefs were weaker; also, Christians and Jews showed less concern (at least by one measure) than non-Christians/non-Jews.

We should point out, however, that even though Eckberg and Blocker (1989) used regression analyses to control for the effects of age, gender, and so on, there may have been *yet other* variables (e.g., respondents' ethnicity or race) that Eckberg and Blocker failed to control via the questions they asked and the statistical analyses they performed. Therefore, their results cannot prove that belief in the literal truth of the Bible and being Judeo-Christian actually *cause*

low concern about the environment. (All correlational research is limited in this way; a correlation between two variables—even a partial correlation derived from regression analyses—never proves that one variable has a true causal effect on the other.) Indeed, as we discuss below, Greeley (1993) argues that Eckberg and Blocker failed to control for the effects of certain key variables.

Andrew Greeley (1993), a social scientist and a Roman Catholic priest, took Eckberg and Blocker's (1989) research plan and data analysis a statistical step further; his results suggest that Eckberg and Blocker drew the wrong conclusion. Greeley examined the relationship between religious beliefs and environmental concern in a representative sample of Americans who took part in a large (well over a thousand respondents) public opinion survey in 1988—the annual General Social Survey (GSS) conducted by the National Opinion Research Center in Chicago. This survey, which covered a broad range of topics, contained several different questions about the respondents' religious beliefs (including one question on biblical "literalism" identical to the one used by Eckberg and Blocker, 1989), and also a single question about respondents' levels of environmental concern. That question was: "[Are we as a nation now] . . . spending too much money, too little, or about the right amount . . . on improving and protecting the environment. . . ? [p. 22]" As the left-hand data column of Table 3-2 shows, Greeley found small but statistically significant correlations between religious affiliation and literal belief in the Bible, and environmental concern: Respondents who were Christian and respondents who believed in the Bible's literal truth were less concerned about improving and protecting the environment than were respondents who were non-Christian and respondents who did not believe in the literal truth of the Bible; (note that Greeley used the category "Christians," rather than "Christians and Jews;" however, the number of Jews in the U.S. population is relatively small, and inclusion of them in the same category should not have greatly affected the results). Greeley's "raw," or zero-order, correlations were, thus, similar to those obtained by Eckberg and Blocker (1989) in Oklahoma.

However, Greeley argued that the biblical literalists in his study were more likely to be *political conservatives* than political liberals, and that it was political conservatism, not literalism, that accounted for the lower levels of environmental concern found for literalists. Put another way, Greeley argued that there are some biblical literalists who are politically liberal, rather than conservative—Vice-President Albert Gore, for example—and these literalists are as concerned about the environment (and about other liberal causes) as are nonliteralist liberals. Greeley further postulated that Christians and biblical literalists were also more likely to be *morally rigid* and have *harsher religious mental imagery* (e.g., a mental conception of God as a father, master, judge, and king, versus as a mother, spouse, lover, or friend) than were non-Christians and non-literalists. Therefore, Greeley used regression analyses to statistically control for the effects of political liberalism/conservatism, moral rigidity, and religious imagery, and also the effects of such sociodemographic variables as age and level of education. The right-most column of Table 3-2 shows the results of his analysis. Note that the standardized beta weights, which can be interpreted as partial correlations, relating Christian/non-Christian and biblical literalism/nonliteralism to environmental concern were *not* statistically significant. These results therefore contradict those of Eckberg and Blocker (1989).

TABLE 3-2 Results from the Greeley (1993) Study

CORRELATIONS BETWEEN RESPONDENTS' BELIEF/NONBELIEF IN THE LITERAL TRUTH OF THE BIBLE, AND THEIR CHRISTIANITY/NON-CHRISTIANITY ON ONE MEASURE OF ENVIRONMENTAL CONCERN

	ZERO-ORDER CORRELATIONS	PARTIAL CORRELATIONS (STANDARDIZED BETA WEIGHTS)[†]
Not enough money spent on the environment		
Christian vs. non-Christian	−.10[*]	−.02
Belief in literal truth of Bible	−.11[*]	−.04

Source: Adapted from Greeley, A., Religion and attitudes toward the environment. *Journal for the Scientific Study of Religion*, Volume 32, pp. 19–28. Copyright 1993. Society for the Scientific Study of Religion. Used with permission.
Legend: [†]These are standardized beta weights from a regression analysis, and can be interpreted as partial correlations.
[*]significant at the .05 level or better

Though Greeley's (1993) conclusions were the opposite of Eckberg and Blocker's, we believe that Greeley's results are not definitive for two reasons: First, Greeley used a single and ambiguous measure of environmental concern. A respondent's judgment about whether the government is spending the right amount of money on improving and protecting the environment actually measures both the respondent's environmental concern as well as his or her assessment of the adequacy of government spending. (In contrast, Eckberg and Blocker, 1989, used twelve different questions in an effort to tap their respondents' levels of environmental concern). Second, Greeley may have gone too far in his attempts to statistically control for his respondents' religious imagery and moral rigidity. More specifically, his questions on these dimensions may actually have included aspects of biblical literalism. For example, one question on moral rigidity asked how important was it to the respondent ". . . to follow [his/her] . . . conscience even if it means going against what the churches or synagogues say and do. . . ." If this question duplicates or overlaps the biblical literalism question, then using it to statistically control for moral rigidity could spuriously mask a real correlation between biblical literalism and environmental concern.

A Look at More Recent GSS Data. For our theoretical and empirical analyses of how values influence environmental attitudes and actions and for the analysis of the 1993 General Social Survey data on religion and environmental concern, we are especially indebted to Thomas Dietz (e.g., Dietz and Stern, 1995; Stern, Dietz, and Guagnano, 1995; Stern and Dietz, 1994). As we point out above, Greeley's analysis suffers from its use of a single, ambiguous measure of environmental concern. His analysis would be more convincing if measures of religious belief were related to more trustworthy measures of environmental concern. Fortunately, the 1993 General Social Survey contains both a series of items on religion and a series on environmentalism. We now describe and analyze results from the 1993 survey, which involved approximately 1,600 randomly selected Americans.

Among the environmental items in the survey, in addition to the one Greeley used, were a series of items that we have grouped into *three* scales of proenvironmental action and willingness to support environmental protection financially. One scale, measuring consumer behavior, is made up of three items such as "How often do you make a special effort to buy fruits and vegetables grown without pesticides or chemicals?" The political behavior scale consists of three items such as "In the last five years, have you taken part in a protest or demonstration about an environmental issue?" The third scale measures willingness to make financial sacrifices for the environment, and includes three items such as "How willing would you be to pay much higher prices in order to protect the environment?"

Table 3-3 presents regression coefficients that represent the strength and direction of the relationship between measures of Christianity and religious belief and behavior and the four measures of environmentalism. The first two columns, which report results on the indicator of environmental concern that Greeley used, show that the 1993 data replicate Greeley's findings. The two indicators of religion that Greeley reported (see Table 3-2) have a weak but significant relationship to the government-spending measure of environmentalism, but the relationship disappears when age, education, income, gender, and political liberalism-conservatism are taken into account (boldface/italicized entries).[1] For other indicators of environmentalism, however, measures of Judeo-Christian religious belief do sometimes make a difference, and usually, the effect is negative, as White's (1967) thesis predicts. The significant relationships between religion and the political behavior and financial support scales are almost uniformly in the direction predicted by White's thesis.

The consumer behavior scale seems to present a different picture, however. Although Christians are less likely than non-Christians to change their behavior specifically to protect the environment, among Americans of all faiths, those who are stronger in their religious practices (frequency of praying and self-reported strength of affiliation) are more likely to engage in such behavior.

TABLE 3-3 Analysis of 1993 GSS Data

REGRESSION COEFFICIENTS REPRESENTING ASSOCIATIONS BETWEEN INDICATORS OF ENVIRONMENTALISM
AND OF RELIGIOUS BELIEF AND BEHAVIOR

	SPENDING ON ENV'T.		POLITICAL BEHAVIOR		FINANCIAL SACRIFICE		CONSUMER BEHAVIOR	
	A	B	A	B	A	B	A	B
Christian	−.17***	−.05	−.50***	−.30**	−.45***	−.24*	−.36***	−.33**
Belief in God	−.03**	−.01	−.17***	−.07**	−.14***	−.05	−.04	−.01
RELIGIOSITY								
Frequency of prayer	.03*	.01	.04	.02	.05	−.02	−.09***	.09***
Strength of affiliation	−.04	−.02	.02	.04	.17**	.20**	.17***	.17***
FUNDAMENTALISM								
Biblical literalism	−.10***	−.05	−.30***	−.16*	−.29***	−.12	−.04	−.02
Fundamentalist	−.07***	−.05*	−.20***	−.09*	−.27***	−.14**	−.12**	−.09*
SACREDNESS OF NATURE								
Nature sacred in itself	.30***	.21**	.58***	.41***	.61***	.41***	.40***	.26**
Nature sacred; made by God	−.08*	−.04	−.33***	−.16*	−.07	.16	−.02	.03
Nature import. not sacred	−.14***	−.12**	−.09	−.15*	−.39***	−.48***	−.28***	.23**

Columns A present regression coefficients uncontrolled for other variables.
Columns B present them **controlled for age, education, income, gender, and political liberalism.**
*p < .05; **p < .01; ***p < .001

This finding seems self-contradictory and needs further explanation. The data in the table suggest at least one possibility. Adherence to religion is sometimes understood in terms of two separate dimensions, religiosity (typically measured by frequency of praying and self-reported strength of affiliation) and fundamentalism (typically measured by items like the belief that scripture was literally written by God and other indices of strict religious ideology—this survey used a measure that rated each respondent's religious denomination as fundamentalist, moderate, or liberal). This distinction makes a bit more sense of the data. The effects of two measures of fundamentalism on proenvironmental behavior are uniformly negative, as White would have predicted, although the effect may be weaker for consumer behavior than for the other indicators. The two measures of religiosity have positive effects on environmentalism—opposite to White's

thesis—but these effects are restricted almost entirely to consumer behavior. This pattern suggests that although strict adherence to Judeo-Christian religious ideology (fundamentalism) has antienvironmental implications, particularly for political behavior, something different applies for religiosity and for personal behavior. One interpretation is that both religious attendance and proenvironmental individual behavior are part of a broader pattern of altruism and good citizenship among Americans who identify with organized religion. This hypothesis, however, is not strongly supported by the data, and so deserves further testing. In fact, a different form of analysis of the same data found no relationship between strength of religious affiliation and consumer behavior (Dietz, Stern, and Guagnano, 1998).

The data clearly suggest, however, that the relationship between religion and environmentalism is more complex than White's thesis implies. It seems to depend not only on beliefs in broad religious ideology but on other factors associated with religious adherence—and it may be that religious influences affect different kinds of proenvironmental behaviors in different ways. We should emphasize, however, so as not to lose sight of the main point, that all the relationships represented in Table 3-3 are rather weak, suggesting that religious belief by itself has a limited influence on behavior among present-day Americans. Furthermore, as we noted above, a significant correlation between two variables (or a significant regression coefficient) never proves causality. Similarly, if biblical literalists tend to be more antienvironmental than nonliteralists, this does not demonstrate that biblical scripture (or, for that matter, the teaching of certain religious denominations) is correctly characterized as being antienvironmental.

One other finding about religion and environmentalism from the 1993 General Social Survey is worth further mention. Respondents were asked to choose the one of the following statements that most closely represented their personal beliefs:

- Nature is spiritual or sacred in itself.
- Nature is sacred because it is created by God.
- Nature is important, but not spiritual or sacred.

The lower part of Table 3-3 presents the relationships between these beliefs and the indicators of environmentalism—the three lines under the heading *Sacredness of nature*. The *first line* under the heading gives the regression coefficients when the respondents who chose "Nature is sacred in itself" are compared to subjects who chose the other two alternatives; the entries indicate that respondents who chose "Nature is sacred in itself" were *much more likely* to report proenvironmental concern and action than those who chose the other alternatives, even when age, education, income, gender, and political liberalism were controlled for.

The *third line* shows that respondents who chose "Nature is important but not sacred" were significantly *less likely* to report proenvironmental concern and action than those who chose the other alternatives, even when age, and so on, were controlled for.

Finally, the *second line* shows that respondents who chose "Nature is sacred because it is created by God" were neither more or less likely to report proenvironmental concern/action than the other two groups combined.

These results suggest a surprising conclusion: Individuals whose belief in the sacredness of nature is based on religious teachings are apparently less proenvironmentalist than people who do not tie that belief to God. Thus, it seems that a belief in the sacredness of nature has a significant influence on behavior—both personal and political—but *less so* when that belief is derived from the teachings of organized religion. It is almost as if people who believe nature is sacred because it is God's creation feel that God will take care of nature, and that they need not. We believe that these findings need to be explored further to see which are reliable and to understand their meaning.

The above findings contradict the results of a study that we discuss in the next section, though that study involved a small and nonrandom sample of Americans, unlike the GSS 1993 survey on which our analyses above are based.

Kempton, Boster, and Hartley (1995). We lastly and very briefly describe a study performed by anthropologists Willett Kempton, James Boster, and Jennifer Hartley (1995). Their goal was to try to understand in depth the values and beliefs that underlie people's concern, or lack of concern, about environmental problems. Kempton et al. queried 142 respondents in four U.S. states and in five categories: members of Earth First! (a radical proenvironmental group), members of the Sierra Club (a more mainstream environmental group), the general public, managers of dry-cleaning businesses, and laid-off sawmill workers. Kempton et al. assumed that the environmental group members and some members of the general public would have proenvironmental sentiments, while dry-cleaning managers and laid-off sawmill workers would have antienvironmental sentiments. A key feature of the study was that the researchers asked respondents directly and in an open-ended way *"why* they thought protecting the environment was important [to the extent the respondent felt environmental protection was important]."

Kempton et al. found that a majority of people in all five groups expressed high levels of environmental concern (measured in various ways) and that many of them volunteered that this concern was based on religious and/or spiritual values. When asked, approximately 75 percent of all respondents (in *all* of the five groups) agreed or strongly agreed with the statement: "Because God created the natural world, it is wrong to abuse it." Further, those respondents who agreed most strongly with the statement were significantly more likely to have reported that "they belonged to an organized religion." Kempton et al. did not report *which* religions their respondents belonged to, but to the extent that the religious respondents were predominantly Christian, these results would seem to contradict those of the 1993 GSS that we presented in Table 3-3. Kempton et al.'s sample was very small, however, and their respondents were hardly representative of the U.S. population in general (this was the researchers' intention; the goal of their study was to identify in a general way the values and beliefs that underlie environmental concern). Further, Kempton's analysis did not statistically control for the effects of such potentially confounding variables as respondents' age, gender, income, and so on. Thus, their results must be viewed as interesting but not definitive.

We should mention one other result, a surprising one, that Kempton et al. report. Almost half of the respondents who indicated that they did *not* belong to an organized religion and did *not* even believe in a "spiritual force in the universe" *agreed* with the statement "Because God created the natural world, it is wrong to abuse it." It's as if these respondents had strong spiritual feelings that nature is sacred, but had no other way to express these feelings.

A further relevant result: Kempton et al. found that a number of their respondents reported a strong spiritual experience when they were in natural outdoor environments: Some, respondents who identified themselves as religious, stressed that they felt the presence of God in such environments. Some others, respondents who were not religious in a formal sense,

stressed that they felt a spiritual awareness or consciousness when in such environments.

Overall, though Kempton et al.'s samples were small and nonrepresentative, and though they used simple statistical summaries rather than regression analyses, their data do suggest an important spiritual dimension to people's concerns about the environment: Rather than being a source of antienvironmental feelings and attitudes, organized religions and religious sentiments in general may be potentially major sources of proenvironmental concern. As we discuss later in the chapter under the heading "Ecotheology," many organized religions in the United States are now in a state of flux. Religious leaders are becoming more environmentally conscious and are looking for ways in which religious institutions can encourage proenvironmental concern and action on the part of their congregants. We end our discussion of Kempton et al.'s work by mentioning a significant and relevant example they cite: U.S. Vice-President Albert Gore ". . . justifies his personal environmental values based on his own Southern Baptist faith (1992: 242–248) [quoted from Kempton et al., 1995, p. 39])."

To summarize our discussion in this section of the chapter: There are only a few studies of the relationships in this country between Judeo-Christian affiliation, and also belief in the literal truth of the Bible, and people's level of environmental concern, and the results of these studies, while interesting, are inconsistent.

Historical and Intercultural Evidence

The survey and interview studies that we discussed above found rather weak and inconsistent relationships between religious beliefs and environmental concern and behavior. However, keep in mind that all of the subjects in the studies were Americans and a large majority came from Judeo-Christian religious traditions. As a result, *none of the studies* really addressed the larger and more important issue of whether Western religious/moral beliefs and values are largely responsible for the environmental deterioration found in Western countries, and whether alternative, more proenvironmental values and beliefs would successfully prevent or lessen environmental damage. This major issue *is* addressed in research, which we now review, that examines the environmental records of different cultures throughout the world and at different times of history—cultures that vary in the degree to which their religious teachings, morals, values, and so on were pro- or antienvironmental.

To begin with, several scholars have argued that the environmental records of many non-Western cultures are just as bad as those of Western cultures. Thomas Derr (1975, quoted in Dwivedi and Tiwari, 1987), for example, argues that the ancient Egyptians, Assyrians, North Africans, and Aztecs seriously damaged their environments, in some cases to a degree sufficient to destroy their civilizations.

More powerful evidence against the idea that proenvironmental values and beliefs can avert environmental degradation is the serious environmental damage that has occurred in certain cultures throughout the world that have notably proenvironmental religious teachings. Thus, Cobb (1972, cited in Hargrove, 1986) and Derr (1975, op. cit.) both claim that some Native American tribes had a blemished environmental record, despite their proenvironmental religious and moral codes. They also point out that proenvironmental Eastern religions, including Hinduism in India, and Taoism and Buddhism in China, did not prevent enormous environmental damage from occurring in these countries. To take a closer look at this claim, we devote the next few pages to an in-depth examination of these three Eastern religions and their associated environmental records.

India and Hinduism. O. Dwivedi and B. Tiwari (1987) argue that India's environmental record is extremely poor, despite the fact that Hinduism, the dominant religion in India, is the oldest and the most proenvironmental religion in the world. Hinduism, which predates the birth of Christ by several thousand years, holds that humans, other animals, plants, and even "lifeless" environmental features like stones and mountains are all part of an underlying unity because all are suffused with the same spiritual energy. Humans are thus viewed as an integral part of nature, rather than

as being an exalted species destined to control and exploit nonhuman forms of life.

Further, Hinduism (which has elements of monotheism and polytheism), is *animistic* in that specific gods are thought to be manifested or incarnated in animals and other natural forms (a belief probably derived from prehistoric peoples who worshiped the forms and forces of nature that they could not understand, but which they were at the mercy of). Consider, for example, the following passage from the *Kalika Purana* (1927, quoted in Shiva, 1989):

> *Rivers and mountains have a dual nature. A river is but a form of water, yet it has a distinct body. Mountains appear a motionless mass, yet their true form is not such. We cannot know, when looking at a lifeless shell, that it contains a living being. Similarly, within the apparently inanimate rivers and mountains dwells a hidden consciousness. Rivers and mountains take the forms they wish (Shiva, p. 39).*

Perhaps the best-known animistic feature of Hinduism is cow worship. Cows are thought to be manifestations of a goddess and are therefore sanctified, protected, and allowed to roam at will. The key role of cow worship in Hinduism is made clear in the following words of Mahatma Gandhi:

> *The central fact of Hinduism is cow protection, cow protection is one of the most wonderful phenomena in human evolution. It takes the human beyond his species. The cow to me means the entire subhuman world. Man, through the cows, is enjoined to realize his identity with all that lives. . . . [P]rotection of the cow means protection of the whole . . . creation of God. . . . [C]ow protection is the gift of Hinduism to the world. And Hinduism will live as long as there are Hindu to protect the cows (quoted in Dwivedi and Tiwari p. 45).*

In addition, plants, for example pipal and banyan trees, just like animals, are thought to be the dwelling places of gods and are sacred. Many households worship a pipal branch in a ceremony in the spring. The scriptures, further, contain injunctions and penalties for the cutting of trees and for various acts of polluting the environment.

However, despite the obvious proenvironmentalism of Hindu scripture and practice, the environmental record of India is, as mentioned above, a miserable one. Dwivedi and Tiwari note the ". . . enormous loss of natural resources . . . brought on by the cutting of trees. . . , and by the killing of animals and birds [p. 53]." In addition, heavy pumping of water for irrigation of crops in parts of India has lowered water tables by as much as ninety feet in ten years. In some places the pumping has caused salt water to invade aquifers and contaminate drinking supplies (Postel, 1990). In addition, overirrigation, overgrazing, and deforestation have seriously damaged large tracts of land in India; approximately 35 percent of potentially productive land has been degraded due to the resulting water erosion, wind erosion, and salinization (Postel, 1989). In turn, it appears that deforestation and desertification have worsened droughts and floods (ibid.). And so on.

What explains the above discrepancies between the proenvironmental religious precepts of Hinduism and India's poor environmental record? Several factors appear to be responsible.[2] One, as Dwivedi and Tiwari (1987) point out, is the invasion and occupation of India by Muslim and Western (British) cultures over a period of 700 years and the consequent weakening of Hinduism (p. 90). However, as Hinduism dominates in India today, cultural invasion cannot fully explain the country's current poor environmental record (p. 91).

A second factor is the philosophy of development that has been adopted in India (and many other developing nations) in response to Western ideas—the demands of international markets, and pressure from international lenders whose development loans must be repaid in hard currency. For example, forests are cleared to replace indigenous trees with commercially valuable species that bring in cash for loggers and to repay international debt. Among the results have been a loss of firewood and fodder from the noncommercial trees; hardship for peasants—especially women— who had provided for their families, especially in hard times, by gathering twigs, fodder, and fruit from the forests; and massive downstream floods due to erosion of soils held in place by the indigenous vegetation (Shiva, 1989). Because of these impacts on women,

the poor, and the environment, Vandana Shiva (1989) refers to India's policy as one of "maldevelopment."

A third factor is the unrelenting pressures of population growth, industrial development, and urbanization. The exponential growth of human numbers and the consequent growth in the need to feed, clothe, and shelter them simply overwhelms religious precepts. To provide one example, India's population, which has doubled since 1950, has "outstripped the sustainable production levels of . . . [India's] fuelwood and fodder resources; . . . demand for these resources in the early 1980s exceeded supply by 70% and 23%, respectively [Postel, 1989]." As Dwivedi and Tiwari (1987) put it: "Members of the public, by themselves, will seldom venture into this battle to save the environment if their total attention is always placed on the battle for survival [p. 101]."

China and Taoism and Buddhism. In prerevolutionary China, the pressures of population and industrial and urban growth also overwhelmed proenvironmental religious precepts and contributed to a long record of environmental degradation. As we will see, however, additional factors help account for China's environmental record, factors that also operate in India and other countries that have proenvironmental religious beliefs.

Before the Communist revolution in 1949, there were three dominant Chinese religious, philosophical, and/or moral traditions: Taoism, Buddhism, and Confucianism. (Elements of these traditions are still present in China today, though they are less visible [Kamachi, 1994].) Both Taoism and Buddhism contain many proenvironmental elements. Let's examine the elements in Taoism first. Quoting philosopher Po-Keung Ip (1983, pp. 338–339):

[Basically, the concept of "Tao" is] . . . a totally depersonalized concept of nature. . . . [Tao] is . . . intangible, . . . simple, all-pervasive, eternal, [and] life sustaining. . . . [The word] "Té" [refers to] . . . the potency, the power, of Tao that nourishes [and] sustains beings. . . . Since Té is internalized in all beings in the universe, [all are thereby linked and related].

All beings in the universe, furthermore, are of equal importance. Thus in Taoism, humans are neither superior to nor separate from the nonhuman parts of nature.

Going further, the Taoist doctrine of "Wu Wei" enjoins humans to act in accordance with, rather than against, the laws and processes of nature. Although it is appropriate for people to act to change nature by, for example, building dams and canals, these changes must be designed in accord with the way the hydraulic forces of nature operate, and nature—not people—is always the final arbiter of the success or failure of the project (Goodman, 1980). Finally, the Taoist concept of *reversion* emphasizes the cyclical, nonlinear characteristics of many natural processes, as when a living thing dies, decays, and new life forms from its remains (Po-Keung Ip, 1983; Goodman, 1980). Thus, natural processes are closed processes; garbage or other things that are thrown away by humans never leave natural systems.

The above Taoist principles are very much in line with the basic principles of modern ecology, as Goodman (1980) points out. Indeed, when one first reads the Taoist principles, they seem almost like summary statements taken from a contemporary ecology textbook!

Very briefly, *Buddhism,* like Taoism, stresses the equality of different forms of life and emphasizes the intrinsic value and importance of nonhuman life forms. "Buddhists are taught to love all living beings and not to restrict their love only to human[s]. . . . The Buddha's advice is that it is not right . . . to take . . . the life of any living being since every living being has a right to exist [independent of its utility or value to humans] (Dhammananda, 1982)."

Although Taoism and Buddhism contain many proenvironmental elements, the environmental record of prerevolutionary China was not a good one, as we noted above. For example, Yi-Fu Tuan (1970) documents deforestation on a "vast scale" in northern China and "acute problems of soil erosion on the loess-covered plateaus [p. 248]."

What then explains the discrepancy between the proenvironmental teachings of the traditional Chinese religions of Taoism and Buddhism and the poor environmental record of prerevolutionary China? Yi-Fu Tuan identifies population growth as a major factor (as was the case with India). As the population in China

grew, forests were cut down to provide more land for agriculture, and also wood for construction, and charcoal to heat homes. When faced with the choice between denuding forests in violation of religious teachings, on the one hand, and allowing people to go hungry or to freeze to death, on the other, there is no doubt as to which choice wins. Urbanization and the growth of industries (e.g., metal industries) that consumed wood as fuel also contributed to environmental degradation during certain time periods.

However, the causes of the discrepancy were considerably more complicated than the above paragraph would suggest, and population and urban and industrial growth were not the only factors operating. Possibly as important were the many political, economic, and social factors that—besides religious and moral teachings—determine the behavior of virtually any culture or country, both developing countries as well as advanced industrial states. As Yi-Fu Tuan points out, ". . . China, with her . . . temple compounds, was also a vast bureaucracy, a civilization, and an empire [p. 247]. . . . In the play of forces that govern the world, esthetic and religious ideals rarely have a major role [p. 244]." The egoistic and imperialistic behavior of kings, generals, and other leaders with political, military, and/or economic power can contradict and overwhelm the religious and ethical teachings of a culture. Similarly, groups that have political, military, and economic power are often in a position to override the religious and moral beliefs of other groups that lack such power. National governments sometimes do this by extracting resources from poor or less powerful regions to enrich the rulers and the imperial center, thereby disrupting what may have been sustainable local relations between people and their environment.

An example of this "override" phenomenon is the large-scale deforestation that is now taking place in the Amazon River basin of Brazil. The efforts of indigenous tribes to protect their sacred ancestral homelands, and also the efforts of nonindigenous rubber tappers to prevent deforestation, have been swept aside, sometimes violently, by cattle ranchers and others who—spurred on by such government policies as providing access roads deep into the forest, granting ownership to anyone who clears a piece of forest, and providing tax incentives for farming—have ruthlessly pursued their own economic agendas by clearing land and creating large cattle ranches. International lending institutions long supported these policies by funding large-scale development projects that cleared forest land and made Brazil more dependent on cash crops, such as timber and beef from Amazonia, to repay international loans (Stern, Young, and Druckman, 1992).

National governments may also disrupt sustainable local relations between people and their environment in the name of development, as we have seen in the case of India. In postrevolutionary China, the government—in addition to suppressing the traditions of Taoism, Buddhism, and Confucianism—has sacrificed the environment to development, for example, by pursuing a policy of heavy industrialization that relies on burning highly polluting soft coal, without emission controls (Smil, 1988).

Many of the political, social, and economic forces and processes that we outlined in the last few paragraphs above play a significant role in *any* country, and most certainly help explain the poor environmental record of India. Some of these forces and processes also operate in advanced industrial nations like the United States today, nations in which population growth is much slower than in developing nations. The strong influence of nonreligious forces and processes in any country is a major reason that religious/moral/and so on efforts by themselves are insufficient to solve major environmental problems.

PROENVIRONMENTAL RELIGIOUS/MORAL MOVEMENTS—CURRENT DEVELOPMENTS AND POSSIBLE FUTURE TRENDS

The experiences of India and China warn us of the limits of religious and moral controls alone as a way to solve environmental problems. Nevertheless, we believe that a stronger moral consciousness of the environment will help, and that religious and moral changes will play a role—most likely an important one—in any successful, permanent solution to global environmental problems. With this in mind, we briefly review in this section four current movements in the United States and other advanced industrial societies that are trying to raise this sort of moral/religious consciousness. We then, in later sections, examine these four movements in terms of their common threads and their prospects for success. The religious/moral movements are: contemporary Christian and Jewish *ecotheology*, which emphasizes proenvironmental aspects of traditional scripture; Catholic theologian Thomas Berry's proenvironmental religion, which includes elements of Eastern religions and of modern scientific ecology; the *deep ecology* movement, led by Arne Naess, Bill Devall, and George Sessions, a movement that provides a new worldview and that urges major changes in Western lifestyles and values; and *ecofeminism*, which claims that there is an intrinsic moral and practical linkage between solving environmental problems in Western cultures and ending gender-role stereotyping and discrimination against women. Note that these four religious/moral movements overlap with each other to some extent and that they are not mutually exclusive (i.e., people could embrace more than one of the movements at the same time). Note also that the four differ in the degree to which they are complete religions: Both ecotheology and Berry's work involve full-blown religions, whereas deep ecology and ecofeminism are each moral/ethical/values movements that have some religious elements.

Ecotheology

As we mentioned earlier in this chapter, many modern theologians argue that the "multiply and subdue the earth" quote from the Bible (Genesis), above, is misinterpreted, taken out of context, and viewed simplistically. There is little in traditional Judeo-Christian scripture, these theologians argue, that endorses environmental exploitation, and much that supports a *stewardship* philosophy, one that stresses our responsibility to respect and care for the Earth, its ecological systems, and nonhuman forms of life (Gelderloos, 1992; Whitney, 1993).

There is now considerable change occurring in many Western religions, even religions traditionally viewed as very conservative (Gelderloos, 1994). Indeed, a proenvironmental "Statement by Religious Leaders. . ." stressing the concept of stewardship was endorsed in 1991 and again in 1992 by a large panel of prominent Catholic, Greek Orthodox, Jewish, and Protestant religious leaders representing a total of 330,000 different congregations in the United States (Anderson et al., 1991; Moehlmann, 1992). Further, a National Religious Partnership for the Environment has been formed by organizations of Protestant, Catholic, Jewish, and evangelical Christian denominations. These movements tend to confirm Kempton et al.'s (1995) conclusion that, contrary to White's (1967) thinking, religions in this country may be a major source of, rather than an obstacle to, proenvironmental sentiments and actions.

Many ecotheologists base their arguments for the stewardship concept on a close examination of the wordings of original biblical and other scriptures, as distinct from the teachings and writings of more recent religious leaders and movements (Gelderloos, 1992; Whitney, 1993). Thus, Gelderloos (1992) distinguishes between *Christianity* and its Judaic roots on the one hand, and *Christendom* on the other. The former refers to actual Judeo-Christian Holy Scripture, the latter to commentaries, teachings, sermons, encyclicals, and practices of members of religious organizations over the past two millennia. Gelderloos argues that the concept of stewardship appears clearly in the Jewish Torah and Old Testament, and even in some New Testament scripture, but that Christendom has greatly deviated from these scriptures in recent centuries. Gelderloos attributes this deviation to several causes, including the influences of the prominent Christian theologian Saint Augustine several centuries after the death of Christ, the Protestant Reformation in the sixteenth century, the mechanistic worldview that emerged in the Western Enlightenment during the eighteenth century, and the industrial revolution.

Gelderloos (1992) and others base their claim that the stewardship ethic is clearly present in the Jewish Torah and the Old Testament (centuries before the deviation described above) on several lines of reasoning. For one thing, these scriptures repeatedly emphasize that God is the creator of ". . . the heavens, . . . the earth and all that is on it, the seas and all that is in them [Neh 9:6, quoted on p. 13], and that [God] . . . gave life to everything"; things created

by God, in turn, are sacred and must be protected. In addition, the Book of Genesis (the book containing the "multiply and subdue" passages) also describes the role of humankind as the "tillers" and "keepers" of God's creations, with the relevant Hebrew words clearly implying that humankind is to care for and keep these creations forever. Similarly, Barr (cited by Whitney, 1993) argues that the Hebrew word appearing in Genesis and translated as "to have dominion" is more correctly translated as implying responsible leadership, and the word translated as "to subdue" specifically refers to the physical act of plowing soil, and not to the domination and exploitation of animals.

Gelderloos (1992), further, points out that support for the stewardship concept is found in Jewish rabbinical writings (which are separate from the Old Testament and therefore not shared with Christians) as well as the Old Testament itself. Gelderloos cites the writing of Ehrenfeld and Bentley (1985) on the (Old Testament) Jewish law of *bal tashhit* (meaning "do not destroy"), and also specific rabbinical injunctions against ". . . overgrazing of the countryside, the unjustified killing of animals or feeding them harmful foods, the hunting of animals for sport, species extinction and destruction of cultivated plant varieties, pollution of air or water, overconsumption of anything, and the waste of mineral and other resources [quoted in Gelderloos 1992, p. 16]."

In summary, Gelderloos and other ecotheologians argue that Christians and Jews should now return to the stewardship "paradigm" that is clearly present in the original Judeo-Christian religious tradition. (Note that Gelderloos uses the word *paradigm* to encompass worldviews, ethics, and values, and we will adopt this usage for the remainder of the chapter.) To quote Gelderloos: "Today we are looking for new paradigms to lead us out of the most severe planetary crisis we have faced since the end of glaciation. . . . [A] new look at one of the oldest paradigms in history, the Judeo-Christian [religious tradition reveals a clear paradigm] . . . of stewardship or earthkeeping [p. 7]." Given that Judaism and Christianity have been the main Western religions for centuries, and given that radical changes in peoples' religious beliefs are not likely in the time we have left to avert environmental problems that threaten human survival, the ecotheology movement, by emphasizing the proenvironmental teachings of the dominant religious tradition, is a promising approach for moving Western values and beliefs in a proenvironmental direction.

We end this discussion of ecotheology with a moving quote from Psalm 104 from the Old Testament that has some of the flavor of the passages ascribed to Chief Seattle in the Chapter Prologue (as quoted in Gelderloos [1992], pp. 29–30):

> [God] . . . makes springs pour water into the ravines;
> They give water to all the beasts of the field;
> the wild donkeys quench their thirst.
> The birds of the air nest by the waters;
> they sing among the branches.
> [God] . . . waters the mountains . . .
> the earth is satisfied by the fruit of [God's] . . . work. . . .
>
> The trees of the Lord are well watered,
> the cedars of Lebanon that [God] . . . planted.
> There the birds make their nests;
> the stork has its home in the pine trees.
> The high mountains belong to the wild goats;
> the crags are a refuge of the coneys [small mammals].
> The moon marks off the seasons. . . .
>
> How many are your works, O Lord!
> In wisdom you made them all;
> the earth is full of your creatures.
> There is the sea, vast and spacious,
> teeming with creatures beyond number,
> living things both large and small. . . .
>
> May the glory of the Lord endure forever. . . .

The work of ecotheologians, which we discussed in this section, overlaps with the work of Thomas Berry, which we describe in the section that follows.

Thomas Berry's Work

Thomas Berry, a Catholic monk and an "historian of cultures," believes that the survival of Western civilization hinges on its ability to create and gain adherence to a radical new religion. The new religion must include, Berry emphasizes, an environmentally sound worldview and *cosmology* (story of the creation of the universe and the role of humans in it).[3] While this religion does not yet exist in final form, it will, Berry (1988) proposes, include the following elements: The new religion will be Earth- rather than human-centered. It will give us a sense of the Earth's sacredness, and we will respect, revere, and feel gratitude toward our home planet. We will fully understand the interrelatedness and interdependency of all living things. Berry emphasizes that these elements are found in the religious beliefs of some Native American and other traditional cultures. Berry (1988) writes:

> [The] story [will present] the organic unity and creative power of the planet Earth as they are expressed in the symbol of the Great Mother; the evolutionary process through which every living form achieves its identity and its proper role in the universal drama as it is expressed in the symbol of the Great Journey; the relatedness of things in an omnicentered universe as expressed by the mandala; . . . and finally, the symbols of a complex organism with roots, trunk, branches, and leaves, which indicate the coherence and functional efficacy of the entire organism, as expressed by the Cosmic Tree and the Tree of Life. [p.34.] (This quote and ones below are from The Dream of the Earth. Copyright 1988 by Thomas Berry. Reprinted with permission of Sierra Club Books.)

Berry's new religion, however, would augment the spiritual beliefs of native peoples with cutting-edge, scientific knowledge about the ecology of the planet and the functioning of the entire universe. This would make religious beliefs and, thereby, resulting human actions, consonant with the actual processes and limits of the natural world, so that human actions will be eternally sustainable. (Note that many ecotheologians, e.g., Gelderloos, 1992, also see a need for inputs from scientific ecology.)

Berry suggests an additional reason for giving modern science a major role in creating a new religion: He argues that scientists in physics, astronomy, biology, meteorology, and other disciplines are now coming to a radically new understanding of the Earth as a living organism, as well as an understanding of the "oneness" of all things in the universe, and even of an intelligence that is manifest in the universe's design. Thus, contemporary scientists, Berry argues, are arriving at the same kind of worldview and cosmology (or creation story) that native religions, as well as Eastern religions such as Taoism and Hinduism, have held to for thousands of years. This worldview and cosmology is, therefore, one that people can now accept both emotionally and intellectually. Berry (1988) writes:

> . . . [O]ur sustained [scientific] inquiry into the inner functioning of the planet [has] . . . brought us [to an] . . . awareness that the entire planet . . . [may be] a single organic reality. . . . [D]esignation of the earth as "Gaia" [referring to the "Gaia hypothesis" a scientific theory that views the Earth and all its biological and physical processes as a single integrated organism] is no longer unacceptable in serious [scientific] discussion. . . . Here the ancient mythic insight and our modern scientific perceptions discover their mutual confirmation. . . .
>
> [Furthermore], science is [now] providing some of our most powerful poetic references and metaphoric expressions. . . . We are more intimate with every particle of the universe and with the vast design of the whole. . . . We experience an identity with the entire cosmic order within our own beings. This sense of an emergent universe identical with ourselves gives new meaning to the Chinese sense of forming one body with all things [as in Taoism]. . . . That some form of intelligent reflection on itself was implicit in the universe from the beginning is now granted by many scientists.

The inclusion of new developments in science in the creation of a new environmentally sound religion is ironic. Earlier science and technology, shaped so strongly by the human-centered worldview and values of Western culture, may be seen as responsible for the industrial and postindustrial revolutions that have led to the serious environmental problems the world now faces, including the alteration of global climate and the erosion of the Earth's protective ozone layer (White, 1967; Berry, 1988, p. xii).

One major point of Berry's argument needs further clarification: Berry is talking not just about creating environmentally sound rules or codes to guide human action, based on modern scientific ecology, but also about creating a full new religion, one that has, as do all religions, significant subjective, spiritual, intuitive, and emotional components. Berry sees the subjective and spiritual components of religion as necessary for two reasons. First, from a practical point of view, he argues that the spiritual components of a religion are much more effective in changing and regulating human actions than are

mere rules or codes of conduct. Thus, for example, he writes:

> Without a fascination with the grandeur of the North American continent, the energy needed for its preservation will never be developed. Something more than the utilitarian aspect of fresh water must be evoked if we are ever to have water with the purity required for our survival. There must be a mystique of the rain if we are ever to restore the purity of the rainfall [p. 33].

Secondly, the spiritual, emotional components of a religion are, Berry argues, necessary to satisfy basic spiritual and emotional needs that all humans have, but that have been given short shrift in Western culture. Berry believes that many Westerners are emotionally starved and unfulfilled in our supremely secular and materialistic society—a belief he shares with several other writers (e.g., Theodore Roszak, 1973). Berry writes:

> . . . [O]ur secular society remains without satisfactory meaning or the social discipline needed for a life leading to emotional, aesthetic, and spiritual fulfillment. Because of this lack of satisfaction, many persons are returning to a religious fundamentalism. But that, too, can be seen as inadequate to supply the values for sustaining our needed social discipline [p. 124].

The kind of new religion that Berry envisions would fulfill our spiritual and emotional needs, while at the same time it would guide us into environmentally sound, permanently sustainable behavior.

The Deep Ecology Movement

Although it has spiritual and religious elements and some overlap with the work of Thomas Berry, the deep ecology movement mainly involves an overriding philosophy and worldview, and a prescribed lifestyle. The concept of deep ecology and its philosophical foundations derive from the work of Norwegian philosopher Arne Naess (e.g., Naess and Rothenberg, 1989). The concept has been further articulated by sociologist/ecologist Bill Devall, philosopher George Sessions, and others (e.g., Devall and Sessions, 1985).

Deep ecologists emphasize the major differences that exist between their basic philosophy, values, and worldview (i.e. their *paradigm*) and the philosophy, values, and worldview currently dominant in Western countries. Table 3-4 summarizes these differences. The dominant Western paradigm (left column) is human-centered and materialistic, and derives, deep ecologists argue, from the basic Judeo-Christian worldview criticized by White (1967), from traditional scientific orthodoxies, and also from capitalism (Devall and Sessions, p. 45). This paradigm, deep ecologists claim, is responsible for current global ecological crises, is factually and scientifically incorrect, and is spiritually impoverishing.

The deep-ecological paradigm (right column), in contrast, is nature-centered and stresses the *intrinsic* value of nonhuman forms of life, rather than their value defined only by their usefulness to humans. Deep ecologists identify several sources of their beliefs and values, including: the new research findings in physics, astronomy, and so forth (discussed above in connection with Berry's work); modern scientific ecology; Eastern religions, such as Taoism; Native American religions; contemporary feminism (discussed in the next section, Ecofeminism); certain aspects of Christianity (especially the teachings of Saint Francis of Assisi); and the writings of ecologist Aldo Leopold, authors Theodore Roszak and David Ehrenfeld, and others.

Deep ecologists generally advocate: A decrease in the Earth's human population; less human interference with the nonhuman natural environment; a lifestyle of "voluntary simplicity" (that is, one that minimizes resource consumption and environmental pollution and that avoids superfluous material possessions); and frequent communion with nature, for example, by hiking in natural environments—a communion that reconnects people with the Earth, provides spiritual fulfillment, and is the only way, deep ecologists believe, to truly understand and appreciate the deep ecology worldview (Devall and Sessions, 1985; Devall, 1988).

The literature of the deep ecology movement draws clear distinctions not only between the dominant Western paradigm and the deep-ecological paradigm (as in Table 3-2), but also between "shallow" and "deep" levels of ecological consciousness. The literature claims that a great many of the people who are truly concerned about environmental problems and are motivated to solve them (this includes a majority of Americans, according to public opinion poll results) still do not realize the degree to which their worldviews and values are frozen at a not sufficiently radical level. In Table 3-5, adapted from Miller (1990) and Devall and Sessions (1985), we outline three different levels of environmental consciousness, going from shallow to deep.

TABLE 3-4 The Deep Ecology Paradigm Versus the Dominant Western Paradigm

DOMINANT WESTERN PARADIGM	DEEP ECOLOGY PARADIGM
Dominance over nature	Harmony with nature
Natural environment as resource for human use	All nature has intrinsic worth; biospecies equality
Material/economic growth for a growing human population	Elegantly simple material needs (material goals serving the larger goal of self-realization)
Belief in ample resource reserves	Earth "supplies" limited
High technological progress and solutions	Appropriate technology; nondominating science
Consumerism	Making do with enough/recycling
National/centralized community	Minority tradition/bioregionalism

Source: Adapted from Devall, B., and Sessions, G., *Deep ecology: Living as if nature mattered,* p. 16. Copyright 1985. Gibbs Smith, Publisher.

Deep ecologists urge all Westerners to progress by reading the deep ecology literature and communing with nature to the third, or deepest, level of ecological consciousness and follow the tenets of their movement. Nothing short of this, they argue, can save our species and planet.

Ecofeminism

Because there are several different versions of ecofeminism and a diversity of ecofeminist beliefs (Cuomo, 1992), it is difficult to briefly characterize the ecofeminism movement. However, most ecofeminists are concerned *both* about solving global and regional environmental problems and about eliminating sexism, that is, eliminating discrimination against women and gender-based stereotyping of social roles. Ecofeminists argue forcefully that there is an underlying *linkage* between major environmental problems and sexism, and that both problems can only be solved together. Both problems, they argue, reflect the West-

ern male paradigm, one that stresses dichotomy, hierarchy, discrimination, domination, and exploitation (Shiva, 1989; Salleh, 1992; Cuomo, 1992). Thus, the Western view of nature, historically developed by men, sees nonhuman forms of life as separate from and inferior to human life, and Westerners have exploited and subjugated nonhuman life for centuries; similarly, the Western view sees women as inferior to men, and Western men have subjugated and exploited women for centuries. (Some ecofeminists argue that the linkage in the Western male paradigm between women and nature derives from the paradigm's view of *nature as feminine;* in other words, women are identified with and symbolic of nature [see Ortner, 1974].) Further, ecofeminists argue, both the subjugation of nature and the subjugation of women are strongly linked together in male-centered Western religions and religious institutions, in Western science, and in everyday Western life.

Concerning Western religions, we have already reviewed the claims of White (1967) that the Judeo-

TABLE 3-5 Levels of Environmental Consciousness

Level One—Shallow ecology: Concern about pollution and resource depletion: Acute and visible cases of environmental degradation are a cause of serious concern and action. Different environmental problems are seen as largely unrelated, and are to be corrected on a case-by-case basis. Natural resources should not be squandered, but should be consumed efficiently. Nature exists for humans and human use, but it is in our own self-interest that nature be managed wisely.

Level Two—"Intermediate-depth" ecology: The spaceship-earth analogy: Nature exists for human use, but humans are polluting and despoiling it badly. We must more fully understand the high degree of human dependence on nature, the finite ability of nature to absorb pollution and yield resources, the interrelatedness of all life forms, and the complexity of global ecosystems. With advanced scientific knowledge and management (for example, sophisticated computer models of ecosystems), and with the proper laws, regulations, and other existing societal institutions, we should be able to manage the planet wisely so that humans will continue to prosper materially into the foreseeable future.

Level Three—Deep ecology: Bioequality: Morally speaking, humans are not more important than nonhuman life forms and are not fundamentally different from or separate from them; all forms of life have a basic right to exist. Humans and human science and technology will never be able to fully understand and manage global ecosystems; to assume so is merely human arrogance. Nature can't be bent to the ways of humans. Pursuit of material comforts is intrinsically unrewarding; a lifestyle of voluntary simplicity is rewarding. We must replace our Western worldview with that of Eastern and native religions, and develop a spiritual/religious bond with the Earth and all its creatures, including fellow humans of all cultures and countries.

Source: Adapted from Miller, G. T. *Living in the environment*, pp. 612–613. Copyright 1990. Wadsworth Publishing Co. Also adapted from Devall and Sessions, 1985, Chapters 3–5, (see citation in Table 3-4).

Christian tradition is human-centered and demeaning to nonhuman forms of life. Though many scholars and theologians have questioned these claims (as we discussed above), it is difficult to deny that Western religions are patriarchal and male-centered. Judaic and Christian scripture both portray God as male, and emphasizes the male progenitors of the human species more than the female progenitors. In Christianity, the child of God, Christ, is a male, and he has greater importance than Mary, the mother of Christ (or than any other female figure). In most organized Jewish and Christian religious institutions and traditions, furthermore, women have until recently played a less important role than men, both in participating in services and rituals, and in serving as priests, ministers, or rabbis. Western science is, similarly, male-centered, ecofeminists argue (Shiva, 1989). Francis Bacon and the other (male) founders imbued science with the distinctly male hierarchical and patriarchal worldview and values, a worldview and values that justify the exploitation of nature[4] and that at least implicitly justify sexism and the exploitation of women. Finally, it is apparent that just as deep ecology is not yet widely accepted in Western nations, there is residual sexism in most Western nations (this is not to deny the considerable progress made in the last two to three decades).

Going further, some ecofeminists argue that the differences between the male paradigm (one that stresses dichotomy, hierarchy, discrimination, domination, and exploitation, as we discussed above) and the female paradigm (which we discuss in more detail below) are mainly culturally rather than genetically determined and can be traced back to the gender-based division of labor found in a nomadic hunting and gathering society, the main form of human living arrangement for the first 30,000 years of our species' 40,000 or so years of life on this planet (Maria Mies, 1986, [cited by Shiva] and Shiva, 1989). Men generally hunted animals, whereas women, besides bearing and nurturing new human life, gathered foods such as berries, nuts, and tubers. Shiva argues that the role of hunter intrinsically involves the use of tools that destroy life rather than produce life, a basically exploitative, dominant stance toward nonhuman

nature, and a life-or-death power over other living things—in other words, the basic male paradigm that still appears to dominate today.

In contrast, Shiva argues, the female role as food gatherers led women in Stone Age cultures to a different, female, paradigm, one that stresses: the interconnectedness and interrelatedness of all life forms; the sanctity of all forms of life; the diversity and complexity of life forms and natural processes (which cannot be understood by examining small, individual parts); and nature as productive, creative, and bountiful (Shiva, 1989). Shiva (1989) also argues that, in Stone Age hunter/gatherer cultures, women's gathering activities actually generated as much as 80 percent of the food consumed. Furthermore, women who live today in developing countries, Shiva points out, still have a disproportionately large role in food production, both as gatherers and in small-scale farming activities.

Some ecofeminists and others further claim that although the male paradigm has predominated in Western society for tens of thousands of years, there was a brief period (6500 B.C. to 3500 B.C.) during which certain cultures in Europe followed feminine principles (Gimbutas, 1974; Berry, 1988). These cultures were, according to at least some archaeologists, egalitarian, democratic, and peaceful. However, according to these accounts, the cultures were swept aside by an invasion of Aryan peoples, the male-centered forebears of contemporary Western cultures (Berry, 1988, pp. 139–140). These archaeologists argue that a key element in the replacement of feminine-oriented cultures by masculine-oriented cultures was the replacement of the *Earth Mother*, worshiped by the feminine cultures, with the *Heavenly Father* conception found today in Judeo-Christianity. The masculine-oriented cultures, by these accounts, were apparently responsible for 5,000 uninterrupted years of warfare, brutality, and environmental destruction. As political philosopher Herbert Marcuse (1974, [quoted in Shiva, 1989]) wrote, "Inasmuch as the male principle has been the ruling mental and physical force [in Western civilization and has produced such negative consequences], a free society would be the 'definite negation' of this principle—it would be a female society."

The ecofeminist solution to global environmental problems and to sexism, then, is for Western society to readopt the feminine worldview and religious/moral codes that some ecofeminists claim existed in European cultures between 6500 and 3500 B.C. Again, ecofeminists stress that since environmental problems and sexism have the same underlying cause—the male paradigm—they can only be solved together and through a fundamental and radical change from the male to the female paradigm. Note, finally, that the female paradigm Shiva describes is highly similar to the basic worldview and perspective of modern ecology (as we discuss in greater detail later in the chapter). In fact, the contemporary ecological theory that views the Earth as a single, living, and self-regulating organism is called by its proponents the "Gaia Hypothesis," Gaia being the name of the (female) Earth goddess in ancient Greek mythology.

COMMON THREADS IN RELIGIOUS/MORALLY BASED ENVIRONMENTAL MOVEMENTS

Each of the four proenvironmental movements or positions we just discussed—ecotheology, the teachings of Thomas Berry, deep ecology, and ecofeminism—has an explicitly religious/moral/ethical base, and each claims to be a departure from a dominant Western paradigm its proponents hold responsible for environmental degradation. What do these environmental movements have in common, aside from what they oppose? We contend that the four movements share two key components: 1) *A worldview (or system of beliefs) consonant with contemporary scientific ecology,* and 2) an *ecocentric value orientation.* These key components overlap and are tightly intertwined. We now discuss the components one at a time.

Shared Ecological Worldview

All four religious movements share to varying degrees the basic worldview or belief system that many ecologists and environmental scientists have been advocating for the last few decades. A look at any recent text on human ecology, environmental sciences, or environmental studies reveals a set of interrelated beliefs

about: the relationship between humans and the rest of nature, the workings of global and regional ecosystems, the disruptive impact that human activity is now having on those systems, the potential for catastrophic consequences, and the changes in human activity needed to avert those consequences. The ecological worldview advocates harmony with nature, emphasizes the finiteness of natural resources, the limited resilience of ecological processes, and the necessity of controlling human population and material growth.

Table 3-6 contains a more detailed statement of the major beliefs that comprise the worldview of scientific ecology, as we understand it, based mainly on a close examination of the best-selling undergraduate text in environmental studies/science, G. T. Miller's *Living in the Environment,* 1994. Again, we maintain that all four movements—ecotheology, Berry's religion, deep ecology, and ecofeminism—share many of the beliefs in this table. Items in the table aren't necessarily listed in order of importance, nor is the list intended as exhaustive. A look back at the discussion earlier in the chapter will verify that the four movements embrace some or many aspects of the ecological worldview. For example, compare Table 3-6 with the right-hand column of Table 3-4 on the deep ecology movement.

Consider how different the ecological worldview, described in Table 3-6, is from the Western worldview (compare items in Table 3-6 with items in the left-hand column of Table 3-4 titled "Dominant Western Paradigm"). The Western worldview holds that there is an essentially unlimited supply of natural resources and that science and technology have essentially unlimited powers. Similarly, it assumes that continual material growth and growth in human numbers are both possible and desirable. The Western and the ecological worldviews are, thus, in many ways polar opposites.

To repeat and summarize, we contend that all four of the cutting-edge environmental movements described earlier in the chapter share a rejection of the Western worldview and embrace, to varying degrees, the contrasting ecological worldview in Table 3-6.

Before we leave the topic, however, we should briefly qualify our discussion of the ecological worldview. All worldviews, including this one, are sets of *beliefs* that provide cultures and individuals with general perspectives or vantage points. Worldviews are not God-given truths. Thus, it can be argued that no worldview is valid in an absolute sense, including this one. On the other hand, the ecological worldview is a conceptual framework accepted by most ecologists and environmental scientists. The framework is implicitly if not explicitly endorsed by national and international scientific panels that have voiced warnings about global and regional environmental problems. For example, a group of over 1,000 scientists, including dozens who received Nobel prizes, signed a "World Scientists'

Warning to Humanity," issued by the Union of Concerned Scientists (1992) that basically accepts the framework of Table 3-6. This degree of scientific agreement at least suggests that the Western worldview is flawed and that the ecological worldview must be given serious consideration.

Shared Ecocentric Values

All four proenvironmental movements share not only a highly similar set of beliefs, or a worldview, but also

TABLE 3-6 The Worldview of Modern Scientific Ecology

1. There are complex, multiple interactions and linkages between the different forms of plant and animal life on the planet. The forms of life are highly interconnected and interdependent.
2. Because of these complex interconnections and interdependencies it is difficult to change one thing in a natural or environmental system (e.g., to increase or decrease the population of a species) without creating other changes, often ones that are remote and unanticipated. This principle is sometimes summarized by the saying "You can't do just one thing."
3a. Human survival is highly dependent on services provided by nonhuman forms of life and by global and regional ecological processes.
3b. There are limits to the resiliency of the ecological processes upon which human life and activity depend (for example, the oxygen cycle, and global climatic processes).
4. The Earth's supply of natural resources upon which human activity (especially technologically advanced human activity) depend are finite and exhaustible.
5. The global impacts of human activity (affected by the size of the human population, the nature of the technologies humans use, and the intensity with which they use them) have recently begun to disrupt key ecological processes, disrupt the "balance of nature," deplete natural resources, and cause the extinction of plant and animal species at unprecedented rates.
6a. Growth in human population and industrial activity, and hence in resource use and pollution generation, cannot continue indefinitely.
6b. Exponential human population growth must be slowed and stopped, along with the growth of human industrial activity. Permanently sustainable levels of population, resource use, and pollution generation must be reached.
7. Natural/environmental systems and processes are closed or circular ones (e.g., the waste materials and dead bodies of a species become the food and sustenance for others), not linear ones. There is no "away" to which human garbage or pollution can truly be thrown.
8. Global and regional environmental systems are highly complex, and humans may never be able to fully comprehend them. Human efforts to "manage" nature—even aided with such technology as computer models—could well lead to catastrophic failure.
9. "Upstream" solutions to environmental problems (for example, limiting the material introduced into the solid-waste stream to begin with) are generally more beneficial than "downstream" solutions (for example, trying to clean up a pollutant after it has been widely distributed). A summary of this principle is "An ounce of prevention is worth a pound of care." [We further discuss this and related principles in Chapter 10.]
10. The ecological integrity and diversity of the Earth's life-support systems must be maintained.
11. There is some evidence that the Earth, its biota, its atmosphere, and so on can be conceived of as a single, complex, integrated organism (i.e., the "Gaia hypothesis").

a single major value and ethical orientation. We can best describe this orientation in relation to Carolyn Merchant's (1992) analysis of the values and ethical bases that underlie human views on the environment. Merchant argues that controversies about human-environment relations center on three different values/ethics. One, which she calls the *egocentric* ethic, judges acts against the standard that the pursuit of self-interest is to be placed above other values. People who uphold this ethic tend to oppose environmental policies that would lead individuals to take actions against their desires. A second ethic, that Merchant calls *homocentric*, holds the good of the human species above other values. People who uphold this ethic support environmental policies that constrain individual choice if the effect is to promote a greater good for a greater number or to advance humanistic ideals such as equality or justice. They oppose environmental policies that promote the well-being of nonhuman species if the policies cause human beings as a species to make sacrifices or if they create injustice. The third ethic, termed *ecocentric*, judges acts according to their effects on the biosphere. People who uphold this ethic favor environmental policies benefitting ecosystems even if human beings must sacrifice.

Of course, few people hold any of these values/ethical positions in the extreme. Even the most egocentric agree that there are limits to selfishness set by the good of society, and even the most ecocentric would not place unlimited burdens on human beings to achieve small benefits for other species. Nevertheless, it is useful to distinguish these three ethical positions because doing so makes clear the moral basis of the positions of environmental movement groups and their opponents.

The four radical proenvironmental movements we discuss in this chapter all share an ecocentric value/ethical orientation. Their proponents consider environmental quality to have intrinsic value and believe on ethical grounds that people should protect the environment even at some cost to themselves and their societies. Their positions can be contrasted with those of environmentalist organizations that do not challenge the human-centered basis of ordinary ethics. These environmental groups may organize people to reduce pesticide residues in food on the grounds that

they are carcinogenic, to oppose nuclear power because of the risk of accidents that can cause human illnesses and fatalities, and to control toxic waste disposal because of the threats air and water pollution pose to human health or to aesthetic or recreational values. These arguments are all homocentric, in Merchant's terms, and in that way contrast sharply with the positions of the ecotheologians, Thomas Berry, the deep ecologists, and the ecofeminists, as well as with the teachings of Native American religions and Hindu, Buddhist, and Taoist scripture. Homocentric and ecocentric environmentalists often take the same positions on policy issues, but on different value and ethical bases.

Plan for the Rest of the Chapter

We devote the remainder of this chapter to exploring the potential for these radical religious/ethically based proenvironmental movements to succeed. In Merchant's terms, the goal of the movements is to put

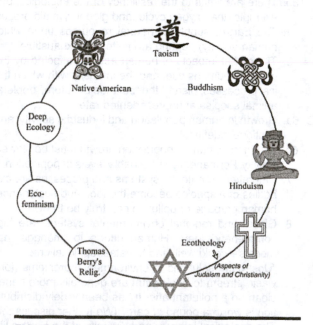

FIGURE 3-1 Religious/Ethical/Moral Traditions/Movements That Share an *Ecocentric* Value Orientation

human-environmental relations on a more ecocentric ethical basis. In addition, the movements aim to disseminate and make dominant the contemporary ecological worldview, as we discussed earlier. These changes are major and represent dramatic shifts in Western values and beliefs. To judge whether or not these environmental movements—and the religious/moral/ethical strategy—are likely to succeed, we must consider three important and related issues. *First,* we need to examine more closely the values and worldview that now underlie public support for environmental protection and the potential that exists for making that support more ecocentric and more based on the ecological worldview. In the next section of the chapter, we carefully review relevant public opinion data on these issues. We'll see evidence that people's basic values and beliefs in Western and other nations have already begun to shift in the direction that radical ecologists want them to shift. *Second,* we need to consider whether or not the value and belief shifts are likely to persist and become permanent. *Third,* we need to assess the probable effects of value and belief shifts on people's actual behavior, such as their taking energy-saving actions in their homes and cars, purchasing proenvironmental goods and services, participating in recycling programs, choosing to limit the size of their families, supporting proenvironmental government policies, voting for proenvironmental political candidates, and so on (there are analogous changes in the actions of government policy makers). Major shifts in values, morals, and beliefs must translate into such changes in people's behavior if the shifts are to have a significant positive impact on global and regional environmental problems.

ISSUE ONE: ARE ENVIRONMENTAL VALUES AND BELIEFS CHANGING?

Explicit in the arguments of ecotheologians, deep ecologists, ecofeminists, and other radical environmentalists are the claims that Westerners are now primarily *homo*centric or *ego*centric in their value orientations and primarily adherent to the Western worldview. But what do we really know about the values and beliefs of the general public in Western countries? Are the ecologists' claims about the *non-*

*eco*centricity of people's values and about their worldviews backed by empirical data? As it turns out, there are extensive public-opinion data concerning environmental values and beliefs, and we devote this section of the chapter to a review of these data. The data, indeed, confirm that an *ecocentric* orientation is *not* now common among the Western general public. However, the data do show surprisingly strong, widespread, and increasing environmental concern in Western and in many developing countries as well. There is also some evidence that this concern is linked to deep, underlying values and beliefs, suggesting the initial emergence of a more ecocentric orientation and of the contemporary ecological worldview, and possibly of an equally fundamental shift in general societal values. These trends may reflect in part the influence that liberal and radical environmental movements have already had on the values and beliefs of the public and of policy makers.

Strong Public Support for Environmental Protection

Opinion polls show that public proenvironmental sentiments in the United States are very strong—stronger, indeed, than they have ever been (Mitchell, 1990; Dunlap, 1991). Recall from Chapter 1 that public concern about environmental problems and support for proenvironmental measures first increased in the late 1960s and early 1970s, and then decreased during the mid- and late 1970s, though it was still considerable (Dunlap, 1991). In the 1980s, public environmental concern and support once again increased, leading up to the very high levels of environmental concern and support in recent polls. Evidence of the strength of public proenvironmental sentiments in the United States is shown in Table 3-7. The table displays people's responses to questions asked in three different public opinion polls taken in 2000 and 2001. These polls involved random samples of about 1000 Americans each; the data should be accurate within plus or minus three percentage points. All the surveys show strong support for the environmental movement. Figure 3-2 shows graphically the strong levels of public environmental concern since the 1970s, as well as the fluctuations in this support

TABLE 3-7 Public Opinion Poll Results: Proenvironmental Sentiment in the U.S. General Public, some results from 2000 and 2001. (Figures shown indicate the percentage of people polled who held the corresponding view.)

National Opinion Research Center Poll, 2000 (*Source:* Dunlap, 2002)
The United States is spending on improving and protecting the environment:
   too little: 62%
   too much   7%

Gallup Poll, March 2001 (*Source:* Dunlap and Saad, 2001)
Right now, do you think the quality of the environment in the country as a whole is:

Getting better	Getting worse	Same	No opinion
36%	57%	5%	2%

For each of the following, please say whether you think they are doing too much, too little, or about the right amount in terms of protecting the environment:

	Too much	Too little	About right	No opinion
The federal government	11%	55%	31%	3%
U.S. corporations	4%	68%	23%	5%
The American people	2%	65%	30%	3%

Would you be willing to pay more each year in higher prices do that industry could reduce air pollution?

	Yes	No	No opinion
$100 more per year	74%	24%	2%
$500 more per year	63%	35%	2%

Gallup Poll, Earth Day 2000 (*Source:* Dunlap, 2000)
Do you agree or disagree with the goals of the environmental movement?

Strongly agree	Somewhat agree	Somewhat disagree	Strongly disagree	No opinion
43%	40%	10%	5%	2%

from time to time. It displays results from two questions that have been asked in the same poll repeatedly over many years. The top line represents responses over time to the same question on the National Opinion Research Center's poll that appears at the top of Table 3-7. It shows the percentage of respondents who responded "too little" when asked about how much the United States is spending on improving and protecting the environment. The bottom line represents the percentage of respondents in the Roper poll who believed that "environmental laws and regulations have not gone far enough" when asked if they had gone too far, not far enough, or achieved the right balance. Both questions show strong and long-standing concern about environmental quality, and an increase in concern during the 1980s, when the administration of President Reagan propounded the view that governmental regulations, including environmental regulations, were excessive. During the 1990s, under the Clinton administration, which was seen as more favorable to environmental regulation, concern about under-regulation for environmental protection declined somewhat, while about the same size majority of Americans continued to believe the country was spending too little on environmental improvement and protection.

The evidence of strong and growing support for environmental protection among the U.S. public is matched by evidence from other countries that indicates widespread environmental concern worldwide.

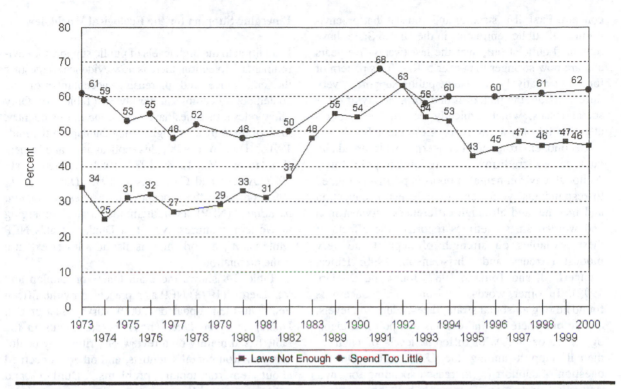

FIGURE 3-2 Public Views of Environmental Protection 1973–2000. Top line: Percentage of U.S. public agreeing that "Environmental laws and regulations have not gone far enough." (National Opinion Research Center polls). Bottom line: Percentage of U.S. public agreeing that "The U.S. is spending too little on improving and protecting the environment." (Roper polls).
Source: Dunlap (2002).

Dunlap, Gallup, and Gallup (1993) reported results of a huge survey-research project involving approximately 1,000 respondents in each of twenty-two countries. The countries ranged widely in level of wealth and economic development and in geographic location. In twenty of the countries, a majority of the respondents gave environmental protection first priority when asked to rank the importance of environmental protection relative to economic growth. In sixteen of the countries, a majority even indicated a willingness to pay higher prices for goods and services if necessary to achieve environmental protection. In addition, "[p]eople in the poorer, developing countries . . . clearly recognize[d] the impact of population growth on their environment and . . . [did] not place

all of the blame for global environmental problems on richer nations. Likewise, residents of wealthy nations . . . [did] not attribute the world's environmental problems primarily to overpopulation [in poorer countries] [p. 2]."

Of course, people's responses to public opinion poll questions and their true values and beliefs are not necessarily the same. Further, people's values and beliefs do not necessarily consistently affect their actual behavior, a fact that behavioral and social scientists have long emphasized (and which we discuss in detail in Chapter 4). Although for these reasons it is possible to infer too much from public opinion data, there are still several reasons to be impressed by the extent of public expressions of environmental concern. First, it is striking and unusual that proenvironmental public sentiments in the United States have remained quite strong over the last twenty-five years and are now stronger than ever before. This pattern of long and relatively consistent public support is very different from the pattern found for a variety of other social issues, which typically occupy the public's attention for only a few years or even months.

Second, environmental concern is widespread and cuts across traditional sociodemographic lines. Although environmental support is positively related to respondents' youth and their levels of education and income, and although differences between men and women are sometimes reported, the effects of these variables on strength of support are very modest (Arcury and Christianson, 1990; Hines, Hungerford, and Tomera, 1987; Jones and Dunlap, 1992). In other words, environmental concern is found among all social strata, races, education levels, and so on. There are indications that political ideology is associated with different levels of environmental support among the U.S. public. On the question of support for increased spending for environmental protection, Democrats have been consistently more supportive than Republicans, and self-described liberals more supportive than conservatives (Dunlap, Xiao, and McCright, 2001). Still, even among Republicans and conservatives, majorities have supported increased environmental spending in almost every year since the mid-1980s. Moreover, at the international level (as we noted above), high levels of environmental concern are found not only in wealthy countries but also in developing countries and countries of the former Socialist bloc (Dunlap, Gallup, and Gallup, 1993). The claim that is sometimes made—that environmental concern is a luxury of the rich and well educated—seems to be untrue both within the U.S. population and at the international level.

Third, research data suggest that proenvironmental public sentiments are strong enough to affect such actual behaviors as voting for proenvironmental political candidates and purchasing proenvironmental goods and services, as we discuss in detail later in the chapter. In sum, environmental concern appears to be widespread, relevant to action, and gaining in strength.

Emerging Support for the Ecological Worldview

In addition to the high levels of public support for environmental protection, there is now evidence suggesting the emergence and increasing acceptance of the contemporary ecological worldview (Table 3-6). Only a few relevant studies have been done in this country, most of them by sociologist Riley Dunlap. In the mid-1970s, Dunlap (1978), his colleagues, and others (Dunlap and Van Liere, 1978, 1984; Lovrich et al., 1987; Arcury and Christianson, 1990; Dunlap et al., 1992) proposed the concept of a *"new environmental paradigm"* (NEP), a paradigm, they argued, emerging in Western countries. Note that Dunlap et al.'s NEP while *mainly* a worldview, is also to a lesser extent a value orientation.

Table 3-8 shows the main tenets of Dunlap and Van Liere's (1978) NEP as represented by the fifteen items in their updated (2000) survey instrument. Dunlap and Van Liere chose the statements in the table based mainly on a review of writings by ecologists, environmental scientists, and others concerned about environmental problems. Dunlap and colleagues asked their research subjects to rate, on a questionnaire form, the degree to which they agreed or disagreed with each of the statements. Note that some of the statements are worded so that a respondent's *disagreement* with the statement signifies belief in the NEP; this wording is used to detect if a respondent did not really read the statements and only agreed or disagreed with them uniformly.

The first twelve items in the table assess respondents' beliefs about the relationship between humanity and nature—how it is and how it ought to be. These items duplicate several aspects of the ecological worldview that we discussed earlier (Table 3-6). The final three items of the NEP scale focus on the value of nonhuman versus human life. Although Dunlap et al. did their initial research before Merchant (1992) published her three-part classification of ethics/values, disagreement with these final two items is quite consistent with Merchant's *ecocentric* value orientation. Dunlap and colleagues' NEP scale is like the programs of the proenvironmental movements we discussed earlier in the chapter, in that it embraces both the ecological worldview and ecocentric values.

TABLE 3-8 Dunlap and Colleagues' New Ecological Paradigm Scale (Subjects Rated Their Level of Agreement with Each Statement

---

WORLDVIEW

We are approaching the limit of the number of people the earth can support.

When humans interfere with nature it often produces disastrous consequences.

Human ingenuity will ensure that we do NOT make the earth unlivable.

Humans are severely abusing the environment.

The earth has plenty of natural resources if we just learn how to develop them.

The balance of nature is strong enough to cope with the impacts of modern industrial nations.

Despite our special abilities humans are still subject to the laws of nature.

The so-called "ecological crisis" facing humankind has been greatly exaggerated.

The earth is like a spaceship with very limited room and resources.

The balance of nature is very delicate and easily upset.

Humans will eventually learn enough about how nature works to be able to control it.

If things continue on their present course, we will soon experience a major ecological catastrophe.

VALUES

Humans have the right to modify the natural environment to suit their needs.

Plants and animals have as much right as humans to exist.

Humans were meant to rule over the rest of nature.

Source: Dunlap, R.E., Van Liere, K.D., Mertig, A., and Jones, R.E. Measuring endorsement of the New Ecological Paradigm: A revised NEP scale. *Journal of Social Issues*, Volume 56, p. 433. Copyright 2000. The Society for the Psychological Study of Social Issues. Used with permission.

In several research studies, Dunlap and his colleagues found high levels of public agreement with items of the NEP. Thus, Dunlap and Van Liere (1978) found that an average of 74.5 percent of 806 randomly selected residents of the state of Washington "agreed" or "strongly agreed" with the NEP items (items scored in a proenvironmental direction to compensate for differences in item wordings). Further, Dunlap and Van Liere found that approximately 25 percent of the respondents adhered "coherently" to the NEP items. That is, 25 percent of the respondents responded to the scale in a highly consistent proenvironmental way. They "strongly agreed" with most, if not all, of the proenvironmental statements, suggesting a coherent ecological worldview (and an ecocentric value orientation), and not merely a tendency to support environmental protection measures on a piecemeal, case-by-case basis.

In subsequent research, Lovrich, Tsurutani, and Pierce (1987) found significant numbers of "coherent" new-environmental-paradigm supporters in Japan as well as the United States. Further, research data suggest that public acceptance of the new environmental paradigm has been increasing. In 1990, Dunlap and his colleagues (Dunlap et al., 1992) presented eight of the original NEP items to a representative sample of residents of the state of Washington and compared the results with those from the similar sample, noted above, surveyed in 1976. Overall, the 1976 respondents endorsed each of the eight items at an average level of 71 percent—a high level of acceptance. By 1990, acceptance had increased somewhat further, to an average of 78 percent endorsement. This evidence suggests that it is not only support for environmental policies that has been increasing but also support for the ecological worldview (and also ecocentric values).

A Direct Search for Ecocentric Public Values

In related and more recent research, Paul Stern and colleagues surveyed a representative sample of residents of Fairfax County, Virginia, a large suburban area of

Washington, D.C., in a direct attempt to detect the emergence in the public of a *ecocentric* value orientation (Stern, Dietz, Guagnano, and Kalof, 1994; Stern and Dietz, 1994). Their data failed to confirm a coherent ecocentric orientation distinct from a homocentric one (in their terms, a "social-altruistic" one). However, the Stern et al. data revealed surprisingly strong public support for several of the individual value items that comprise the ecocentric ethical position.

In their research, Stern et al. first compiled a list of specific values—including both homocentric and ecocentric values—from the work of Schwartz (1992). The list is summarized in Table 3-9. Note in the leftmost column of Table 3-9 the cluster or group of values that Schwartz calls self-transcendent values. In Merchant's (1992) terms, these values include some that appear to be ecocentric (unity with nature, protecting the environment, and a world of beauty) and others that appear homocentric (a world at peace, equality, social justice, and helpfulness). Egocentric values are to be found in the two other value clusters that Schwartz calls self-enhancement and openness to change.

Stern et al. added two environmental value items to Schwartz's self-transcendent values to see if lengthening the list would make it easier to statistically identify in their respondents' value systems a coherent ecocentric value cluster separate from a homocentric cluster. Stern et al. presented each respondent with the list of value items and asked him/her to rate the importance of each item on a scale, along with other items about environmental beliefs and actions. Stern et al. found that most of their respondents strongly endorsed many of the homocentric value items and many respondents strongly endorsed several of the individual ecocentric value items. However, so far, Stern et al. have failed to find convincing evidence that a separate coherent ecocentric set of values is emerging in their representative U.S. population. As Table 3-9 indicates, homocentric and ecocentric values, though they are distinct in the literature of the environmental movement, could not be disentangled in this cross-section of citizens. That is, individuals who endorsed the ecocentric items tended also to endorse the other items in the self-transcendence cluster, all of which are homocentric. (It is possible, though Stern et al. have not yet examined the possibility, that a separate ecocentric value orientation is emerging in certain subgroups, such as environmental activists or youth.)

The Stern et al. data thus show that the ideological struggle concerning proenvironmental values is not

TABLE 3-9 Four Classes of Values

SELF-TRANSCENDENCE	SELF-ENHANCEMENT	OPENNESS TO CHANGE	TRADITIONAL[a]
Unity with nature	Authority	An exciting life	Honoring parents
Protecting the	Social power	A varied life	and elders
environment	Wealth	Curiosity	Honesty
Preventing pollution[b]	Influence	Enjoying life	Obedience
Respecting the Earth[b]			Self-discipline
A world at peace			Family security
Equality			Cleanliness
Social justice			Politeness
Being helpful			Social order
A world of beauty			

Source: Stern, Dietz, Guagnano, and Kalof, 1994. The values are listed in order of the strength of relationship between individual items in a cluster and a scale that represents the entire cluster as a single factor.
[a]Schwartz's (1992) term is "conservation," in the sense of wanting to conserve existing institutions. Stern et al. (1994) used the term *tradition*, because in the literature on environmentalism, the term *conservation* has a very different meaning. Traditionalists are not necessarily conservationists.
[b]These items were added to those originally used by Schwartz (1992).

nearly over. At this point, the homocentric and ecocentric ethics seem to be combined in many people's minds, rather than competing. Stern et al. believe the progress of radical ecologists' efforts to change the ethical basis of environmental concern can be gauged by monitoring changes in the structure of human values in particular populations. If, for instance, the increasing environmentalism of younger cohorts of the population reflects a value shift toward ecocentric values, that may be an early indicator of a change in the typical value structure in the society. If radical ecologists can succeed in socializing youth in a new value structure, the result might be significant for the future of mass environmental concern and possibly for action.

The Emergence of "Post-Materialist" Values

The political scientist Ronald Inglehart, in his book *Culture Shift in Advanced Industrial Society* (1990), argues that a profound shift of basic values has been occurring in Western countries over the past two decades, a shift that is broader and more general than one toward ecocentrism. This shift, Inglehart suggests, may play a role in increasing public environmental concern and support for proenvironmental policies. Based on an extensive analysis of data from public opinion surveys in as many as twenty-five different countries between 1970 and 1988, Inglehart concludes that the public has become less interested in the pursuit of money and material objects ("Materialist" values) and more interested in pursuing nonmonetary goals ("Post-Materialist" values), such as esthetically pleasing surroundings (including environmental protection), self-expression (for example, more say in how things are decided in workplaces), and self-esteem. Inglehart argues that although this trend had its origin in the decades of economic prosperity after World War II, the trend toward postmaterialism has not weakened during major economic recessions in the mid-1970s and early 1980s.

One significant facet of Inglehart's (1990) work concerns the values of young people. He writes: "Everyone has heard that youth . . . [in the late 1980s turned] conservative, and that they . . . [were] mainly interested in preparing for lucrative careers so that

they . . . [could] become Yuppies and devote their lives to conspicuous consumption." Inglehart's data, however, do not support this perception. Although more students went into business careers in the 1980s than before, Inglehart sees that trend as a rational response to job availability. His general conclusion is this: ". . . [T]he overwhelming bulk of the evidence indicates that the basic values of . . . youth [in the late 1980s were] not more materialistic than those of their counterparts a decade or two earlier. Nor . . . [were] they politically conservative in any basic sense [p. 12]."

Inglehart's data indicate that materialists in Western countries outnumbered postmaterialists by approximately four to one in the early 1970s, but by 1988 the ratio was about four to three. He predicted that because materialists are older than postmaterialists, by 2000 they will be about equal in numbers. Inglehart concluded that even by 1990, the postmaterialists represented a large group of potential votes for proenvironmental political parties. He noted that in eleven European countries, an average of 47 percent of the adult population expressed a willingness to vote for "ecologist" political parties, such as the West German "Greens."

Despite this finding on people's willingness to vote for ecologist political parties, the impacts of postmaterialist values on public concern about the environment remain uncertain. The results of several research studies that attempted specifically to identify these impacts are so far equivocal. Inglehart presented some additional evidence to support a link between environmentalism and postmaterialist values (Inglehart, 1992), but other preliminary studies have failed to support this relationship (Dunlap et al., 1993; Brechin and Kempton, 1994).

Looking broadly, then, at the above research on postmaterialist values and the research on the new environmental paradigm and on ecocentric values, we conclude that increased public support for environmentalism is due at least in part to increasing acceptance of new beliefs about human-environmental relations, such as Dunlap's NEP. However, we cannot tell with certainty whether environmentalism also reflects a shift in basic human values, such as increasing postmaterialism or ecocentric (biocentric-altruistic) values, because not enough studies have been done. Postmaterialism is increasing, but its connection to environmentalism has not yet been demonstrated, and data on Schwartz's value clusters have not been collected for a long enough time to tell whether a value shift is occurring.

ISSUE TWO: WILL CHANGES IN VALUES AND BELIEFS PERSIST?

If, as Inglehart and Dunlap claim, the developed countries (at least) are experiencing a shift toward postmaterialist values and an initial emergence of a new ecological paradigm for understanding human-environmental relationships (and assuming that postmaterialist values do have a significant positive impact on people's environmental concern), will these value and belief changes be permanent? The permanence of shifts in basic values or in basic orientations towards the natural environment depends, of course, on what is causing the shifts.

One theory, implicit in some of the work of both Inglehart and Dunlap, is rooted in the satisfaction of human needs. The argument has its roots in the work of the psychologist Abraham Maslow (1954), who postulated that human beings have a hierarchy of needs beginning with so-called basic needs, such as the needs for food, air, and protection from danger, and moving up through "higher" needs, such as respect from others, social position, self-esteem, and self-expression. According to the theory, higher needs are not expressed unless the basic needs are supplied to an adequate level. Since postmaterialist values and concern with the environment both represent higher needs, they can be expected to be significant in people's consciousness only as lower, material needs are satisfied. It follows from this line of reasoning that as people's basic material needs are met, they begin to express postmaterialist values. This is why postmaterialism first appeared as a major social phenomenon during the period of Western prosperity after World War II. If this analysis is correct, postmaterialist values are likely to continue to be expressed as long as individuals expect to have their material needs adequately met. Thus, barring economic collapse in the West, postmaterialism can be expected to last a long time. Moreover, economic development in other countries can be expected to lead to increasing postmaterialism there as well, and thus to increased concern about environmental quality.

This theory is plausible but hard to test without letting several decades pass. We should note, though, that the widespread environmental concern now being expressed in developing countries suggests that material satisfaction is not always necessary for environmental problems to be taken seriously.

A second theory is that the value shift toward environmentalism has been influenced by scientific information demonstrating the interconnectedness of all life and by graphic representations, including photographs from space, of the finiteness of the planet Earth. Dunlap's new ecological paradigm is in large part a set of beliefs that have been supported by decades of ecological science, and it may be that science has played an important part in changing people's understanding of the world, which in turn has led to a change in values (again, it is hard to disentwine ecocentric values and a scientific-ecology worldview). If this theory is correct—and it is also very difficult to test—the new ways of thinking are likely to become more prevalent as older people with older beliefs are replaced in the population with new generations raised with the new scientific understanding. Under this theory, the change is also likely to be very long-lasting.

A third theory is that value changes occur not in whole populations but in cohorts, groups of people of similar age who share common formative experiences. Under this theory, postmaterialism is historically rooted in the experience of a generation that saw privation (the Great Depression of the 1930s) replaced by prosperity. The children of the Depression were materialist because of their experience, and the children of the postwar period were postmaterialist because of their experience. Similarly, the children of Earth Day adopted the beliefs (and values) of the new ecological paradigm, while their forebears, who were not raised with the environmental movement, did not. Under this theory, the future of ecocentric values and ecological thinking is much more uncertain. As events bring new issues to the forefront of public concern, people may shift their priorities between values—materialist versus postmaterialist, ecocentric versus homocentric or egocentric, and so forth. In fact, much of the debate over environmental policy can be seen as an effort to bring one set of values or another to the fore. When the U.S. Environmental Protection Agency accepted the language of risk analysis in its policy thinking during the 1980s, it signaled a subtle change of values. *Environmental protection* seemed to put ecocentric values first, whereas *risk management* implies that the environment needs to suffer some insult, and the question is how to balance it against other risks, particularly risks to economic growth. The results of such political and ideological debates express a society's values at a particular time, and it may be that individuals raised in a period of such debate have their values shaped for a lifetime by the winning paradigm. If this is how environmental values and thinking changes, the proenvironmental developments of the late 1960s and the 1970s may be reversible.

Not enough is known to tell which of these theories is most accurate, but one thing is common to all of them. It is that values, and basic ways of thinking about human-environment relations, are hard to change in adults. Shifts in the dominant values or ways of thinking in a society are therefore slow, generational processes, so that whatever effects they have on behavior will be long-lasting.

ISSUE THREE: HOW DO VALUES AND BELIEFS INFLUENCE PEOPLE'S ACTIONS?

In prior sections, we described four Western, radical proenvironmental religious/ethical movements, reviewed survey data on people's values and beliefs toward the environment and changes over time in these values and beliefs, and discussed the likelihood of such changes becoming permanent. However, we have not yet discussed the effects of people's values and beliefs on their actions. As we noted earlier, if proenvironmental shifts in people's values and worldview don't translate to changes in their behavior (e.g., taking energy conservation actions, voting for proenvironmental political candidates, and so on), then shifts in values and beliefs will have relatively little impact. We now, therefore, take an in-depth look at the links between people's values and beliefs and their willingness to support environmental protection policies or to take proenvironmental actions. Our focus will be primarily on the impacts of *values*.

To understand the value and ethical basis of environmental concern and action, we need first to step back and consider the *full range* of human values (i.e., not just environmentally relevant ones). An ongoing international research project, organized by the social psychologist Shalom Schwartz at Hebrew University in Jerusalem, is examining just this question (Schwartz, 1992), (see Table 3-9). The project is searching for universals in the structure of human values by asking respondents in many different countries to rate the importance to them of a list of fifty-six different values, including almost all those in Table 3-9. It has found that although the importance of particular values varies from one national population to another, the relationships between the values are quite constant. For example, people who value wealth highly tend also to value power and influence highly, regardless of the country they live in. People who value variety in their lives tend also to value excitement and enjoyment. People who value honesty also tend to value obedience, cleanliness, and family security. And people who value social justice also tend to value equality, beauty, and protecting the environment. Schwartz found that in every country studied, the fifty-six values fell into almost exactly the same clusters. Table 3-9 lists the clusters and some of the values in each.

As we noted earlier, several of the self-transcendent values in the left-most column of Table 3-9 are clearly relevant to protection of the environment: These are the ecocentric values (e.g., protecting the environment, a world of beauty); the other self-transcendent values are homocentric (e.g., world at peace, social justice). Egocentric values appear in two other value clusters, self-enhancement and openness to change. (The relationship between Inglehart's postmaterialist values that we discussed earlier in the chapter and the broad value clusters identified by Schwartz has not yet been systematically investigated. It seems on the surface, however, that Inglehart is claiming a shift away from traditional and self-enhancement values to self-transcendent values and openness to change, which seem to be postmaterialist. In one study

(Stern et al., 1999), a measure of postmaterialist values did not add any predictive power beyond what could be achieved by measures of the Schwartz value clusters. This finding suggests considerable overlap between the concepts.)

In recent research, which we partially described earlier in the chapter, Stern and colleagues have begun to look for direct evidence of links between value clusters and individuals' willingness to take political action for environmental protection (Stern, Dietz, and Kalof, 1993; Stern, Dietz, Guagnano, and Kalof, 1994; Stern and Dietz, 1994). Recall that Stern et al. compiled an enhanced list of values from the Schwartz (1992) work and presented the list, along with other questions about environmental beliefs and actions, to their sample of Fairfax County, Virginia, residents. As we already noted, Stern et al. failed to find convincing evidence that a separate ecocentric set of values is emerging in the population of their study. The results indicated that homocentric (social-altruistic) and ecocentric (biospheric) values could not be disentangled in their sample of citizens. However, while the Stern et al. results failed to show a distinction between ecocentric and homocentric sets of values, the results did clearly suggest an influence of values on behavior. The cluster of self-transcendent values (Stern et al. termed it "biospheric-altruistic") was strongly predictive of people's self-reported willingness to take politically significant actions such as boycotting the products of a company that pollutes, signing a petition for tougher environmental laws, and refusing to invest in or work for a polluting company. The self-enhancement value cluster, which consists of egocentric values, also predicted willingness to take action, but in a negative direction—people with strong self-enhancement values were less supportive of proenvironmental actions. The other two value clusters were unrelated to action. Thus, the values or ethics that writers like Merchant link to environmental concern on theoretical grounds in fact predict support for environmental protection; other human values, though also universal, are generally unrelated to environmentalism. In a study with a national sample of respondents, Stern et al. (1999) found that self-transcendent values also predicted self-reports of a set of pro-environmental consumer behaviors.

A Closer Look at How Values (and Beliefs about the Consequences of Environmental Problems) Influence Actions

Stern et al. take the position that people's values are likely to be especially strong determinants of their proenvironmental actions because people often need to react to environmental conditions or problems on the basis of very *limited knowledge or experience.* People are continually faced with environmental issues appearing in the media that are newly discovered or reinterpreted by scientists, and they must find ways to determine whether to take these issues seriously enough to do something about them. Stern et al. reason that values provide a basis for those choices. People consider what they know about an environmental condition or problem and ask themselves whether that condition or problem is likely to have harmful consequences for anything they value. If it is, and if they can do something to prevent the harm, they are inclined to take action. In fact, they may feel a sense of moral obligation to take action. (We discuss the role of moral obligation in more detail in Chapters 4 and 6.)

In the Stern et al. formulation, then, the keys to people's response to any environmental problem are people's values as well as their *specific beliefs about the consequences of environmental* problems to the things they value (note that a person's beliefs about harmful consequences are just one part of his/her set of environmentally relevant beliefs or worldview.) As we already described, Stern et al. found that respondents' values were significantly correlated with their actual political behaviors. Stern et al. further found that respondents' beliefs about the adverse effects of environmental conditions also affected their willingness to act. Effects on nonhuman species and the biosphere made the most difference, but effects on oneself also mattered, and in one of the studies (Stern et al., 1993), effects on people in general also motivated action. These beliefs, however, were in turn affected by values. A striking finding was that people with strong biospheric-altruistic values were more likely than others to believe environmental changes had adverse effects on all kinds of valued objects: nonhuman species, other human beings, and oneself.

Stern et al. interpreted their results as indicating that values can affect proenvironmental action both directly and indirectly, through beliefs about consequences. Values may affect these beliefs by simply motivating people to pay attention to information about how environmental conditions affect things they value, or they may shape the beliefs directly by leading people to believe what they want to believe. For instance, individuals who hold strong traditional or self-enhancement values, which environmentalists often claim will need to be sacrificed to preserve the environment, may deny that human activities are harmful to nature because such a denial allows them to believe they need not give up what they value; similarly, individuals who value the biosphere for its own sake may believe with minimal evidence any claim that a human activity is threatening a natural system. The Stern et al. data so far do not make clear whether one or the other, or both, mechanisms are operating to influence environmental beliefs. The data do show, however, that basic values are relevant both to individuals' beliefs about the consequences of environmental problems and to their actions.

The research suggests that biospheric-altruistic values predispose people to accept worldviews like the NEP and to believe that environmental conditions present negative consequences for other people and species. These beliefs, in turn, predispose people to accept pro-environmental personal norms—that is, to have a sense of personal obligation to do something to reduce those negative consequences, in accordance with their values. This formulation, which Stern and colleagues (1999) refer to as the value-belief-norm (VBN) theory, offers an explanation of the propensity of some people to engage in a variety of behaviors that they believe to be pro-environmental.

What do these studies of environmental values and action have to say about the prospects for social change through movements like deep ecology and ecofeminism? The studies demonstrate that basic values can have far-reaching implications for individuals' willingness to support proenvironmental actions. This means that the ideological debates about homocentric versus ecocentric ethics, "deep" versus "shallow" ecology, and the like, which center on whether nonhuman aspects of the environment should be valued in their own right, are potentially important.

Factors That Can Limit the Effect of Value and Belief Changes

As we have just discussed, changes in people's values can have significant effects on their willingness to support proenvironmental policies and to take proenvironmental actions. However, a personal norm that makes someone feel obligated to take pro-environmental action does not automatically lead to any particular action, as Stern and colleagues have emphasized (Stern et al., 1999; Stern 2000). When environmentally related public morals, values, and beliefs change, the effect on the environment may be sharply limited by several factors. Assume, for example, that as of tomorrow morning, all citizens of the United States considered it their moral duty to conserve energy and reduce pollution by reducing gasoline consumption. When morning came, a great many of them still would be living in suburban areas with separated single-family houses, and still would be dependent on automobile transportation. These structural conditions would greatly limit what individuals could do for the environment by acting on their new values and beliefs. Markets would further constrain their options. Anyone who wanted to trade in the family automobile for one that could drive 70 miles on a gallon of gas would not be able to do so because, though such cars can be built, they are not now on the market. These sorts of structural and market factors are difficult to reverse. Moreover, they change very slowly, and the unorganized actions of individuals can do very little to change them.

There are other barriers to action that may be easier to remove. Many people, for instance, do not know which of their daily actions are most responsible for energy use or toxic waste production. Without this information, they are unlikely to act effectively on their values and beliefs, but with it, they might change.

CONCLUSION

We have argued in this chapter that Western culture is strongly human-centered, considers nonhuman forms of life as having importance only in direct proportion to their usefulness to humans, and embraces a nonecological set of beliefs about the environment. These basic Western values and beliefs have several possible origins, though the Judeo-Christian religious tradition is probably not primarily responsible. Regardless of their source, however, these values and beliefs cannot be helping Western countries solve major environmental problems that ultimately threaten human survival.

While it seems quite plausible that changes in Western values and worldview—changes that have already begun—are necessary to lessen or avert environmental problems, a look at cultures that have strongly proenvironmental religions and belief systems suggests that even radical changes are not likely to do the job by themselves. The influences of religious and moral beliefs and institutions are often severely limited by the influences of political and economic forces and also the force of population growth. Furthermore, basic values and beliefs usually change slowly—in entire populations, it may take a generation (or more) for major changes to be achieved. In addition, value and belief change does not lead in any straightforward way to behavior change and environmental improvement: Structural conditions and other barriers often keep values and beliefs from being enacted as behavior, and when this occurs, it will require changes in the structure of society and in people's funds of knowledge for changed values and beliefs to be carried into action. In the following chapters, we discuss ways of eliminating barriers to action by changing attitudes, providing information, changing incentive structures, and creating new institutions.

But even though proenvironmental shifts in moral/religious/value/belief systems can bring about little change in behavior and environmental improve-ment by themselves, this sort of change can still be a critical part of the solution to the problems of the commons. We will see in subsequent chapters, and especially Chapter 7, that the four main solution approaches (of which the religious/moral approach is one) can reinforce each other, creating more effect in combination than the sum of their individual influences. Further, removing the barriers that often block proenvironmental individual action, as discussed in this and subsequent chapters, often requires action by governments, which, in turn, may require an aroused and concerned public that actively promotes the necessary changes in government policy. Public concern and political action, in turn, may require that the public embrace proenvironmental values and an ecological worldview. Thus, change in values and worldviews may be an important stimulus for public concern, arousal, and political action; similar changes on the part of government policy makers may be critical as well.

Finally, we have reviewed in this chapter four new and "radical" religious/moral/ethical movements that all share an ecocentric value orientation and a worldview consonant with contemporary ecology. These movements—ecotheology, Thomas Berry's religion, deep ecology, and ecofeminism—share major elements with Asian religions, and the religions of some Native American tribes. If these new movements, or parts of them, are widely adopted in Western cultures, they may have a positive influence in averting global and regional environmental problems, as well as in helping to satisfy our spiritual needs.

NOTES

1. For the analysis of the 1993 General Social Survey data on religion and environmental concern, and for our analysis later in the chapter of how values influence environmental attitudes and actions, we are indebted to Thomas Dietz, Gregory Guagnano, and Linda Kalof (e.g., Dietz and Stern, 1995; Dietz, Kalof, and Stern, 2002; Stern, Dietz, and Guagnano, 1998; Stern and Dietz, 1994; Stern, Dietz, and Guagnano, 1995; Stern, Dietz, Kalof, and Guagnano, 1995).
2. In addition to the factors described below, some argue that Hinduism's otherworldly orientation and emphasis lead to the neglect of real-world problems.

3. Berry and others want to retain many elements of our basic Judeo-Christian heritage, such as the Ten Commandments, that proscribe murder, and so on.
4. This claim must be qualified. In Bacon's day, more than now, nature was a source of danger and threat, poorly understood, and much feared. Disease was rampant and nature was a prime killer of people. It is not surprising that humans wanted to understand and master nature.

CHAPTER 4

EDUCATIONAL INTERVENTIONS: CHANGING ATTITUDES AND PROVIDING INFORMATION

CHAPTER PROLOGUE

This chapter begins with brief passages from two of the best-known and most widely read books ever written about environmental problems. The books were written to educate people about the problems and, thereby, change their behavior toward the environment. Such efforts to educate usually have two main thrusts, which the passages below illustrate: changing people's attitudes and providing them with information.

The first passage comes from Rachel Carson's classic book on the dangers of pesticide use, *Silent Spring* (1962):

[Insecticide and herbicide] sprays, dusts, and aerosols are now applied almost universally to farms, gardens, forests, and homes—nonselective chemicals that have the power to kill every insect, the "good" and the "bad," to still the song of birds and the leaping of fish in streams, to coat the leaves with a deadly film, and to linger on in the soil—all this though the intended target may be only a few weeds or insects. Can anyone believe it is possible to lay down such a barrage of poisons on the surface of the earth without making it unfit for all life? They should not be called "insecticides," but "biocides."

. . . Future historians may well be amazed by our distorted sense of proportion. How could intelligent beings seek to control a few unwanted species by a method that contaminated the entire environment and brought the threat of disease and death even to their own kind? (pp. 7–8)

[Carson goes on to argue for the use of biological pest control, a system that controls pests with predators, diseases, and other natural enemies.]

Carson wrote *Silent Spring* to *change attitudes* about pesticides. She tried to alert people to an environmental problem and convince them that the problem was so important it needed urgent action—either by the reader or by government agencies, industrial organizations, or others in a position to do something. Carson tried to develop in readers strong beliefs about the seriousness of threats to the environment and a strong attitude about the pesticide problem in question—a predisposition to do something about it or to encourage others to do something. She was part of a growing movement of scientists who sought to alert the public to threats to the environment resulting from human actions.

The second passage is from the popular book *50 Simple Things You Can Do to Save the Earth* (Earth Works Group, 1989), published in connection with the U.S. observance of Earth Day 1990. The passage is part of a discussion of water-saving shower heads from number 24 of the 50 things.

Shower Facts:
- *Showers usually account for a whopping 32% of home water use.*
- *A standard shower head uses about 5–7 gallons of water per minute (gpm)—so even a 5-minute shower can use 35 gallons!*
- *"Low-flow" shower heads reduce water use by 50% or more. They typically cut the flow rate to just 3 gpm—or less. So installing one is the single most effective water conservation step you can take inside your home.*
- *. . . [In addition, with] a low-flow shower head, energy use (and costs) for heating hot water for showers may drop as much as 50%. (pp. 50–51)*

   [The section goes on to explain how to tell if the shower head in your bathroom is a standard or a low-flow model, how low-flow models work, and where they can be purchased.]

The passage above from *50 Simple Things . . .*—and the entire book—aims to change the way people treat the environment by *providing information*. The authors don't try to change attitudes; they assume that the reader already wants to save the earth. The authors believe, however, that the reader needs to know exactly what to do and how to do it in order to take effective action.

INTRODUCTION

The books *Silent Spring* and *50 Simple Things You Can Do to Save the Earth* were written on the assumption that educating people—changing their attitudes and beliefs and providing them with information—would change their actual behavior. The kinds of educational efforts we discuss in this chapter are much more focused than the kinds of moral and ethical-religious appeals we discussed in Chapter 3. This is because people's beliefs about particular environmental issues, such as the effects of pesticides on bird populations, and their related attitudes, such as about the widespread use of pesticides, are much more specific and less deeply rooted than their morals and basic values (such as a religious reverence for nature) or their general ideas about how the environment responds to human intervention.

The assumption about the efficacy of education that underlies *Silent Spring* and *50 Simple Things* is not one confined to environmentalists who write books. It is shared by many public officials, doctors, educators, and ordinary citizens who are concerned about societal problems. Indeed, it is almost common sense that education is essential for solving a wide range of social problems, and many also believe that a good educational effort will be sufficient to do the job. Consider, for example, this brief quotation from the Saline, Michigan, hospital newsletter: "Today, marijuana use is not uncommon in junior high schools, and is creeping into elementary schools. How can it be stopped? As with any behavior, the most effective way. . .is through education." Following the same logic, people propose sex education as the way to prevent the spread of AIDS and other sexually transmitted diseases, education on smoking and diet as ways to prevent heart disease, and environmental education as the way to get people to be more respectful of wilderness areas and other fragile environmental systems (see Figure 4-1).

Behavioral and social science research, however, indicates that this conventional wisdom—that education is enough to solve social problems—is oversimplified and misleading. The research shows that education can help but that education is rarely sufficient. For example, decades of careful study of

FIGURE 4-1 Some Written Educational Materials Intended to Promote Healthy Behavior

health promotion campaigns show that it is possible to get people to stop smoking tobacco or to eat healthier foods, but not with education alone. In the 1960s and 1970s, a large number of programs were conducted in schools to keep children from developing the smoking habit. These programs, which operated mainly by providing information on why smoking is bad for health, changed some of the students' beliefs and attitudes, but rarely reduced the onset of smoking behavior (Thompson, 1978). Other educational programs for health promotion—to improve eating habits, cut alcohol consumption, and the like—have been plagued by problems of limited success and frequent relapses into old behavior patterns.

Clearly, there have been major strides in health promotion in the United States over the past 30 years. Since the 1950s, when the link between cigarette smoking and cancer first came to light, the proportion of American adults who smoke has decreased, and after decades of publicity about the health effects of

fiber and cholesterol in the diet, sales figures have shown increased consumption of whole grains and fresh produce and decreased consumption of red meat. But these successes are based on more than just education. We identify the other key elements of success later in the chapter.

Chapter Overview. We devote this chapter to a careful examination of efforts to encourage proenvironmental behavior via education. We focus on interventions that aim to change people's behavior in the relatively short run. (We do not address general environmental education programs, such as those in some schools, targeted specifically at children, that attempt to produce changes in the long run by changing children's basic environmental understanding so they will believe differently as adults. The long-term effects of such education are very difficult to measure, but we believe that these effects can be significant, and we return to this theme at the end of the chapter.)

We find that, as with health promotion, education is helpful but not sufficient for promoting the desired behaviors. We look first at educational efforts that try to change fairly specific environmental attitudes and beliefs and then at efforts that offer information about how to act on proenvironmental attitudes. We see that education can change attitudes and beliefs, but that many barriers, both within individuals and in their social and economic environments, can keep proenvironmental attitudes from being expressed in action. Some internal barriers can be overcome with informational programs, but only if the programs are carefully designed to take advantage of psychological principles of communication. The chapter presents those principles and some illustrative examples of successful and unsuccessful information programs. But even the best educational programs cannot overcome external barriers to action, such as financial expense or serious inconvenience. The chapter details what environmental education can and cannot accomplish, and tells what must be done to take the educational strategy as far as it can go. In later chapters, we show how even greater success can be achieved by combining education with other approaches.

EDUCATION TO CHANGE ENVIRONMENTAL ATTITUDES AND BELIEFS

Education can change attitudes and specific environmental beliefs, but it cannot quickly or easily change ethics or values. Furthermore, education is not likely to work if it promotes attitudes that clash with people's basic ethics or values. Educators like Carson know this. If an educator tells people that in order to have a clean environment they must sacrifice financial security, fresh food, or time with their families, people who value those things highly will reject the educator's message. But if the message is that environmental quality does not require people to reorder their basic values, it will go down easier. Carson's message can work partly because she explains how giving up pesticides does not mean giving up fresh food. It is not necessary to choose between environmental values and fresh produce, because one can have both by rejecting pesticides in favor of biological controls. (Of course, major shifts in Western values may also be needed to permanently solve environmental problems, as we discussed in Chapter 3.)

Changing environmental attitudes can make a difference. It is no coincidence that the increased awareness and concern about environmental problems in U.S. public opinion beginning in the 1960s was followed by a burst of new legislation in the 1970s. And many scholars and writers believe that this shift in opinion was strongly influenced by Carson's *Silent Spring*. Also, when the word first came out in the mid-1970s that the chlorofluorocarbon propellants used in aerosol cans might harm the earth's ozone layer, Americans quickly reduced their purchases of the cans and the government instituted a ban (Morrisette, et al., 1990). This could not have happened without widespread public concern. People who strongly favor environmental protection are more likely to join environmental movement organizations (Mitchell, 1979) and vote for environmental protection in public referenda (Gill et al., 1986), so attitudes can lead to action. But environmental attitudes are not always correlated with behavior, and attitude change does not always lead to behavioral change. These facts greatly limit what the attitude-change strategy alone can accomplish.

Controlled studies show that educational efforts to change environmental attitudes and beliefs generally have little effect on behavior. The most careful studies focus on consumer behaviors—recycling, energy conservation, and other things individuals can do on their own to directly change how environmental resources are used. (Researchers have not conducted experiments on changing people's political attitudes and beliefs—what *Silent Spring* tried to do—probably because doing this as an experiment poses serious ethical questions.) The following examples are typical. They focus on energy conservation in the home, an important way of reducing environmental problems such as air pollution and global warming and one on which there is considerable research.

In 1977, a year when natural gas shortages caused some businesses and schools to close down to preserve heating fuel, state agencies in Virginia conducted three-hour workshops in various communities to educate people about energy conservation in the home. The workshops, which consisted of lectures, slide shows, discussions, and demonstrations, were designed to convince people that they could save substantial amounts of energy in their homes and to show them how. Scott Geller and his colleagues at the Virginia Polytechnic Institute and State University evaluated the effects of the workshop approach with surveys and follow-up visits to participants' homes to look for behavioral change (Geller, 1981). The workshops were effective in changing attitudes and beliefs, as measured by before-and-after surveys. After the workshop, participants expressed increased concern about the energy crisis, increased awareness that simple changes in the home can yield substantial energy savings, and stronger beliefs that they could do something about the energy crisis and that they had not yet done enough to insulate their homes. The surveys also revealed stronger expressed commitment to change "residential lifestyle for energy conservation." But these attitudes, beliefs, and commitments did not translate into behavioral change. Follow-up visits to participants' homes six weeks after the workshop revealed that only one of forty workshop participants had lowered a water heater thermostat, as the workshop had recommended, and that the only two with insulated water heaters (another workshop recommendation) had insulated them before the workshop. The only behavioral change was in installing low-flow shower heads. Eight of forty workshop participants had installed them, compared with two of forty nonparticipants in nearby homes. But this change was not produced by education alone. The workshop leaders also gave participants water-flow restrictors and explained how to use them. By doing this, they removed a barrier to energy conservation—the effort of obtaining the flow restrictor people may have come to want as a result of the workshop. Such barriers between attitudes and behavior impede educational efforts, as we see throughout this chapter. In sum, although the workshops changed people's attitudes, beliefs, and even their plans to act (at least for a while), education alone did not lead to any observable action.

A similar result was observed in a government pilot program conducted in 1977 in Denver, Colorado. The purpose of this program was to change people's attitudes about appliance purchases so that instead of trying to get the lowest price, they would want the model with the lowest "energy cost of ownership." This concept is that the true cost of a household appliance such as a refrigerator includes not only the purchase price but also the cost of the energy used to operate it. The U.S. Department of Energy believed that if consumers developed energy-wise attitudes about appliance purchases, they would begin to buy models that achieve great energy savings, even if they cost a bit more to buy. The program used paid radio, television, and newspaper advertisements, as well as signs in appliance stores, using the slogan, "Products That Save Energy Pay for Themselves." In addition, displays were placed in shopping malls and in bank lobbies to show how much could be saved, carrying the message, "Products That Save Energy Finance Themselves." The program, which ran for seven

FIGURE 4-2 Some Barriers to Making Major Energy-Conserving Home Improvements

months, produced increased awareness and some attitude change. At the end of the project, people knew more about the cost and expected savings from specific conservation measures and were more likely to believe they could personally help solve the energy problem. They also expressed greater willingness to pay 10 to 15 percent more for energy-efficient appliances. But their actual behavior changed very little (Hutton, 1982).

Why did these efforts fail? One likely explanation is the gap between attitudes and behavior. There are many good reasons people may not take actions that reflect their values and attitudes. Consider the example of someone who wants to cut his energy bills. He may not know how much he can save in his particular home by upgrading insulation or installing an energy-efficient furnace, may not have the necessary money or credit, may not want to change a heating system that is functioning adequately, may want to spend the money on something else, may not trust a local contractor to do the work, or may be unable to act because, as a renter, he does not have the right to alter the building (see Figure 4-2). The more of these barriers that exist, the less difference a strong attitude in favor of saving energy will make in terms of behavior. Box 4-1 reports the results of a statewide survey of Massachusetts households that demonstrates this point. It shows that attitudes predict simple, low-cost energy-conserving behaviors such as resetting thermostats, but the more difficult or expensive the behavior, the weaker the relationship to energy attitudes.

Barriers to action also prevent other kinds of environmentally responsible behavior. Raymond DeYoung (1989) interviewed thirty-two participants and fifty-nine nonparticipants in a long-established community recycling program in Ann Arbor, Michigan, to try to understand their behavior. He found that the groups had about equally strong prorecycling attitudes. On a scale that represented strong antirecycling attitudes as 1 and strong prorecycling attitudes as 5, recyclers' responses averaged 4.13 on items such as "I like it when stores carry recycled products" and "recycling is good because it helps reduce imports." Nonrecyclers' attitudes were not significantly different at 4.02. What differentiated the groups was their beliefs about barriers to recycling, particularly difficulty. The two

groups were a full half-point apart on these items: "It's a big nuisance to keep everything separated for recycling," "A problem with recycling is finding a place to put the stuff," and "I'm never exactly sure what I'm supposed to do to recycle." Recyclers' opinions were about neutral at 3.14, but nonrecyclers, who averaged 2.65, saw significant difficulties with recycling.

In a similar vein, Georg Prester and his colleagues (1987) examined the differences between participants and nonparticipants in a local political debate about extending a high-speed railroad line in a residential area of Mannheim, Germany. People who became politically active in the controversy had a slightly higher level of general environmental awareness and were more likely to believe that the project would decrease local environmental quality. However, two of the strongest determinants of political involvement were knowledge about how to participate and interest in politics. When it came to action, political skills—interest and know-how—were more important than environmental attitudes.

The studies of environmental attitudes and behavior indicate that although the right attitudes are conducive to environmental action, they are only predictive of action under certain conditions. Attitudes are more likely to lead to behavior when strong barriers to action are removed. (A recent study of recycling attitudes and behavior suggests that attitudes have their strongest effects when external conditions—barriers or inducements—have moderate strength, and that both strong barriers and strong inducements limit the effects of attitudes [Guagnano, Stern, & Dietz, 1995].)

The conceptual framework of Table 4-2, which is based on an analysis of numerous studies of proenvironmental behavior, makes clearer what the barriers are. The table shows a long causal chain of factors influencing environmentally relevant behavior, which is at the bottom of the chain (level 1). Note that any variable at a higher level in the chain is able to influence any variable at a lower level. For example, owning one's own home rather than renting (level 6 in the table) may affect one's attitudes toward energy efficiency (level 4). This is because, to a homeowner, an energy-efficient furnace and well-insulated attic mean more than just lowered utility

Box 4-1 _____

Attitudes versus Barriers to Action:
Energy Conservation in Massachusetts, 1980

In the summer of 1980, a period of serious national concern about energy conservation and rapidly rising energy prices, one of the authors and his colleagues conducted a statewide survey of energy conservation activities in Massachusetts (Black, Stern, and Elworth, 1985). We surveyed a random sample of the households served by the state's five major electric utility companies, and received responses from 478 households across the state. We tried to explain why households differed in what they had done to conserve energy and particularly, to see how much internal psychological factors, such as attitudes and beliefs, mattered in comparison with external factors such as income, home ownership, household size, and the like. We examined four classes of energy-saving activities: major investments (such as insulating walls and ceilings, adding storm windows, or making improvements to furnaces), low-cost investments (such as caulking, weather stripping, or fixing leaky hot water faucets), minor curtailments (such as turning off heat in unoccupied rooms or lowering the temperature of home hot water), and changes in indoor temperature. Households were asked which energy-saving activities they had undertaken, and they also responded to numerous questions about their energy attitudes and beliefs, household composition, income and energy expenditures, and the structures of their homes and heating systems (such as number of rooms, heating fuel used, and ability to control heat room-by-room). The attitude questions tapped respondents' feelings of personal obligation—given U.S. energy problems at the time—to use less energy and to use it more efficiently.

When we analyzed our results statistically, this is what we found: Generally speaking, as the kind of energy-saving activity went from easy and inexpensive (changing temperature settings) to difficult and expensive (insulation and major furnace repairs), attitudes and beliefs became less and less important as predic-

tors of behavior. The key results are shown on the bottom line of Table 4-1 on page 78. The numbers are based on a statistical technique known as regression analysis, the details of which are beyond the scope of our discussion here. In intuitive terms, however, the numbers on the bottom row indicate the strength of the relationship between people's proconservation attitudes and the number of conservation actions they took (technically, the boldface entries are the percentages of total explained variance in conservation actions accounted for by attitudes and beliefs). A high number indicates that respondents who had strong proconservation attitudes tended to take more energy conservation actions than respondents with weaker proconservation attitudes; a low number indicates little relationship between the strength of respondents' proconservation attitudes and the number of conservation actions they took. As the bottom row of the table shows, strength of proconservation attitudes correlated highly with the number of temperature-change actions taken, less highly with the number of minor curtailments made, less still with the number of low-cost investment actions taken, and least with the number of major investments made in energy efficiency. Relevant attitude and belief items from our survey are shown on the second row of the table. To restate our main findings: The more difficult and expensive the conservation action, the less people's attitudes and beliefs related to whether or not they performed the action.

These findings strongly suggest that external barriers and constraints set limits on what can be accomplished by changing peoples' attitudes. The higher the barriers—expense, inconvenience, technical difficulty, and so on—the less effect proenvironmental attitudes have on behavior. It follows that inducing proenvironmental attitudes will have little effect on expensive or difficult behaviors unless the external barriers can be lowered.

expenses. Having an efficient, modern furnace and good insulation may be an important part of a homeowner's attitudes about taking good care of her home. For this reason also, homeowners may

become more knowledgeable than renters about how to install insulation (level 3), and more committed to making this kind of home improvement (level 2). Note that it is possible for factors lower on the chain

TABLE 4-1 Factors Most Closely Associated with Four Types of Energy-Saving Actions Among Massachusetts Households, 1980

	TEMPERATURE CHANGES	MINOR CURTAILMENTS	LOW-COST CAPITAL IMPROVEMENTS	MAJOR INVESTMENTS IN ENERGY EFFICIENCY
		TYPE OF CONSERVATION ACTION		
External factors involving home and household	Some residents are home during the day Number of rooms in home Elderly in household	Number of rooms in home Residents own the home	Residents pay heating bill Winter heating cost	Residents own the home Number of people in household
Attitudes and beliefs	Feel personal obligation to cut back energy use	Feel personal obligation to cut back energy use Feel personal obligation to use energy efficiently	Feel personal obligation to use energy efficiently	Expect family to benefit from efficiency
Percent of explained variation in energy-saving action attributable to attitudes and beliefs (see text)[a]	59%	50%	44%	25%

Source: Black et al., 1985. The table lists only those external factors and only those attitudes and beliefs most closely associated with each type of conservation action. See text of box for further explanation.

a. External factors, attitudes, and beliefs together could explain more of the total variation in behavior for temperature settings than for the more difficult or expensive actions. For the four types of behavior listed, the total percentage of variation explained was 17%, 8%, 9%, and 8%, respectively.

in Table 4-2 to influence those higher up. For example, behavior (level 1) can change attitudes and knowledge (levels 4 and 3) through a process of learning from experience or a psychological process of justifying one's past efforts by adopting attitudes consistent with them—the phenomenon of cognitive dissonance reduction.

The framework shown in Table 4-2 indicates that there are two main types of barriers that can keep people from acting on proenvironmental attitudes. First, the framework implies that any break in the chain between attitudes (level 4) and behavior (level 1), such as absence of appropriate knowledge (level 3) or of attention or commitment (level 2), can keep proenvironmental attitudes from generating action (see examples in the table). Such barriers exist *within individuals,* so they can be addressed with interventions aimed at individuals. Information programs, which we discuss in the next section, are designed to remove knowledge barriers at level 3. Other programs

can increase levels of attention and commitment, as we discuss in a later section.

Second, the framework in Table 4-2 identifies barriers that lie *outside the individual.* These external barriers, which appear at levels 7 and 6—the individual's socioeconomic background, available technology, social and political institutions, economic forces, and inconvenience—precede attitudes in the causal chain and so can prevent proenvironmental attitudes from forming. For example, opinion polls show a weak but consistent relationship between socioeconomic factors such as level of education (level 7), and concern with the environment (level 4) (Hines, Hungerford, and Tomera, 1987). External barriers can also inhibit the expression of proenvironmental attitudes. Attitudes in favor of recycling produce no action when recycling is too inconvenient, and attitudes favoring energy conservation lead nowhere when action is costly, difficult, or blocked by the rules of property ownership.

TABLE 4-2 A Causal Model of Resource-Consumption Behavior with Examples from Residential Energy Conservation

LEVEL OF CAUSALITY	TYPE OF VARIABLE	EXAMPLES
7	Household background	Income, education, number of household members
6	External incentives and constraints	Energy prices, size of dwelling, owner/renter status, available technology, difficulty and cost of energy-conserving action
5	Values and worldviews	New Ecological Paradigm, Biospheric-altruistic values, Postmaterialism (see Chapter 3)
4	Attitudes and beliefs	Concern about national energy situation, belief households can help with it, belief neighbors expect you not to waste
3	Knowledge	Knowing that water heater is a major energy user, knowing how to upgrade attic insulation
2	Attention, behavioral commitment, etc.	Remembering to install weather stripping before heating season
1	Resource-using or resource-saving behavior	Decreased use of air conditioner, purchase of high-efficiency furnace, lowering winter thermostat setting

Source: Adapted from Stern and Oskamp, 1987.
Note: For practical purposes, it is important to remember that resource-using behavior does not completely determine resource use. For example, someone who buys a high-efficiency air conditioner may take advantage of it by keeping the home cooler, so some of the benefit in energy savings may be lost.

As we mentioned in Chapter 3, external barriers can also impede the expression of values (level 5). The proenvironmental values of Indian Hindus and Chinese Taoists were not strong enough to overcome the pressures of poverty, tyranny, and competition for scarce resources (factors at level 7). As we noted in Chapter 3 and see again in this chapter, such external factors are very difficult to change at the individual level. In Chapters 5, 6, and 7, we examine the effects on individual behavior of interventions that alter some of the external economic and social forces shaping people's treatment of the environment.

To summarize: When can one expect attempts to change attitudes and beliefs to induce proenvironmental behavior? The simple answer is: When the barriers to action are low. In the case of consumer behaviors, barriers are low for inexpensive actions that are ready at hand. These include participating in well-designed, convenient recycling programs, making simple and low-cost changes in household energy use, and the like. The barriers are higher when the actions are inconvenient, complex, or when they have costs to the individual in terms of money, time, or opportunities foregone. Note further that some political actions are relatively easy to take. The most obvious one—voting—is the one where attitude-behavior relationships are easiest to demonstrate. In contrast, joining organizations takes more time and sometimes money, and becoming an environmental activist, which takes considerable effort, requires much more than just a proenvironmental attitude.

What can educational efforts aimed at attitude change accomplish when the external barriers to action are high? In the short run, they can do very little by themselves. But interventions need not be restricted to attitude change. As a few of the examples above have already shown, efforts to change attitudes and beliefs, combined with a lowering of the external barriers to action (for example, providing a flow-restricting shower head to install), have real potential. We discuss combination approaches to behavior change later in the book, especially in Chapter 7. In the short run, the most promising role for education is to help overcome internal barriers to action, particularly the barriers of ignorance and misinformation. We turn now to this use of education.

EFFORTS TO CHANGE BEHAVIOR WITH INFORMATION

Lack of information can be a serious internal barrier to action because it is not always obvious to an individual how to act effectively on his or her attitudes. This is especially the case for environmental protection, because the connections between behavior and its environmental effects can be impossible to discern from personal experience. Only expert analysts can tell which behaviors have the greatest effect on global warming or the extinction of species in distant tropical forests, so nonexperts cannot be expected to know what to do without some assistance. Even with a relatively simple problem, such as reducing energy use in the home, many people do not know which conservation actions are most effective, as we show in Chapter 10.

How much can be done to protect the environment by informing consumers? The best evidence comes from careful studies of deliberate interventions—studies that compare the behavior of people who have been informed with similar people who serve as a comparison group. In this section, we review several of these studies. We find first that simply providing straightforward information can make a difference, but mainly with easy, low-cost actions. We then look at other ways of providing information, methods based on principles of psychology and communication. These methods are much more successful, and illustrate what can be accomplished by information alone. We begin with studies involving simple, straightforward information, starting again with energy conservation examples.

Information, Plain and Simple

In the 1970s, in the early days of excitement of the modern environmental movement, researchers and governments began to put "conventional wisdom" into practice: They assumed that if concerned people were only told what to do, they would act to preserve the environment. This approach had very limited success, as the following examples illustrate.

Shortly after the Arab oil embargo of 1973 shook the faith of many Americans in the perpetual availability of fossil fuels, a number of U.S. gas and electric utility companies began preparing and distributing glossy informational brochures on how to save energy in the home. Some of these brochures targeted relatively simple, cost-free measures such as resetting thermostats on furnaces and air conditioners to use less energy in winter, and setting them even lower at night and when the home is unoccupied. The companies typically distributed the brochures by inserting them in the envelope along with the regular utility bill, a so-called bill stuffer. Note that there are few external barriers to making these simple changes, and that the American public in the late 1970s had a positive general attitude toward energy conservation. Thus, the main barrier to action seemed to be lack of information about which behaviors effectively save energy—the barrier that bill stuffers attempted to overcome.

Despite all this, the few reported studies of the effects of these bill stuffers on actual energy use yielded disappointing results. Thomas Heberlein (1975) conducted a small experiment just before the 1973 energy crisis in which he mailed a utility-produced brochure on electricity conservation to fifteen households in a Wisconsin apartment complex. His research team read electric meters throughout the complex for about twelve days before and after the brochures were received and found no change in electricity use by the control households and a small *increase* in use, though not a statistically significant one, among households that received the brochure. In a larger study, Samuel Craig and John McCann (1978) monitored the effect of a utility-produced pamphlet on how to cut electricity use by air conditioners. In early August of

1977, they sent pamphlets to about 800 apartments in New York City where the pattern of electricity bills indicated that air conditioners were in use. By 1977, energy was a national concern, so a strong effect might be expected. Nevertheless, a month later, the apartments that received the brochure along with a letter from Consolidated Edison, the local utility that produced the brochure, showed no change in their energy use compared to a control group that received no information.

The study also had a curious and more hopeful finding. Other apartments, randomly chosen to receive the same brochure along with a letter from the commissioner of the state's public utility regulatory commission, cut electricity use by 7 percent compared to the controls and the Con Edison group. Since air conditioners use only about 40 percent of household energy in the summer, the savings in terms of air-conditioning use was approximately 17 percent. This study shows that something in addition to the information itself—something about the way information is provided—can determine whether information works. In the next section, we return to the question of what makes some information programs effective when so many others are not.

Some information programs, carried out both by gas and electric utilities and by government agencies, targeted much more difficult-to-take conservation actions, such as adding insulation to attics and walls or replacing energy-inefficient heating equipment. Such actions are often costly and many require major modifications to one's home. Put another way, these are actions for which there are major external barriers (limitations outside the individual). But there are also internal barriers, because people do not always know which actions are most important or how to take them.

Some of these information programs featured bill-stuffer brochures, while others featured home "energy audits." As an example of the latter, consider a program started in 1977 by the Canadian government. The program, called ENER$AVE, offered all home-owners a free computerized "energy audit." Participants filled out a questionnaire about their home, giving its age, size, form of construction, and other information. By return mail, each received a computer analysis with recommendations for home insulation, weather stripping, and other energy-saving actions, complete with estimates of the cost of each action, the energy and money that would be saved, and the "payback period"—the time it would take for the savings to repay the cost.

In late 1980, a group of Canadian professors of business administration (McDougall, Clayton, and Ritchie, 1983) surveyed a sample of homeowners, most of whom had completed the ENER$AVE survey about two years before. They asked whether the household had undertaken any of six energy-saving actions that were sometimes recommended by the ENER$AVE program: adding insulation in attic, walls, basement, or over unheated areas, installing weather stripping and caulking, or installing storm windows. If the household had taken any of these actions, the respondent was asked whether the action occurred within the past two years. The researchers assumed that if ENER$AVE was effective, the households that had participated would have taken more of these actions in the last two years than the comparison households. After excluding actions that people reported they had done more than two years before, and which would therefore probably not have been recommended by the ENER$AVE audit, they found that households that had not participated in ENER$AVE reported having taken 45 percent of the energy-saving actions over the previous two years. The households who participated said they had taken 46 percent of the actions.

This is not much of a difference, and is too small to be statistically reliable. Of course, the study is not definitive. The ENER$AVE participants may have been more likely to have forgotten what changes they made in their homes (although there is no particular reason to expect this), and it is possible that the people who participated in the program—or their homes—differed from nonparticipants in some important respect that the study did not measure. However, this study finds about the same thing as studies of other computerized home-energy audits that used different research methods and asked different questions. This sort of information program appears to have little overall influence on how people use energy at home.

Why did ENER$AVE's computerized energy audits have so little effect? One possible answer that occurred to many conservation advocates was that the audits did not offer good enough information. When a homeowner says there is insulation in the attic, the computer cannot tell how much. Neither can it tell how well caulking or weather stripping has been applied. But if the energy audit is done personally, by a trained energy analyst, the computer can get better information. Moreover, the analyst can explain the recommendations and answer the homeowner's questions.

Following this logic, in the late 1970s U.S. gas and electric utility companies began offering customers free or low-cost on-the-spot energy audits. Soon afterward, the federal Residential Conservation Service program required the states to see that these audits were available to households at a minimal cost. Were these programs effective? Table 4-3 reports the results of two early evaluations.

These two programs appear to have been partially effective. They increased the frequency of a few energy-saving actions, but had no effect on most of them. More specifically, the energy audits increased the frequency of relatively low-cost behaviors (caulking, weather stripping, and modifying water heaters),

but not expensive ones (insulating walls, ceilings, and floors). Apparently, the energy audits removed the information barrier to action, but not the external barriers that prevent householders from taking expensive energy-saving actions. Consequently, the only behaviors that changed were the ones for which information was the only significant barrier. The conclusion is hopeful in that it shows that detailed, accurate information can make a difference. But it is also discouraging in a larger sense. Success was only partial, and it required a significant investment of money and the time of trained personnel in interacting one-on-one with householders. Moreover, this effort failed to change the behaviors that have the greatest energy-saving potential, because these were precisely the ones with major external barriers (see Chapter 10).

Better Ways to Provide Information

We have seen that simply providing people with straightforward information has weak effects on only a limited set of behaviors. This section shows that behavior-change programs can be much more successful if they pay attention to the way they provide the information. The successful programs we

TABLE 4-3 Actions Reported by Participants and Nonparticipants in Two On-the-Spot Energy Audit Programs, about 1979

ACTION		SEATTLE (WASHINGTON) CITY LIGHT	NORTHERN STATES POWER (MINNESOTA)
Caulk and/or	Participants	18%	55%
weatherstrip	Nonparticipants	18%	45%
Insulate and/or reduce temperature of water heater	Participants	30%	N/A
	Nonparticipants	20%	N/A
Install attic	Participants	7%	34%
insulation	Nonparticipants	7%	35%
Install wall and/or	Participants	9%	14%
floor insulation	Nonparticipants	6%	18%
Install storm windows	Participants	20%	27%
and/or doors	Nonparticipants	14%	28%

Source: Energy, Volume 6, Hirst, E., Berry, L., and Soderstrom, J., Review of utility home energy audit programs, 621–630, copyright 1981, with kind permission from Elsevier Science Ltd., The Boulevard, Langford Lane, Kidlington, OX5 1GB, UK.

describe in this section found ways to deliver information that caught people's attention and made the information credible.

Feedback. One approach to making information more effective is to tie it directly to people's behavior. Beginning in the 1970s, psychologists began experimenting with a method that, instead of telling people what to do to save energy, offered higher quality information about how much they were already using. The experiments provided regular, usually daily, feedback on how much energy a household was using and on what that rate of energy use would cost by the end of a month. Some studies used simple technology, for example, students reading electric or gas meters every day and leaving a note on the front door. Other studies used electronic monitoring devices, installed in a prominent place in the home such as on a wall near the kitchen sink, with the information made available automatically. Such devices are capable of providing feedback by the hour, minute, or second, but most of the early devices were not so advanced. Feedback systems provide information much more easily than reading a utility meter, and in a form that is personalized and easy to understand.

The theory of feedback is a simple application of operant learning theory from psychology (Skinner, 1938). If people are motivated to save energy, or to lower their energy bills, they will repeat whatever behaviors produce that reward. But it is difficult for people to tell which behaviors work because energy savings are not directly visible, and money savings are only realized once a month when the utility bill arrives—much too infrequently to help them learn what they have done to lower the bill. Feedback devices let people teach themselves how to save energy. In terms of learning theory, feedback acts as a signal of a reinforcer—financial savings—that is slow in coming. Feedback provides much more specific and valid information than a general brochure or even an expert's energy audit because it is directly related to the householders' actual behavior and because it tells what people actually *have* saved, not only an estimate of what they might expect to save.

The effect of energy-use feedback depends on several factors. To change everyday behavior, it needs to be sufficiently frequent, and it is probably most effective if it is available immediately before and after people have done something to try to save energy (Seligman et al., 1981; Shippee, 1980). It must be related to behavior in understandable ways. For example, feedback about energy used for home heating and cooling should be corrected for variations in weather (Winett and Neale, 1979). Otherwise, the large, weather-related changes in the need for heating or cooling can hide the effects of people's actions. It should also use units of measurement the householders can easily understand, such as dollars saved. And feedback is more effective when it concerns an energy source that is a large portion of the household budget (Winkler and Winett, 1982). That is, information works better when people have a strong financial motive to learn from it.

Overall, feedback experiments demonstrate under controlled conditions that real households during the late 1970s cut their energy use by around 10 percent immediately after feedback started and that the savings continued for at least several months, with feedback still being provided. The immediate savings indicate that the change was accomplished by altering behavior rather than by installing energy-saving equipment such as more fuel-efficient furnaces or appliances.

Although frequent feedback works, its effect is of limited magnitude and staying power. Because it operates mainly by getting people to use less, rather than by encouraging people to install equipment that can give the same comfort for less energy, the energy savings from feedback will sooner or later be perceived as sacrifices. (An argument has been made that annual or semiannual feedback may encourage people to install energy-efficient equipment, whose benefits can be seen most easily if they are averaged over a long period of time [Layne et al., 1988].) And feedback only works if the participants are strongly motivated. If the experiments were repeated in the mid-1990s, when there is no talk of a national energy crisis and when energy prices are no longer such a large portion of most people's incomes, feedback might be much less effective than it was in the late 1970s.

Modeling. One can also make information more effective by using a presentation that combines concepts from behavioral psychology and communication research. Richard Winett and his colleagues (1982) demonstrated a program that effectively reduced people's energy use without having them invest in new equipment or sacrificing comfort. The program featured twenty-minute videotapes of a young couple, much like most of the people in the Virginia apartment and townhouse complex where the experiment was conducted, demonstrating ways to save energy. For example, the tape on saving energy in the summer showed how to use fans and natural ventilation in the evening to save on air-conditioning, how to dress in lightweight clothing, how to shift the time and place of eating and cooking, and so forth. The script was carefully designed to present energy saving as a positive action. It used the visually compelling medium of television to demonstrate the desired behavior, and it employed the behavioral technique of modeling: the demonstrations were by people the audience could readily identify with and imitate. Participants in both the experimental and control groups in the study by Richard Winett and his colleagues (1982) also attended a forty-five-minute meeting in which they were instructed on the proper use of window fans, the insulating value of different items of clothing, and how to use a hydrothermograph installed in their homes to monitor temperature and humidity. Some of the participants were also given daily energy-use feedback for thirty days.

Compared with the group that only attended the meeting, the group that saw the videotape used 10 percent less household electricity immediately, and 19 percent less three weeks later. The savings for air-conditioning, which was the target of the program but is only a fraction of household energy use, were obviously much larger. Participants who also received feedback saved even more. The savings were accomplished with little or no change in indoor temperature, and the participants in the different groups reported the same levels of comfort. A companion experiment

in the winter produced similar results. People saved more than 25 percent of the electricity used for heating. They did this mainly by lowering indoor temperatures, but because they were instructed in how to make the change slowly and to adapt with warmer clothing, they reported a level of comfort equal to that of the comparison group.

Winett's experiment demonstrates energy savings of over 20 percent from a carefully constructed information program. It is reasonable to ask, though, whether this sort of intensive effort, with meetings, feedback, and a specially created videotape with demonstration by models, is cost-effective. To answer this question, Winett's research group conducted another experiment in July 1982, this time using a local-access television channel to broadcast twenty-minute videotapes (Winett et al., 1985). People in the experimental groups were telephoned and asked to watch the program, which was broadcast four times over a five-day period. Their energy reduction was around 10 percent for the rest of the summer, compared to control groups (a reduction of about 23 percent of the energy estimated to be used for cooling). In a follow-up the next summer, the experimental group was still using 5 percent less energy, compared with the controls. The researchers concluded that this method could be cost-effective on a large scale, because once the videotape had been paid for (about $40,000), the cost of the program would be about $1 per household, for the telephone contact. If one million households could be reached and each saved $14 in a summer, as these households did, a $1 million program would save $14 million in energy.

As with feedback, this program achieves reductions in energy use by behavioral change rather than by improving technology, so the results may be hard to duplicate when people have lower levels of motivation, such as when energy prices or environmental concern are low or people are affluent enough to use electricity rather than sweaters to keep warm at home.

"Framing" Messages. Another way to make information more effective involves paying close attention to how proenvironmental behaviors are described. The

program developed by Richard Winett's group provides an example: It referred to energy "efficiency" instead of "conservation" because Winett and his colleagues believed that their audience would perceive energy conservation as sacrifice, but would think of efficiency as a desirable goal. Another example is the experiment Suzanne Yates (1982) conducted in Santa Cruz, California, for her Ph.D. dissertation in psychology. She provided householders with information about the benefits of insulating their water heaters. When she presented the information in terms of how much money they were wasting by *not* insulating them, people became much more willing to insulate them than when she presented the information in terms of how much money could be saved. Of course, the two amounts were the same. Yates's experiment was based on the principle developed by Kahneman and Tversky (1979) that people are more sensitive to the prospect of losing something than to the prospect of gaining something of equal value. We discuss these ideas further in Chapter 9.

What the methods of feedback, videotaped modeling, and framing have in common is that they present information in ways that are particularly personalized, attention-getting, or motivating for the audience. Such methods can make educational programs appreciably more powerful. But even these methods do not overcome all the internal barriers that can prevent the expression of proenvironmental attitudes. The next section describes ways to tighten the links among attitudes, information, and behavior in order to make education yet more effective.

TIGHTENING THE LINKS FROM ATTITUDES TO BEHAVIOR

Table 4-2 shows that even in the presence of favorable attitudes, knowledge does not lead directly or automatically to proenvironmental behavior. People do not always do what they are predisposed to do, even if they know how and there are no external barriers. An example is people who save recyclables for a long period but never "get around to" taking them to the recycling center. Another example: Homeowners who want to use the city's collection service for compostable yard wastes but forget to put the wastes near the curb on the proper day. Or shoppers who prefer environmentally friendly products but feel too preoccupied with getting through their shopping lists to fully attend to their environmental concerns. We are referring here to level 2 of the table—"attention, commitment, etc." In order for people to express their proenvironmental attitudes in actual behavior, they must pay attention to environmental issues in their everyday lives, overcome the laziness or "behavioral inertia" that tends to oppose any new behavior, make a commitment to act even in the face of competing demands on their time, and remember to take action at the proper moment. In this section, we discuss ways of promoting proenvironmental behavior that remove these internal barriers to action. These methods remind people to do what they are predisposed to do or encourage them in various ways to act on proenvironmental attitudes or information they already have. Such methods can help get the most out of the educational strategy.

Reminders and Prompts. The simplest way to get people to act out their attitudes is to ask them. All of us are familiar with environmental slogans and reminders, such as "Only You Can Prevent Forest Fires," "Keep America Beautiful," "Every Litter Bit Helps" (on a trash can), and the like. These messages are designed neither to change attitudes nor to give information, but simply to remind readers and listeners to do things that they presumably are already predisposed and knowledgeable enough to do. These messages are intended to overcome internal barriers to action such as laziness or forgetting.

Research indicates that nonspecific reminders like these generally have very little effect on actual behavior. But timely and specific reminders can be effective. For example, Scott Geller and his colleagues (1971) handed out one-page flyers outside grocery stores asking customers to purchase their soft drinks in returnable bottles and giving reasons for the request. They counted the proportion of customers purchasing most of their soft drinks in returnables when they were or were not distributing the flyers. At the two large supermarkets that were leafleted, the request—what behavioral psychologists call a *prompt*—had no effect; but at the one small convenience store, the percentage of returnable-bottle customers increased 32 percent when leaflets were handed out. A likely inference, which is supported by other studies, is that to be effective, a request must be very close in space and time to the behavior people are being asked to perform. If you want people to turn out lights on leaving a room, it is most effective to put the message near the door; if you want people to invest in insulating their homes, it makes sense to have posters or flyers available in offices where people apply for home-improvement loans. Similarly, in the convenience store, people bought their soft drinks soon after receiving the flyer, whereas in the supermarket, they did so, on the average, only after many other purchases.

In another experiment on resource recovery, Harvey Jacobs used *reminders* to improve participation in a residential recycling program in Tallahassee, Florida (reported in Geller et al., 1982). Four neighborhoods of different socioeconomic levels were monitored after the residents had been initially informed of a weekly curbside pickup of newspapers and cans. The level of participation correlated strongly with socioeconomic level. It was 3 percent in the lowest social class neighborhood and 25 percent in the highest. After four to six weeks, all the residents were given a flyer reminding them of the program, to see if this would increase their participation. This prompt added no new information, but only reminded people of past information. In the middle and upper-middle income neighborhoods, where participation was already higher, participation immediately increased by ten to twelve percentage points—but there was no change in the lower and lower-middle class neighborhoods. This finding again demonstrates that a request—or, for that matter information such as the initial notice about the recycling program—can help, but it also suggests that messages must be designed to fit the audience. A message that is delivered in the wrong way or by the wrong messenger is likely to be ignored or even mistrusted. Numerous studies on energy conservation as well as recycling show that written communications tend to be ineffective with U.S. audiences of lower socioeconomic status.

Public Commitment. It is also possible to increase proenvironmental behavior by getting people to make a public or quasi-public commitment to taking an action. A public commitment appears to strengthen people's private, personal commitment to the action. Recall that in the framework in Table 4-2, a personal commitment to take action despite competing demands on one's time is part of level 2—a main link between attitude and behavior; therefore, a publicly made commitment, freely given, should make a proenvironmental attitude lead more reliably to action by creating a personal commitment. The principle, derived from cognitive dissonance theory (Festinger, 1957), is that when people undertake an action in the absence of any obvious external force or reward, they see that action as something they have chosen themselves. People who see their behavior as based on their own internal motives are likely to persist in the behavior even after the commitment has lapsed.

A simple experiment by Anton Pardini and Richard Katzev (1984) on recycling behavior shows the power of public commitment. Pardini and Katzev asked twenty-seven households in a middle-class neighborhood of Portland, Oregon, to participate in a feasibility study of neighborhood recycling. Nine households were asked impersonally: Informational brochures were left at their doors to explain how the program worked and give the dates of the first two weekly pickups. Another nine were asked in person to make a minimal public (or quasi-public) commitment. They were approached by one of the researchers, who explained the program, gave them a piece of paper listing the two pickup dates, and asked, "Will you commit your household to participating in this recycling project for two weeks?" All agreed. Nine were asked to make a "strong commitment." Instead of the oral commitment, they were asked to sign this statement: "In the interest of conservation, I commit my household to participating in this newspaper recycling program for two weeks." Again, all agreed. After two weeks, all households were recontacted, and urged to participate for two more weeks. As Table 4-4 shows, public commitment was more effective than mere information, and stronger commitments led to more recycling than weaker commitments. Over the first two weeks, the two commitment groups recycled about three times as often, providing about three times as much paper as the households receiving only information. For the strong-commitment households, but not the weak-commitment households, the effect continued for two more weeks, after the commitment had ended.

Personal commitment—besides being a link between attitude and behavior—is also a link between knowledge and behavior (see Table 4-2). Therefore a stronger personal commitment caused by public commitment should make information more effective as well. An experiment on energy-use feedback by Lawrence Becker (1978) demonstrates this kind of effect. Becker asked participants in the experiment to make a quasi-public commitment to saving a specific amount of energy—either 2 percent of what they had been using, or 20 percent. In this experiment, what was stronger about the commitment was not the way it was made (e.g., on a signed document, or orally), but the difficulty of the behavior people were committing themselves to. When people received feedback, those who made the stronger (20 percent) commitment used 9 percent less energy than those who made the weak (2 percent) commitment. When they did not receive energy-use feedback, commitment had essentially no effect. (For more detailed review of research on prompts, reminders, and commitment effects, see Katzev and Johnson, 1987).

Highlighting Attitudes and Norms. Yet another way to break down internal barriers to action is to call people's attention to attitudes and beliefs that they already have, but that they may not connect to the situation they are in. The following experiments show that people sometimes need to be reminded that they are in situations in which it is appropriate to exercise their proenvironmental attitudes or in which other people expect them to do so.

Robert Cialdini and his colleagues at Arizona State University (Cialdini, Kallgren, and Reno, 1991) conducted a series of experiments demonstrating that subtly calling people's attention to the social norm against littering decreased their littering behavior. In one study, visitors to a municipal library, on their way back to the parking lot, saw a passerby (who was in reality working for the researchers) do one of three things: put a fast-food restaurant bag in the trash can, *pick* up a littered bag and put it in the trash can, or simply walk by. On returning to their cars, they found a handbill on automotive safety attached to their windshields, and the researchers watched to see if they littered it.

The researchers reasoned that the passerby's simply depositing the trash in the can called attention to a "descriptive norm," that is, it told observers that people usually don't litter in the area. Picking up a littered bag called attention to an "injunctive norm,"

TABLE 4-4 Effects of Public Commitment on Participation and Paper Collected in an Experimental Recycling Program

CONDITION	NUMBER OF HOUSEHOLDS	FREQUENCY OF PARTICIPATION		POUNDS OF PAPER COLLECTED	
		FIRST 2 WEEKS	SECOND 2 WEEKS	FIRST 2 WEEKS	SECOND 2 WEEKS
Information	9	3	4	70	57
Minimal public commitment	9	10	4	210	54
Strong public commitment	9	13	11	247	166

Source: Pardini, A., and Katzev, R. The effect of strength of commitment on newspaper recycling. *Journal of Environmental Systems,* Volume 13, 245–254. Copyright 1984. The Baywood Publishing Co., Inc.

that is, it reminded the observers that others *disapprove* of littering. They predicted that reminding people of either norm would influence littering, but that the descriptive norm would influence behavior only in the location where people were reminded of it, while the injunctive norm would generalize outside the immediate geographical area. To test this prediction, they ensured that half the respondents in each group encountered the passerby on a grassy, landscaped section of the property while the others saw the passerby in the parking lot. The results confirmed the predictions (see Figure 4-3). The strongest effect, in fact, was observed when the injunctive norm was evoked in a different environment from the one where people were given the opportunity to litter.

What this experiment and the others in Cialdini's series imply is that behavior can be affected by efforts to bring into people's awareness things they already know about how people normally behave, what is expected of them, or (as another experiment in the series showed) what they themselves believe they should do. When people's attitudes and values, and the expectations of others, support protecting the environment, it helps to remind people of that fact. The experiments suggest that the reminders have to be subtle so as not to seem coercive. In particular situations, it is left to the ingenuity of those who would change behavior to find effective ways to implement the principle.

The results of a final study support the findings on reminders about attitudes, and also help clarify how several strategies for linking attitudes, information, and commitment can work together. Beginning in the summer of 1982, Joseph Hopper and Joyce McCarl Nielsen (1991) experimented with three strategies for increasing participation in an ongoing, but rather ineffective, residential recycling program in a stable, middle-class neighborhood of Denver, Colorado. They randomly assigned blocks in the neighborhood to one of four experimental conditions. One group received only information, in the form of a flyer that described how the program worked and what could be recycled, and that listed the next seven monthly pickup dates. The flyers were distributed twice over the seven-month study. The second group received

the flyers plus a prompt, in the form of bright yellow flyer announcing each pickup date one to three days in advance. The third group received the information and the prompt, and in addition were contacted by volunteer block leaders who had been instructed to talk with every household on the block about the program and to encourage their neighbors to recycle. The fourth group, a control, was not contacted at all. All the households had been monitored over the previous seventeen months, and less than 1 percent of them had left recyclables for pickup in an average month. Over the seven months of the study, the participation rate rose to 2 percent for the control group, 10 percent for the information-only group, 21 percent for the group receiving information and monthly prompts, and 28 percent for the group with block leaders.

These results show that lack of information was a barrier to participation in the program, and that forgetting was also a barrier (that the prompts helped overcome). In addition, the volunteer block leaders added something to the program. One thing they may have added is models whom their neighbors could imitate, as with Richard Winett's videotapes on energy conservation. But there is no evidence that the neighbors actually saw the block leaders demonstrate recycling. A second possibility is that by talking to every household on the block, they may have reminded people of their attitudes about recycling, with the result that prorecycling neighbors followed their attitudes more closely.

There is a third possibility that is even more promising for promoting proenvironmental behavior. It may be that information given in the course of social interaction in a community helps create a shared norm in favor of recycling that changes behavior both by creating perceived social pressure and by modifying internal, personal motives so as to promote behavioral change. Hopper and Nielsen present some evidence that supports this interpretation. Participants in the experiment completed questionnaires both before and after the seven-month experimental period. Two sets of questions concerned norms. One set, about what Cialdini calls injunctive social norms (see above), asked whether people's friends and

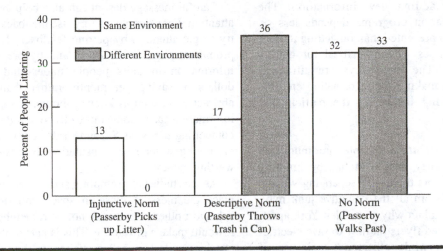

FIGURE 4-3 Percentage of People Littering as a Function of the Type of Norm Evoked and the Environment in Which It Was Evoked

Source: Cialdini, R., Kallgren, C., and Reno, R. A focus theory of normative conduct: A theoretical refinement and re-evaluation. *Advances in Experimental Social Psychology,* Volume 24, 201–234. Copyright 1991. Academic Press.

Shaded bars signify different environments for littering and evocation of the norm; white bars signify the same environment.

neighbors expected them to recycle and whether they expected their friends and neighbors to recycle. The other set concerned personal, internalized norms, that is, people's expectations for their own behavior. Hopper and Nielsen asked how much it bothered the respondent to throw away recyclables and how much personal obligation they felt to recycle. Over the course of the experiment, both types of norm became stronger in households living where there were block leaders, but not on other blocks in the neighborhood. The survey results thus suggest that talking with block leaders actually changed both social and personal norms. Going one step further, if this were true, one would expect groups with block leaders to continue recycling at a high level even after information and prompts are withdrawn. Although Hopper and Nielsen did not follow the experimental groups beyond seven months, they do report on four blocks in the neighborhood that had had block leaders for two years before the experiment began. People on those blocks were already recycling 21 percent of the time when the experiment started and over the next

seven months, without experimental intervention, their recycling rate increased to 34 percent. It appears that information given in the right social context at the community level can change behavior more effectively, and maybe also more permanently, than information given to individuals without supporting social interaction. We examine this possibility in more detail in Chapter 6.

WHEN DOES INFORMATION WORK?

What makes some informational programs succeed where others fail? Successful programs are not necessarily ones that offer more or better information. Richard Winett's videotapes on energy conservation presented essentially the same information that the participants could get from the meeting they attended, and daily energy-use feedback presents exactly the same information that people could get if they read their own utility meters. Similarly, when information programs use prompts or try to raise participants' commitment levels, they become more effective without adding new information. The success of information programs depends less on getting information presented than on getting it used. This section discusses what is critical for getting information used. The main keys are attracting people's attention, making the information credible to the audience, and increasing the participants' involvement.

Getting People's Attention. People are inundated with information. They deal with this by ignoring most of what confronts them—by separating what is important to them from all the cognitive junk mail. This process may explain why some New York apartment dwellers ignored flyers on how to save electricity in the summer, while others cut their use of air-conditioning by 17 percent. The flyers that were ignored came from Consolidated Edison, the local electric utility. All these people had received mail from Con Edison before, so they knew what to expect. Most often, that mail contains a bill—sometimes along with other written material that most people ignore. They probably learned to operate under the rule that with mail from Con Edison, if it's not a bill, you can throw it out. But New Yorkers have much less experience getting letters from the state Public Service Commission. Most people probably opened these and many probably read them. The information worked only when it could get from the flyer into people's awareness.

There are many techniques to attract people's attention. One is with a personal approach, such as Pardini and Katzev used when they asked people to commit themselves to a recycling program and Hopper and Nielsen used with block leaders for recycling. Word of mouth has often proved the best form of advertising for energy conservation programs. Making the invisible visible also attracts attention, as shown by energy-use feedback programs, which convert electricity or gas use into a daily message. A compelling medium of presentation also helps—for example, Winett's use of television. Video presentations, in addition to being inherently attention-getting, can use demonstrations, which present information more vividly than verbal descriptions can.

Careful message design can also help get people's attention. For example, energy-use feedback programs try to get attention by putting feedback devices in a prominent place in the home and by presenting the information in units people understand, such as dollars of saving per month, rather than in more abstract units such as kilowatt-hours. Also, the same information can become more effective if it is stated in compelling terms, as Suzanne Yates demonstrated by promoting water heater insulation as a way to stop wasting money.

As the studies of prompting show, it is important to place the message close in space and time to the behavior; otherwise, it may not be remembered when it would make a difference. This is part of the logic of attaching miles-per-gallon stickers to the windows of cars in dealers' showrooms and bright yellow labels to major household appliances to tell prospective buyers what energy costs to expect when operating them.

And as we have already noted, what gets people's attention depends on the audience. It may depend on socioeconomic differences, as Jacobs found with the Tallahassee recycling program, but there are many other variables. Evaluations of home insulation programs typically conclude that working with local groups—churches, neighborhood associations, and the like—is the best way to promote a program (Stern et al., 1986). As one example, when utility companies in Minnesota used their own personnel to conduct home energy audits, they reached 4 percent of the eligible homes; other utilities, which hired community groups to do the job, reached 15 percent of homes—and did it for one-third the cost (Polich, 1984)! The community groups were locally known and trusted, so messages from them got serious attention. Moreover, because of their commitment to helping their neighbors, they probably worked harder at marketing the program than the utility companies' employees did.

Credibility. Information must be credible to be effective. Part of credibility lies in the source of information. This may be why a message from the New York State Public Service Commission was more effective than one from the electric utility, and why community groups were more successful than utility representatives at encouraging people to have energy audits. Electric utilities may be highly credible for some purposes, but people may not take them seriously when they offer advice on how to use less of their product. By far the most important factor affecting the purchase of solar-powered equipment in a study of California homeowners in the late 1970s was the number of people they knew who already owned solar equipment (Leonard-Barton, 1980). This fact and other information from the study clearly suggest that the word of trusted friends and neighbors was more important in the decisions than the word of solar energy experts.

Credibility also depends on people's ability to validate the information they are given. With energy use, which is generally invisible, this can be a serious problem. It is nearly impossible to tell whether a home insulation contractor has done a thorough job inside one's attic or walls, so people are understandably suspicious. One or two horror stories in a community can kill a program, because people are more likely to trust a neighbor's experience than the word of someone who is promoting a product. For this reason, energy conservation programs have often provided independent inspections of contractors' work or even performance guarantees as a way to become more credible.

Involvement. Information becomes more effective with people who have made a commitment to act on it. This has been demonstrated experimentally by studies of commitment such as those of Pardini and Katzev, Becker, and others. The block-leader approach also seems to depend on getting people involved by talking with their neighbors about recycling, and Cialdini's efforts to call people's attention to social or personal norms can also be considered a way of increasing involvement. Crisis can also increase involvement. For example, in periods of severe drought, people have made major efforts to conserve water simply on the basis of requests from local authorities and concern for the community (Agras, Jacob, and Lebedeck, 1980). Of course, it helps if the requests are made credible by photos of low water levels in the local reservoir.

These examples suggest some general rules about how to make information more effective; however, the

specifics depend on the kind of behavior one intends to change. For informational approaches to reach their potential, they need to be designed creatively to maximize their credibility and the audience's attention and involvement. To do these things, it is important to make a concerted effort at the outset to understand the audience's perspective. This may be done either by systematically surveying the audience group or, what is often better, by involving representatives of the audience group in designing the program. The latter approach, one of *community* involvement, is suggested by the use of block leaders, and has been used successfully on a larger scale in a number of cities and towns, as we show in Chapter 6.

Using Social Networks to Diffuse Information. One of the most effective strategies for spreading information is to take advantage of existing networks of communication. The tendency of California homeowners to buy solar collectors if they knew other people who had it is an example of a broader principle, that innovations diffuse through a population along the lines of social influence. Agricultural extension programs have used this principle for generations to spread new and improved farming practices in farm communities. They identify individuals who are well known and respected in the community, and focus their efforts on getting a few such "opinion leaders" to adopt the new technology. Once they have benefited from it, the technology tends to spread with little additional effort.

It is easy to see why information coming from individuals someone knows and trusts is particularly effective. Such information automatically gets attention and has high credibility because of its source. And it tends to increase involvement as well, because whatever someone does with information from a trusted friend or neighbor is likely to be of subsequent interest to that person, and may affect the future relationship between the giver and receiver of the information. The experience of community energy conservation programs has repeatedly validated the diffusion-of-innovation approach, which relies on sending information through existing social networks (Darley and Beniger, 1981; Stern et al., 1986). And as we will see in Chapter 6, diffusion of information is not the only important function that existing social networks can serve in promoting proenvironmental behavior.

SUMMARY AND CONCLUSION: WHAT CAN EDUCATION ACCOMPLISH?

We have discussed educational interventions aimed at promoting fairly specific proenvironmental attitudes and beliefs among individuals and overcoming internal barriers, such as lack of knowledge or commitment, that keep them from acting on those attitudes. Education can make a difference in people's behavior, but there are serious limits to what it can accomplish. The chapter supports the following general conclusions:

In the short term, educational approaches work only when the main barriers to action are internal to the individual. As we have seen, education is effective mainly with relatively simple, low-cost behaviors, such as depositing cans in curbside recycling bins or altering home thermostat settings. Such actions help, but they typically have smaller effects on the environmental problems they are meant to lessen than more permanent actions such as purchasing an energy-efficient vehicle or appliance (see Chapter 10). Information has also been effective in getting people to request home energy audits, an action that has the potential to lead to larger and more permanent energy savings and environmental benefits by changing heating and cooling equipment. Nevertheless, when protecting the environment requires great effort or expense, as it often does, there is no experimental evidence that education alone will do the job. Under such conditions, behavior change requires interventions to reduce the external barriers to action. We examine those interventions in the next chapters.

Education may have important indirect effects over the long term. Though external barriers to individual action limit the effectiveness of education in the short run, education may have important positive, though indirect, effects in the long run. For example, the block-leader approach to recycling (discussed in

Tightening the Links) had indirect beneficial effects by changing community norms. A longer-term and possibly more important indirect effect—one we have not yet discussed in this chapter—can occur when education changes people's political behavior; this behavior, in turn, changes government policy so as to lower the external barriers to proenvironmental behavior. The history of smoking reduction illustrates this process. Over the several decades since the health hazards of smoking became established and widely publicized, the proportion of smokers in the United States has slowly decreased. During that time, scientists, physicians, and other individuals who became convinced of the dangers became politically active and brought pressure on governments and other powerful actors to bring down the barriers to behavior change and alter some of the incentives that govern smoking. Since 1964, cigarette advertising has been restricted, tobacco taxes have been deliberately raised, no-smoking rules have been applied in airplanes and many public buildings, life insurance companies have made smokers pay more than nonsmokers for coverage, and employers have implemented antismoking programs. These changes are fair because governments, employers, and insurance companies incur higher costs for smokers than for nonsmokers. At the same time, these changes have made it easier for individuals to act on antismoking attitudes. People who intend to stop smoking find more justifications and social support, and nonsmokers find it easier to speak their minds to smokers. Some of these changes, of course, even influence people whose attitudes are not antismoking.

By a similar process, changes in environmental attitudes may come to affect behavior over the long term. A generation of voters and environmental activists, influenced by the writings of Rachel Carson (1962), Paul Ehrlich (1968), Barry Commoner (1970), and other scientist-educators, has pressed government agencies, corporations, and other important actors to implement new policies on air and water pollution, energy development, and land use, and thus change the way they treat environmental resources. Some of these policies also remove barriers to individuals' acting on their own proenvironmental attitudes, and thus change individual behavior. For example, they have helped bring more energy-efficient automobiles and appliances to market, so that environmentally conscious consumers can buy them. If education about environmental problems has been indirectly responsible for these advances in environmental policy over the last few decades, this would be a highly significant accomplishment. Although it is difficult to conclusively demonstrate the causal role of education over such long time periods, improved public awareness and understanding are among the most plausible causes of the policy changes. This sort of long-term effect of attitude change provides a key rationale for environmental education programs in the schools.

Education is only likely to induce behavior that is compatible with people's deeper values. As we note above, environmental values and ethical beliefs are broader and more deeply rooted than environmental attitudes or the specific beliefs addressed in this chapter. They are also more difficult to change. Therefore, educational efforts aimed at attitude change are unlikely to succeed if they go against people's ethics and values. An example may be the repeated efforts of a coalition of nuclear energy industries through the 1980s and early 1990s to change public attitudes toward nuclear power with multimillion dollar advertising campaigns emphasizing the benefits of the technology. After well over a decade, public opinion is even more strongly opposed than before. The bases for public opposition to nuclear power are discussed in detail in Chapter 9.

Educational programs are more effective when they are designed according to psychological principles of communication and also directly address the links between attitudes and behavior. As the chapter repeatedly shows, making information available is not the same as getting it used. Even when people are being asked to act according to their attitudes, and so are predisposed to use information, it is essential to make special efforts to get their attention, use information sources the audience trusts, and involve the recipients of the information in the effort. It may also be neces-

sary to remind people that their proenvironmental attitudes apply to the situation at hand, and to tell them what to do to enact the attitudes. These things need to be done in different ways for different behaviors and audiences, and the chapter illustrates a number of useful tools for the purpose.

Education works best when combined with other strategies of intervention. We have seen how external barriers such as cost and difficulty keep educational programs from reaching their goals. We have also seen that programs work best when they do more than just educate. For example, recall that when an energy conservation program provided water-flow restrictors along with information on how to use them and on what they could save, it achieved its only behavioral success (Geller, 1981). These observations support a more general conclusion, that education and other strategies can act in synergy: The effects of both together are greater than one would expect from their separate effects.

The general point has been demonstrated by decades of research on health promotion, showing that educational campaigns are not enough to change individuals' smoking, drinking, dietary, and exercise behavior without supplementary efforts. However, education plus other changes can make an important difference over time. In the words of one review of the literature (Green, Wilson, and Lovato, 1986):

> . . . [H]ealth promotion has been occurring and health practices have been changing. . . . The changes have been more notable since the advent of official policies supporting nonsmoking with more than information alone. . . . Organizational changes, such as smoking restrictions on airplanes, restaurants, and other public places have helped. Economic supports, such as excise taxes on tobacco and alcohol, insurance incentives for driver training, not smoking, and blood pressure control, have helped. Environmental supports for behavior conducive to health, such as regulations on marketing food products as healthful and availability of fitness facilities in worksites and public parks, have helped. The combination of these supports with health education appears to have made a substantial dent in social norms of health-related behavior (p. 513).

Green et al. (1986) then go on to point out the important role of education in helping to bring about the above supports and changes. They conclude:

> . . . [F]ew of the organizational, economic, and environmental changes would have been possible without the support of an enlightened, or at least willing, public (p. 513).

The experience with health promotion suggests that although education may look like an ineffective strategy if it is judged over the short term and in isolation, it can be an essential part of effective intervention. In fact, the next few chapters show that the success of incentives and other methods of changing environmentally relevant behavior sometimes depends critically on the quality of the information provided and on the level of public concern and willingness to support the incentives or other interventions. Short-term educational interventions are important sources of information, and long-term environmental education strategies can be critical in building the public support necessary for a variety of environmental policies to be effective. No single strategy is sufficient by itself. Thus, the key issue is not how much can be accomplished by education alone, but what the place of education is in a comprehensive strategy of behavior change. We return to this question in Chapter 7.

CHAPTER 5

CHANGING THE INCENTIVES

CHAPTER PROLOGUE

Like many other environmentally concerned Americans, I (Paul Stern) want to minimize my personal contribution to environmental degradation. I know that one of the most effective things I can do to reduce air pollution and the threat of global warming is to avoid using a car, especially for traveling alone. I could do a lot by not driving to and from work. But my choices are limited.

The place I live, just outside Washington, D.C., is about 13 miles from where I work in the city—much too far to walk. Commuting by bicycle would be possible but still time-consuming—and also dangerous. There are no bicycle paths on my way to work, and lots of rush-hour traffic—fast-moving in some areas, and tightly packed with frustrated drivers in others. I do not want to risk life and limb. What's more, if I biked, the extra time in traffic and the exertion would leave me breathing far more than my share of auto exhaust, while the auto commuters, who are

not doing their part for the environment, breathe more easily. This would not only endanger my health (I suffer from asthma, and would be one of the first encouraged not to exercise on Washington's hot, polluted summer days), but it also would make me resentful and angry toward those who are harming both the environment and me while enjoying the comfort of an air-conditioned ride. So I don't bike.

Joining a carpool or vanpool might be possible, but then again, it might be very inconvenient. Although there may be people living near me who work near where I work, I don't know them. There is no easy way to identify them, because neither my neighborhood nor my employer keeps the sort of records I could check to find them. And if I did find them, I might have to change my work hours to join a pool, and I would probably still have to drive a car to the place the pool leaves from. So I don't even try to find out whether I should carpool.

I could move closer to work. But housing near my office costs at least two or three times what it does where I live now. There is lower-cost housing in Washington, but it is in areas nationally famous for their high crime rates, and I would not be safe traveling there by foot or bicycle, especially after dark. Besides, my wife and I like the rural feel of the area immediately around our home. So I don't move.

I could look for another job where commuting wouldn't be so difficult. Although I often wish this solution would work out, it hasn't so far. The challenges and enjoyments of my work are a strong tie to it, so I would need a stronger inducement than just making a personal contribution to environmental quality to get me to give up my job. Besides, I tell myself that by working on environmental policy issues in Washington, I am doing good for the environment to compensate for the harm I do by transporting myself.

It would be possible, although slow and inconvenient, for me to take public transportation. I would walk five minutes to the nearest bus stop to catch a bus that runs every thirty minutes in rush hour and takes thirty minutes on a circuitous route to get to the Washington Metro, three miles away. The Metro trains run often and stop within 1-1/2 miles of my office, where I can catch a city bus or my employer's interoffice shuttle van. The 13-mile trip would take between 1-1/2 and 2 hours each way, depending on whether I made the bus connections.

What do I do? I compromise by driving a car to a parking lot at the nearest Metro stop, and catching the Metro and then the bus or van. Instead of driving 13 miles each way, I drive only 3 miles each way. The trip takes about sixty-five minutes in each direction—more time than I would like, but much less than it would take if I gave up the car. Compared to driving all the way, I save 20 miles of car travel a day, or 100 miles a week, and reduce my contribution to air pollution and global warming accordingly.

In honesty, though, that is not why I don't drive all the way to work. The main reason isn't the environment but the traffic. I have found, on occasions when I need to take a car into the city, that the traffic is slow and frustrating. The trip usually takes fifty to sixty

minutes, which doesn't save much time over my present route, and fighting the traffic would put me in a foul mood when I got to work or got home. In addition, I enjoy the 25-minute Metro ride, on which I can usually read, write, or think without interruption or frustration. In fact, I wrote most of this account while riding on the Metro. I take my present route to work because it is the best of the alternatives for me. It is better for the environment than driving all the way to work, but that is just a bonus. My proenvironmental attitudes are not the main cause of my proenvironmental behavior. If I could drive to work in twenty-five minutes, as I once did when I commuted 13 miles to work in the small city of Elmira, New York, most likely I would do it in preference to a two-hour trip on public transport or my present sixty-five-minute route.

What is the significance of this story for saving the environment? It illustrates the many factors other than saving the environment that determine people's transportation choices and the importance of external barriers to proenvironmental action. For me, and I think the great majority of other American workers, the environment is not the deciding factor in how we travel to work. Time and inconvenience are major barriers: Few people would sacrifice over an hour a day from time with their families to spend it on a more environmentally benign but slower trip to work. Unavailability of alternatives is an important barrier for many people. When I worked in Elmira in the 1970s, for example, there was no public transit alternative there. This is effectively the case for most Americans, because they either live or work in places that are far from public transport. Cost is a major barrier. I can afford a car to get to the Metro, but many of my neighbors cannot, and find themselves being even more proenvironmental than I am—but not by choice. They ride the bus when they would rather drive.

The same considerations of time, convenience, available alternatives, and cost dominate most choices about travel—for shopping, visiting friends and family, and taking vacations. Even among the environmentally concerned, I suspect that it is the configuration of barriers like these that determines how far people travel to shop and whether they visit relatives on holidays or stay home and use the mail or phone to stay in contact.

The point is that people make transportation choices mainly as a function of their immediate, personal consequences, because these are often more important to us when it comes down to action than our commitment to the environment. It would be a mistake to conclude that someone with proenvironmental attitudes is being hypocritical by driving to work in the face of all the barriers to other modes of transportation, and given that so few others are giving up their cars. The cause of behavior that damages the environment is not necessarily a lack of the right attitudes. If I could join an organized national or even citywide effort to reduce car travel, for example, as part of a campaign for cleaner air, I might well change my behavior and give up my car ride. For one thing, the organized effort would make behavior change easier. Someone might make a centralized effort to set up car pools, for example. The antidriving movement would offer social support, the good feeling of being part of a group working for a better earth, and something to talk about with people on the bus or in the carpool. It would lower the barriers to behavior change. But as it is, my attitudes are outweighed by the society's incentives. If I acted on my own to help the environment, any contribution would be vanishingly small. And if I chose to help in certain ways, such as by riding a bike, I might come to see myself as either a masochist or a fool. I would have to breathe an increased volume of polluted air and thus threaten my health. I would also resent the comfortable drivers who leave me in their exhaust while they benefit from my contribution to lighter traffic and continue to harm the environment. Who but a masochist or a fool, I might ask myself, would go out in summer heat, exercising in dangerously polluted air, while everyone else is driving in comfort and staying within health guidelines?

In Garrett Hardin's classic analysis of the tragedy of the commons, he encompasses all of this in a simple theory of why people destroy environmental resources: It pays. His theory does not imply that people are crass or amoral. Rather, the tragedy lies in

human nature—we have no choice. According to Hardin, whenever human beings have free access to a valuable but depletable resource, we are in a situation that ensures that by acting to promote our own well-being and the well-being of our families, we inevitably destroy the resource base.

As we explained in Chapter 2, Hardin shows how a resource user on a commons—we used the example of the crab fisher—is better off by taking more of the resource so long as the additional effort, with costs taken into account, brings in more food or money than not making the effort. Anyone who, out of religious belief or proenvironmental attitudes, refrains from taking more crabs does nothing to help the situation because someone else will catch them. In fact, the result may well be to punish the conserver because the increased supply will drive down the price of the conserver's catch. I am in just this kind of situation when I decide whether to drive to work. If everyone else is polluting the air, I get punished for riding a bike.

According to Hardin, the tragic flaw in the tragedy of the commons is people's desire to better themselves as individuals. That characteristic, combined with a free but finite resource and unlimited access to it, is a potent barrier to conservation and results in destruction of the environment. What solution is possible? It is impossible to make a finite resource infinite. Adopting environmentalist religions or changing attitudes is unlikely to work because those who do not change their morals or attitudes can get rich off other people's restraint, while they destroy the resource. According to Hardin, there are only two solutions. One is to restrict access; the other is to make the resource costly. What these approaches have in common is that they change the individuals' incentives—that is, the positive and negative conditions surrounding their behavior—so that it pays them to get as much as they can out of a limited amount of the resource rather than to harvest as much as possible. This chapter concerns the theory and practice of changing the external conditions. We call these incentives although they can be both positive and negative; that is, they include what are sometimes called disincentives as well. Changing incentives is

Hardin's recommended approach to environmental problems. We emphasize, however, that some of the premises of Hardin's argument—particularly that human behavior is by nature egoistic—are highly controversial. We return to the question of egoism and altruism in human nature in Chapter 8.

THE THEORY OF INCENTIVES FOR ENVIRONMENTAL PROTECTION

There are innumerable ways to end common access or make a resource costly. A crabbing area can be divided and fishers given rights to identified areas; government can charge fishing fees, sell licenses, ration access, or rent or auction fishing rights; and so on. All these approaches, Hardin argues, require an authority strong enough to keep individuals in line. Because of the need for an authority, Hardin calls the general strategy *coercion.* He proposes, with a nod to democracy, that it be "mutual coercion, mutually agreed upon." The implied threat of physical force will not need to be carried out if individuals abide by the rules, but in Hardin's view there must be some authority that has the right to use force, if necessary, to protect the commons. (As we show in Chapter 6, the argument that coercion is necessary has also been called seriously into question.)

Hardin's solution is familiar from political theory, as noted in Chapter 2. It is the solution political philosopher Thomas Hobbes offered in the seventeenth century to the eternal problem of government: How to protect the common good from the acts of bad individuals. People allow kings and democratic governments the right to use force to protect them from criminals, invading armies, and other threats to the common good. Hardin argues that environmental destruction is such a threat, and the same solution should be applied.

Hardin's solution also has a psychological basis in B. F. Skinner's theory of operant behavior. Skinner argued that except for a small number of biologically predetermined ("unconditioned") and classically conditioned (or Pavlovian) reflexes, behavior is learned by a process in which people (and other animals) repeat behaviors as a function of their consequences. Whatever is rewarding to an individual is repeated until it is no longer rewarding or until the individual finds a more rewarding behavior. Skinner and his followers demonstrated in hundreds of carefully controlled experiments that animals repeat behavior that is rewarded, stop repeating it when the reward is removed, stop doing things that are punished, and so forth. Careful analysis of the ways behavior responds to its consequences has proved to be a powerful model for predicting animal—and human—behavior.

Let us see how it explains tragedies of the commons, and what solutions it suggests. For a crabber, harvesting an additional crab is rewarding (that is, it has a positive immediate consequence) so long as it increases the total value of the individual's catch. But there is also a long-term negative consequence: Enough of this rewarding behavior harms the environment so much that it depletes the crab population, with the result that it takes more and more work to earn the same amount and with the eventual result that there are no more crabs. Early in this process, catching more crabs still benefits the fisher, although at the expense of others. Some continue to catch crabs out of greed, and the others do the same out of self-protection. They are aware that if the greedy catch more crabs, prices will fall and their incomes will suffer. If uncontrolled, the process goes on inexorably until catching another crab is so difficult that it is not worth the extra time or money of operating the boat. Only when conditions get that bad—at a time when overfishing may have already ruined the crab grounds—does a self-interested crabber stop fishing.

In Skinnerian terms, there are two reasons for tragedies of the commons. One is that the rewards for using the environment's resources go to the individual who uses them, but most of the costs are paid by others. The tragedy occurs, according to Skinner, because behavior changes only as a result of consequences to oneself. People do not stop doing things that reward them just because those things also harm others. They stop only when the behavior stops benefiting themselves. With open access to a common-pool resource, it follows that they overuse the environment. The other reason for the tragedy is that the rewards are much closer in time to the behavior than the costs are. Skinner's experiments show that the effect of a consequence on behavior drops off rapidly as the consequence is removed in time from the behavior. Immediate consequences shape behavior much more effectively than delayed ones. A consequence that is delayed by years or decades, such as depleting a crab fishery, is likely to have almost no effect on behavior. The tragedy of the commons is what John Platt (1973) called a social trap: Free access entices fishers to keep taking crabs, but they do not see—until they have gone too far into the trap—the larger punishment that is the ultimate result of their behavior.

This analysis implies a strategy for solution. If the shared, long-term costs of resource use could somehow be charged to the individuals responsible and brought closer in time to the behavior, people would not do things that harm the environment. To put it positively, if the rewards for environmentally appropriate behavior accrued to the environmentally responsible individual immediately, instead of requiring an initial sacrifice followed by a long waiting period, and instead of being shared with people who may not have done anything to help, people would take care of the environment by taking care of themselves. (A dominant theory in economics offers much the same analysis, although it uses different language. The economic version is summarized in Box 5-1.)

Consider, for example, an impoverished country where small farmers raise large families that overtax the ability of the land to provide food and firewood for cooking. In countries such as India, Nepal, and Madagascar, this pattern, driven by rapid population growth, is one of the causes of deforestation. Skinner would presume that the families are large because children are rewarding to parents. As discussed in Chapter 2, children provide more hands to work in the fields, and they are valued because they will care for their parents in sickness or old age. Families are even larger where the local peasants experience a high rate of infant and child mortality, because this prospect gives parents a reason to "invest" in extra children, just in case. The result, sooner or later, is that people produce a greater population than the

country can support. Although people do not want to impoverish the country, their only alternative is to put their own well-being at risk.

How would Skinner recommend that such a country change the incentive structure and lower its birth rate? There are many possibilities, involving both positive and aversive consequences. It might legally limit childbearing, applying financial and other penalties against violators. This approach has been a cornerstone of Chinese population policy in recent years. It might provide benefits to families so long as they have two or fewer children. It might create social security programs to care for people in sickness or old age, so that they no longer need large families for that purpose. It might invest in rural education and economic development programs so that people have an alternative to living off the land. This policy makes children less of an economic asset because they have to be supported while in school, but it gives each child a better chance to earn enough in adulthood to support aging parents. Government might aim education programs at women, so that families are better off if mothers work than if they have additional children, as we discuss in Chapter 12. Creativity can suggest almost endless possibilities. Among them, behavior theory prefers rewards to punishments on the grounds that they are more effective. A reward increases the frequency of a specific behavior, while it is highly unpredictable what behavior will result from punishment. People who are punished for having children may instead engage in all sorts of other behavior—including evading or changing the policy that threatens to punish them. When India tried to control population growth with a coercive sterilization policy in 1975 (one province, for example, ordered teachers to be sterilized or lose a month's salary), one demographer remarked that the policy was more likely to bring down the government than the birth rate. And indeed, the government was defeated and the policy reversed (Visaria and Visaria, 1981).

In this chapter, we look at changing incentives as a strategy for promoting proenvironmental behavior. We show that for any kind of proenvironmental behavior there are many barriers that can be lowered

Box 5-1

The Economic Theory of Externalities

An economic analysis of the causes of environmental problems begins with an account of how markets work. In market transactions, buyers are willing to pay only for the value they expect to receive; the price of a good or service therefore depends on the values individual buyers place on it. If a transaction has effects beyond the buyer and the seller, those effects—known as externalities—are not reflected in the price. Externalities can be positive or negative. A homeowner who buys flowering shrubs beautifies the neighborhood; the neighbors benefit even though they do not pay. Environmental problems, however, involve negative externalities. Someone who drives a car pollutes the air but pays no more for pollution control than someone who rides a bus or bicycle. Someone who buys groceries in nondegradable packaging causes solid-waste problems but pays nothing extra for municipal waste services.

Markets have difficulty solving environmental problems because the environment is a public good. Because no individual can own the clean air, no one can charge polluters for using it. The same is true for clean water, beautiful views, endangered species, the ozone layer, and so on. Moreover, it is unrealistic to ask people to make voluntary contributions to preserve the environment because of the "free rider problem": any individual is better off by letting other people make the contributions because no one can keep a noncontributor from enjoying the benefits of a public good.

Economists offer a number of solutions to environmental problems, all of which are based on the principle of "internalizing" the externalities. The idea is that if people who benefit from environmentally damaging goods and services can be made to pay individually for the environmental damage they are indirectly causing, they will have an incentive to maintain environmental quality.

One way of internalizing externalities is to establish property rights. This approach can work with grazing lands—a commons can be divided into family plots, giving each family an incentive not to overgraze. In the arid western United States, the national government grants water rights to ranchers, farmers, and municipalities and leaves them to manage their own allotments. This approach is not practical for some problems, such as managing ocean fisheries or preventing air or water pollution.

Another approach is for government to auction rights to use the environment up to a limit that is considered safe. For example, it could auction hunting licenses for threatened species or licenses to release waste materials into a river. In either case, a public decision would have to be made about how much hunting or waste the environment could stand, and enforcement would be required to prevent unlicensed or excessive use. The purpose of using an auction rather than, say, a lottery, is to guarantee that the resources being allocated are put to their most highly valued use.

A third approach is for government to charge people and organizations for the use of resources in excess of what a supplier would charge, so as to include the value of the negative externalities in the price. An example is the idea of a carbon tax on the use of coal, oil, and natural gas to discourage their use, encourage the use of substitutes, and thus help solve the problem of greenhouse warming, which is caused in large part by the burning of those fuels. The theory is that if the social costs of greenhouse warming—for example, to future generations that will live with its effects—were estimated and added to the price of fuel, people and businesses would use fuel more sparingly. If the price is set right, people would decrease use enough that the taxes collected would compensate future generations for the burdens they may face but not so much that the present generation is unfairly penalized.

It may occur to you that the economists' solutions are often hard to implement. You can't make some parts of the environment private (the climate system, for example), proposals for auctions and taxes often meet strong political opposition, and finding the right price for damages to future generations may be a task beyond the ability of any economist. We agree that it is much easier to state the principle of internalizing the externalities than it is to make it practical. Nevertheless, the principle provides a very useful way of thinking about environmental problems. And as the chapter shows, incentive systems that work are those that put the principle into practice.

and many ways to lower them (and for environmentally dangerous behaviors, many barriers that can be raised). We consider in some depth three examples of behavior changes that could significantly benefit the environment: increasing the use of carpools and mass transit, recycling and waste reduction, and reducing energy use in homes. We show that incentives can be effective in encouraging these behaviors, but that they are not effective automatically. Not all incentives that seem appropriate can be implemented, and not all of those are effective. Incentives can also have unexpected side effects, both positive and negative, and effectiveness sometimes depends critically on what is being done simultaneously with other strategies, particularly information.

INCENTIVES FOR RIDE SHARING AND MASS TRANSIT USE

We look first at the problem raised at the start of the chapter—reducing use of the automobile. To use incentives to this end, it is necessary to understand the incentives that lead people to use automobiles. We have already seen the incentive structure for Paul Stern's trip to work; Peter Everett and Barry Watson (1987) have offered a more comprehensive list of the rewarding and punishing aspects of driving and of using mass transit for a typical American (see Table 5-1). Everett is a psychologist who has devoted his career to applying behavioral insights to transportation planning and management.

It is clear from the table why most people prefer driving to mass transit: The benefits outweigh the disadvantages. The same would be true if we compared driving with ride sharing (using carpools or vanpools), although some of the items in a table of incentives would change. The imbalance of incentives explains behavior and also offers many ideas for changing it. In principle, one could weaken any of the rewards for driving or strengthen the punishments, strengthen any of the rewards for using transit or weaken any of the punishments. One could also invent new rewards for using transit or punishments for driving. Let us look at some examples of incentive approaches that have been tried in practice.

Everett and his colleagues have experimented with rewarding patrons of city buses with a token for each trip, exchangeable for discounts in participating city stores. In a pilot experiment in a university bus system, the reward increased bus ridership by 27 percent (Deslauriers and Everett, 1977). The system was later adapted for use in municipal bus systems in Spokane and Seattle, Washington, and a few other cities (Everett and Watson, 1987). In Spokane, although the tokens were widely used, the system produced only a small increase in bus ridership. The program was considered a success because it induced businesses to market the bus system, and it may also have prevented the decline in bus ridership that most U.S. cities experienced in the early 1980s.

In Seattle, people who bought a monthly "flash pass" for the bus system got, in addition to unlimited bus rides for the month, discounts at some of the best restaurants in the city, movie and performing arts theaters, health spas, and several retail establishments. These incentives were chosen to attract middle- and upper-income residents, who normally did not ride the buses. By 1985, sales of passes had increased 37 percent under the system, and the program was declared a success. We do not know, however, whether the sales were mainly to new bus riders or to people who had previously paid by the ride.

It is important to note that these incentive programs were carefully designed to reward everybody involved: Bus patrons saved on their consumer purchases, the bus system gained ridership, and the participating businesses attracted new customers. Peter Everett believes that a reward system has to have this character if it is to stay in operation.

But do incentives for bus riding reduce automobile use? The studies suggest that they may have only limited effect in the short run, even though they seem to pay for themselves and to have secondary benefits to local businesses. Helping downtown businesses, however, may indirectly benefit the environment. In many U.S. cities, downtown businesses face bankruptcy because of competition with suburban malls that offer easy access and convenient parking. As downtowns decline, fewer people travel there to work or shop, and eventually there are not enough travelers

TABLE 5-1 Rewarding and Punishing Aspects of Car Driving and Mass Transit Use

	REWARDING	PUNISHING
Car Driving	Short travel time Prestige Arrival/departure flexibility Privacy *Route selection* Cargo capacity Predictability Delayed costs Enjoyment of driving	Traffic congestion Gas and maintenance costs
Using Mass Transit	Making friends Time to read	Exposure to weather Discomfort Noise Dirt Surly personnel Long walk to stops Danger (crime) Immediate costs Unpredictability Small cargo capacity Limited route selection Crowded Limited time flexibility Low prestige Long travel time

Source: Everett, P., and Watson, B. Psychological contributions to transportation. In Stokols, D., and Altman, I., (Eds.), *Handbook of Environmental Psychology*, Volume 2, p. 999. Copyright 1987 by John Wiley & Sons, Inc.

to fill the buses. Bus lines become increasingly uneconomic and local governments phase them out, with the result that bus riders are forced into less energy-efficient auto travel. So, the kinds of incentive plans Everett describes do benefit the environment by slowing the broader trend toward sprawling and totally auto-dependent cities.

Some municipalities have tried to get commuters out of their cars by making car driving less convenient compared to the alternatives. One way to do this has been to reserve lanes on heavily traveled commuter roads for high-occupancy vehicles and buses, so that people who give up driving can shorten their commuting time. This strategy has increased bus ridership and ride sharing in a number of cities,

although there are enforcement problems. Where it is easy to use a carpool lane without being caught, such as when it is a center lane of a multilane highway, violations are frequent. And people sometimes ride with mannequins or pick up riders at bus stops so that they can use the fast lanes. Of course, these evasive strategies undermine the purpose of the programs—to reduce the number of cars on the road. They can be countered with lane designs that discourage evasion and with increased enforcement.

Some companies have tried to induce ride sharing by offering their employees the service of matching them with neighbors who might be able to pool with them, by reserving the best spaces in their parking lots for carpools and vanpools, or by a combination of these methods. Both approaches reduce the time and inconvenience that are otherwise part of ride sharing. In a number of efforts during the energy crisis period of the 1970s, matching services added from 7 to 30 percent to the proportion of employees ride sharing; incentive systems that featured priority parking added from 22 to 55 percent (Geller, Winett, and Everett, 1982). The evidence indicates that in the companies that used them, parking benefits were much more attractive to employees, and were sometimes sufficient to induce carpooling even when matching services were not offered.

The experience with ridesharing and transit incentives suggests that incentives can work, that some incentives are much more effective than others, and that it is not obvious in advance which ones will be best. It is also hard to predict which incentive systems will be easy for people to evade. Finding an effective incentive package seems to require a certain amount of trial-and-error learning.

Despite the successes with some incentive programs, the big picture is still discouraging. As of 1990, only 5 percent of U.S. workers traveled to work on mass transit and 13 percent shared rides, while 73 percent drove alone. These figures are down from 1980, when they were 6 and 20 percent, with 64 percent driving alone (data are from the U.S. census, reported in Davis and Strang, 1993, Table 4.7). Why has the overall progress been negative? The answers, we think, lie in forces that are much stronger than these incentives and that have the opposite effect. Consider three factors that combine to put city buses and subways on the endangered technology list. First, the price of gasoline. For generations, U.S. gasoline prices have been much lower than those of Western Europe, and since the energy crisis of 1979, they have been declining in real terms (that is, gasoline has become less expensive after accounting for inflation). In Western Europe, where consumers pay at least three times as much for gasoline as we do in North America, there is a very strong incentive to avoid driving and to use small, fuel-efficient cars when driving is necessary. This incentive structure helps keep people riding the railroads in Europe, where the average citizen travels four to eight times as far by rail in a year as the average American, despite the

shorter distances (data from Davis and Strang, 1993, Table 1.15, converted to per capita rates). Second, highways. Since the 1950s, the United States has invested billions of dollars annually ($7 billion in 1991, for example; Federal Highway Administration, 1991) in an interstate highway system that shortens travel time for long commutes and makes it easier for people to live far from where they work. Third, the income tax deduction for home mortgage interest, which encourages the American dream of owning a detached house on a plot of ground.

Many Americans benefit from relatively low gasoline prices, good highways, and incentives for home ownership, but these incentives hurt mass transit. They have made it convenient for Americans to move out of central cities by the millions into dispersed suburban settlements from which commuting by mass transit is highly inconvenient or impossible. Once people live in such places, it takes large incentives indeed to get them on a bus. Moreover, they are likely to oppose policies, such as sharp increases in gasoline taxes, that would provide the needed incentives but that they would see as punishing them for living where they choose.

These barriers to ride sharing and mass transit use can be called *structural* barriers because they are literally built into society (for example, by the locations of buildings and roads). They cannot be changed quickly by any policy because they are shaped by the history of past decisions that necessarily take a long time to reverse. The past decisions to build highways and support home ownership have created incentives for people to oppose policies that would change them, and powerful institutions as well: The interests of commuters and homeowners are institutionalized in lobby groups and in the person of legislators who depend on their votes. In this sense, past decisions have been built into the social structure as well as the physical. For these reasons, structural barriers are more difficult to change than other kinds of incentives. It usually takes slow historical processes to remove them.

The structural barriers to reducing energy use in transportation are so strong that transportation analysts have focused most of their proenvironmental efforts on encouraging use of less-polluting and more

A Large Suburban Residential Area Near San Francisco Bay
This sort of dispersed housing makes ride sharing and mass transit impractical for most residents because of
the long distances they would have to go to get their rides.
(Louisa Preston/Photo Researchers)

fuel-efficient automobiles rather than on alternatives to the private car. Incentives, including regulations that attach financial costs to noncompliance, are the predominant policy tools for this goal as well. In the United States, the major policy options have been raising the Corporate Average Fuel Economy requirement on auto manufacturers, raising taxes on motor fuel, introducing "gas guzzler" taxes or "gas sipper" incentives for new automobiles, and recently, a proposal to offer more than the market value to buy and scrap old, inefficient, and polluting vehicles. We will not review the evidence on the effectiveness of these incentives here. But we emphasize that most of

these incentives directly affect the behavior of manufacturers and oil companies, and only affect individual consumers indirectly. Consequently, they have the potential to make a huge difference when they can be enacted, but they also meet fierce and well-organized opposition from the political lobbies that represent the present structure of transportation.

INCENTIVES FOR RECYCLING AND WASTE REDUCTION

The United States produces 1,900 pounds (864 kg.) of municipal solid waste per person per year (1986 data are from World Resources Institute, 1992). The resulting waste stream pollutes water, land, and air and taxes the resources of local governments responsible for disposal. It is a problem facing rich and poor, rural and urban. Although the greatest potential for reducing waste lies in preventing it (see Chapter 10)—for example, by using less packaging material for consumer goods—this strategy usually requires changing the behavior of corporations. It is possible to make significant advances, however, by changing individual behavior. Getting people to purchase products that have minimal packaging or that are made from used materials, or encouraging repair rather than disposal of products that stop working, are among the ways that individuals can reduce waste by preventing it. Recycling waste material is usually a second-best choice. We discuss it here because this is the approach most commonly studied by behavioral researchers.

An early example of such behavioral research was the work of E. Scott Geller and his students in the 1970s, who offered rewards to students in dormitories at Virginia Polytechnic Institute and State University and James Madison University for recycling wastepaper (see Geller et al., 1982, 133–136). The researchers tried two kinds of rewards. Students in some dorms who brought wastepaper to the recycling location at the appointed time received raffle tickets, usually one per pound of paper delivered, for a weekly drawing with prizes donated by local merchants (valued at between $1 and $30). Other dorms were paired in a contest in which the dorm that produced more paper per resident received a $15 cash prize. Both rewards were effective in increasing recycling, and Geller considered the program cost-effective because of the low cost of the program and its positive advertising value to participating merchants. However, Geller was disappointed at the low proportion of residents who delivered paper and the low yield of recyclable paper from the drives (see Table 5-2). He concluded that it would be more worthwhile to try incentives in homes and offices, which produce more wastepaper than dorms.

Harvey Jacobs (1978, cited by Geller et al., 1982) tried the same kinds of incentives on households in Tallahassee, Florida. He offered some of the households one cent per pound for newspapers left at the curb for recycling every other Saturday. (One cent per pound was the approximate salvage value for newspaper at the time.) Other households were offered a chance for a $5 prize given by lottery to one participating household on the day after the trash pickup. These households were compared with others that received only reminder notices announcing the dates of the pickups and the procedures for participating. Before the experiment, each group contained 3 to 4 percent recyclers; afterward, 8 percent of the households receiving only information put recyclables at the curb, compared to 9 percent of those offered payment and 14 percent of those offered a chance in the lottery. The increase in the amount of recyclables collected was too small to make the program practical. No group provided enough additional material to pay more than one-third of the cost of the program.

TABLE 5-2 Responses to Contests and Raffles as Rewards for Recycling Wastepaper in University Dormitories, 1970s

	PAPER RECOVERED (LBS./WK.)			PERCENT PARTICIPATION		
STUDY	BASELINE	CONTEST	RAFFLE	BASELINE	CONTEST	RAFFLE
Geller et al., 1975	141	237	253	2.2	3.7	7.3
Ingram and Geller, 1975	58	—	113	2.9	—	4.9
Witmer and Geller, 1976	49	544	820	2.6	5.9	12.2
Couch et al., 1979	134	—	763	—	—	—

Source: Geller, E., Winett, R., and Everett, P. *Preserving the environment: New strategies for behavior change.* Copyright 1992. Adapted by permission of Allyn & Bacon.
Note: Participation is defined as delivering at least one sheet of 8-1/2 × 11-inch paper during a one-week period.

These incentives were relatively ineffective, even though the penny a pound essentially returned the salvage value of the wastepaper to those who recycled it. Apparently, this was not enough money to make a large difference for householders or college students. Incentives might still make a difference if a different and more effective kind of incentive could be found, or if a larger financial reward could be justified.

To find a more effective incentive, one should begin by asking why people fail to recycle. As we saw in Chapter 4, inconvenience is a major barrier to recycling. So, it is no surprise that successful recycling programs have been built on making the behavior more convenient. In Jacobs's work in Tallahassee, increasing the frequency of collecting recyclables had as much effect on participation as offering one cent per pound of wastepaper. Another good example of successful efforts based on convenience are the programs that encourage recycling of high-quality office paper by providing specially designed receptacles at each worker's desk. In an early program of this type, the U.S. Environmental Protection Agency collected 12.5 tons of high-grade wastepaper per month from its 2,700 Washington employees beginning in 1975 (see Geller et al., 1982). Such programs collect large amounts of paper nearly uncontaminated by other kinds of waste because of the receptacle design, and they work without offering financial incentives.

For recycling household wastes, it is possible to maximize convenience by using curbside pickup rather than recycling centers, picking up recyclables and trash at the same place and time, using identifiable receptacles (e.g., different colors or shapes for different types of waste), and so on. We believe much can still be done to raise rates of recycling by concerted efforts to make it easier for people to do what they want to do. The way to find effective convenience interventions is to ask people what would make recycling easier, try out some of the suggestions, and keep those that work.

One of the most familiar incentive systems for recycling, and one of the most effective, is the so-called bottle bill—a system of legislatively mandated deposits on containers for soft drinks and sometimes other beverages as well. In this system, people pay a deposit at the time of purchase, usually 5 cents, for each container covered under the law. The deposit is returned when the empty container is returned to a designated location (usually, a store that sells soft drinks). Bottle bills have benefited the environment by reducing littering, saving space in landfills, and saving the extra energy that would be needed to make glass and aluminum containers from raw materials rather then recycled ones (see Chapter 10). They sometimes also encourage reuse of containers, which is even more environmentally beneficial than recycling. Despite these environmental benefits, bottle bills have been vigorously opposed by the soft-drink industry wherever they have been introduced. They now operate in nine U.S. states.

Bottle bills work by providing a rather large incentive—much greater than the one cent per pound used in the early recycling experiments—for recycling materials that have special environmental benefits if returned to the same industry that produced them. In effect, people who fail to recycle are made to pay the cost of litter and trash pickup and disposal, and people who recycle—even if they did not purchase the containers in the first place—are rewarded for reducing these costs. The bottle bill is thus a good example of incentive theory in action.

Even better for the environment than recycling are methods that reduce the quantity of waste that needs to be recycled or disposed of. Incentives can also be used to promote waste reduction, as an increasing number of communities are learning. Many cities, towns, and counties are now paying for trash disposal by charging per-trash-can fees to those who put out trash, rather than using older methods financed by property taxes or flat monthly fees per household. The effect is to create an incentive for each household to produce less trash.

In Seattle, where a pay-per-can system has been in effect since the early 1980s, households pay $10.70 per month for pickup of a 19-gallon "mini-can" or $13.75 per month for a standard 32-gallon can, plus $9 for each additional can. Seattle households put out an average of three and one-half cans per week in the early 1980s, but only one can per week by 1992

(Cohn, 1992). People accomplish the reduction in many ways, from buying fewer throwaway products to doing the "Seattle stomp," a dance on top of trash to get more of it to fit in a can. Both methods reduce demand on garbage trucks and landfills. Illegal dumping of household waste, which some feared would be a side effect of the program, has not become a serious problem. The Seattle Solid Waste Utility attributes success in part to the strong environmentalism of city residents.

Over 200 U.S. communities now offer incentives of various types for reducing household waste. Some charge by the can, some by the pound, and others charge for stickers that must be affixed to a trash can or bag for it to be picked up. Seattle and other communities offer a discount for low-income households so that the system is not unfair to the poor. And in some communities, pay-per-can is an option rather than a requirement, and is enthusiastically accepted by small households.

Pay-per-can is another practical application of incentive theory. It directly rewards any behavior that reduces trash volume (or, in pay-per-pound, trash weight) and punishes increases in trash. It leaves the household the choice of finding ways to reduce waste that fit family needs, but it creates a more or less constant reminder to think about how to reduce waste. This sort of system may lead people to buy fewer disposable products and choose products that use minimal packaging, and thus put pressure indirectly on manufacturers. Of course, the system is not perfect. Wealthy people who feel they can afford to make trash are completely unconstrained by the system, a fact that could lead to objections on the ground of unfairness. Pay-per-can is often resisted by people who see it as adding a new cost to daily living. And the possibility of illegal dumping is always there for people who want to evade the system or cannot afford the fees. In addition, there are implementation problems, ranging from trash stickers that fall off and cause complaints from people whose trash has not been collected to the difficulty of working out a system of contracts with the dozens of small, private trash haulers that serve some municipalities. But the system is working in a growing number of communities. Success is partly due to the large incentives that can be offered. A saving of $9 per trash can not filled is enough to make a real difference to many families, and the large incentive is justified by the avoided cost of trash hauling and disposal.

These few examples show the significant potential that exists for reducing household wastes by offering incentives. We should not leave the topic, however, without commenting on why incentive approaches were adopted so slowly for decades and why they came to look so encouraging in the early 1990s. Much of the reason lies in the changing physical, legal, and social context of municipal waste disposal in the United States. A generation ago, landfills were easy to locate and most had plenty of room to expand. By the early 1990s, many old landfills were getting full and cities had expanded so far that it was hard to locate new ones within reasonable distance. Moreover, citizens had come to place a higher value on environmental quality, particularly in the area of toxic waste disposal, and had caused laws and regulations to be enacted. Because people know that hazardous wastes can be dumped in landfills, no one wants a landfill for a neighbor. Alternatives, such as trash incineration plants, also meet public opposition because of concerns about air pollution. All these changes greatly increased the financial cost of waste removal, and all indications are that these costs will continue to increase. When waste removal is expensive or new disposal sites are politically unacceptable, even fairly large incentives for waste reduction and recycling look like bargains. In short, incentives became more practical because the structural conditions changed over a twenty-year period. New physical, legal, and social conditions made the environmental costs of waste disposal more visible to those who managed waste disposal—that is, they have helped internalize the externalities. The incentives for local governments changed, so they became more willing to pass those incentives on to individuals. And people grew more aware of the environmental costs of disposal. In short, incentives became more effective because environmental attitudes changed in ways that made larger incentives justified.

REDUCING ENERGY USE IN HOMES

Reducing the use of fossil fuels and electricity can do great things for environmental quality. Burning coal, oil, and natural gas pollutes the air and contributes the majority of all greenhouse-warming gases produced by human activity. Nuclear-powered electricity does not cause these problems but is responsible for environmental threats from long-lived radioactive waste. Households account for about one-third of all energy consumption in the United States, and most of that—amounting to almost 20 percent of the national total—is used in homes (most of the rest is used in vehicles; see Chapter 10). So, residential energy conservation can do a lot to improve the environment.

Three kinds of incentives have been used to promote energy conservation in homes: energy price changes, financial rewards for desired behavior, and methods that simplify the task of conserving energy and thus make conservation more convenient. This section discusses examples of each approach.

Energy Price Changes

Economic theory holds that people find ways to economize on things if they are sufficiently expensive. It follows that if Americans use too much energy, it is because the prices consumers experience are too low. Here are a few things that have been tried to address the problem in homes.

Many apartment houses include heat and electricity costs in the rental charge, either because local authorities require landlords to provide heat or because a large building has only one furnace or electric meter. If instead of getting energy at no apparent cost, each household paid directly for the energy it used, it would have an incentive to economize. One way to internalize the costs is to provide electric or gas meters for each household. This may require each apartment to have its own heating and cooling system. When the individualized approach is tried, it usually results in reduced energy use in the building, but the change can be very expensive. It often results in a change to electric heating, which is simpler and cheaper to install in apartments but can be less energy-efficient than

having one heating plant for a large building. Less expensive alternatives include submeters on the steam or hot water flowing through each apartment and an allocation system that divides the total energy cost for a building among the apartments proportionally to the amount of floor space in each. This last system creates an incentive to economize for the group of apartments rather than for each individual, but has the advantage of requiring no new equipment. Even this simple system, called the Residential Utility Billing System, has resulted in building-wide energy savings of 5 to 8 percent (McClelland, 1980).

It is also possible to build conservation incentives into the energy price system. Electricity, which accounts for over half of all the energy used in U.S. homes (Hirst et al., 1986), has traditionally been billed under a "declining-block" rate system that rewards overuse. The system essentially gives volume discounts: The more electricity you use, the less each additional kilowatt-hour costs. This system made sense in terms of marketing electricity, but when excessive energy use became a national concern, several state utility regulatory agencies began to change the rate structures, either introducing flat rates (the same charge for each kilowatt-hour) or increasing-block or "lifeline" rates, which offer enough electricity for household necessities at a lower rate, with a higher rate for usage beyond the basic level.

Another form of electricity price incentive is called time-of-use pricing. Although most people are unaware of this fact, electricity costs different amounts to produce at different times of the day and year. Every electric company is responsible for having the capacity to meet demand at the day and time when it is greatest—in most of the United States, on the hottest summer afternoons; in a few of the coldest areas, on the coldest winter mornings. To meet this demand, many companies have power plants that they operate only at the high-demand or "peak" times. They use their most inexpensive power plants all the time and leave the expensive ones for infrequent operation, so peak power is by far the most expensive to produce. It is also often more polluting, because the peak plants tend to use older technology. To give people an incentive to use less power at peak times, many utility companies now charge a higher rate at those times—sometimes as much as eight times the usual rate, but only for a few hours a day in the peak part of the year. They install new meters in homes that record the times when power is being used.

The idea of time-of-use pricing is, of course, completely new to most people who are first exposed to it, so utility companies have needed to explain it to people, much as we have done in the previous paragraph. As one might expect, a large incentive induces people to shift some of their electricity use to off-peak hours. But nonprice factors can have a larger effect than the size of the incentive. Thomas Heberlein and Keith Warriner (1983) analyzed energy use among participants in an experiment in the state of Wisconsin that set the cost of electricity during peak periods at rates between twice and eight times the cost of off-peak power. The price differential between 2:1 and 8:1 had a significant but small effect on the amount of electricity people used in peak periods: It accounted for two percent of the variation across households. Behavioral commitment—a measure of how important the household considered it to be to reduce peak-period electricity use and of whether doing this was considered a moral obligation within the household—accounted for 11 percent of the variation. One reason that the price effect was so weak was that many people in the experiment did not know or did not believe that the price differential was as large as it was. They knew there was a sizable incentive to use electricity at off-peak times, but beyond that, the exact size of the incentive mattered relatively little.

These results suggest that the effectiveness of the price incentive will depend on how it is explained to people. Heberlein showed this in another study (Heberlein and Baumgartner, 1985) that compared two different ways of explaining time-of-use rates. The electric company's usual information package consisted of notification letters, a brochure, and two bill-stuffer notices. The enhanced communications package added frequent reminders about the rates, letters from the state Consumer Advisory Council, detailed information about rates, advice on how to monitor home energy use, and other information.

Consumers receiving the enhanced package reduced peak-period energy use 16 percent below the level attained with the utility's information package.

Financial Rewards

Since the 1970s, psychologists have experimented with systems of rewards to encourage households to use less energy. Early experiments, for example, offered financial payments to households that reduced energy use by a certain percentage on a weekly basis (e.g., Hayes and Cone, 1977) or that had their air conditioners set above 74 degrees Fahrenheit when an inspector stopped by (Walker, 1979). Such rewards changed behavior, but they are inherently limited because they only affect people's daily behavior and do not change the inefficient equipment many people have in their homes. In addition, people who use less often perceive themselves as sacrificing comfort because their homes are colder in winter or warmer in summer (see Chapter 10). People usually resist making major cuts in energy use when it involves sacrifice, only doing so when they perceive a general emergency or when the household is financially strapped. Sacrifice-type responses are often only temporary. (A significant exception to this rule is that people sometimes adapt to changes they have made that they may have considered temporary. People can and do adapt to lower indoor temperatures in winter [Winett et al., 1982], and there is evidence that average indoor winter temperatures in the United States have decreased since the early 1970s [Kempton, Darley, and Stern, 1992].)

The limitations of changing daily behavior do not affect energy-saving approaches that change household technology so that people have the same comfort with less energy use. Improving the energy efficiency of buildings with measures such as added insulation in attics and walls can yield substantial savings in household energy used for space heat—30 to 50 percent, according to many estimates—while holding indoor temperatures constant (Hirst et al., 1986, see Chapter 10). Replacing old, energy-inefficient furnaces can save almost as much (the precise figures vary widely, depending on the condition of the house, the fuel, the climate, and so on). Of course, these changes are expensive, so money is a barrier to action. Here is where incentives come in.

Two kinds of incentives that governments and utility companies have used are loan subsidies—offers to lend money for energy efficiency at below the usual rate of interest—and partial rebates that effectively reduce the cost of new household equipment. In the United States in the 1980s, the incentive programs typically worked as follows: A utility company, usually acting on the insistence of a state regulatory agency, would offer incentives to households for upgrading the energy efficiency of their homes. To qualify, a householder would have to call the company to request a home energy audit. An energy auditor would arrive and assess the status of the home's insulation, storm windows, and so forth, and recommend measures that met the utility's standard of cost-effectiveness. Then the householder could have recommended measures installed by a qualified contractor under a low-interest loan contract with the utility company or, if the utility offered a rebate program, could pay for the contractor's work and exchange the receipt for a partial reimbursement from the utility. Some utility companies also made the incentives available for do-it-yourself work.

Several conclusions emerge from the experience of these incentive programs (Stern et al., 1986; Berry, 1990). One unsurprising conclusion is that the stronger the incentive, the greater the percentage of eligible households that use it. In addition, households vary in their preference for different types of incentive (such as grants or rebates versus loan subsidies), even when the incentives have the same monetary value. On average, households prefer grants or rebates to loans, but many higher-income households and people skilled in managing budgets prefer loans. The difference seems to depend on willingness to go into debt for energy conservation and people's expectations about being able to maintain enough income to repay the debt. The repayment issue may explain the great success of one type of loan subsidy—a loan that does not need to be repaid until the home is sold.

What is clearest in the reviews is that the size of an incentive is not the most important factor affecting the proportion of people who use it. We looked closely at three regional programs in which several utility companies in the same geographic area offered exactly the same incentive package to encourage households to invest in insulation and other major efficiency improvements (Stern et al., 1986). The evidence shows that a strong financial incentive was necessary for a highly successful program, but far from sufficient. In New York State, where nine utility companies offered loans for energy-efficiency improvements at slightly below-market interest rates, no company got more than one-half of 1 percent of its customers to take advantage of the incentive within a year. In the Pacific Northwest, when eleven companies offered interest-free loans, the most successful program attracted over 10 percent of the eligible households per year, but the average rate was only 4 percent per year. When seven northwestern utilities offered an even stronger incentive, a rebate that covered 93 percent of the cost of the recommended energy improvements, the most successful program involved almost 20 percent of its customers in a year—but the average program reached only 5 percent. Figure 5-1 shows these results. It shows that with small incentives, no program was very effective, and that large incentives made much greater success possible. But success depended on much more than the incentive, as shown by the tremendous variation in effectiveness between programs offering exactly the same incentive. In fact, the stronger the incentive, the more difference nonincentive factors make.

We gained some understanding of how the nonincentive factors work by examining the two steps consumers went through during these programs: requesting energy audits, and deciding to invest in what the auditors recommended. Figure 5-2 shows that larger incentives affect household behavior primarily after they receive energy audits. Once people get their energy audits, the size of the incentive has a strong effect on household decisions—a sufficiently strong incentive, such as that offered in the 93 percent rebate program, means almost certain action among those who received audits. So, once people are seriously considering their energy choices, a large incentive strongly influences behavior. But the size of the incentive makes little difference in attracting people's attention. As the white bars in Figure 5-2 show, strong-incentive programs did not do very much

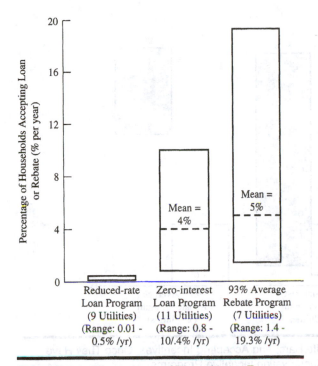

FIGURE 5-1 Effectiveness of Three Home Energy Conservation Programs.
Source: Stern et al., 1986.
Dotted line signifies mean percentage across programs; bar covers range from most successful to least successful program.

in Canada and Western Europe. After considering various possible explanations, we concluded that the most likely reason was the procedures required for people to take advantage of the incentives. The U.S. programs all used the same two-step procedure. To get the incentive, consumers first had to request a home energy audit and then, after waiting for the audit to be scheduled and conducted, act on the auditor's recommendations and file a claim for the rebate or loan. The U.S. utility companies used this procedure to ensure that people were not getting rewarded for installing uneconomic energy improvements. The non-U.S. programs did not require audits. They simply listed the improvements that were covered and paid the incentive on receiving proof (a receipt) that the improvements had been installed. They took the chance of paying for some ill-advised improvements in return for getting more homes improved. The more convenient procedure made the programs more successful, apparently because it required consumers to take one less action. In the terms of Chapter 4, it lessened a barrier to final action and also tightened the link between attitude and action by requiring less of the consumer's attention.

Making Conservation Convenient

Convenience is very important for home energy efficiency because, as we noted in Chapter 4, it is so difficult to make and carry out wise decisions about making one's home more energy efficient. One must select from many possible improvements—adding insulation, installing storm windows and doors, sealing cracks around windows, maintaining or replacing furnaces, and so on. Most of these measures are expensive, and it is difficult to judge in advance how much each measure will save in one's home. Moreover, many of them require experts, such as heating or insulating contractors, to install them. The consumer is faced with a range of choices and may have to rely for information and installation on experts whose trustworthiness is unknown. So, lack of knowledge, uncertainty, and the need to devote significant attention to the choice are major barriers to action. A consumer may understandably

better than weak-incentive programs at getting people to request energy audits. This is why so many of the strong-incentive programs had only moderate success in changing behavior. Attracting people's attention is a job for information (sometimes called marketing), and it is an absolutely critical job—especially when a very strong incentive is being offered. It seems clear that once an incentive is fairly large, it may be more effective for an energy incentive program to invest in information than to increase the incentive. In terms of Figure 5-1, it is often more cost-effective to move a program up one of the columns than to shift it to a column farther right.

Our review uncovered another interesting fact: Incentive programs in the United States were systematically less effective than those we reviewed

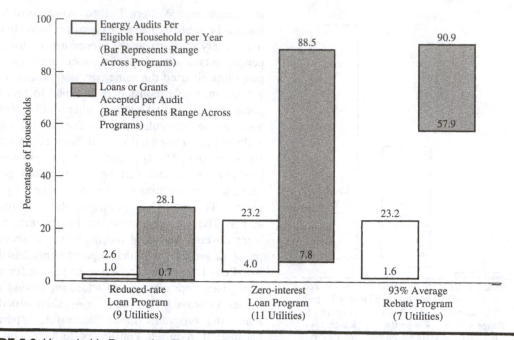

FIGURE 5-2 Households Requesting Energy Audits (White Bars) and Accepting Incentives Once They Have Received Audits (Shaded Bars) in Three Home Energy Conservation Incentive Programs
Source: Stern et al., 1986.

take the attitude that "if it's not broken, don't fix it," and do nothing, even if energy improvements would save money in the long run. Doing nothing may not save energy, but it certainly saves time, planning, and effort.

When home energy conservation programs make efforts to emphasize convenience, it helps. The Canadian and European incentive programs that dispensed with energy audits were noticeably more successful than programs that required audits. Many programs in the United States have provided other services to increase convenience and consumer confidence in the program. For example, an early program in the state of Rhode Island offered free energy audits, an approved list of contractors to perform the recommended work, assistance with low-cost bank financing, inspections of the completed work, and follow-up to make sure the contractors corrected any problems the inspection revealed (Stern, Black, and Elworth, 1981).

A few programs have gone even further, by allowing people with one simple action to learn what energy improvements they need, and to have them installed and paid for. One such program was implemented in the small city of Fitchburg, Massachusetts, in the fall of 1979 (Stern, Black, and Elworth, 1981). Fitchburg is a northern city of old, oil-heated houses that were poorly insulated when built and that, as they aged, developed cracks around windows and doors. There was a large population of older and low-income people who had neither the money nor knowledge to make the improvements their homes needed, and the price of home heating oil was rising rapidly. A crash program was needed to improve energy efficiency in the city's houses and protect the people's income and comfort.

The city government, working with an advisory council of prominent citizens and the labor of city workers and volunteers, created FACE (Fitchburg Action to Conserve Energy), an ad hoc group that held instructional workshops and distributed home weatherization kits free to anyone earning less than 80 percent of the median income in the city (the federal government provided grants for the kits, which cost under $20 each). People were encouraged to install the kits themselves after receiving training, but for people who could not do this, volunteers came around to install them. In six weeks, one-sixth of the homes in the city were weatherized through the program.

The strategy of giving away energy-saving equipment recalls the educational program discussed in Chapter 4 (Geller, 1981) that achieved its only measurable success when it distributed water-flow restrictors for shower heads, along with information on how to install them. The strategy of giving away equipment combines the tangible reward of cost-free equipment with the convenience of avoiding the considerable effort of deciding what equipment to buy and shopping for the lowest price. Installing the equipment, as was done for the Fitchburgers who could not do it themselves, increases the incentive further. (We should emphasize that the Fitchburg program, successful as it was, was a low-cost, relatively low-effect program. Given the status of older homes in Fitchburg, much larger investments in energy efficiency would have been warranted, and would have saved much greater amounts of energy and money, cost-effectively. However, the money to make larger investments was not available.)

We close this section by reporting on what is probably the ultimate in the use of incentives for energy conservation. An experimental program in metropolitan Minneapolis, Minnesota, in 1984 offered homeowners a free energy audit, free installation of the recommended conservation measures (which were estimated to save an average of almost $1,300 per home in energy costs over the first three years), and a guarantee that as a result, their monthly utility bills would decrease from that time on. A wary reader or consumer may well say, "What's the catch? I know there's no such thing as a free lunch." So before reporting the results of the program, it is worth explaining how it was possible to offer such incentives.

Energy analysts have long realized that in most U.S. homes, it is possible to make investments in improving energy efficiency that offer a far greater economic return in terms of money saved than people could get by putting their money into a bank account, a money-market fund, or even the stock market. If this is so, it should pay people to invest in home energy efficiency rather than the alternatives. It should even pay people to borrow money for energy efficiency. Yet, this is not what most people do. There are many good reasons. For one, as we have already pointed out, people would need to develop a great deal of specialized knowledge—about how to choose the best energy investments; where to find low-cost, reliable contractors; how to get the best loan rates; and so on. For another, people may not expect to live in the home for the full five to ten years it will take for the investment to pay off. For such people, the investment is not worth making unless they are confident that it will increase the resale value of the home. Also, because most people do not think of energy efficiency as one of the options for financial investing, they may not reason out the alternatives as we have done here.

There is a way to overcome all these barriers at once. Imagine a private, profit-making company that developed the specialized knowledge on how to do energy audits, calculate the best energy investments for each home, install the improvements, and borrow the capital to pay for the work. Most of this knowledge applies to a large number of homes, so the company could gain the knowledge at lower cost per home than a homeowner could. Because of such cost-cutting economies of scale, the company should be able to make a better investment than the average homeowner. It should therefore be possible for the company to install the energy improvements at no cost to the consumer and earn a profit by collecting part of the money that the customers save on their utility bills. The company and its investors profit, and the homeowner saves money, probably ending up with a home that is more salable because of its low energy costs. Everybody wins. This is the concept of the shared-savings conservation program. It relies on the efforts of an energy services company—a firm that, in effect, sells home heating, cooling, lighting, and the other services energy provides at lower cost than the local utility company. And the environment benefits because people get their energy services without burning fossil fuels or using nuclear power.

The reason energy services companies have not already covered all of the North American market is that it is impossible to put a meter on the gas and electricity people no longer use. With no way to measure the savings actually achieved, it is not easy to divide their value equitably between the homeowner and the energy service company. This is a problem for which a simple and generally accepted solution has not yet been worked out. The Minneapolis program worked this way: The energy service company, after completing its energy audit, calculated the expected energy savings from installing the package of improvements it recommended. It offered the homeowner a contract in which the homeowner would agree to pay the company 75 percent of the projected savings for the first three years and 50 percent for the fourth and fifth years. After three years, the projected savings would be compared with the actual savings so that the consumer's cost could be adjusted to reflect actual savings, and after five years, the contract would end, with the homeowner retaining the improvements and all future savings they produce (Miller and Ford, 1985). The plan guarantees homeowners a negative cost for the investment in energy efficiency, and enables an efficient company to make a profit at the same time.

Although this arrangement is not fully convenient—it gives consumers an additional bill to pay—it proved quite attractive to the Minnesota homeowners. The program was initially marketed in early summer of 1984, and despite that being a time of low interest in energy conservation in a cold climate, the most successful of the marketing strategies attracted interest from 20 percent of the eligible households, and resulted in signed contracts with 6 percent. For a program in operation only a few months, this is a very high rate of participation.

The Minnesota experiment showed that a program that removed all the financial barriers to energy efficiency could rapidly attract participants. But it also demonstrated that incentives are not the only key to success. Within the experiment with incentives, the

program conducted an experiment in marketing technique. Anticipating that consumers would be skeptical of a company that seemed to be offering something for nothing, the program tried three different ways of introducing the program to homeowners. In one method, the energy service company sent letters on its letterhead to the eligible homeowners, explaining the program and encouraging them to request a free home energy audit. In the second method, the same letter included mention of the fact that the county government was cosponsoring the program. In the third marketing approach, the invitation letter came from the chairman of the county Board of Commissioners, on his letterhead. It contained the same information, and introduced the energy service company as the county government's selected contractor.

As Figure 5-3 shows, the effect of marketing techniques was dramatic. The letter from the county government was over five times as effective as the letter from the private company that did not mention the government, both in encouraging energy audits and in getting contracts signed. It is easy to see why the letter made such a difference. People tend to be suspicious of unknown private companies making offers that sound too good to be true. They are well aware of consumer fraud and deceptive sales techniques, and the energy service company's offer must certainly have looked suspicious to many consumers. But they probably did not believe their local government would sponsor a fraud. On receiving an invitation letter from the county government, people were much more likely to accept the claim that the program was in the public interest and their personal interest. (County officials, well aware that public trust and well-being—not to mention their own careers—were at stake, made sure the energy service company was offering a good program.)

The Minnesota experiment demonstrates once again that incentives do not work automatically and that they need an effective informational component to work well. Figure 5-3 also reinforces the lessons of Figure 5-2 about the different roles of information and incentives. It shows that regardless of the marketing strategy, about 25 to 30 percent of the people who received energy audits went on to sign contracts. That

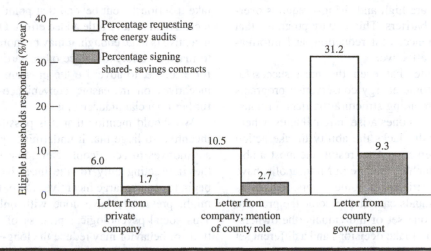

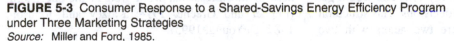

FIGURE 5-3 Consumer Response to a Shared-Savings Energy Efficiency Program
under Three Marketing Strategies
Source: Miller and Ford, 1985.

proportion is what this particular package of incentives—financial savings, convenience, and so forth—could accomplish. Where the letters mattered was in getting people to make initial contact with the program by requesting energy audits. The invitation letters worked by attracting people's attention—getting them to think seriously about what the program offered. Once they were paying attention, the incentive sold many of them on taking action. Information and incentives addressed different barriers to action; an effective program needs both.

The experience with energy conservation programs shows how different kinds of incentives can complement each other. Programs work better when they combine convenience with financial savings. And needless to say, they are likely to work still better when high or rising energy prices give people an added reason to improve efficiency and bring a higher payoff from any particular improvement. It is important to point out, however, that price increases by themselves have only moderate effects on consumers' decisions to improve home energy efficiency. In the 1970s, when energy prices increased rapidly and were expected to rise even further, reinsulation of attics and walls in existing homes, and other major energy-

saving improvements, proceeded rather slowly, compared to what one would expect if everyone who could save money by taking action did so. One of the major effects of the price increases of that period was on public policy. Seeing the slow pace of energy improvements in homes, local and national governments implemented incentive and information programs to encourage people to take energy-saving actions they were not taking but that would be in their own economic interest.

A look at the evidence in this chapter shows why price increases had to be supplemented by policies that supplied information and additional incentives. As we have noted, improving the energy efficiency of a home is not a simple matter. Someone in the household must learn about the energy condition of the home and how to improve it; decide what actions are best for the household; shop for the needed materials and services; find, interact with, and wisely judge the providers of these services; make the relevant financial choices; and keep track of the whole complex process. These things require considerable knowledge, time, and effort and thus present barriers to action that a price increase does nothing to remove. To effectively promote energy conservation—even when fuel prices are high and rising—requires overcoming all these barriers. This is why programs that combine convenience, cost reduction, and information are the most effective.

We should note that even the most successful incentive-based home energy conservation programs have trouble overcoming structural barriers. In housing, the most serious ones arise in rental units, where the occupant usually lacks the ability to take action on home improvements. As a result, the most ambitious programs, such as the one in Minneapolis, have generally been restricted to homeowners. Incentives directed to individuals cannot overcome the problem that arises when one set of individuals (the owners) has the authority to make decisions and a different set (the occupants) pays the energy bills and would get most of the benefit of those decisions. This structural problem, in which there are two actors with two different incentive structures, is among the most difficult in the area of residential energy conservation. The effect is to hold back progress in the homes of low-income people, who tend to rent rather than own.

PRINCIPLES FOR DESIGNING EFFECTIVE INCENTIVES

The most general principle is probably that of the economists: Internalize the externalities. But as the great variety of effective incentive programs suggests, there is no one sure way to apply this principle. There are many ways to match costs to environmental damage, but no set rule for deciding when it is best to offer rewards, charge fees for services like waste disposal, make it more convenient to do things that provide environmental benefits, or employ some other method. Experience shows that there is an art to designing effective proenvironmental incentives. Good design follows these principles:

Make the incentive large enough. This is not the same as saying that with incentives, the bigger the better. Rather, as the experiments with time-of-use electricity pricing and with energy efficiency suggest, there is a sort of law of diminishing returns for incentives. An incentive must be large enough for people to

take it seriously, but beyond that point, increasing the incentive may have little added effect. Once a financial incentive is large enough, it may be more cost-effective to make an effort to reduce other barriers to action—for example, to advertise the program more or focus incentives on increasing convenience—than to add further financial inducements.

We should mention that it is possible to make an incentive so large that it undermines people's intrinsic motives to act. People can come to believe that they are acting only for the incentive, so that they begin to require large incentives to do things that they might previously have done with only small ones. This social-psychological process of "overjustification" of behavior may reduce the long-term effectiveness of incentives that work well in the short term (Lepper and Greene, 1978; Katzev and Johnson, 1987; DeYoung, 1993).

In some situations it is easy to get the incentives large enough, but in others it is not. Energy efficiency presents a good opportunity, because even though it is difficult for a homeowner to learn what improvements to make and to get them done properly, an entrepreneur such as an energy service company can earn money by providing incentives that remove those barriers. Recycling is presenting an increasingly attractive situation as the cost of waste disposal rises. Municipalities can afford to offer larger incentives than ever before, and still save money. Providing large enough incentives is much more difficult, however, for getting commuters out of their automobiles and into alternative transport. There, the combination of a lack of convenient alternatives to the car, the relatively low price of motor fuel, and the various structural elements of an automotive society combine to make it extremely costly to provide strong enough incentives. Structural barriers also stand in the way of improving energy efficiency in rental housing.

Match the incentives to the barriers that prevent action. We have seen that the external barriers to proenvironmental behavior are different for each behavior, and also that different people respond well to different kinds of incentives. An incentive that changes one behavior may have very little effect on another. A small, well-placed financial inducement for recycling, such as a deposit on returnable drink containers, can work wonders to reduce that part of the solid waste stream. But fairly large tangible rewards to get commuters out of their cars and into buses have had only moderate effects, while improved convenience—access to preferred parking spaces in large industrial parking lots—addressed a more significant barrier and produced a surge in carpooling.

Similarly, an incentive that works well for some people may have little effect on others, even when the behavior is the same. In home energy conservation, some people prefer grants while others prefer loan subsidies, because of differences in household incomes, living situations, and attitudes toward indebtedness. In general, the most effective incentives are the ones that attack the most significant external barriers for the individual. But it is not always obvious in advance which barrier is the greatest or which incentive will best surmount it.

Because people's situations vary, an incentive that looks appropriate in advance may turn out to be ineffective or even counterproductive. A single incentive can have positive and negative effects at once, though on different individuals. Consider the short-term effect of an energy price increase—say, a large increase in gasoline taxes—intended to reduce energy consumption in automobiles. Among people who can afford new cars, there would be a shift toward more fuel-efficient models, with the result that more of these models would be produced and total gasoline use would decline. For most people, who have no immediate plans to buy cars, the price increase would not get them into more efficient cars immediately, though it might induce them to make fewer trips. For lower-income people, especially rural people who commute long distances to work, the price increase would be an economic hardship, and it might even force them to keep fuel-inefficient cars longer because they would have less disposable income available for replacing them. The reason the incentive has these different effects is that the barriers to owning fuel-efficient cars are different for different people. Where income is a significant barrier, price increases may only make matters worse in the short term.

There is no formula for finding the best incentive for each behavior and each individual, but there is an effective process for designing incentive programs. We discuss it at the end of this section.

Get people to notice the incentives and the behaviors they are meant to change. We saw in Chapter 4 that information had no effect unless it was noticed. The same is true for incentives, because incentives do not always advertise their own existence. Some are easier to notice than others. Bottle deposits and trash-bag fees are hard to ignore because they require people to take some additional action every time they engage in a target behavior. Gasoline taxes are somewhat less noticeable because they are included in the price, so they do not require special action. They can still have powerful effects, though, because people are likely to notice higher gasoline prices whenever they take long trips or select between models when buying a car. Changes in electricity prices are less noticeable because the bill comes only after a month or more of assorted electricity-using behaviors (and nonbehaviors, such as leaving a basement water heater on even when a household is on vacation). It can be hard to notice electricity price increases because usage tends to fluctuate with the seasons. And because the bills are infrequent and cover many appliances, it is extremely hard to tell which specific behaviors could significantly lower the bill. It should follow that electricity price changes will have greater effect if they are supplemented with information about how the rates work and how to keep track of home energy use so as to keep costs down. This is exactly what was learned from Heberlein's experiments with time-of-use electricity pricing.

Among the least noticeable kinds of incentives are the loan subsidies and rebates that utility companies have offered for energy-efficient home improvements and appliance purchases. With these incentives, if people who go on behaving as usual, nothing changes at all. Someone must notice the incentive and actively connect it to behavior, for example, by requesting a home energy audit or filing a claim for a rebate, in order to benefit. Because such incentives do not advertise their own existence, program managers use direct-mail advertising, telephone banks, community advisory groups, and many other methods of getting them noticed, just as they do with information programs. As we saw with the home energy conservation incentive programs, marketing can be increasingly important as incentives get larger.

In sum, the experience of incentive programs teaches that *incentives work better when combined with information* or when designed so as to have useful information built in, as with trash-bag fees. This is an instance of a broader point we make throughout the book: that any single solution approach is likely to work better in combination with other approaches.

Make the incentives credible. Sometimes credibility can be a problem for incentives. It has been a problem for utility companies that sponsor programs to get people to use less of their product and with private companies that offer incentives that may seem too good to be true. As the Minneapolis shared-savings program demonstrates, such programs may need to find a credible sponsor to overcome consumer doubts that the program is genuine and that they will actually receive the savings promised.

Find politically acceptable forms of incentive. It is tempting to conclude that incentives that are inherently noticeable are always preferable to those that must be advertised. This is not necessarily so. Some of the most noticeable incentives, especially those that impose new costs on identifiable political interests, are not feasible politically precisely because they are noticeable. Those who would pay the costs anticipate them and organize political opposition to prevent their ever being put into practice. For two decades, proposals for higher gasoline taxes in the United States have been blocked by opposition from the oil industry, advocates for low-income people, and other affected groups. Efforts to require automobile companies to produce a more fuel-efficient fleet of cars by strengthening the Corporate Average Fuel Economy regulations, which include stiff fines for noncompliance, have met strong opposition from the automobile industry.

Generally, positive incentives are more acceptable than regulations, price increases, or other mecha-

nisms that impose new costs on individuals or organizations. Regulations, which generally limit behaviors that had previously been unlimited, draw opposition from those who would be regulated. For this reason, consumers oppose limits on driving to control air pollution and prefer regulations that would make car manufacturers adopt cleaner-exhaust technology; car manufacturers, however, tend to oppose regulations on their behavior. Despite the prevalence of opposition, regulatory solutions that apply across whole industries have been among the most effective of environmental policies. In the U.S. transportation industry, for example, these have included abolition of leaded gasoline, required installation of catalytic converters on automobile exhaust systems, and auto emissions inspection programs, as well as the Corporate Average Fuel Economy regulations. These and related regulations have drastically reduced emissions of nitrogen oxides and carbon monoxide from motor vehicles. Note that most of these regulations affect the behavior of corporations. Although they are mostly out of the awareness of motorists, they alter the set of choices available for individuals.

Regulations and other new costs are more likely to be accepted if they are perceived as fair, either because they are shared evenly by everyone or because greater costs tend to be borne by those who benefit most. Also, policies that are politically impractical in normal times because of the costs or restrictions they impose are sometimes enacted during a crisis situation that creates strong pressure to act quickly. For example, pay-per-can waste removal systems, which are often perceived as increasing consumers' costs or limiting their freedom, are more acceptable in cities facing a waste disposal crisis because they have run out of landfill space and have no easy choices.

Design the incentive system to discourage evasion. People can evade both positive incentives and penalties. For example, several early experiments used rewards to encourage the pickup of litter in public places such as parks, stadiums, and the yards of public housing projects (Geller et al., 1982). When experimenters offered payment for bags of litter, the public spaces became significantly cleaner—but it also turned out that some people were bringing trash from home to get the rewards. By trial and error, the researchers discovered that offering rewards for clean yards was much more cost-effective than offering them for bags of trash. Drivers who evade carpool-lane restrictions by putting mannequins in their cars are another instructive example.

These few examples can serve as reminders of the ingenuity of people and organizations that are motivated to take advantage of incentives and avoid penalties and that can make aspiring social engineers look very foolish. Designing an evasion-proof incentive system can take tremendous ingenuity. There is no formula for success, but the best single principle is to look for ways to make the desired behavior coincide with narrow self-interest, so that people have an interest in using the incentive instead of evading it. Container deposit laws are a good example of this principle in action. A deposit gives people an incentive to return their own containers and, if they discard them anyway, gives other people who need the money an incentive to pick up after the litterers. It is possible to design such self-enforcing incentive systems to fit many environmental problems, but experience shows that systems that look good in theory are often outwitted in practice. The best rule is to try systems that look good and then closely watch what happens, being prepared to make adjustments.

The above principles are much more specific than the advice to "internalize the externalities" (see Box 5-1), yet they can still be very difficult to put into practice. Finding an incentive that is highly noticeable but not politically objectionable can be a hard task. So is finding an incentive that cannot be easily evaded. So, although the principles above provide useful guidance for creating effective incentive programs, they fall far short of a recipe for foolproof program design. We believe it is a mistake to look for such a recipe. Changing human behavior is not like baking a cake or fixing a piece of machinery, because the "materials" one works with—people and their interactions—are not interchangeable parts, and they are always in flux. As we have noted, different people respond to different incentives. Also, what people notice is always subject to change, as is the range of interventions that are polit-

ically acceptable. People can defeat an incentive program at any time, either by ignoring it or, in the case of government programs, by actively opposing it. Consequently, anyone who would design an incentive program should plan for individual variation and be prepared to modify the program to meet changing conditions. Program designers should follow two additional principles for the process of program design.

Interact with people to understand the barriers to environmentally desirable behavior. As we have noted, the barriers—and the most attractive incentives for overcoming them—vary with the behavior and the individual. Sometimes the best way to understand the barriers is to observe people's behavior, but it is usually far better to ask people—with questionnaires, or in conversation—why they do not behave differently and what might induce them to change. The most common form of structured interaction is a sample survey in which people from the group whose incentives are to be changed are asked questions about the incentive structure facing them. Another valuable method is the detailed interview, using ethnographic techniques from anthropology, to get a deeper understanding that questionnaires normally give or to reveal things that survey researchers may not even have thought to ask (see, e.g., Kempton et al., 1995). The goal of such interactions is to help program managers find effective incentive packages for changing a particular behavior in a particular place and time. Although they can get good ideas from experience in other places and times, they need to determine what will work in the current, local situation. This means that to some extent, incentives must be reinvented in each situation.

The most effective way to learn what incentives can work is often to *involve some of the people who are the targets of behavior change in actually designing the program.* After all, many people would like to engage in proenvironmental behaviors but are prevented from doing so by the incentive structures they face. Who is in a better position to understand the barriers than the people most frustrated by them? And who can know better what incentives will work than the people they will be offered to? Often, a community has organized groups that would gladly send representatives to help design effective environmental programs. They may be general civic groups, like the school and church groups that sometimes mobilize to clean roadsides or riverbeds, or groups that represent a specific constituency. For example, advocacy groups for low-income housing have helped develop effective energy conservation programs, and groups that oppose waste incinerators can be encouraged to help design programs to make the incinerators unnecessary.

Involving the target population in program design has advantages that go far beyond research. Nothing is more politically acceptable than a system designed by the people who will be asked to follow it. In the case of regulations and other costly or restrictive external conditions, opposition is almost guaranteed unless the people who will pay the costs or face the restrictions come to a decision that the goal justifies them. Moreover, the process of public involvement can attract volunteer help in implementing a program and building a sense of community support for the program and its environmental goals. It can thus combine the incentive strategy with the community-based strategy for promoting proenvironmental action that we discuss in Chapter 6.

Continually reassess the program. Throughout this section, we have been offering advice like this: "Try systems that look good and then closely watch what happens, being prepared to make adjustments." More than advice, this is a general principle for designing effective programs, especially incentive programs. What fails may need only slight modification to succeed, and what succeeds at first may not continue working. Program design is unavoidably a process of trial and error, and proenvironmental programs are a kind of experiment that must be monitored.

Sometimes program managers resist evaluation out of fear that their programs may be evaluated negatively and be discontinued. But evaluation can be used to improve a program's design rather than to give a simple passing or failing grade, and systematic evaluation efforts used for this purpose can greatly benefit a program. Specialists in program evaluation are constantly improving quantitative techniques for

monitoring how people respond to programs, how much their behavior changes in response, their attitudes toward the programs, and the reasons the programs have the effects they do. Even when the resources for a formal evaluation are lacking, managers can monitor programs informally by continued interaction with the target population or by seeing that it is well represented in a group created to advise on program reassessment. This procedure, like that of public involvement in program design, has the advantage of strengthening community support for programs.

CONCLUSION:
WHAT CAN INCENTIVES ACCOMPLISH?

Incentives present a powerful strategy for promoting proenvironmental behavior, but as with information, there are limits. The chapter supports the following conclusions:

Incentives can overcome specific external barriers to action. They are especially effective at removing financial barriers such as cost and access to money, as well as a variety of barriers we have grouped under the general heading of inconvenience. In short, incentives can reduce the cost or the effort involved in following one's proenvironmental attitudes. They can also make it attractive for a person who lacks such attitudes to engage in proenvironmental behaviors. In addition, incentives can create barriers to actions that harm the environment. Sometimes, a single well-designed incentive can do both, as with the container deposit law. Since behavior is so dependent on the incentive structure facing individuals, changing the incentives is a powerful approach.

Incentives for individuals fail when significant barriers to action lie in the larger social system. Energy conservation in the U.S. automobile sector faces several of these structural barriers. Any car owner could switch to a more fuel-efficient vehicle if there was the political will to offer incentives large enough to compensate for losses of comfort or performance that may be associated with the change and for the cost of switching cars before one is ready. But it is much more difficult to conceive of incentives that would get people out of cars entirely. As we noted at the start of the chapter, the geography of American homes, workplaces, and shopping would make the shift tremendously inconvenient and time-consuming for millions. Incentives to individuals can have little effect on where corporations build their offices or developers build housing or shopping developments. Similarly, incentives will not get people to buy 70-mile-per-gallon cars or superinsulated homes, because they are not yet on the market. And incentives to individuals cannot stop manufacturers or retailers from using nondegradable packaging for their products.

Incentives can sometimes be most effective when aimed at organizations. When the key decisions are made by corporations, government agencies, or other large institutions, incentives aimed at individuals do not work, but incentives that affect the actual decision makers can be very effective. We saw this in the example of municipal solid waste, where the increasing financial and political costs of waste disposal in landfills changed the incentives for the managers of municipal waste disposal programs and led many cities and towns to sponsor recycling programs and other initiatives that made it possible for individuals to recycle, where it would have been virtually impossible without action by local government.

In fact, the great majority of government environmental programs are aimed at changing the incentives for organizations rather than individuals. These include regulations on average automobile fuel economy; emissions from automobiles, power plants, and factories; toxic waste disposal; sewage disposal; and so on. In the future, government programs may increasingly replace regulations with financial incentives. An example is the idea of issuing tradable permits to release air pollutants up to a limit set by government (see Box 5-1). The design of regulations and incentives for corporate polluters is beyond the scope of this book because our focus is on what individuals can do and how to change individual behavior. Many of the same principles that apply to individuals also apply to corporations, but the incentives are sometimes different, and it is probably necessary to find different ways to apply them. We offer an intriguing example at the end of the chapter.

Under some conditions, incentives can be counterproductive. When a new incentive effectively limits people's freedom to act, it tends to be experienced as punishment. An example is price increases for energy that force low-income people to make hard choices, such as between heating their homes in winter and buying food or clothing. Regulatory restrictions are often experienced as punishing. People might react in this way to restrictions on driving automobiles because of air pollution or rules against disposal of toxic household wastes in ordinary trash. Sometimes, incentives that are seen as punishments generate enough political opposition to prevent their being put into practice; and even if they are put into practice, the result may be that their proenvironmental goals become distasteful to people who suffer under the incentive system. It is very difficult to tell in advance, however, whether a particular new regulation will be seen as a punishment, or as something that must be done for the common good. Changing ethical and value systems are one factor that can alter a population's willingness to support changes in incentives.

Incentives work best in combination with other influence techniques. We have said throughout the book that no single influence strategy is optimal by itself for promoting proenvironmental behavior, and this general point is true of the incentive strategy. This chapter identifies many situations in which incentives are much more effective when accompanied by well-designed information: in promoting energy efficiency rebates, shifting electricity use to off-peak periods, and increasing the use of curbside recycling programs, for example. We have also seen that moral and ethical concerns about the environment can help provide the public support needed for incentives to be acceptable. And the chapter also shows that incentive programs sometimes need to draw on resources in the communities where they operate—credible local institutions to respond to consumer skepticism, and representatives of the target population to help in program design, evaluation, and redesign. These features take advantage of social support systems—either preexisting ones in the local community or new ones created to improve the incentive program. They are examples of the community-based strategy of changing behavior. We devote Chapter 6 to a discussion of this strategy.

EPILOGUE: HOW PEOPLE CHANGED THE INCENTIVES FACING A CORPORATION

Most of our discussion of incentives has been from the top down—we have presumed that some high-level entity, such as a government agency or a utility company, creates programs to change the incentives facing individuals or businesses under its jurisdiction. But governments and utilities also face incentives, and these can be affected by the organized actions of individuals. So, incentives can flow from the bottom up. A well-known example is public opposition to nuclear power in the United States, which has made it increasingly difficult and expensive for utility companies to get all the necessary regulatory approvals to operate a power plant and has for over fifteen years discouraged the American electric utility industry from ordering any new nuclear plants.

Another example, less well known but also very instructive, concerns the decision by the McDonald's Corporation in 1990 to stop using plastic packaging for its hamburgers and other products. The story of this victory for environmental activism holds important lessons about how individuals can help the environment by changing the incentives that affect corporate behavior.

Before November 1990, anyone who ate McDonald's Big Mac hamburgers got them in "clamshell" packages made of polystyrene (better known by the brand name Styrofoam). Polystyrene is environmentally damaging in at least two ways. Its manufacture (about 5 billion pounds per year in the United States) left millions of pounds of toxic waste products, chiefly benzene and toluene, in the environment, and the process of blowing polystyrene into foam used chlorofluorocarbon (CFC) gases that deplete the earth's ozone layer and contribute to the greenhouse effect. Many environmentally concerned individuals

were offended by the clamshells, which also took up large amounts of landfill space and were vivid symbols of a wasteful, environmentally destructive, throwaway society. Although concerned individuals could stop buying Big Macs, there was nothing an individual could do to make a larger difference.

During the late 1980s, a small, focused social movement changed all that. A key event occurred in the summer of 1987 when a statewide protest was held by the grassroots group, Vermonters Organized for Cleanup, in which parents and children organized boycotts and picketed McDonald's restaurants around the state of Vermont. This was the opening event in the national McToxics campaign, organized by the Citizens' Clearinghouse on Hazardous Waste (CCHW), a national organization that provides technical assistance to local groups organizing to reduce hazardous wastes. CCHW's network spread the word around the country, and local groups organized to ban polystyrene from local landfills and to conduct boycotts and demonstrations at McDonald's restaurants. The sight of children on picket lines generated lots of news coverage, and McDonald's—and the plastic packaging industry—took note. They announced a goal of phasing out CFCs in packaging by the end of 1988.

The story does not end here, though. The movement's goal was to eliminate polystyrene packaging entirely (not only the CFCs), but McDonald's and the packaging industry resisted a change to paper and cardboard. They engaged in several countering actions. One involved switching the material used to blow the foam. Instead of CFC-11 and CFC-12, the compounds that had been used, McDonald's switched to CFC-22, a compound estimated to be about 95 percent safer to the earth's ozone layer (although no different in the amount of benzene and styrene waste is produced). Because CFC-22 contains a hydrogen (H) atom in addition to the chlorine (Cl), fluorine (F), and carbon (C) atoms that give the name CFC to CFC-11 and CFC-12, the leading manufacturer, the DuPont Corporation, began calling it HCFC-22, and the U.S. Environmental Protection Agency soon accepted the name change. By early 1989, the packaging trade industry was able to make the misleading announcement that it had totally eliminated CFCs in food service products, and McDonald's was making the same claim on its place mats.

Three national environmental organizations other than CCHW had a role in this change in company policy—the Natural Resources Defense Council (NRDC), the Environmental Defense Fund (EDF), and the Friends of the Earth. They had advised McDonald's to switch to CFC-22, and then defended the partial measure as a great step forward and the most they could reasonably accomplish. David Doniger of NRDC said, "I didn't see that we had any leverage on them to say, 'You all ought to go to cardboard.'" CCHW, other activist groups, and a *Washington Post* article criticized McDonald's—and the three environmental groups—for trying to deceive the public.

The industry also responded with a major campaign for plastics recycling as an alternative to eliminating production of the clamshells. Recycling could improve the industry's environmental image and at the same time circumvent a growing number of local ordinances that banned polystyrene from the solid waste disposal system unless it was recyclable. Starting in late 1988, with contributions of $16 million from the major manufacturers and a supply of Styrofoam trash from McDonald's, polystyrene recycling plants began opening, with a goal of recycling 250 million pounds per year (5 percent of the national output) by 1995. But the program had problems. Though McDonald's set up recycling bins at its restaurants and publicized them, half the clamshells continued to be sold to take-out customers, and the remainder had enough food waste mixed in to make the recycled product unacceptable to potential buyers. Moreover, the recycling program did not stop the consumer pressure. As one protestor shouted when the program was unveiled at Boston Children's Hospital, "Why do you produce so much trash in the first place?" CCHW encouraged people to "help" McDonald's recycle by sending loads of Styrofoam trash to corporate headquarters.

Third, the industry engaged in public relations campaigns. In addition to making the misleading claim about phasing out CFCs and advertising its commitment to recycling, McDonald's sought partners within the environmental movement. It offered

the Sierra Club a $700,000 grant for its youth activities at Earth Day 1990 (which was refused), and it got EDF to join it in a task force on solid waste management. And in late 1989, in a confidential memorandum obtained by CCHW, the president of the Society of the Plastics Industry put out a call for a $150 million campaign to counter "the image of plastics among consumers," warning the manufacturers that "Business is being lost. Product growth rates are being dampened. And stock analysts are beginning to take notice." Clearly, the stakes had grown beyond hamburger packages—public pressure had significantly changed the incentive structure for the entire plastics industry.

Through all this, consumer pressure continued. Church groups organized to stop using Styrofoam and initiated a resolution at McDonald's annual stockholders' meeting. School children organized boycotts of school cafeterias that used Styrofoam, and received prominent news coverage. More communities banned Styrofoam from dumps. Finally, in November of 1990, McDonald's announced that it would end nearly all Styrofoam packaging at U.S. restaurants within 60 days. Other fast-food chains quickly followed suit, and Burger King, which had been using paper and plastic all along, claimed credit for its environmental foresight. The decision by McDonald's sharply affected the nascent polystyrene recycling industry.

McDonald's at first denied any environmental reason for its actions. Its president, Ed Rensi, said, "Our customers just don't feel good about" Styrofoam. Both EDF and CCHW claimed victories, with EDF getting most of the media attention. And the EDF-McDonald's partnership continued. With advice from EDF, McDonald's set a goal of an 80 percent reduction of solid waste, and by 1991, the corporation was being hailed in *Advertising Age* as "a leader in environmentally sensitive marketing."

What are the lessons in this story? First, small-scale, grass-roots action can make a difference by changing corporate incentives. Second, the target must be carefully chosen. The activists chose to focus on a product that was ubiquitous, that had symbolic value as an icon of waste and environmental degradation, and that was also nontrivial in its local and national impact. Polystyrene food packaging produced huge amounts of solid waste a year and large quantities of toxic benzene and styrene. Third, the corporate target was easy to organize against because nearly any community that wanted to could find a nearby McDonald's restaurant—and involving children assured media interest. Fourth, the incentives that the movement used were meaningful to the corporation. Not only are children an environmentally concerned group, but they are major customers of McDonald's, so the corporation was especially sensitive to the kind of publicity the movement generated. McDonald's was a large enough and prominent enough customer that even the polystyrene industry became concerned.

Another lesson is that changing corporate behavior is different from changing individual behavior. Not only do corporations respond to different kinds of incentives (boycotts, bad publicity concerns about attracting and keeping investors), but they can also organize very strong resistance, including working to divide the environmental movement and influence public opinion. But the McDonald's story shows that activity by collections of citizens can be effective nevertheless.

The McDonald's story is also, in part, a victory for environmental education. It shows that people who understand the negative environmental consequences of using a product can produce change, if they organize effectively, of a much larger order than they could ever hope to achieve by altering their behavior as individual consumers. Education was an important element in a strategy that also involved political organizing to change the incentives facing the producers of the target product.

Of course, the story is full of ironies. McDonald's had dragged its feet and wound up with kudos for environmentalism. EDF, after settling for a half-measure, took credit for the full success and gained the opportunity to work with a newly greened McDonald's to make even further environmental improvements. Burger King got little or no credit for its decades of paper packaging, and CCHW was barely mentioned in mass media accounts despite its strong efforts, so that many citizens may have come to believe that McDonald's changed its long-standing practices simply because of its foresight and its cooperative discussions with EDF and NRDC. Nevertheless, the victory proved that individuals can, through political action to change the incentives for larger social organizations, make changes they could never make on their own. They can create a sort of structural change, in the sense of changing the possibilities for future individual action. Now, nobody who buys a Big Mac gets foam packaging, and the success has had ripple effects across the fast-food industry.[1]

NOTE

1. This account was pieced together from the following sources: *Everyone's Backyard,* Dec. 1990; Lipsett (1990); Moore (1989); Holusha (1990); Hume (1991); Hamilton (1991), James (1989); Citizens' Clearinghouse for Hazardous Wastes.

CHAPTER 6

COMMUNITY MANAGEMENT OF THE COMMONS

CHAPTER PROLOGUE

The village of Törbel, in the Swiss Alps, has managed its communally owned forest and grazing lands successfully for centuries, sustaining the resources and avoiding the tragedy of the commons. Written rules for resource management go back at least to the thirteenth century, and on February 1, 1483, the villagers established an association to regulate Törbel's communally owned lands. The villagers had previously decided that the forests and certain low-productivity lands, such as the high Alpine meadows, should belong to the community rather than private owners, and they had set the boundaries between private and community land. The community gave villagers rights to graze their cattle on the mountain in summer, and to use timber from the forest, and it established rules to protect the resources.

Cattle Grazing on an Alpine Pasture
Some Swiss villages prevent overgrazing by establishing community ownership and control of such pastures.
(Georgia Engelhard/Monkmeyer Press)

The grazing rules, established in 1517 and still enforced, state that "no one is permitted to send more cows to the alps than he can [feed in] winter" (Netting, 1981, p. 61). The cows are sent to the alp all at once, and are immediately counted, because each household is allocated cheese in proportion to the number of cows it owns. A local official is authorized to levy fines on those who violate the rules against overgrazing and other kinds of misuse of the common resource. The official, in turn, is elected at the annual meeting of the association of local cattle owners, who have the power of removal. Cattle owners contribute an annual fee, proportional to the number of cattle they own, that pays a staff to maintain roads and paths on the mountain and rebuild corrals and huts damaged by avalanches.

The forestry rules work as follows: Once a year, the village forester marks the trees to be harvested. The families eligible to harvest logs form work teams and equally divide the work of cutting and stacking logs. The households are then assigned stacks of wood by a lottery. This procedure assures a fair distribution of work and of logs, and since trees can be harvested only once a year, it is easy to spot rule violators.

These systems have operated in Törbel for centuries, and similar systems were developed in many other Swiss villages. The villagers' methods for avoiding the tragedy of the commons have stood the test of time. In many villages, though not all, the rules have adapted well to changes such as population growth and increases in the value of the villagers' labor in the outside economy. (The resource management system of Törbel was described in detail by the anthropologist Robert Netting, 1976, 1981.)

The coast of central Maine is the leading lobster-producing area in the United States. Unlike many fisheries, where valuable species have been fished almost to extinction, the Maine lobster fishery has produced a nearly steady yield for many decades. The State of Maine's government is partly responsible for this success by setting legal limits on the size and sex of lobsters than may be caught, requiring licenses and patrolling lobstering areas. But most of the credit belongs to the lobstermen themselves. There are not enough fishery wardens for adequate enforcement, and it is so easy to get a license that licensing is no barrier to overfishing. The lobstermen sustain their livelihoods and the resource by their own informal means of control, based on assigning lobstermen to particular territories, and in the places where the strictest territoriality is maintained, both the resource and the lobstermen do best. The anthropologist James Acheson (1975, 1987) has studied the fishery in detail, and this account is taken from his work.

Along the Maine coast, lobstering is done in small boats by fishers who drop wooden traps or "pots" into the water and pull them up regularly to collect any legal lobsters inside. The law requires that each trap and buoy be marked with the owner's license number, making it possible to check that lobstermen are licensed. (This law also makes it possible to for a lobsterman to know who is fishing where.) A typical lobsterman owns 400 to 600 pots and uses them in a very small area near a home harbor, moving them farther out in winter, close to the shore in summer, and to intermediate areas in spring and fall when the lobstering is best. Legally, any licensed lobsterman can fish anywhere, but because of the effort of pulling pots, it is more efficient for them to work in contained territories near their homes.

Lobstermen have developed strong, unwritten rules for governing these territories and defending them from outsiders. As Acheson describes it,

> To go lobster fishing, one must be accepted by the men fishing out of a harbor. Once a new fisherman has gained admission to a "harbor gang," he is ordinarily allowed to go fishing only in the traditional territory of that harbor. Interlopers are met with strong sanctions, sometimes merely verbal but more often involving the destruction of lobstering gear. This system . . . contains no "legal" elements (Acheson, 1987, p. 40).

Along some parts of the Maine coast, the lobstermen enforce territories strictly, keeping all outsiders away and reserving certain areas for particular individuals or families, for example, on the basis of their ownership of nearby land on islands. Territories are marked by landmarks on the shore and are defended, sometimes to the yard, by individuals who silently

> sanction the interloper. First, the violator may be warned, usually by having his traps opened or by having two half-hitches tied around the spindle of his buoys. If he persists, some or all of his traps will be "cut off." That is, his traps will be pulled, the buoy, toggles, and warp cut off, and the trap pushed over in deep water where he has little chance of finding it (Acheson, 1987, p. 41).

Far from shore, where the boundaries are harder to define, lobstermen from neighboring harbors may share the same area.

Along other parts of the coast, it has become more difficult to defend well-defined territories. With the advent of motorized boats and depth-finding equipment, lobstermen who could purchase larger boats increased their effective range. But to pay for the boats, they had to fish in all seasons and to invade other lobstering territories, especially far offshore. The men in the invaded territories generally let mixed fishing occur rather than imposing sanctions because they knew how much the interlopers had at stake and feared a full-scale "war," with large financial losses for all.

Both types of territoriality establish forms of property rights in an area where the law does not recognize private property, and the effect has been to prevent overfishing. The stronger the enforcement of territoriality, the greater the benefit. In the tightly "perimeter-defended" areas we described first, there were fewer boats per area of fishing grounds, with several salutary results. Each lobsterman could make a living with less effort, there were more lobsters per area of ocean, and larger lobsters were caught. It is only in the perimeter-defended areas that lobstermen have been successful in imposing conservation measures such as closed seasons and limits on the number of traps a lobsterman can use. Such measures benefit every lobsterman in the area, and put less stress on the lobster population, but they do so only if the rules are observed. Lobstermen in the perimeter-defended areas did the best job of enforcing the territories, so they were best able to prevent the tragedy of the commons and benefit themselves. But even in the territories where control was much looser, the fishery and the lobstermen have survived.

INTRODUCTION

The systems of resource management used in Törbel and on the Maine coast are different from the systems we have discussed in Chapters 4 and 5 because they were created and operated by the resource users themselves—they are self-organized systems. We refer to them as systems of community management. The resource users devise their own management rules, accept the rules voluntarily, and have the power collectively to change them. An important characteristic of community management systems is that when they work well, the self-imposed rules become shared social norms that most people adhere to because they believe they are doing what is right, or at least necessary to keep the system working. When most people internalize the community's norms as their own, minimal policing is needed and individuals do not feel coerced. Communities sometimes impose incentives like those discussed in Chapter 5—they levy fines and even physically interrupt behavior that violates community rules—but successful systems are marked by how lightly such coercive means of

behavior control are used, compared to individual self-control and informal social pressure.

The examples of Törbel and the Maine coast are special in that they involve small groups of people who depend on local, renewable resources for a significant part of their livelihoods. We will see that successful community management is most often observed in settings that have these and other characteristics in common. We should emphasize that small communities do not always maintain their natural resources over the long term. As we will see, smallness is an advantage only in combination with other qualities of the communities, the resources, and the ways the two interact.

The relevance of the success stories from Törbel and Maine to the world's environmental problems is limited by the fact that most of the world's serious environmental problems arise on too large a scale to be managed by villages, local fishers, and other small community groups, no matter how careful they are with their natural resources. But as we will see, some of the management principles that work so well in Törbel and the Maine lobster fisheries can be extended to other settings.

This chapter first examines the conditions that enable some local groups like those in Törbel and Maine to manage their renewable resources successfully, while other groups suffer the tragedy of the commons. We discuss the implications of successful community management for Garrett Hardin's model, which predicts that successes like those of Törbel and the Maine coast cannot be achieved. We then ask how applicable the techniques of community resource management are to the major environmental problems of modern societies. We conclude, as in other chapters, by discussing the conditions that are favorable for using these techniques, and the limits of their applicability.

HOW DOES RESOURCE MANAGEMENT WORK IN SMALL COMMUNITIES?

In Hardin's formulation of the tragedy of the commons, the only alternative to "ruin," brought on by the remorseless working of individual self-interest, is "mutual coercion, mutually agreed upon"—that is, the establishment of rules, backed by the coercive power of government, that force individuals to do what is good for the group. The success of the Törbel villagers and the Maine lobsterers is inconsistent with Hardin's ideas in that members of these communities seem to have put group needs ahead of narrow self-interest—in Törbel, over many lifetimes—without depleting the resource, and without coercion. Community management does not depend on central governmental regulations or the other sorts of externally imposed incentives we discussed in Chapter 5. Neither does it depend on organized programs of persuasion or information, such as we discussed in Chapter 4. And it does not seem to require a deep religious or moral commitment either (Chapter 3), although in some societies, systems of community management are built upon shared religious beliefs (Rappaport, 1970).

A strict adherent of Hardin might argue that these examples are merely rare exceptions to a general rule, but in fact, they are not isolated instances. It is true that small communities often suffer tragedies of the commons, but there are hundreds of documented cases like these two, in which communities have are maintained their important natural resources over very long periods without the use of coercive governmental institutions, and there are probably innumerable undocumented cases throughout human history. Various researchers since Hardin have examined large numbers of cases of success and failure (e.g., McCay and Acheson, 1987; Berkes, 1988; Ostrom, 1990; Bromely et al., 1992; Ostrom, Gardner, and Walker, 1994; Wade, 1994; Baland and Platteau, 1996). Their results are converging on a general understanding of why the "drama of the commons" (National Research Council, 2002, where a recent analysis of research on commons situations can be found) sometimes ends tragically but at other times has happier endings. A particularly useful and influential analysis was presented by Ostrom (1990).

Ostrom focused on the sustainability of what she defines as "common-pool resources." A common-pool resource, as we noted in Chapter 2, is one that is large enough geographically to make it difficult, though not impossible, to exclude individuals from benefiting from its use. For example, a water well can be controlled easily by its owner, but the large under-

ground aquifer that provides its water is a common-pool resource. Sustainability is a mark of successful management because renewable resources, such as grasslands, forests, aquifers, fisheries, and many others, replenish themselves at a limited rate. If the average rate of resource use exceeds the average rate of replenishment, the resource cannot be sustained. Sustainability is an issue because common-pool resources can be depleted by overuse.

Ostrom looked at renewable resources where substantial scarcity existed, where relatively small numbers of individuals (no more than 1500) depended heavily on the resource, and where resource management choices would not produce major harm to outsiders. Thus, she excluded very large-scale resource systems and problems like air and water pollution that cannot be geographically contained. We return later in the chapter to the question of whether the strategies that work on the scale Ostrom studied can also work in these other situations.

Ostrom found that success in developing long-lasting, sustainable community management systems depends on the characteristics of the resource, the group using the resource, the rules they develop, and the actions of government at the regional and national levels (see Table 6-1).

Characteristics of the Resource

When community management works well, the resource is always one with fairly clear boundaries, so that it is possible to define who has rights to use it and to exclude outsiders if necessary. Also, the resources need to remain within their boundaries. For example, commercial marine species that are caught in bays and near shore, like the Maine lobster, are much more likely to be sustainably managed by fishers than those caught in the open ocean, such as whales. Whale stocks tend to be overexploited unless strong agreements are made between national governments, which then impose the rules on their fishing fleets. Fishers by themselves can sometimes (though not always) succeed in controlling the harvest of species like lobsters, but they virtually never succeed with species like whales.

TABLE 6-1 Conditions Conducive to Successful Community Resource Management

1. Resource is controllable locally
   a. Definable boundaries (land is more controllable than water; water is more controllable than air)
   b. Resources stay within their boundaries (plants are more controllable than animals; lake fish more than ocean fish)
   c. Local management rules can be enforced (higher-level governments recognize rights of local control, help enforce local rules)
   d. Changes in the resource can be adequately monitored
2. Local resource dependence
   a. Perceptible threat of resource depletion
   b. Difficulty of finding substitutes for local resources
   c. Difficulty or expense attached to leaving area
3. Presence of community
   a. Stable, usually small population
   b. Thick network of social interactions
   c. Shared norms ("social capital"), especially norms for upholding agreements
   d. Resource users have sufficient "local knowledge" of the resource to devise fair and effective rules
   [(a) facilitates (b), and both (a) and (b) facilitate (c). All three tend to make it easy to share information and resolve conflicts informally.]
4. Appropriate rules and procedures
   a. Participatory selection and modification of rules
   b. Group controls monitoring and enforcement processes and personnel
   c. Rules emphasize exclusion of outsiders, restraint of insiders
   d. Congruence of rules with resource
   e. Rules contain built-in incentives for compliance
   f. Graduated, easy to administer penalties

Success also requires that the resource be of a type that makes it apparent to most of the people who use it that they will be harmed, usually by the results of resource depletion, if they do nothing to control their collective behavior. Ostrom offers the example of groundwater supplies in the area around Inglewood, California, a semiarid region near the Pacific Ocean where water wells began showing signs of salinity in the early 1940s. All the wells in an area of 170 square miles (435 km2) drew their water from an underground natural reservoir called West Basin. At first, many people believed that the salinity was due to a temporary condition affecting the wells nearest the ocean, and that the wells would soon return to normal. But when nine city governments in the West Basin commissioned a study, they learned that overuse of the underground water by the rapidly growing human population was drawing salt water from the ocean into the whole basin. Once it became clear that

overuse would ruin the water for the entire basin, the local water users quickly got together to establish a set of rules to ration water pumping and to enforce the limits (Ostrom, 1990, pp. 114–123).

Characteristics of the Group

Groups of people that devise and maintain successful systems of community resource management typically have certain characteristics. Successful groups are rather stable, with limited population growth and relatively few members moving in and out, and with most members of these groups placing high value on maintaining the resource. These two characteristics often go together. When members of the group have opportunities to meet their needs in the larger economy, they are less dependent on the local resource. They have decreased incentive to follow local rules for resource management and often leave the community when resources become scarce, rather than working to create better management systems. Similarly, when outsiders enter a community at rapid rates, local management systems are faced with the problem of getting newcomers to accept rules they did not help create and whose importance they may not readily understand. Population growth, however, has sometimes been a stimulus for a change from an open-access system, where there are no rules governing resource exploitation, to a community management system. Such seems to have been the history of the English commons, where for hundreds of years, population increases were associated with more effective local resource management systems, which decayed when population decreased. The changes depended on community cohesion, which increased as population rose (Levine, 1986).

Stability is important for an additional reason. Stable, geographically tight social groups are characterized by thick networks of social interaction. People interact with many different neighbors for a variety of purposes: with some around issues of child care, with others in business, with others for food production and consumption, and so on, so that through one network or another, everyone in the group is linked to virtually everyone else. Groups with these sorts of thick, ongoing social networks build up shared expectations of the behavior of members—that is, norms of interaction—around such matters as keeping promises, following rules, and reciprocation.

Community management is more likely to be successful in groups where there are widely shared norms before the resource management problem arises. If people in a community already know whom to trust and what to expect from each other, it is easier to arrive at rules that individuals believe will work. If people in a locality know that their neighbors can be trusted to keep their promises, they can be confident that the neighbors will abide by resource management rules they have agreed to. Ostrom describes shared norms as a kind of "social capital," with which a group can build institutions that can maintain resources with minimal expense for enforcement. Groups that already share social norms find it easier to create new norms that group members will follow. The term *community* is sometimes reserved for

groups that have the social characteristics just described: relative stability of population, long-term direct social interactions, thick social networks, and a body of shared norms (Singleton and Taylor, 1992). To the extent that a group has these characteristics, we can describe it as a strong community in a sociological and psychological sense. We use the term *community management* to reflect the fact that resource management is much easier to organize and maintain when community is strong in this sense.

Successful groups are also those in which there are easy, low-cost ways to share information, enforce rules, and resolve conflicts. For example, the anthropologist Fikret Berkes (1986) has reported the experiences of several coastal fishing areas in Turkey that were threatened by overfishing in the 1970s. In Alanya, the local fishing cooperative was able to devise an effective system for managing the harvest. At the beginning of the fishing season, each fisher was assigned a fishing area by lot for the opening day of the season. After that, fishers changed areas daily according to prearranged rules, so that each fisher had an equal chance to fish in the better areas. Any fisher who was assigned a good fishing ground on a particular day would automatically monitor behavior in the area, and would quickly know if someone else was fishing there. In addition, the fisher could easily confront the interloper and if that failed, could get the support of the community at the end of the day when the fishermen met at the local coffee shop. The system worked smoothly for as long as it was studied. Farther to the west, in Bodrum, the fishing cooperative was unable to prevent overfishing. In Bodrum, outsiders, including commercial fishing trawlers and charter boats from a booming tourist industry, were increasingly fishing in the area. Even when the locals could identify the interlopers, it was difficult to address the conflicts directly or to enforce the local rules. The outsiders did not respond to informal social pressure because they were not part of the community, and moreover, the government did not effectively enforce the laws, such as a three-mile limit, that existed.

When individuals in the group know each other well and have frequent occasion to interact, they also find it easy to get information that may be needed to modify the rules. For example, fishermen in Alanya

For a small Turkish community, a coffee shop like this is a good place for informally sharing information, enforcing community norms, and resolving conflicts.
(Paul Conklin/Monkmeyer Press)

would easily know from daily conversation whether total fish catches were declining, making it necessary to reconsider the management rules. Cattle owners in Törbel would quickly learn if the local official in charge of levying fines had become corrupt or unfair. In this way, the close contact of community members helps groups keep their management systems working well.

Finally, successful management tended to occur in small communities. Smallness facilitates interaction, monitoring of violations, and enforcement of norms, all of which are important to community management. But it is worth noting that Ostrom found some community management systems that served large numbers of people. These systems worked by building larger units out of smaller ones, in what she called "nested enterprises." West Basin and the other water basins Ostrom studied in Southern California provide good examples. In each basin, there was a relatively small number of wells, but some of them, particularly municipal wells, served large numbers of people. Each municipality had already established a system for allocating its own groundwater, so it was possible for the municipalities to act as if they were individuals for the purpose of agreeing on rules for pumping water. The principle of nested enterprises can allow for community management of resources even when the number of dependent people is large—but Ostrom concluded that it is effective only when rules are developed from the bottom up, when rules for small-group management are already in place before the small groups get together into a larger agreement.

Characteristics of Effective Rules

Successful community management is characterized by rules that limit resource exploitation by excluding outsiders and controlling the level of resource use by insiders. Rules that work are, first of all, the product of participatory choice: Most of the people who must abide by them have had a say in making and modifying the rules. Effective rules are also perceived as fair, or equitable—group members are convinced that the rules have more or less the same effect on all those who are asked to abide by them. Participation and fairness increase the likelihood that people will internalize the rules and obey them without coercion.

Effective rules must be congruent with the resources they are designed to manage, and changeable when conditions change. For example, it was fair to assign each Maine lobsterer a fixed territory because the lobsters were distributed more or less evenly, but in Alanya, where the fish ran in certain areas in particular parts of the season, the only way to arrive at a fair allocation was to rotate fishers through all the areas of the fishery. The Alanya fishing cooperative found its system after a process of trial and error, changing the rules until it arrived at some that seemed fair and enforceable. It is important to note that no one is in as good a position as the fishers themselves to develop fair rules, because they have the best understanding of the tides, fish behavior, and other factors that determine where the good fishing spots are from day to day. Their local knowledge may be essential for devising fair and effective rules.

This is one reason Ostrom emphasizes that the best way to develop rules that are congruent with local conditions and flexible enough to change with those conditions is for the people most knowledgeable about those conditions—the users of the resource—to make the rules. She cites numerous sad cases in which well-meaning outsiders, including public officials who believed, like Hardin, that only government action could restrain individual selfishness and prevent the tragedy of the commons, imposed management rules to the detriment of the resource. For example, the government of Nepal nationalized its forests to protect them from overexploitation. The result was to override management rules established at the village level, so that individuals, who saw their villages as having lost control of the forests, began to act as egoists, cutting trees without any restraint to meet their needs and wants. The national government could not afford to police the forests adequately, so deforestation accelerated. The government eventually repealed its nationalization law, making village control the national policy. The new system, in which government provides a legal status for community management, seems to work much better (Arnold and Campbell, 1986).

Successful rules also tend to build in incentives for compliance, so that following them has benefits that counteract the temptation to overexploit. In strong, tightly linked communities, members who comply with rules establish reputations as reliable community members, and so may find it easier to ask their neighbors for favors or to make exchanges on trust, because they are known to be trustworthy. Successful rules may also have built-in material benefits. In the perimeter-defended lobster areas in Maine, lobstermen could spend the poor-fishing months on land, repairing their equipment, secure in the knowledge that because no one was poaching in their areas, they could catch larger lobsters when they returned to the water and make as good a living as if they had spent those slow months fishing. Territorial systems also build in incentives for careful resource management by rewarding people for investments in the productivity of "their" resources. Robert Repetto (1986) offers this dramatic example:

> Oyster grounds in Connecticut that are leased to individuals are ten times as productive as those in Maryland that are fished in common, because on individually leased beds, fishermen will seed with shellfish spat for higher yields, thin and transplant the growing crop, take steps to eliminate predators, and make other improvements.

Oystermen on public waters do not, because the returns are not assured. The world harvest of aquaculture products [Repetto claims] could be expanded thirtyfold or more, if constraints on the leasing of coastal areas could be overcome and investment opportunities realized (Repetto, 1986, p. 30).

Individual territories are not the only way to get these benefits. Small communities can benefit from investments in resource productivity as well. For example, in Törbel, the Alpine grazing lands were kept as communal property, probably because they were too unproductive to support families if used as private property (they were useful for only ten weeks of the year). Even so, the community made investments in the productivity of the alp through collective maintenance projects such as rebuilding walls and trails, and spreading manure to improve fertility. Whether the resource is owned privately or communally is not so important as whether the management rules can keep outsiders from capturing the benefits of the owners' investments in it. In Törbel, the highly productive land was privately owned and the forests and meadows were communal, but in both cases, the owners had built-in incentives for sustaining the resource because they could benefit for many years from their efforts, and would suffer if they did not manage the resource wisely. In both cases, the management rules internalize the externalities of resource management.

Successful community management also requires accurate, accountable, and relatively inexpensive systems for monitoring the state of the resource and individuals' compliance with the rules. Accountability means that the people who enforce the rules should be subject to control by the resource users, so that they can be controlled or removed if they become corrupt or unfair. In the simplest and most effective form of accountability, the monitors are the resource users themselves. A management system has the lowest cost when the resource users do the monitoring automatically in the course of their everyday activities, as with the fishers in Maine and Alanya, who would automatically notice declines in the catch or the presence of poachers. It can also be effective, however, to delegate the monitoring job, for example to village officials in Törbel or to "watermasters" in

the California groundwater basins, who were given the jobs of ensuring that all the wells have accurate water meters and making annual reports on withdrawals of water.

Community management also requires quick, convenient, and inexpensive ways to resolve conflict and deal with violations of the rules. These can include informal procedures, such as the discussions in the Alanya coffee shop, and also formal sanctions. In either case, the system works best if administered by the resource users or an accountable party. Successful systems typically use graduated penalties—small ones for small or initial violations and more serious penalties for persistent violations. When most of the penalties are small and monitoring is built into the system, the cost of enforcement is low, and when most people adhere to the norms most of the time, small penalties are usually sufficient. The low cost of monitoring and enforcement is a major advantage of local management over central management, as the case of the Nepalese forests shows.

The Role of Central Government in Community Management

Ostrom found that the effectiveness of community management also depends on factors outside the community, particularly in government. Sometimes, as with the Nepalese forests, government officials take the attitude that they know best how to manage a local resource, and override community management rules. The typical results include overexploitation of resources, as in Nepal, extraordinary efforts to circumvent the rules, and public protest. In addition, Ostrom reports numerous instances in which central government officials who were responsible for resource management accepted bribes or political favors in return for allowing some individuals to take more than their share of a resource. This rewards the selfish at the expense of the resource and the people who depend on it. Such corruption arises most easily when central officials who have limited ties to the local community are in charge, and where local resource users do not have enough political power to exercise control over central government officials.

Central government can also help community management, for instance by affording local rules the legal status of contracts enforceable in the courts, and by providing support for monitoring the condition of the resource. The State of California was helpful to the West Basin and other regional water users in both these ways. For local water users, the likelihood of expensive litigation over water rights created a strong impetus to negotiate agreements to restrict pumping, and the courts approved the agreements as legally binding. The state helped further by agreeing to pay part of the cost of monitoring the agreements. The concept of *comanagement,* in which local communities manage resources under rules that they develop with the support of government and where they and the government share power and responsibility, is one of the promising new ideas in environmental resource management (McCay, 1993). (For a recent analysis of comanagement and other strategies for linking community management to larger layers of government, see Berkes, 2002.)

The Psychology of Community Management

In organizational terms, the keys to community management systems such as those in Törbel and Maine are participatory decision making, monitoring, social norms, and community sanctions. But as Hardin convincingly showed, the success of resource management ultimately depends on controlling the behavior of individuals. How does a set of community management rules change individual behavior? To ask the question another way, what makes individuals follow the rules when they can gain something by breaking them? The key is that most people do what is good for the group and the resource because they *internalize* the group's interest, rather than acting out of *compliance* with a set of external incentives. This is a subtle but important social-psychological distinction (Kelman, 1958). Compliance—the method of control most closely associated with regulations—works only when people expect to be punished for a violation, but internalization works all the time. A compliant motorist will stop at a red light only if he or she fears punishment in the form of a fine or a traffic accident. A motorist who internalizes

the red-light norm will stop even with no police or traffic in sight. Obviously, a system that runs on internalization can be effective with much less policing than one that relies only on compliance.

Community resource management systems run mainly on internalization, but they always include an incentive structure as well, consisting of built-in incentives where they can be devised, and monitoring and formal and informal sanctions where built-in incentives are insufficient. The incentive structure is very important even when most people internalize group norms so that it is unnecessary to threaten them very often with sanctions. Effective incentives are necessary to control the few who do not internalize the norms and others who usually follow the rules but may sometimes be tempted to stray. The ability to penalize the few violators assures the many that they will not suffer by controlling their own behavior. People need to know that violators will be discovered and dealt with in order to be comfortable doing what is good for the group. When a system is effective and most people internalize the norms, penalties are rarely imposed and the costs of maintaining the system are low. Effective systems usually use graduated sanctions. Few individuals break the rules, and most rule breakers comply with mild sanctions, making more severe ones almost unnecessary to use. But without any incentives against overexploiting the common resource, some individuals could take advantage of other people's self-restraint with impunity, and the system's whole basis in trust would begin to unravel. Enforcement costs would climb, and people would become less willing to exercise self-control, leading to a vicious cycle ending with the tragedy of the commons.

Why do people internalize the group norms? Because they have participated in creating them, because they see their value for themselves and their community, and because the norms become part of the meaning of the community they share with others, with whom they have ongoing and trusting relationships. It is necessary to have a system of sanctions to protect group members from anyone (including themselves) who might be tempted to violate the rules for personal gain. But the reason most people act on the norms most of the time is that doing so is what it means to be a member of the community—as members, they have a sense of responsibility to follow the rules. Most people see following the rules as the right thing to do, rather than as coerced behavior. In short, the incentive structure that a community creates (including monitoring and enforcement) and the processes it follows (participation and social expectations of rule-following) create the psychological conditions for self-control (internalized norms). It is self-control that ultimately makes community management different from the incentive strategies we discussed in Chapter 5, and it is belonging to and feeling responsible to the community that shape self-control. This is why a strong community is so important to successful resource management at the local level.

We should be careful not to glamorize the psychological and social climate in the strong communities that manage their resources successfully, or to conclude that because these social systems work well for the environment they necessarily work well for the people. Small, cohesive communities, even if they are participatory in making decisions about the natural environment, do not always distribute resources equally or fairly, and the presence of the thick social networks that help make community management work does not necessarily guarantee that community members are happy or that their relationships are harmonious. Strong communities are often good at exercising informal social control, a skill that helps them provide public goods such as natural resource management, crime control, and the like, but these communities' norms sometimes repress individual community members, and there is an inevitable tension between the demands of community and such widely held modern values as freedom of individual expression and procedural justice.

The social downside of community management can be seen in some of the success stories we have cited. In the Maine lobster fishery, the rules of territoriality favored landowners over others, and therefore families that had lived in the area for generations over relative newcomers. There was not full equality of access to the resources. Further, the rules were enforced by a kind of informal justice that included illegal destruction of private property and that lacked avenues of appeal if the enforcement was unfair. In the Alanya fishery, disputes were settled informally at the coffee shop—an environment that, like the fishery itself, excluded women. These examples suggest that small communities that achieve social control by the use of norms and informal sanctions can be quite repressive in their own ways. This is why for generations, some people, particularly those who felt repressed or out of place in their small social systems, have migrated from rural communities to the cities or to other countries.

Community Management and Hardin's Model

According to Garrett Hardin's formulation of the problem of common-pool resource management, successful community management cannot occur. Hardin's formulation assumes that the overriding human motives are always self-centered. If this were the case, whenever there is a valuable, depletable resource to which individuals cannot be denied access (that is, a common-pool resource), some individuals would selfishly exploit it and others would follow suit out of the need for self-preservation. Tragedy would inevitably follow.

Successful community management shows that under some conditions, it is possible for other motives to win out over selfishness so that a common-pool resource can be managed over long time periods with very limited coercion. Without putting up fences or stationing armed guards, communities can get individuals to control themselves well enough to protect shared resources. Ostrom refers to the systems of rules, norms, social pressures, participatory decision making, and sanctions that are responsible for this achievement as resource management *institutions*. These social inventions help people act in the collective interest when they recognize that doing so will also benefit them as individuals.

Hardin offered one essential insight. He realized that an individual's awareness of a common fate with a larger group—even when combined with willingness to sacrifice for the group—is not enough to solve the resource management problem because the structure of the situation can create irresistible pressures on individuals to take more of the resources than the pool can sustain. Put in our terms, education, even combined with the right values, is insufficient when the incentives are wrong. It only makes sense for an individual—even one who cares about the group—to exercise self-restraint if there is reasonable assurance that others will do the same.

What Hardin did not recognize was that coercion is not the only way to provide that assurance. Individuals can and sometimes do create noncoercive institutions that give people the assurance they need. It is important to repeat that these institutions do not depend on individuals' willingness to sacrifice themselves. They depend on individuals' seeing how self-interest and group interest can reinforce each other and creating community norms and sanctions that they believe will restrain some people's impulses to take advantage of each other. To the extent people have faith in a community and its management institutions, they are willing to comply with the community's rules, help to enforce them, and help modify them as needed to maintain the resource and the group's collective well-being.

Hardin's blind spot about the potential value of noncoercive social institutions is important because it is shared by many other individuals, and even whole intellectual disciplines. Hardin's idea of self-interest implies that community management institutions cannot work, so he teaches that government coercion is the only way. He is not alone in this way of thinking. Both Skinnerian psychology and neoclassical economics view individuals as acting in isolation and do not often consider how social institutions and relationships can shape individual behavior. (Recently, some neoclassical economists have begun to address the question of institutions.) We believe that it may make a great difference in terms of humanity's ability to solve environmental and other problems whether people think of the problems only in individualistic terms or also consider solving them by creating social institutions and making use of social relationships. The following true story suggests how individualistic ways of thinking can get in the way of resource management and why we sometimes call Hardin's scenario the "Tragedy of the Economists."

During the 1970s, a number of researchers began studying behavior in the commons by creating small-group laboratory simulations. One of us (Paul Stern) developed a four-person game that presented people in a schematic form with the choices they would make if they were deciding whether to join a carpool. Following public concerns of the time with the "energy crisis," the simulation focused on the depletion of oil supplies. Each player got a "salary"—a small amount of real money—before every round of the game and then spent money to get to work. Every round, the players could discuss and then choose one of two alternatives that amounted to driving alone or joining a carpool. Driving alone had a known cost, representing the cost of fuel and maintenance for a car. Those who chose to join the pool then decided on one member (the driver) who would pay a higher cost, representing not only the cost of fuel and maintenance but also the extra time and inconvenience of driving the group. The others (passengers) paid a low cost that did not include fuel. The game was repeated, allowing the carpoolers to take turns driving if they chose. At the start of the game, the cost of driving alone was less than the average cost of being in a carpool, creating an incentive to drive alone, much the way one exists in reality: for most workers in the United States, after all the financial and convenience issues have been considered, driving to work alone is preferable to carpooling. The game simulated resource depletion (an energy shortage) by having the cost of fuel rise at a rate determined by the total amount of driving that had been done during the game. The result was that incentives for carpooling increased over time as the resources were depleted. The more solo drivers there were early in the game, the faster the resource was depleted and the faster the costs of driving increased later on. Thus, the strategy that was best for the group in the long run (and for the environment) was to carpool from the beginning, in order to delay resource depletion.

A number of student groups played the game at Elmira College, in Elmira, New York (Stern, 1976). Generally, when the players were given detailed information about how the costs would be affected by their behavior, they fairly quickly agreed to carpool consistently and earned a few dollars each before the resources ran out. (When they lacked detailed information, they used the resources much more rapidly, and earned much less.) The findings were consistent with Ostrom's conclusion that people are more likely to manage resources well when they can see that all will suffer unless they collectively control resource use.

Once, professors at a faculty seminar were invited to play the game. The four volunteers were a philosophy professor, an English professor, a chemist, and an economist. They were given complete information about how the game worked, and then began a long discussion about strategy. The philosopher suggested (quite correctly) that it would be best in the long run for everyone to join together in a carpool from the beginning, but the economist pointed out that at the start of the game, everyone would be better off if each drove alone. He reasoned, therefore, that everyone should drive alone until the cost of driving got to the point where it would pay better if everyone joined the carpool, and then everyone should join the carpool. The argument convinced all his colleagues. The result was that the group used resources very fast at the start of the game, exhausted the resources much more quickly than they might have, and did not do nearly so well as the student groups.

We take this as an object lesson in the limits of the kind of shortsighted analysis that adds up what is good for individuals in the short run rather than choosing what is good for communities in the long run. The economist convinced his colleagues to engage in a sort of egoistic thinking that Garrett Hardin believes to be universal among human beings, and that is in fact practiced and advocated by many economists. The argument was powerful enough to convince the others, even though one of them had initially suggested a better solution. The tragedy of environmental management is a certainty if everyone takes the individualistic view this economist advocated. But fortunately, people, aided by the institutions they create, do not always think and act this way. (We do not mean to accuse all economists of being shortsighted egoists. However, we are not the only ones to report evidence that training in economics is associated with an increase in egoistic behavior [see Marwell and Ames, 1981; Frank, Gilovich, and Regan, 1993]).

Other Benefits of Community Management

Community-based institutions for resource management may have value beyond their effect on the natural environment. Ostrom points out that groups that have an abundance of "social capital" in the form of norms of reciprocity and shared expectations of behavior start out at an advantage in building resource management institutions. By the same logic, building and operating successful resource management institutions provides social capital that can be valuable for solving other social problems that require cooperation. That is to say, strong communities succeed at resource management, and success in resource management strengthens communities. Groups that have built up familiarity, shared expectations, good communication, and trust working on an environmental problem may find it easier to address other community problems that involve providing public goods and controlling selfishness, such as neighborhood safety, drug abuse, and school truancy. We return to this possibility later.

APPLYING COMMUNITY MANAGEMENT PRINCIPLES BEYOND SMALL GROUPS

Most of the examples of community resource management we have discussed so far come from rural communities that depend economically on locally available natural resources. Life in such communities is far different from the life of most citizens of urbanized, developed countries, whose economic survival does not depend nearly so much on natural resource supplies in their immediate vicinity. This section considers whether the principles of community management we have described are applicable under the conditions of modern developed economies. We find that they are applicable in two kinds of situations: in which modern communities still depend on local resources, and in which some of the principles can be applied even in the absence of significant resource dependence.

Local Resource Dependence in Modern Societies

Few people in modern, developed societies depend for much of their livelihood on fishing, hunting, cutting trees, or grazing cattle. In fact, because of global markets for food, fuel, and raw materials, people get what they need from suppliers all over the world. People eat fruits and vegetables that are out of season locally, and even some that must be imported throughout the year. Most people who buy in the global economy do not even know where the raw materials come from that make up their automobiles, home appliances, or other household goods. Global markets ensure that people with cash incomes can almost always escape the pain of local resource shortages by simply buying products from elsewhere.

There are, however, a few important exceptions to this rule—situations in which people, even in modern societies, are still largely dependent on local, common-pool resources. In such situations, Ostrom's principles for community resource management should be widely applicable. We briefly examine two of these situations—water supply and waste disposal—and find that community management principles can be effectively applied.

Water Supply. Most modern communities depend on nearby rivers and reservoirs or local aquifers for their water supplies. Transporting water over long distances is uneconomic for most purposes, though there are exceptions such as the canals that move water hundreds of miles to Southern California and an emerging international market for bottled drinking water. Most communities depend on the availability and quality of local water, and community institutions are an obvious management strategy. Even in Southern California, where much water comes from long distances, Ostrom found water supply institutions based on community management principles.

The California water management institutions that Ostrom studied are "nested" institutions. The members of the water associations are city water departments and other major water pumpers, rather than individuals, and the smaller units to which individuals belong operate by the norms of businesses or public utilities—service in exchange for payment—rather than those of community management. Consequently, even where community institutions manage the water supply, individual Californians do not normally internalize norms for careful use or experi-

ence a sense of responsibility to the community to husband water resources.

The relationships of individuals to water change, however, during the periodic droughts that hit California. During those periods, local governments have called on citizens to cut water use, and people have responded. During the serious drought of 1976–77, a number of water districts in Southern California sent educational brochures to their residential water users explaining the need to conserve (which was widely covered in the local news as well) and advising on ways to save water. Even though the brochures were usually enclosed with the water bill—a procedure that has been generally ineffective when applied to energy conservation, as we saw in Chapter 4—the typical consumer response was to reduce water use by 10 percent for the duration of the drought (data are from Berk et al., 1981, Appendix 3).

More remarkable was the response to the water rationing programs instituted in many Northern California communities. For example, in the East Bay Municipal Utility District, which supplies water to the cities of Oakland and Berkeley and sixteen neighboring communities, communities restricted water use for landscaping, prohibited decorative fountains, and allocated water to each household based on a formula that estimated water needs. The goal was to reduce water use by 35 percent, and in fact reductions averaging about 40 percent were achieved (Berk et al., 1981). What was most impressive, however, was the level of community acceptance of the restrictions. For example, the city of San Leandro in the East Bay area was included in a 1979 study of drought response. The conservation programs in San Leandro were even more successful than planned. Instead of the 35 percent planned savings, residential water use declined 60 percent. When residents were asked to rate the conservation program on fairness and effectiveness, using the A–E grading system common on school report cards, fully 73 percent gave the program an A for fairness, and 81 percent gave it an A or a B for effectiveness (Bruvold, 1979).

The studies show that information was effective and that rationing was even more effective. Combinations of programs saved more than single programs (Berk et al., 1981). Moreover, people seemed to willingly accept restrictions on their water use during the drought emergency, and in the areas where the restrictions were the most stringent, the most frequent complaints about the program were not that the restrictions were too severe but that they were not enforced strictly enough and that more education was needed to increase their effectiveness (Bruvold, 1979). In short, Californians strongly supported their communities' water conservation programs.

It appears that the drought created conditions that were conducive to shared norms in favor of restricting water use. The research is not conclusive about the psychological processes that were involved, but between the research results and anecdotal accounts, the following picture emerges. The water shortage was clearly visible to Californians: Grass turned yellow, lawns needed more water, and pictures of the low levels in water-supply reservoirs appeared prominently in the newspapers and on television. The reservoirs, in particular, may have been especially graphic evidence of a very unpleasant common fate that awaited unless the rains came or consumption decreased. Water districts were a sufficiently local entity to take collective action, and typically, they were accountable to citizens as public agencies. Where water restrictions were in force, violations were easily identified because the largest water uses in homes—watering lawns and filling pools—were highly visible to the neighbors. Anecdotal accounts suggest that Californians got angry and used all sorts of informal social sanctions to control the behavior of neighbors whose green lawns showed them as violators of the drought norms. In short, the drought situation had many of the characteristics, and brought out many of the psychological mechanisms, that Ostrom identified in small communities of resource-dependent people in rural communities.

Waste Disposal. The waste products of modern industrial society are almost always disposed locally, into landfills, waterways, and the air. Air and water often carry wastes outside the local area, thus weakening pressures for community management, but not always. Communities often face public health threats due to the actions of local motorists, whose automobile exhaust produces smog, or local manufacturers, whose liquid chemical wastes seep into water supplies. These conditions hold the potential for management by community institutions because they present communities with a common fate and clear evidence that continuing the status quo will become unacceptable to most citizens.

The clearest examples of common fate in waste disposal, however, concern solid waste. Solid wastes in most U.S. communities are deposited in central locations, usually landfills, either directly or after part of the waste has been changed into other forms (e.g., by incineration). In many communities, shortages of landfill space, concern about toxic materials, increasing disposal costs, and public opposition to siting new waste facilities have made old waste disposal practices untenable and have forced local governments to consider new policies to reduce the volume of waste. Many have responded with community-based recycling programs.

The block leader approach described in Chapter 4 is a good example of how principles of community management work for recycling. Hopper and Neilson's (1991) experiment in Denver suggests that much of the effectiveness of adding block leaders was based on increasing the strength of norms—both social norms (the behaviors people expect of each other) and internalized personal norms (the things people feel a personal obligation to do without considering what others may expect). We have already seen that norms are a key to the success of community management. Let us now examine how the block leader approach affects norms.

One thing block leaders do is turn recycling from a private, individual activity to a social one, in which social norms can influence behavior. People become aware that recycling is a neighborhood effort, and they expect that what they do in that effort will be monitored by their neighbors. Since most Americans are predisposed in favor of recycling and believe it is a good thing for their communities (Dunlap and Scarce, 1991), they can expect that their neighbors will not only notice, but judge. Thus, the presence of block leaders can give people the expectation that they face the social disapproval of their neighbors if they do not contribute to the recycling effort. Block leaders provide a monitoring system and, as Ostrom noted, monitoring and norms reinforce each other. In the case of recycling, the prospect of monitoring puts teeth into preexisting social norms.

To understand how block leaders affect personal norms, we need to understand the relationship between personal norms and prosocial behavior. The social psychologist Shalom Schwartz (1977), whose research on values we discussed in Chapter 3, has developed the concept of *norm activation* to describe a process by which, under certain conditions, individuals experience a sense of personal obligation to act in a prosocial way, and a sense of guilt if they do not. Personal norms for prosocial behavior are activated under two conditions: First, people must believe that an existing condition poses a threat of harm to others (Schwartz calls this Awareness of Consequences, or AC); second, they must believe that their personal action or inaction has the power to prevent that harm (Schwartz calls this condition Ascription of Responsibility to self, or AR). When a person holds both beliefs, he or she experiences a sense of obligation to act to prevent the harm. Thus, if someone believes that failure to recycle is harmful to the community, and believes that his or her own action can make a difference, that person will feel obligated to recycle, independently of social pressure. Note that the absence of *either* AC or AR will keep the person from feeling a personal obligation—it is still possible to get such a person to recycle, but not on the basis of internalized norms. In Hopper and Neilson's experiment, having block leaders led to stronger personal norms, and to more recycling behavior. We do not know whether the block leaders helped spread the awareness of the negative consequences of waste disposal, and thus increased AC. But it is quite likely that they increased people's sense that their own behavior would matter (AR), because having the whole block in a sense committed to recycling made personal behavior part of a larger effort in which people were responsible to each other and collectively could make a difference.

We should add that Schwartz's norm-activation model has been shown to explain a number of environmentally relevant behaviors. For example, people are more likely to reduce energy use (Black, Stern, and Elworth, 1985) and refrain from burning trash in their yards (Van Liere and Dunlap, 1979) if they believe these activities threaten the well-being of people in general and if they also believe their personal actions or those of people like them can make a noticeable difference. They are also more likely to support government policies for environmental protection if they believe that environmental conditions are harmful to people and that the policies are directed to changing the behavior of the responsible parties (Stern, Dietz, and Black, 1986). There is evidence that for some people, harmful consequences to ecological systems and nonhuman species has the same norm-activating consequences as harm to people (Stern, Dietz, and Kalof, 1993; Stern, Dietz, Kalof, and Guagnano, 1995).

A city block is different in some important ways from the communities that Ostrom studied. Like those communities, residents may interact frequently, and they may monitor each other's behavior, but their relationship to their resource (in this case, the landfill) is not as close as the Maine lobsterers' relationship to theirs. Although each individual would find it very inconvenient to use a different landfill, the neighborhood does not control the landfill, because it is shared with the rest of the city. Consequently, one block's reduction in solid waste may have little effect on the resource as a whole. People's awareness of that fact may eventually reduce their enthusiasm about block-level recycling programs. Of course, a nested approach using block leaders around the city might solve that problem. We report in Chapter 7 on a successful citywide recycling program that uses such an approach.

The facts that block-level programs can work at all in a large city, and that citywide conservation programs can also be effective, are significant. They suggest that the techniques of community management can be useful even when some of the optimal conditions for community management are absent.

Community Management without Resource Dependence

The success of the block leader approach to recycling relies on some of the key social-psychological mechanisms that make community management work. It uses face-to-face communication among people who already have social contacts and some degree of trust, it establishes and reinforces shared social norms, it activates personal norms, it makes people aware they may be monitored, and it makes monitoring simple. The following examples show how some of the same mechanisms have been used effectively even when the basic condition of community resource dependence is absent.

Although few U.S. communities depend on local resources for their energy supplies, a number of them have operated successful community-based conservation programs. As we have seen in Chapters 4 and 5, carefully designed information and wisely chosen incentives are elements of successful programs. As we show here, the most successful programs also use principles of community management.

In Chapter 5, we mentioned the FACE program in Fitchburg, Massachusetts, that in only six weeks succeeded in getting low-cost weatherization materials installed in one-sixth of the homes in that city. We emphasized the way the program removed barriers to action by providing information, low-cost or free materials, and, if necessary, installation, all in one step. But the notable success of the program also depended on its creative use of some of the principles of community management.

FACE was conceived from the start as "a local collective" (quotations are from Fitchburg Action to Conserve Energy, 1980). The city planning coordinator's office, which organized the program, immediately created an Advisory Council that represented a cross-section of the community and appointed coordinators from each neighborhood of the city to provide "direct access and feedback on a neighborhood level." It opened centers in each neighborhood to increase participation and provide convenient locations for people to get training in how to weatherize their homes. FACE relied heavily on volunteer labor from college students and other interested individuals.

The program made special efforts to provide weatherization training in settings that were comfortable and familiar to the citizens. The two most successful techniques were training sessions arranged by appointment at workplaces and so-called Tupper-

ware sessions that were held for any group of neighbors that could gather ten or more people in a place of their choosing. The informality of these sessions and their reliance on existing social groups made it easy for participants to ask questions and to encourage and learn from each other. Encouragement was especially important to many women who initially believed weatherization was too technical or difficult to accomplish on their own. The program used the popular rhetoric of "self-help" and a "hands-on approach" to training to make conservation an active and rewarding experience. "People were constantly reminded at training sessions that this was *our* program and that we had to spread the word to relatives and neighbors if it was going to work. The sense of ownership and identity with FACE became very strong as the program progressed" (emphasis in original). The final report claimed that "a sense of shared commitment resulted" and that "there is now an increased atmosphere of community and cooperation in Fitchburg."

In short, the FACE program successfully applied and adapted community management principles, among them reliance on face-to-face communication, shared commitment to a common activity, interdependence within the group, and the creation and reinforcement of shared norms (spreading the word, helping oneself and others). The community approach allowed the successes of individuals and the group to reinforce each other: "Individuals proud of the work they had done apparently made a point of showing their neighbors," thus spreading the word and adding a kind of credibility that even a well-liked local government can only hope for. The use of face-to-face groups made it easier to elicit the level of individual commitment necessary for this program's success.

The FACE program did not have all the above elements at the start. Rather, they evolved over the program's six-week duration because the program used a participatory approach—another key principle of community management. The final report mentions a number of outreach techniques that were tried but failed, and several neighborhood centers that were closed because they added little to the program's effectiveness. The explicit effort to get continuing feedback from all the neighborhoods allowed FACE to find particular methods that worked well for Fitchburg and to discard ideas that looked good in the abstract but that did not work. The process is similar to the one used by the Alanya fishermen, who spent years tinkering with their system for allocating fishing areas until they found one that worked.

The experience with residential energy conservation programs in the 1970s and 1980s repeatedly confirms the value of using existing community institutions (a form of what Ostrom calls social capital) to help in advertising, marketing, and modifying the programs where necessary. Community institutions are especially critical for reaching groups of people who tend not to respond to mass-media campaigns: low-income and low-education groups, renters, speakers of foreign languages, and so forth. Such people tend to be especially skeptical or unresponsive to programs brought to them by large, established institutions, but they are much more responsive if the church, civic, or community groups to which they belong act as intermediaries.

Experience shows that the community approach can add greatly to the efficiency of programs, in the sense of getting the most benefit for a limited cost. An example is the experience of the Residential Conservation Service program in the state of Minnesota, already mentioned in Chapter 4, which aimed to promote energy efficiency by conducting energy audits in homes. The participating utility companies used three different methods to perform their energy audits. Some companies had their own employees perform the audits, while others hired private companies or community groups as subcontractors. Utility companies that hired private companies spent half as much per audit as those using their own staff, while maintaining the same quality of audits and reaching more homes (6 percent versus 4 percent of those eligible). But the utility companies that used community groups got the best results. Their audits were of the highest quality of all, reached 15 percent of the eligible homes, and cost one-third of what companies paid to their own employees (Polich, 1984). The community groups apparently benefited from their credibility with consumers, which made marketing easier and

gained them easier access to homes. It may be that when accepted community groups are the bearers of the program, people tend to act on the norms of community membership: It is a valued thing to participate in programs for the good of the community.

The best way to use the strengths of a community to design an effective environmental program cannot be known from theory, and they can only be predicted in rough approximation from an examination of the community. An indispensable key is to establish a participatory process that can, by trial and error if necessary, find ways to involve the right groups, address any emerging problems, and keep the environmental program on track. The experiences in Fitchburg and other communities show that community-based programs can have positive effects on community morale, as well as achieving their environmental goals: There is a connection between community resource management and the strength of community feeling.

COMMUNITY MANAGEMENT AS A WAY OF LIFE

For some people, community management institutions are desirable not only because they can sustain natural resources, but also because they support a much-desired way of life. Community management works best in relatively small, relatively stable social groups characterized by common geographic location; frequent personal contact; economic and social interdependence among group members; shared norms; and informal means of monitoring, enforcing norms, and resolving conflicts. This is roughly the social organization of a nomadic tribe, an agricultural or fishing village, or a small rural town. It is a form of social organization that has been the norm throughout most of human history, as we note in Chapter 8, and one with strong cultural resonances even for many modern urban and suburban dwellers. The positive images associated with small communities include friendly and helpful neighbors, trusting relationships, freedom from crime, a leisurely pace of life, and closeness to nature. There are also negative images, of course—backwardness, boredom, loss of privacy, restrictions of free expression, and intolerance or even repression of differences—but nevertheless, the ideal of village life holds strong attractions for many modern human beings, particularly in the United States. We do not mean to advocate life in small, cohesive communities as a better way of living—that, after all, is a value judgment on which people differ. But for people who are living happily in such communities or who desire to do so, community resource management has important social benefits as well as environmental ones. Such people often prefer community management even when other management methods are beneficial in economic terms.

This section explores the link between community resource management and community social organization from two angles. It examines the efforts of communities to maintain their own systems of resource management—and their ways of life—in struggles against development policies imposed from outside that they see as threatening both. It then examines the idea that people can organize the way they interact with the natural environment so as to make resources more manageable at the community level.

Community Management versus Development Policy

Recent decades provide numerous stories of economic development efforts that have negative environmental consequences. There are many reasons for these effects; one is that the development process often disrupts local community management institutions.

An illustrative story comes from the foothills of the Indian Himalayas, where for centuries people have relied on the forests for cooking fuel, fodder, and food. The forests also helped control the floods that sweep through the region every monsoon season by holding rainwater in the soil and releasing it slowly. For centuries, most of the forest land has been the communal property of the local villages and hamlets, with the poor and hungry having rights, recognized by their communities, to gather firewood, mushrooms, fruit, and fodder from the many tree species growing there. These rights were a matter of life and death for landless people and for everyone

when droughts reduced the productivity of the fields. Gathering in the forest was mostly the work of the village women.

Beginning in the late nineteenth century, the forests entered the cash economy, first through commercial felling of native trees and, since the 1950s, through the replanting of forest lands to a single, fast-growing commercial species, usually eucalyptus, that produces lumber for market but no fruit, little fodder, and relatively few twigs for cooking. This sort of forest development was the policy of the British colonialists and then of the Indian government, both of which wanted the cash. The resulting loss of tree cover in the higher elevations meant that by the 1960s, floods in the Ganges valley were becoming increasingly dangerous. Moreover, because the commercial plantations were private property, communities lost control of the resources, and villagers (except for a few landowners) lost their right to turn to the forests for sustenance in hard times.

In this historical context, a social movement began growing in the late 1960s and early 1970s to save the multispecies, common-property forests because of their value for flood control, subsistence, and the local way of life. The story has been told in detail by Vandana Shiva (1989), an Indian physicist who, out of concern for the Indian environment, became an expert on the ecological and human benefits of India's ancient subsistence systems and the human and environmental costs of many modern development plans.

The movement, known as Chipko and organized originally by women who had worked in Mohandas Gandhi's nonviolent resistance to British colonial rule, used nonviolent methods to resist development plans that involved cutting trees from the remaining common lands. Recognizing the multiple threats inherent in replacing forests that provided diverse resources for all with tree plantations that provided cash incomes for a few, the leaders were able to organize an increasing number of grassroots groups for action when government agencies decided to give forest lands over to commercial forestry. Women from local communities organized to protect trees from foresters, sometimes literally hugging trees in the presence of men with axes, and staying in place until the scheduled tree harvest was postponed or canceled. The word spread from isolated successes until people were mobilizing across the region to protect threatened forests.

The Chipko movement drew on deeply held beliefs and values, as well as on very real concerns for survival. The following statements made by movement women to axmen suggest the depth of feeling:

> *The forest is our mother. When there is a crisis of food, we come here to collect grass and dry fruits to feed our children. We dig out herbs and collect mushrooms from this forest. You cannot touch these trees.*
>
> *Stop cutting trees. There are no trees even for birds to perch on. Birds flock to our crops and eat them. What will we eat? The firewood is disappearing: how will we cook?* (Shiva, 1989, pp. 74–75)

These statements show both the spiritual and material value of the indigenous forests to the people, as well as their awareness of the ecological connections between forests and croplands. The burden of deforestation fell especially heavily on women, who had the responsibility to gather firewood for cooking, and whose workload increased as the forests receded.

Shiva shows how community institutions and community well-being suffered under a government policy that treated common lands as "wastelands" and that sought to convert them as quickly as possible to marketable uses. She shows how the Chipko movement was an attempt to reassert community values and local control by people who realized that the community's management institutions were the only proven way to meet basic needs.

The Chipko story makes two things clear. First, whatever benefits may come from commercial development of natural resources, market-oriented development can be incompatible with community management. This is not because it replaces common property with systems of private property—in Törbel, private and communal property have coexisted for over 500 years under a set of rules chosen by the community itself—but because commercial considerations put decisions about resource use under the control of forces outside the local community. Indian forest policy relied on cash cropping (in this case, lumber) to repay development loans to outside lenders, who did not have to bear the costs that development imposed on local people and resources. Commercial systems place little importance on the noncashable benefits that natural resources provide and that local people value highly—in the Himalayas, food for the poor and hungry, fodder, flood control, and low-quality cooking fuel. Nor does market-based management place any value on the social capital present in community institutions that manage local resources and provide other desired public goods. The local benefits of the indigenous system of forest management became, under the cash-crop economy, externalities in transactions with outside purchasers who, because they were outsiders, did not have to take them into account.

Second, community management can sometimes yield social, environmental, and even economic benefits far in excess of what development experts and central government officials recognize. Shiva shows in her book that indigenous systems often provide greater economic value, when all the forest's products (not only the lumber) are taken into account, than the modern development systems that replace them. The same argument has been made in studies of sustainable development around the world. It appears that communities that have depended on local resources for generations sometimes learn to use them more efficiently than any modern development plan.

The Chipko movement had numerous successes in defending community management systems against the inroads of national and global markets, but the struggle remains a difficult one. India is increasingly part of a global economy in which cash is needed for fuel, transportation, medicines, and manufactured products. People who live in subsistence economies lack the cash to benefit from modern advances of all kinds, and their governments often lack the cash to provide for them, so a strong desire for money is understandable in terms of meeting local needs. In the Himalayas, needs for cash often created conflict between the women of the Chipko movement and their own husbands, who, in order to get money to support their families, worked at jobs in the forest products industry. The story of Chipko suggests that there may be a fundamental incompatibility between community resource management and the global market economy that provides so many of the benefits of modern life. It hints that there may also be an incompatibility between the global economy and the global environment. That unhappy possibility brings us to another approach to community management.

Making Resources More Manageable

A number of social thinkers have advocated reorganizing societies and economies into smaller units that they believe would be better able to manage environmental problems. The *bioregionalist* movement, for example, advocates drawing political boundaries to match ecological ones, such as watersheds (Sale, 1991). Under this system, a problem like water pollution could be handled by one regional government. Under today's political arrangements, water pollution tends to be a source of perennial conflict between upstream governments that can stop pollution but have little incentive to do so because the pollution flows downstream, and downstream governments that have the incentive but no power to act upstream, where the pollution originates. Bioregionalists argue that more "natural" political boundaries would reduce social conflict and also give people identities tied to their environments rather than to nations or other political entities that can come into conflict over claims on the same territory. A problem with the bioregionalist approach is that even after all the effort that would be required to alter political boundaries and institutions, there would still be many environmental problems—air pollution, for example—that cross any boundaries that might be drawn.

Some argue that the key to environmental management is to bring social institutions down to "human scale" (Sale, 1991; Schumacher, 1973). They claim that smaller units are socially easier to manage and that they can depend on technologies that are easier for relatively untrained people to understand, produce, and maintain. This argument is often made in the context of development policy, using the concept of "appropriate technology." Proponents say that large hydroelectric projects, for example, transfer political power to the banks, governments, and engineers

responsible for the project and away from the people who are supposed to benefit, whereas smaller-scale projects can be financed and controlled locally, so are more likely to bring local benefits. The argument about human scale is consistent with the evidence that community management techniques are easier to implement and use when the community is small.

The physicist Amory Lovins (1977) made an important argument along these lines about energy policy in the 1970s, when the United States was facing expected shortages of petroleum supplies and was looking to coal and nuclear power technologies as the chief alternatives. Lovins argued that either of these technologies would take the country down a "hard energy path" that required centralized control and regulation and large-scale financing. He contended that the result would be bad for the environment, and would erode democratic principles of public control. Only experts would understand the technologies, but they would not have incentives to protect the environment or the local residents; the people who might be most affected would not have the knowledge or power to decide. Lovins saw the hard path as leading to technocratic control by technological experts. His favored alternative was "soft energy paths," relying on energy efficiency and renewable energy technologies based on wind, water, and solar power that were relatively small-scale and local in character. Lovins argued that soft paths are environmentally preferable and also that they would better preserve American political values of democratic control and individual autonomy.

"Soft-path" technologies tend to be inexpensive, small-scale, and designed for local energy production. An example would be a small city that, instead of getting energy for public and private uses from large electric and gas utilities, produced gas from its landfill and electricity from windmills supplemented by hydroelectric power from a nearby reservoir. It is in the nature of soft paths that each community would be free to find its own combination of supplies to meet its needs. In terms of social organization, however, soft-path solutions tend to have some things in common. They rely on the initiative of local households, businesses, and governments and they are managed by those same actors. Some technologies may be adopted privately (for example, a solar water heater for a house) and some may be communal (for example, a hydroelectric power station); similarly, the management rules are likely to be a mixture of private-property rules and community-based ones.

The choice of a soft path makes community management a more viable option. A community that depends for its electricity on its own windmills and dam possesses two necessary conditions for community management: a resource that is controllable locally, and mutual dependence on that resource. If the same community used 1 percent or so of the output of a large coal or nuclear-fired power plant, which in turn purchased its fuel in a global market, it would not have control of its energy supplies, and would probably find it much more difficult to use community management methods to achieve a desired community goal, such as a collective reduction in energy use.

The soft-energy-path idea is an example of a broader strategy, which is to put control over resources, wherever possible, into the hands of human institutions, particularly communities, that can manage them sustainably, and to choose technologies that are manageable at the community level. This strategy turns on its head the usual procedure in modern societies, in which technologies are developed for economic profit or governmental purposes with only secondary regard to their implications for the environment and without considering in advance how any unfortunate environmental consequences will be controlled. Proponents argue that the usual way of choosing technologies has been destructive both to environmental and human values, and that technological choices should take these values more directly into account. They point out that choices about technologies and about the governance of natural resources inherently have a value dimension, and they argue that different values should take priority. They also point out that having resource supplies and management in the control of a group of people that simultaneously benefits from resource use and suffers from any associated environmental damage helps to internalize the externalities of resource decisions.

Renewable energy technologies present many opportunities to use this strategy of promoting

community management, but not the only ones. Water management, as the bioregionalists suggest, can be organized by institutions at the level of watersheds that address environmental problems that arise at the same level, such as water availability and pollution. The management system might be similar to those that Ostrom studied in California that manage water supply at the level of aquifers. And forest management, as the Nepalese experience suggests, can be done at the village level as well as the national, with village-level management having some clear advantages.

The strategy of solving environmental problems by dividing the resources into humanly manageable units is often advocated for its presumed social benefits as well as its environmental ones. A key potential benefit is the strengthening of community. Suppose, for example, that members of a community must interact to manage a reservoir that provides the community's water and electricity. It would be necessary to create a setting—an organization, or a set of town meetings—that would bring community members together to address common needs and goals. This process would increase communication, creating new interactions between individuals who might not have interacted before and thickening the community's social network and thereby making for a stronger community. People would get to know each other better and have the opportunity to build shared expectations and norms, which tends to increase the community's ability to solve other shared, nonenvironmental problems. In addition, situations that create interdependence and a need to solve "superordinate" problems affecting everyone more or less equally are well known to help reduce conflict between individuals or groups (Sherif et al., 1961). As we have already noted, however, strong communities are sometimes repressive of some of their members and inimical to individualist values such as free expression.

WHEN IS COMMUNITY MANAGEMENT LIKELY TO WORK?

Table 6-1 summarizes the conditions under which community resource management is most likely to succeed. These conditions are described earlier in the chapter. The table is ordered roughly from conditions in the natural environment, such as the physical and geographical character of the resources, to social conditions, such as the nature of the community and the rules it devises. The order is only rough, however, because social and environmental conditions can interact. Social conditions, such as national policies, can make a resource that is controllable in principle into one not controllable in practice. Social conditions can also make the environment more controllable, such as when new technology made it possible to perceive and monitor the Antarctic ozone hole and to determine how to control it, or when systems of informal property rights enable the Maine lobsterers to establish and enforce boundaries for "their" resources. Generally, items appearing earlier in the table require larger social forces to change them than those appearing later in the table. This means that if some of the conditions of community management are missing, the chances for success are usually greater if the missing items are toward the bottom of the table.

Large-scale social forces have for generations been weakening two of the major conditions in the table: local resource dependence and the presence of community. The globalization of markets and improvements in transportation have decreased the importance of locally available resources in most people's lives. Fewer people depend on subsistence agriculture or fishing, local energy sources, or locally produced household technology. And when local resources fail, as occurs in periods of drought and famine, people can now be saved by resources shipped from across the world. These trends have brought great benefits to humanity (although large numbers have yet to benefit), but they make environmental conditions less manageable at the local level by decreasing mutual dependence on the local environment. The same social and economic forces have weakened community in many localities, as people seek their fortunes by moving to areas where economic growth provides opportunities. More people migrate, weakening the long-term social interactions and thick social networks that appear to be needed for community management, and also eroding

local knowledge of the properties of the resource base. Economic specialization further thins social networks, as relationships around work, child care, consumption, and so forth become more separate from each other and increasingly based on monetary exchange rather than social obligation. Again, there are undeniable benefits from organizing human life around modern markets, but because this form of social organization weakens community, it makes community resource management more difficult.

Equally threatening to the possibility of community management is the fact that human activity has increasingly altered the environment at the global level, creating new problems such as global climate change, stratospheric ozone depletion, and ocean pollution that are uncontrollable locally because their boundaries are intrinsically global. It is hard to see how community management alone could ever solve such problems, no matter how strong local communities were.

In sum, large-scale social and economic changes have increasingly limited the possibilities for community management. As we have seen, however, there remain significant areas of opportunity, including water supply and solid waste disposal, where resources often meet the conditions of local controllability and resource dependence. There are also many other situations in which the techniques developed by community institutions can be used to supplement informational and incentive-based strategies of behavior change, even when most of the conditions for full-scale community management are lacking.

There is one scenario that might make community management a more viable strategy over time, but it requires some serious shifting of social trends. This is the scenario exemplified by the soft-path approach in energy, in which resource dependence is systematically shifted from global to local supplies, resource-using technologies are increasingly designed and organized to be controllable at the local level, and the decision power is given increasingly to community-level groups. As we have seen, this scenario is imaginable for the management of energy, water, and some other kinds of resources. It would, however, entail a change in social organization that goes somewhat against the grain of recent social trends. It is beyond the scope of this book to try to evaluate all the costs and benefits of this approach. We simply point out that it is a possibility, and that it has the potential advantage of enabling people to use a proven system of resource management that has become decreasingly common over the last century.

CONCLUSION: WHAT CAN COMMUNITY MANAGEMENT ACCOMPLISH?

The community management approach to environmental problems has significant advantages, but like the other major approaches, it is not sufficient by itself. Here we summarize its major advantages and limitations, and draw some conclusions about its appropriate role in preventing tragedies of the commons.

Advantages of Community Management

Community management builds on long-standing social traditions. It is a system of resource management that has served our species well, in an evolutionary sense, for millennia. Whereas formal educational programs and financial incentives are creations of modern societies, which have not yet learned to perfect them, community management systems have been with humanity for thousands of years. We know these systems can work and that they are compatible with human social organization.

Community management internalizes externalities. The problems of externalities—in which some people benefit from the use of resources while creating pollution for others or robbing them of their own resources—cannot occur when the resource and all the consequences of using it are kept within the same group of people. This was the case for a wide range of resources managed communally in the communities Ostrom studied. It is important to note, however, that community management does not always eliminate the externalities. For example, a community upstream in a river valley may manage its water supplies quite well for itself while it lets its polluted water flow downstream. For community management to internalize the externalities, the resource must be a relatively closed system.

Community management can be effective over very long time periods. No other strategy for resource management can claim the centuries-long success that is evidenced in communities like Törbel. The reason is probably that the social system that manages the resources is one that maintains itself. Community management systems contain incentive and educational components, but they are not very noticeable because, rather than being imposed by outside entities such as governments, they are integral to the community. When the controls come from within, the system can last as long as the community, and can evolve to meet community needs.

Community management can encourage people to move beyond selfishness. As we pointed out at the start of the chapter, successful community management is inconsistent with Garrett Hardin's ideas in that people put group needs ahead of narrow self-interest. Yet community management works with the same kinds of people who act as egoists in other social contexts. It is the social character of community management—the participatory process, the creation of a sense of community, and the internalization of group norms—that shapes people to think and act for interests beyond themselves. From many value perspectives, though not all, that is an advantage that puts community management ahead of other strategies that rely on egoism. We return in Chapter 8 to the question of the potentialities for egoism and altruism in human nature.

Community management has low enforcement costs. As we have pointed out, internalized norm-following is much less expensive, in terms of policing, than other strategies of behavior change. Community members in effect police themselves and each other. Such a strategy should therefore be highly attractive in a time when governments experience limited resources.

Community management is the forgotten strategy. Paradoxically, one of the greatest advantages of the community approach to resource management is that it has been so widely ignored. The dominant view in policy analysis is one that looks down on environmental problems, as from on high, and seeks to impose solutions on individuals, groups, or organizations that are presumed to be unable to solve the problems themselves. Garrett Hardin and all other analysts in the Hobbesian tradition presume that some sort of Leviathan is needed to keep people from ruining their environments, and all neglect the lessons from situations in which this assumption was untrue. What this means is that a large number of interventions to solve environmental problems have neglected the principles of community management, and that there may be significant room for improvement by remaking the interventions to be more congruent with those principles.

Limitations of Community Management

Community management works best with a limited range of resource types, such as the coastal fisheries, well-bounded aquifers, grasslands, and forests that Ostrom studied, where resources and pollution can be contained within a small geographic area. Unfortunately, the world's most pressing environmental problems are not like this. Some of them, such as climate change and ozone depletion, are inherently global, so that every community's contribution to the problem is mixed with every other community's. Others are resistant to solution because they present incentives for communities to export their environmental problems to other communities. For instance, many communities can shift their water pollution downstream or their air pollution downwind, take water that is needed downstream, or cut trees in the mountains that provide flood protection in the valleys. In all these cases, the key resources are inherently hard to control locally because community management fails to internalize the externalities. Communities may manage their self-contained resources quite well, yet become involved in a tragedy-of-the-commons dynamic when they share resources with other communities. It seems necessary in these situations for some decision power to lie at a higher level than the community. Although it may be possible to address these problems with nested arrangements among communities or systems of comanagement between communities and larger governmental units,

thus retaining some of the advantages of community management, these strategies are still relatively untested. When a resource affects people outside a community, the community management strategy is insufficient by itself.

Social trends are destroying the conditions for community management. No matter how attractive community management may seem on an abstract basis, it is becoming increasingly difficult because the necessary conditions for it are eroding. People are increasingly dependent on resources that are traded in global markets and are therefore uncontrollable locally, and modernization and migration are making small, stable communities less and less common in the world. As these trends continue, there are fewer locally manageable resources, and fewer groups with the social capital and knowledge to make community management work. What may be desirable is becoming less and less possible, and it may be that in the present era, making community management into a useful strategy for the world's great environmental problems would require nothing short of a social revolution.

Promising Applications of Community Management

The limitations listed above are very serious, indeed. On one hand, the major world environmental problems are not amenable to community management alone; on the other, fewer and fewer communities have the necessary skills for managing the problems that can be handled at the community level. Despite these limitations, we find that community management, or at least principles of it, have great promise for dealing with certain environmental problems and as part of a mixed strategy for dealing with most environmental problems.

Management of locally controllable resources. Land use, water supply, and coastal fisheries, which have been the predominant areas of success for the community management strategy, continue to present important environmental problems, especially in rural areas. Community management remains a highly effective strategy for solving these problems, and governments can do more to provide the conditions conducive to successful community management. In addition, even in modern, urbanized societies, some important resource systems, including water supply and solid waste, are locally controllable and seem particularly suited to community-based approaches. We believe much more can be done to apply the community management strategy to those sorts of problems.

Combining elements of community management with incentives and education. The evidence discussed in Chapters 4 and 5 clearly shows that energy conservation and recycling programs in the United States and other modern, industrialized countries are much more successful when they incorporate elements of the community approach. Incentives and education can be supplemented and made much more effective by relying on word-of-mouth communication and using the resources available to existing community groups, such as their access to audiences and their credibility. Programs that use community institutions and informal social networks are much more effective at spreading their information or advertising their incentives than programs that rely only on mass media or contacts between strangers, and because they are in closer touch with the community, they are better able to identify and make needed improvements. Even though strong communities may be a thing of the past in modernized societies, the kinds of social bonds that existed in those communities still have great influence over individual behavior. In addition, the participatory decision approach that is characteristic of community management turns out also to be effective when decisions must be made on a regional or even a national basis. We believe that for the full range of environmental problems, success will be more likely if the solution strategy used takes advantage of principles such as participation, involvement, creation of norms, and built-in monitoring that proved their value within small communities and are also valuable in larger and more complex social units. Moreover, solving even a few problems with community-based institutions can relieve pressure on the overworked national and international institutions of environmental protection. In Chapter 7, we discuss ways that principles from the community management approach can be combined effectively with the other major solution strategies for environmental problems.

COMBINING THE SOLUTION STRATEGIES

CHAPTER PROLOGUE

A Model Energy Conservation Program

In the spring of 1983, the Bonneville Power Administration, the major supplier of electricity in the northwest of the United States, agreed to sponsor a demonstration project to invest in energy conservation as an alternative to building nuclear or coal-fired power plants. Analysts had projected that within a decade or two, demand for electricity in the region would outstrip supply, and the local power companies wanted to start building new power plants soon, so they could supply the expected demand. Environmental groups argued that it made better sense, both environmentally and economically, to solve the problem by lowering demand than by increasing supply, and they recommended programs to make homes and businesses more energy efficient as the best approach. The electric utility companies did not believe conservation would save as much electricity as the environmentalists claimed, and they argued that because people in energy-efficient buildings could decide at any time to use more electricity, the savings could not be counted on in the same way that the output of power plants can. But neither side had convincing data because no one had yet tried to invest in energy conservation as if it were a power plant.

Bonneville Power, aware that the United States Congress had passed a law in 1980 declaring that power in the northwest must be provided at the lowest possible cost and that conservation must be considered as the equivalent of a power source, agreed to conduct an experiment to see how much a "conservation power plant" would cost and how much electricity it would "produce." The plan was to install improvements in all the electrically heated homes in a single community to make them highly energy efficient, and then to see how much electricity demand was reduced. For the electric utility industry, the experiment was a test of the true cost and reliability of conservation and an aid in planning. For the environmentalists who had called for the experiment, it was a test of how much could be accomplished for the environment if everything possible was done to improve the energy efficiency of homes. For both groups, the experiment would replace theoretical arguments with real experience.

The project was carried out in the town and county of Hood River, Oregon, a community of some 15,000 people about 60 miles up the Columbia River from Portland. Two other towns in Oregon were studied for comparison purposes. The goal was to bring all of the 3,500 electrically heated homes in Hood River up to a very high standard of energy efficiency, with ceiling insulation to the level of R-49 (equivalent to about 15 inches [38cm] of fiberglass insulation), floor insulation to R-38 (about 11 inches [28cm] of fiberglass), triple glazing in windows, and other measures beyond the usual level achieved by well-insulated houses in the region. Bonneville agreed to conduct a community assessment, develop a marketing plan, conduct free energy audits in all eligible homes, and install the recommended home improvements at no cost to the homeowner. In short, Bonneville offered detailed information (the energy audit) combined with a strong incentive (immediate, free installation of water heater insulation and three other minor improvements, and no-cost installation of recommended major home improvements). The project also included a major evaluation component that involved interviews and surveys of the community, analysis of electricity bills, and actual measurements in 320 homes to determine indoor temperatures, appliance usage, and, where wood heat was used, the amount of energy produced by the wood. The evaluation was conducted to get the most accurate information possible about the costs and benefits of the program. It produced some two dozen technical reports and is summarized in a comprehensive report by Eric Hirst (1987), from which this account comes.

The Hood River Conservation Program's great success was to install, within 27 months, major energy efficiency improvements in 85 percent of the eligible homes in the community (93 percent of all eligible homes were contacted by the program, and of these, 92 percent made major improvements). The "conservation power plant" was completed much faster than the ten years it normally takes to build a conventional power plant, and participation levels were far higher than has been documented in other energy conservation programs. The energy savings, however, were less than anticipated. Electricity use decreased only 15 percent in the weatherized homes, compared to the 35 percent savings that the energy audits had predicted. But before we try to interpret the effect on energy use, let us see how the tremendously high level of participation was achieved.

The Hood River Conservation Program (HRCP) took seriously its base in a small community. It was, first of all, a cooperative venture of several local and regional groups, among them Bonneville, the Natural Resources Defense Council, and the two electric utilities—one private, one cooperative—that supplied electricity retail to Hood River homes. Before the program started, Bonneville sponsored a detailed community assessment, consisting of interviews by a consulting sociologist with sixty residents, to gain an understanding of "the community's structure, media channels, local issues, and possible impediments that HRCP might face" (Hirst, 1987). The assessment identified eight major demographic groups in the community that would need to be involved and surfaced some initial concerns of residents about the program, such as a general aversion to "handouts"; a concern that conservation would result in higher electric rates; and a concern about fairness, because the program offered no help to oil-heated homes. It recommended that the program form a Community Advisory Committee, representing a cross-section of the community, that would help educate the community about the program and continue to inform the project staff of community concerns as they developed.

The program began with articles in the local weekly newspaper and the placement of two large billboards with the HRCP logo, proclaiming Hood

River "the conservation capital of the world." Towns-people were also invited to a project Open House in early November 1983. This marketing effort encouraged nearly 1,000 homes to sign up for participation in the first three months—a level of interest that overwhelmed Bonneville's ability to provide energy audits and install the recommended improvements. In addition, over its two-year duration, the program opened a project office in downtown Hood River, enclosed announcements in electric bills, distributed testimonials from satisfied participants, and purchased paid advertising. It also benefited from word-of-mouth publicity from the Community Advisory Committee and other local residents. In short, the HRCP used a multimedia marketing approach that relied heavily on local sources of information. The evaluation showed that more than half the participants learned about the program by word of mouth and that the local newspaper was the second most common source of program information.

Although the Hood River program was tremendously successful in attracting participants, it was not so successful—at least at first—in getting the energy improvements installed. It did not anticipate the need for so many energy audits so soon, and once the energy audits were complete, there were unforeseen problems with the contractors. No local contractor had much experience installing the required levels of weatherization materials, so it took time to develop a simple system for setting prices, and in the first months, half the installations were rejected by the program's inspectors, who called the contractors back to correct their mistakes. Both these situations caused delays, so much so that homes that entered the program early often had to wait more than a year for the recommended work to be done. With time, the program managers improved operations, and the average waiting time dropped to six months. During all this, the program did not report serious dissatisfaction among participants, most likely because the progress of the program was well known in the community, and people understood and accepted the delays.

As the program neared its end, its staff made a concerted effort to reach all the nonparticipant homes and invite them to participate. This was done mainly through direct personal contact, and was quite successful—almost 500 of the nonparticipant homes signed up with the program during that period, bringing the total number of participants to 3,189, of whom 2,989 had major energy improvements installed.

The Hood River program cut electricity use by an average of 15 percent in participating homes. The savings would have been even greater, except for two things. First, most of the participating homes were using wood as a primary or secondary heating fuel when the program began, so some of the saved energy was in the form of wood rather than electricity. Second, households in the community, which had been suffering under rapid energy price increases and locally high unemployment, were able to increase winter temperatures in their homes by an average of $0.6°F$ ($0.3°C$). So, in addition to saving electricity, the program provided the participants with increased comfort, and because wood use declined, it improved local air quality as well.

The program accomplished all this by making a large financial investment—on the average, $4,400 per house. But even analyzed on narrow financial grounds, the money was well spent. Conservation saved 1 kilowatt-hour of electricity for every 7.1 cents it spent, and although this was more costly than the average kilowatt-hour generated in the Bonneville Power system at the time, it was less costly than the new generating capacity it was intended to replace. The program was, therefore, a financial bargain for the utilities compared to the alternative, in addition to its obvious benefits for homeowners and for the environment. If the environmental externalities of building a new power plant were included in its cost along with the dollar expenses, conservation would look that much better by comparison.

In terms of our main concern in this part of the book—changing behavior that is harmful to the environment, the Hood River Conservation Program was an almost unqualified success. The evaluation research attributes the success to a number of factors. First, the commitment of Bonneville Power and Pacific Power and Light to achieving 100 percent participation was important because it provided support and resources for the project staff and gave them a level of autonomy and flexibility that allowed them to build a can-do spirit of teamwork. Also, the recruitment of some households for special projects in advance of the HRCP (such as having their homes monitored for energy use) helped by offering advance word-of-mouth publicity within the community. In addition, creating a Community Advisory Committee before the program began and operating it continuously helped both in spreading the word and in keeping the project staff apprised of community reactions. This participatory feature of the program design maintained close interaction between program personnel and the community and helped the program respond to community needs in a quick, timely fashion and at low cost. The small size of the community may also have helped make the Community Advisory Committee more effective and raise participation rates, because word of mouth is more likely to reach everyone in a small community than a large one. Of course, success would not have been possible without the huge financial incentive provided by the company's willingness to pay all the costs of major weatherization. In short, the project combined a strong incentive with sophisticated use of information sources and an effort to mobilize community participation, support, and solidarity. It drew to some degree on all the major strategies of behavior change described in Chapters 4, 5, and 6, and the result was a level of participation not recorded in programs that rely mainly on only one or two of these strategies.

A Success Story on Recycling

The city of St. Paul, Minnesota (population 272,000), has a private system of trash pickup. Thirty private trash haulers compete for customers, and customers can shop around for the lowest price. This privatized system may save the city and its residents money on trash pickup because of the efficiencies of competition, but it creates a disincentive for recycling, because although recycling would save money for the city by reducing demand on its disposal system, it costs money for the private haulers who would have to collect the trash. Of course, recyclables have cash value, but as of the mid-1980s, sorting and preparing

them for recycling actually cost more in St. Paul than what the haulers would earn. So the private haulers in St. Paul have no incentive to offer curbside pickup of recyclables, much less to promote waste reduction. Yet despite these disincentives, St. Paul has an effective and growing recycling program, thanks to the efforts of a coalition of neighborhood groups.

The Neighborhood Energy Consortium is an umbrella organization representing St. Paul's nineteen neighborhoods and committed to making neighborhoods more self-reliant and increasing their citizens' influence on public decisions in the neighborhood and beyond. It was organized in 1982 to promote energy conservation in the city, but its directors were continually looking for ways to meet community needs. In January 1986, the annual membership developed the idea of offering recycling services, which a previous group had abandoned when the price of recyclables hit one of its periodic lows. The city and regional governments were supportive, and with some seed money from the Metropolitan Council, a seven-county regional planning group, the city subcontracted with the consortium to start a recycling service. Eventually, the city made the system permanent, paying for it through property taxes that raised about $7 per person per year by 1993.

The program started in a few neighborhoods at a time, and spread throughout the city, having modest success simply by appealing to citizens' environmental concerns. It helped that St. Paul and the consortium are organized by neighborhoods. Nineteen elected neighborhood councils pass on zoning, traffic control, and related issues before they are decided by the City Council, and provide a place for people to discuss and organize around issues of local importance. The councils organize block clubs for crime prevention, block nursing services, and other beneficial programs within the neighborhoods, and each council appoints a member to the consortium's board of directors. So, the consortium has close ties to the grass roots. By passing the word through the neighborhood councils and the block clubs, and by announcing the program in its newsletter, the consortium was able to enlist a number of block leaders who were willing to put signs on their lawns advertising the recycling program and to provide information and encouragement to their neighbors by word of mouth. Over time, the consortium also created a telephone service to answer questions about recycling, published and distributed brochures in English and the languages of major immigrant groups in the city, and paid for part-time recycling coordinators on each neighborhood council's staff to promote block clubs, recruit volunteers, and expand the program. Using this system of marketing, the consortium was able to involve 15 to 20 percent of the households in the city in recycling.

To increase the level of participation further, the consortium asked the city to change the incentives for trash hauling by instituting a pay-per-can system for ordinary trash. The city agreed in principle, but was unwilling to impose a fee structure on the private trash haulers. The compromise decision was to require trash haulers to offer different prices for the pickup of one, two, or three trash cans per week or for unlimited pickup—but the city did not dictate what the prices would be. Most of the haulers settled on very small price differentials—for example, if they charged $15 or $20 per month to pick up one can per week—a typical fee—they might charge an extra $1 or $2 per month for two cans weekly. From the trash hauler's point of view, these fees may have made sense because much of the cost to them is for labor and fuel to pick up and move the trash, and not for the amount of trash in the truck.

The pay-per-can system would have provided only a very small financial incentive for recycling, except for one thing. Each household in St. Paul is not required to hire a trash hauler. The consortium started using its network to point out to the citizens that two or more households could combine their trash and only pay the hauler once. That way, a family could save $100 or so per year by reducing the volume of its unsorted trash. The savings were especially easy to achieve, and valuable, for low-trash-producing households, such as single people and senior citizens. Information about how to use the incentive system to best advantage started circulating rapidly through the neighborhood and block networks, and the number of households using the recycling system increased to between 60 and 70 percent of the city. In 1992, the program collected 16,750 tons of recyclable materials, or 123 pounds (56 kg) for every citizen in St. Paul.

The consortium continued working to expand its program by encouraging participants to recycle more of their trash and by extending the service to other kinds of household waste. As of late 1993, it was experimenting with a program of collecting used household goods and clothing and delivering them to Goodwill Industries for resale. Plastics, however, were not part of the recycling program because collecting them was still uneconomic, even for a nonprofit organization.

It is interesting to note that a high level of success was achieved in spite of a certain amount of built-in inconvenience. Although the consortium's haulers pick up trash at the curbside in convenient bins, recyclables are not picked up on the same days as other trash because the haulers are different. People must remember (or consult the lawn signs) to put the recyclables out on the correct days.

It is also worth noting that the Neighborhood Energy Consortium is more than just a recycling program. It is a citywide group with a broad interest in energy and environmental policy, a base in the neighborhoods, and a track record for getting involved in environmental politics at all levels from the city block up to national policy. Through the newsletter, the block leaders, and the neighborhood coordinators, the consortium sometimes alerts citizens to policy issues affecting recycling, and urges political action. For example, in 1992 the consortium participated in a national drive to take the recycling logos off plastic containers that cannot be recycled because there are no buyers for them. (Note that this involved a significant educational effort, because many people mistakenly believed that plastic containers that have a recycling logo on the bottom will actually be recycled if they find their way to a recycling center.) The drive included letters, petitions, and direct action involving mailing plastic containers to the national headquarters of the Society of the Plastics Industry. Such efforts can be controversial, but the consortium's board of directors, with a representative from each neighborhood, ensures that its programs and political actions have wide support in the community. This base of support is vital to the consortium's programs. It is one reason that the pilot recycling program has become a regular service in the city, funded from property tax receipts. People from all over the city have shown their support for the program as a significant public service.

Like the Hood River Conservation Program, the St. Paul recycling program used a combination of strategies to achieve its success. It relied, first of all, on the proenvironmental values and attitudes of the citizens, which generated the idea for the program, helped publicize it, and helped it secure volunteer labor and household participation. It provided a financial incentive—the opportunity to save on trash pickup fees—and it helped citizens achieve substantial savings by pooling their trash. The program used block and neighborhood social organizations to keep in touch with the people whose behavior it aimed to change, to make the program responsive to community needs, and to spread information. Through a coalition of local groups, it was able to achieve community-based action in a city much too large for face-to-face contact. And it engaged in political activity to change the incentive structure for waste disposal in ways that made it easier for people to engage in proenvironmental activities they favored. Like the Hood River Conservation Program, success was based on a combination of the strategies of behavior change discussed in the previous chapters.[1]

LESSONS OF SUCCESSFUL ENVIRONMENTAL PROGRAMS

What can be learned from Hood River, St. Paul, and the other successful efforts we have described, and the partial successes and complete failures of many other programs? What, specifically, can a program manager or an environmental activist do to organize an effective program to achieve a particular environmental goal?

In Chapters 4, 5, and 6, we make a few generalizations that apply to specific types of programs. For example, in recycling programs, inconvenience was a

major barrier to participation and curbside pickup has been a very effective solution in many communities. Similarly, in home weatherization programs, mistrust of contractors was a significant barrier to participation and independent inspection of contractors' work has been a good solution. It would be valuable to have knowledge at a higher level of generality—knowledge that applies across a broad range of behavior change programs. Although there has not been enough research to solidly establish such generalizations, this section offers some general principles that we believe are useful—necessary, though not sufficient—for designing effective programs to change environmentally destructive behaviors.

The most important thing to emphasize, before listing the principles, is that it will never be possible to write a "cookbook" for behavior change that gives specific instructions for intervention. This is true in part because situations are unique—the barriers to behavior change vary, even for a single behavior, from place to place, time to time, and individual to individual. But there is a deeper reason a cookbook can never be written: Human beings are continually responsive to interventions in their lives—even to the point of evading, blocking, or organizing to repeal the ones they find most objectionable. Treating people as inert objects to be moved is a good way to encourage resistance to any sort of effort to change behavior. A good example may be the efforts to solve the municipal solid waste problems of U.S. communities with incineration. Typically, a waste management company proposes building a trash incinerator and tries to convince local governmental officials that it would be both safe and economically valuable to the community. Almost invariably in recent years, suspicious local citizens have organized strong, and frequently successful, opposition fueled in part by indignation that the waste company had tried to avoid public input. The process becomes as important an issue as the merits of the incinerator itself. Because of the inevitable interplay between change agents and those whose behavior they intend to change, some of the most important principles of intervention concern the relationships between these groups. Table 7-1 summarizes our principles for environmental intervention. We explain them below.

TABLE 7-1 Principles for Intervening to Change Environmentally Destructive Behavior

A. Use multiple intervention types to address the factors limiting behavior change
   1. Limiting factors are numerous (e.g., technology, attitudes, knowledge, money, convenience, trust)
   2. Limiting factors vary with actor and situation, and over time
   3. Limiting factors affect each other (interactive principle)

B. Understand the situation from the actor's perspective
   1. Conduct surveys or experiments
   2. Participatory approach to program design

C. When limiting factors are psychological, apply understanding of human choice processes
   1. Get the actors' attention; make limited cognitive demands
   2. Apply principles of community management (credibility, commitment, face-to-face communication, etc.)

D. Address conditions beyond the individual that constrain proenvironmental choice

E. Set realistic expectations about outcomes

F. Continually monitor responses and adjust programs accordingly

G. Stay within the bounds of the actors' tolerance for intervention

H. Use participatory methods of decision making

A. Use Multiple Intervention Types to Address the Factors Limiting Behavior Change

Because there is often more than one barrier to any proenvironmental behavior, programs that combine different types of intervention tend to work better than those based on a "single cause" theory. The Hood River Conservation Program is typical of successful environmental programs in this respect. It combined three major strategies of behavior change—information, incentives, and community management. It successfully employed some of the principles of community management (such as community involvement and word-of-mouth communication) even though certain ideal conditions for community management (such as local control over the scarce resource) were not present. It used information from multiple sources to reach individuals who paid attention to different kinds of people and to the local mass media, and it bolstered its large incentives with efforts to increase the program's credibility and the community's trust. The program's success cannot be reduced to any single part of the package.

The St. Paul recycling program also combined all the major strategies. In the area of community involvement, it even went so far as to use its political power to change the financial incentives for recycling by pressing for a pay-per-can system, and it used its community network to show people how to take advantage of the new incentives by pooling trash to save disposal fees.

We have seen repeatedly in previous chapters that the greatest degree of behavior change occurs when different strategies combine. The research on energy-use feedback described in Chapter 4 showed that this informational technique is more effective when energy costs are a large portion of the household budget—in other words, when there is an incentive to use the information. In the block-leader recycling effort in Denver that was studied by Hopper and Nielsen (1991), success increased with the number of behavior change strategies used. Carefully designed information greatly increased participation in a curbside recycling program that had as its main feature a change in the incentive structure to make behavior more convenient. The addition of block leaders—an application of principles of community management involving the activation of personal and social norms—increased participation even further, possibly also making the behavior change more permanent. The same sort of thing happened in the St. Paul program. Participation increased dramatically when a financial incentive was added to an existing, community-based program. In Chapter 5, we found that time-of-use electricity pricing, an incentive system, becomes significantly more effective when supplemented by carefully designed information, and that major incentives for energy conservation were much more effective when the organizations in charge made serious efforts at marketing. The FACE weatherization program in Fitchburg, Massachusetts, attributed its success to a combination of incentives, information, and community support, much like the combination that worked in the more ambitious Hood River program. And there are other examples, in the previous chapters and elsewhere. Clearly, combining intervention strategies is one of the major lessons of successful environmental programs.

It is worth explaining in some detail why it helps to combine intervention types. First, as we have seen, the barriers to proenvironmental behaviors are almost infinitely varied: expense, inconvenience, government regulations, inappropriate values or attitudes, structural barriers, misinformation, lack of commitment, and so on. To put it another way, people do destructive things to the environment for many different reasons that are compelling to them, and that may override personal concern for the environment in the particular instance. For proenvironmental behaviors to occur where they have not before, a lot must go right.

It helps to think of the problem of bringing about environmentally desirable behavior the way botanists think about plant growth or biochemists describe metabolic processes: in terms of a pathway such that breakdown of the process at any point can prevent the expected product from appearing. Plant scientists talk about limiting factors in plant growth: sunlight, water, the right soil pH, and a number of essential nutrients. Given less than ideal amounts of any one of these factors, plant yield is far below optimum. Providing what is missing will increase yield sharply,

until some other limit is reached. From this point of view, the farmer's tasks are to diagnose the situation—that is, to find the missing factor or factors—and then to supply them. An analogous process occurs for metabolic disorders, in which the absence or shortage of a particular vitamin or enzyme drastically alters metabolism and threatens health and life. The diagnostician's task is to identify the malfunctioning step in the process, and the physician's task is to choose a treatment accordingly.

It is more or less the same for proenvironmental behaviors. Approaches that assume that the absence of a behavior always has the same cause and that the same type of intervention is always needed—whether it be incentives, education, or whatever—are likely to be wrong, or at least partially wrong, most of the time. It is more helpful to begin by presuming that we do not know what sort of intervention is needed. Try first to identify the limiting factors for the behavior—what we have called barriers in previous chapters—and begin by supplying the factor that seems most deficient. If this approach fails, it may be more helpful to add another factor than to use more of the same. This was a lesson of the residential conservation initiatives: Sometimes, more could be gained by improving marketing than by increasing an existing incentive.

Again, we use the term *limiting factors* more or less as a synonym for the term *barriers* in Chapter 5. It serves as a reminder that the key to behavior change varies from one situation to another, just as the needs of a crop may vary from one field to the next. This variation is another reason it is impossible to write a recipe for behavior change. Because of subtle differences in incentive structures and between local communities, the limiting factors for behavior change need to be diagnosed, not presumed. A useful general approach in this sort of situation, especially when the time or money for diagnosis is limited, is to combine several intervention strategies and then observe the response, while continuing to search for the best combination.

In Chapter 4, Table 4-2, we gave a schematic representation of the pathway to environmentally desirable behavior. We reproduce Table 4-2 in modified form as Table 7-2, showing the kinds of intervention that can address barriers, or limiting factors, at different steps in the pathway. The table illustrates that some of the factors affecting behavior are not amenable to intentional change (level 7, household background) and that some can be changed only indirectly and over long periods of time (level 5, values), and it emphasizes that external context (level 6) and social and psychological processes (levels 5, 4, 3, and 2) can act independently as limiting factors on proenvironmental behavior. The following points about limiting factors deserve emphasis:

Limiting factors are numerous. Table 7-2 makes quite clear the variety of barriers or limiting factors at various steps in the pathway to action. Even at a single step, such as level 6 (external incentives and constraints), the limiting factors may be quite varied— including lack of access to technology, legal and regulatory barriers, lack of money or financing, and inconvenience, among others. Energy conservation programs are a good example. Financial cost and the difficulties of getting useful information and finding a reliable contractor are all disincentives for many households, and renters have the additional disincentives associated with the fact that the home is not theirs to improve. Similar variety exists in the limiting social and psychological factors, a fact that explains the success of programs that provided some combination of information, instruction, models to imitate, and efforts to elicit a commitment to change. Because

TABLE 7-2 A Causal Model of Resource-Consumption Behavior with Examples from Residential Energy Conservation

LEVEL OF CAUSALITY	TYPE OF VARIABLE	EXAMPLES
7	Household background	———
6	External incentives and constraints	**Incentive structure** (Chap. 5) (financial incentives, altered regulations, improved convenience, improved technology)
5	Values	**Value change** (Chap. 3) (social movements to change values)
4	Attitudes and beliefs	**Attitude change** (Chap. 4) **Social influences** (Chap. 6) (activate personal norms; build personal and community commitment; build social norms; monitor behavior; participatory management)
3	Knowledge	**Information** (Chap. 4) (match information to consumer's situation; provide it at point of action)
2	Attention, behavioral commitment, etc.	**Information** (Chap. 4) **Social influences** (Chap. 6) (memorable presentations; word-of-mouth; credible sources; framing)
1	Resource-using or resource-saving behavior	———

Source: Adapted from Stern and Oskamp, 1987.

proenvironmental behaviors are often blocked by more than one limiting factor, all the important ones must be addressed for an intervention to succeed.

Limiting factors vary with actor and situation, and over time. Again, residential energy conservation is a good example. Some people (renters) have limited options because they are not allowed to make changes in someone else's building; among homeowners, some are limited mainly by finances, others by know-how, others by conflicting demands on their time, others by mistrust of those available to install the needed improvements, and so on. The picture is very complicated, and it changes with the price of energy, the availability of government programs, the free time available to household members, and many other factors. Moreover, the determining factors in human-environment interactions often have nothing directly to do with the environment and may be very specific to the individual. An example in Chapter 5 was Paul Stern's trip to work. His decision to combine mass transit and driving was affected by the specific locations of his home and workplace, an aversion to driving in traffic, and his asthmatic condition, among other factors—and these factors carried more weight than his environmental concern. This sort of complexity is typical for human interactions with the environment.

Limiting factors affect each other. It is in the nature of limiting factors that the effectiveness of eliminating any one of them depends on the presence or absence of the others—in the language of experimental research, they interact. To return to the example of plant growth, adding fertilizer does no good unless a plant has water and light. Similarly, raising the price of home heating fuel encourages people to conserve, but it will have only a negative effect on renters who are not permitted to reinsulate walls and ceilings, or on people who do not know how to save energy in their homes. They will be forced to choose between paying more for the same comfort and sacrificing that comfort. The effect on fuel consumption in these homes will be limited, and purchased at high personal cost.

A useful way to look at the principle of interactive effects is to consider how the effects of incentives depend on social and psychological processes in individuals or groups. As we pointed out in Chapter 4, the effect of attitudes and beliefs on energy conservation behavior depended on the incentive structure for the behavior. With strong disincentives for action, that is, when the behavior was very difficult or costly, proenvironmental attitudes made little difference. To put this another way, disincentives acted as a barrier to the expression of proenvironmental attitudes. A study of recycling by households in Alberta, Canada, shows what can happen when such a barrier is lifted. The city of Edmonton was the first in Alberta to institute a "blue box" curbside recycling program, intended to make recycling more convenient. Sociologists Linda Derksen and John Gartrell studied the program (Derksen and Gartrell, 1993), focusing on two things: behavior change directly attributable to the program, and behavioral differences as a function of an attitude of environmental concern. They found that the blue box program was responsible for a large increase in self-reported recycling in Edmonton, as it has been in other areas. Environmental concern, however, influenced behavior only for people who had access to the curbside recycling program. People in Alberta expressed strong environmental concern generally, but variations in individual environmental concern were unrelated to recycling behavior in most of Alberta. They made a difference only in those households served by the curbside recycling program—the single-family homes in Edmonton. Environmentally concerned people in those homes reported recycling an average of four different kinds of items, compared to about two and one-quarter for people in similar homes with low levels of environmental concern. Levels of environmental concern made no difference in multifamily buildings in Edmonton or in single-family homes elsewhere in Alberta.

These findings show that the behavioral effect of individual attitudes and beliefs (and probably also values, knowledge, and commitment) can vary dramatically with the incentive structure that provides the context for behavior. In Edmonton, when the barrier of inconvenience was lifted, people readily expressed their proenvironmental attitudes by recycling; when the barrier was in place, they usually did not.

The experience of environmental incentive programs, discussed in Chapter 5, shows another side of the interactive principle: Introducing an incentive often fails if the barriers to action are psychological characteristics of individuals. For instance, when people did not know or trust a home energy conservation program, they usually did not respond even to a substantial incentive, but when the program was marketed properly, they quickly took advantage of incentives that benefited them.

The political sphere provides striking examples of how individual and social processes can interfere so strongly with incentives as to make them impossible to employ. One example is the consistent political opposition to increased gasoline taxes in the United States, which is due to a combination of opposition from the oil and automotive industries, resistance from consumers on grounds of cost, and a widespread positive attitude about automobile driving, which many Americans consider a "right," combined with a negative attitude to government "infringement" of this right. These attitudes, expressed through political activity, have helped keep any major increases in the cost of gasoline from becoming part of U.S. energy policy.

These examples show that the effects of any type of intervention, such as a change in the incentive structure, can be large, small, or nonexistent, depending on other factors. This fact raises some critical practical questions for the design of interventions to change behavior: How can one anticipate whether an intervention that is being considered will have a large or small effect in the conditions given? And how can we tell whether an intervention of one type needs to be supplemented with another type of intervention?

One inference is that when proenvironmental attitudes are strong, removing disincentives can produce large increases in behavior consistent with those attitudes. This seems to be a lesson of the Edmonton experiment in curbside recycling. One might also infer that when proenvironmental attitudes are weak, changing the incentives to favor proenvironmental

behavior will have relatively little effect. An example comes from some rural areas that have instituted fee-per-can recycling incentives. Certain individuals, who want to avoid the fees and who do not have strong environmental concern or do not believe careful disposal is necessary, avoid paying the fees by illegally burying or burning their trash. It would take different attitudes, better knowledge, or stronger incentives to change their behavior, and strong incentives might be politically unacceptable.

Despite plausible generalizations like these, it is not easy to tell in advance whether attitudes are positive enough or knowledge great enough for an incentive of a certain kind and magnitude to work well. And it is not easy to tell in advance whether an incentive is at a level where the best way to make it more effective is to emphasize information and marketing, or at a level where the most effective action would be to increase the incentive. It is easy to survey attitudes or measure the cost of incentives, but no yardstick has yet been developed for comparing the two in order to solve this problem.

We therefore recommend combining solution strategies as a way to deal with the multiplicity of barriers, situations, and individuals, and to help overcome the difficulty of knowing which is the optimal intervention type for any particular situation. If a program supplies a single limiting factor, response soon bumps up against another one, so it is best to design an intervention that provides a mixture of the factors likely to be in short supply. The fact that different individuals face different barriers and have different psychological structures further strengthens the point. Although it is impractical to tailor interventions for each individual, it is sensible to design a program that covers all the barriers that are faced by significant numbers of individuals in the target population. In this way, each individual can take advantage of the program element that is most relevant for his or her situation. It is not necessary to understand each individual's incentive structure or attitudes, beliefs, and level of knowledge if the program is designed to include information for those who need it in forms they find useful and incentives that counter various common types of disincentives.

In the Hood River program, this sort of multi-pronged approach was used to distribute information. The program rented billboards, generated local press coverage, and spread the word of its existence through a community advisory council and by word of mouth. In this way, no matter how an individual preferred to get information about new events in town, he or she would hear about the program. To reach those who still did not respond, program representatives went door to door. Of course, the program also provided strong incentives for participation, which made the information worth having. Hood River covered all the possibilities with its incentives as well. The program supplied a contractor to install the recommended improvements, paid for them in full, and inspected the work to ensure its quality. Other conservation programs with more limited resources have sometimes offered households a choice of a low-interest loan or a rebate of equivalent value, thus allowing them to choose an incentive to suit the household's financial situation.

In applying the principle of combining types of interventions, there still remain the important practical questions of which kinds of incentives, education, community interaction, and so forth to use in any particular situation, and how to combine them. Is there some way, short of offering every imaginable type of intervention at once, to design a package of interventions that will be well matched to the situation? This question brings us to the second principle:

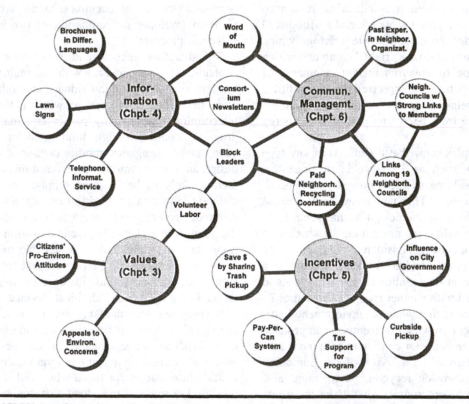

ILLUSTRATION 7-1 Some Ways the St. Paul Recycling Program Combined Strategies of Behavior Change

B. Understand the Situation from the Actor's Perspective

The people whose behavior is to be changed—the actors in the behavior change drama—are in the best position to identify the barriers they face. Although they may not be immediately aware of all the barriers, they are in the best position to know them. Two strategies can reveal to program designers what the actors know. One, a social-scientific strategy, uses surveys and experiments. It is possible, for example, to survey a representative sample of the target group to ask why they do or do not recycle, insulate, reduce the use of water on their lawns, or whatever. And it is possible in a survey or interview format to propose various program designs to them and see which versions are most attractive. The results would be useful for designing the initial version of a program or intervention, but they can be misleading because people do not always know what will affect their behavior until they find themselves in the actual situation. One can usually get more valid data by actually trying the planned intervention—exposing some of the actors to different program designs in a pilot experiment and observing actual responses. This is a common strategy in testing new consumer products, and it was the strategy used in the marketing experiment for the energy service program project in Minnesota (Chapter 5). This experimental approach has the advantages of rigor that come from the ability to use random sampling, quantification, statistical analysis, and experimental control, but it can be expensive, especially if it must be repeated or if the program has a very limited budget.

The other strategy is participatory—it gets the actor's perspective by using social interaction rather than scientific control. It begins by including representatives of the target group in the program design team and continues by relying on informal feedback from the target population, through these representatives and other channels, to tell what is and is not working in the program and to help modify the program to meet consumer needs. This participatory approach to program design is what the Hood River and FACE programs implemented when they created,

and listened to, their community advisory groups. It was built into the St. Paul recycling program, which was created and managed by a community-based organization. The program's staff were, in effect, employees of the target group, and so had the strongest of incentives to listen.

The participatory strategy purchases flexibility and speed of feedback at the cost of scientific rigor. It also potentially gains another benefit. By involving the community, this approach often increases trust, satisfaction, and even participation levels, especially if the target-group members who are involved include important members of significant subgroups within the target population (such as community group officials or well-respected individuals in their neighborhoods, or officers of service clubs). Having such individuals involved serves as a signal of importance and credibility to the wider community and creates an effective channel of communication. (Of course, community leaders and trusted organizations are more likely to support efforts that they have determined will benefit the people they care about.) The participatory approach thus serves two purposes: It gives program designers the knowledge they need to improve the program continually, and it implements one of the principles of community management that can make programs more effective.

The participatory strategy is based on considerable experience with public participation in government program and policy design. A key to effective public participation is establishing effective two-way communication between the officials who run the program and the people it is intended to serve. Another key is that the latter group must have some meaningful influence over the direction of the program or policy for it to be effective. If a program is unresponsive to people's needs, they can simply decline to change their behavior—or, to look on the positive side, behavior change is most likely to occur if the target group is committed to the program designed to help it change. Principles like these have been underlined by the bitter lessons of expensive failures to get public approval for building hazardous waste facilities and nuclear power plants (e.g., Kasperson, Golding, and Tuler, 1992), but they apply to a much wider range of environmental policies as well (e.g., Syme and Eaton, 1989). The participatory principle reaches its highest level when the program managers are directly answerable to a community group, as in the St. Paul recycling program. Because participation has advantages beyond getting the actor's perspective, we discuss it below as a separate principle.

C. When Limiting Factors Are Psychological, Apply Understanding of Human Choice Processes

Chapters 4, 5, and 6 illustrate a variety of ways to make information, incentives, and community management effective by applying knowledge that is fairly well known to social and behavioral scientists but generally unused by environmental policy makers. These insights can be divided roughly into the cognitive and the social.

Possibly the central insight of cognitive psychology is that of bounded rationality (discussed further in Chapter 9). People have limited attention and information-processing abilities. From that insight follows the need to design interventions so as to attract audience attention, to make choices simple and straightforward, and to focus on the most important concerns affecting consumer choice. Doing this decreases the effort necessary for someone to notice and assimilate the information being sent. In energy conservation, these considerations point to the importance of program marketing, one-stop shopping, and efforts to ensure the reliability of anything the program puts into a consumer's home. In recycling, they suggest that pickups should be made convenient and that rules for sorting waste should be as clear and simple as possible. For other proenvironmental behaviors, the implications will of course be somewhat different, but the principle offers useful guidelines for initial design that can then be refined on the basis of experience.

Relevant insights about social influences emphasize the importance of commitment and active participation and the social processes that make communities work: trust, informal communication and conflict resolution, credibility, obligation, and norms. Applying social psychology means remembering that human choice is not a straightforward mechanical process of calculating what is best for the individual. People can be affected more by a friend's experience than an expert's judgment, by neighbors' expectations than by personal inconvenience or cost, by a sense of the common good or the moral thing to do than by self-interest. Although these social influences do not by any means always outweigh other considerations, they can be potent forces, especially when they push people in the same direction as incentives or information. An environmental program that finds creative ways to let social interaction do its work—by feeding information into informal communication networks, appealing to people's desires to benefit their communities, and so on—can be much more successful than one that ignores people's social contexts. A look at Hood River, FACE, St. Paul, and other highly successful programs shows the ways they have harnessed social interaction; the same general approaches can be applied in new programs as well.

D. Address Conditions Beyond the Individual That Constrain Proenvironmental Choice

Sometimes, the best way to change individual behavior is indirectly, by changing conditions that are far beyond individual control, but that limit individual choice. Recall, for example, the difficulties of getting North Americans out of automobiles for their travel to work. Many a committed environmentalist drives a car to work out of experienced necessity: Public transportation is frequently unavailable or extremely inconvenient compared to driving, and the demands of work and family can make flexibility and time saving highly valued attributes of a mode of transport. Such an environmentalist has favorable attitudes and all the necessary information. Obviously, the incentives are wrong, but there is only so much that can be done by changing the incentives at the level of the individual. As Chapter 5 shows, it helps to assign highway lanes to carpools, restrict parking near workplaces, and give bonuses for riding public transport. Even though some of these proposals are difficult to implement because of vigorous opposition from people who see driving to work as a necessity, they are in the realm of the possible. People can choose to ride in carpools or on public transit.

But people are not able to make some of the choices that could have the greatest environmental benefits because choices made by others have limited the options for individuals. Incentives to individuals cannot change such situations. Some examples illustrate. Buying highly energy-efficient cars could save great amounts of gasoline, but manufacturers do not build the most efficient cars technology allows, because they believe the cars would not be sold. Incentives to individuals cannot change this situation, but incentives to get automobile manufacturers to produce highly energy-efficient cars might drastically change individuals' purchase behavior in a proenvironmental direction. Having people live closer to work would also save tremendous amounts of fuel, but many individuals feel they have little choice about where to live, and strongly resist efforts to make them move their homes. Over the long run, changing subsidies for highway construction and suburban housing might eventually reverse the trend toward the geographical dispersal of residences, workplaces, and commercial buildings that increases travel distances and makes cars a necessity for so many people. Again, this effect would occur without directly targeting individuals for change. If, over time, cars become no longer a necessity, people might then be less opposed to increasing the cost or inconvenience of using them.

People can sometimes change incentive structures through political action. In St. Paul, for example, the Neighborhood Energy Consortium convinced the city government to let it organize a curbside recycling program in a situation in which there were no incentives for private trash haulers to offer one and where a previous program had collapsed. To make the program work better, it lobbied for a municipal fee-per-can trash pickup system that changed the incentives for households. Even though the group did not get all it wanted in terms of incentives, it was able to create a strong recycling program to serve people who did not have curbside recycling as an option before.

Another kind of effective political action ended Styrofoam packaging at McDonald's, as discussed in Chapter 5. Because customers had no choice about the packaging material, it is hard to imagine a program of incentives or information targeted at individuals that would have been nearly as effective in reducing the use of Styrofoam as the manufacturer's decision to change its product. It is worth emphasizing that the McDonald's decision shows that people can give themselves new choices by changing the incentives for the large organizations that offer those choices. When consumers and environmentalists organized boycotts and publicity against the use of Styrofoam, McDonald's had a new economic incentive to change its practices, with the result that fast-food customers who wanted paper packaging could find it easily.

Psychologists and others who focus their attention on individual behavior often forget that the behavior of corporations, governments, and other large social institutions is sometimes a major barrier to change in individual behavior and a key to solving environmental problems. Even when individual behavior is the final target of change, the best strategy may be to work on the institutions that select the alternatives available to individuals. The design of interventions needs to result from analyses that ask:

Which actors make the most difference?
Which of their behaviors make the most difference?
Are there important indirect influences—institutional or organizational choices that control the options available to individuals—that can provide handles for behavior change?

We address these questions in more detail in Chapter 10.

E. Set Realistic Expectations about Outcomes

Because it usually takes some time and effort to identify the key barriers or limiting factors affecting a desired behavior and to find effective ways to overcome the barriers, it is a mistake to anticipate that any intervention will accomplish all that is desired from the start. Moreover, as the Hood River program and many others demonstrate, there are often technical problems that arise in program implementation that take time to solve. In Hood River, it took time to find a quick way to set the price of the recommended conservation improvements and for the contractors to learn proper installation methods; in Fitchburg, it took time to determine where to locate neighborhood centers for the greatest effect; for the Alanya fishermen, it took years to arrive at a fair way to divide the fishing grounds throughout the season; and for recycling and carpool-lane programs, as well as many others, constant effort may be needed to keep ahead of individuals who try to evade the system. In short, learning takes time. It is reasonable to expect only modest success in the beginning of an intervention, until the limiting factors are fully understood and the tricks of the trade of implementing the program have been learned. Although it is possible to learn useful lessons from the experience of other programs (for instance, that curbside recycling is cost-effective compared to the use of drop-off centers), experience unfortunately shows that the limiting factors and the tricks of implementation are often uncovered only by trial and error.

This may seem a simple point, but it often goes unrecognized. All too often, environmental programs are judged too early, as if the first version of the program is the best that can be done. Program proponents sometimes make this problem worse when, in order to get political support for carrying out a program in the first place, they make ambitious promises, hoping—unrealistically—that they can quickly find an ideal way to operate the program. Making inflated promises and holding programs to unrealistic expectations in the short term can result in premature termination of a potentially valuable program and can hurt the credibility of environmental activists when they propose future programs. Both environmental activists and those who doubt the value of their proposals need to understand that new ideas take some time to be perfected, and that evaluation should be an ongoing process that accepts the reality that programs learn and that sets reasonable expectations for each phase. It is worth keeping in mind that people-based programs are often expected to make dramatic progress in a matter of months, whereas the hardware-based programs they are intended to replace, such as electric power plants and trash incinerators, take years just in the construction.

F. Continually Monitor Responses and Adjust Programs Accordingly

Some of the best examples of this principle in operation are found in the community advisory processes used by the FACE and Hood River energy programs. In both of these programs, parts of the initial design did not work well, and feedback from the field led the managers to make the necessary changes. In Fitchburg, it was learned that some of the neighborhood offices were working well while others were wasting people's efforts. In Hood River, it was learned that the difficulty of defining the cost of energy improvements was causing major delays in the program. Correcting these problems was a key to the programs' success, and we believe it illustrates a general principle: that those who would promote proenvironmental behavior should remain flexible in their plans.

This principle is consistent with several points we have already made: The difficulty of knowing in advance what is the best combination of intervention strategies for a particular situation, the variability of situations, the need to keep in touch with the situation from the actor's perspective, the fact that interventions must learn in order to become effective, and so on. Flexibility and constant monitoring are made even more important by the tendency of people to respond to efforts to change their behavior and to conditions external to those interventions. It is important to keep in touch with people's reactions to intervention programs, and also to help a program adapt to changes in people's responses to external conditions (for example, changes in the local economy or in the focus of community concern).

All these considerations imply that interventions be thought of as experimental. Each phase of a program, even one that seems to be stable and well established, needs to be considered as a social experiment and to be evaluated to assess whether it should change to meet changing conditions. As with the process of identifying the limiting factors, there are two basic strategies available: standard social science research, and a participatory approach involving social interaction between program staff and the actors whose behavior is to be changed. The social science research approach calls for detailed evaluation studies at periodic intervals, sometimes called process evaluations or formative evaluations, to describe the state of the program, identify its strengths and failings, and suggest ways to improve it. In the participatory approach, a program can use citizens' advisory committees or other similar groups that it has already convened (to help identify limiting factors) to provide continuous feedback on how the program appears to those whose behavior it is intended to change. As we have already noted, the participatory approach sacrifices rigor for flexibility, rapidity, and cost savings. There is a place for both strategies, and they can sometimes be used in the same program—the participatory approach to make rapid adjustments, and the scientific approach to document progress in a rigorous and convincing way and to check periodically on the conclusions reached by the participatory approach.

We showed in Chapter 6 that successful community resource management efforts include ways to continually assess the state of the shared resource and the functioning of the management rules. The St. Paul recycling program shows how the community approach to monitoring and adjustment can work when a group is too large to meet face to face. The Neighborhood Energy Consortium's board represents all the participating neighborhoods and has the power to change the recycling program as needed. Moreover, the recycling program is organized at the neighborhood level. Each neighborhood council can contract with its own hauler for recyclables if it chooses, and using the neighborhood recycling coordinator as the designated monitor, can ensure close monitoring and control of the recycling service at a level that is small enough for face-to-face interaction. With creative "nested" arrangements like this, even large environmental programs can apply the principles developed in small communities.

G. Stay within the Bounds of the Actors' Tolerance for Intervention

We have already mentioned the problem of organized opposition that sometimes arises when an intervention is proposed for approval in a political system. The kind of public opposition that has stymied the siting of nuclear power and hazardous waste facilities can also prevent the enactment of environmental protection policies and programs when the target groups perceive them as overly intrusive, costly, or unfair. In the United States, the problem has arisen with energy taxes, energy-efficient building codes, and some systems of rationing scarce resources in emergencies.

The targeted actors may also resist a policy or program after it has been enacted, thus rendering it ineffective or even counterproductive. We have mentioned the temptation to dispose of waste illegally when the cost or inconvenience of safe and legal disposal are greatly increased. If enough people evade waste disposal regulations by dumping or burning, the environmental effects of regulations or incentives may actually be negative, because the illegal methods are usually much more polluting than the legal ones. Also, information—even helpful information—tends to be ignored or rejected if its source is not trusted or if it tells people something they simply do not believe. This has been a problem for electric utilities telling people how to use less electricity and for chemical companies that announce actions they are taking to improve the environment.

These sorts of problems show the importance of staying within the target groups' limits of acceptance. It is sometimes possible to stretch those limits, for example, by educating the actors about the benefits and importance of following new rules, but it is usually counterproductive to try to achieve changes beyond some outer boundary. But how does one determine the limits of tolerance? This question leads back to the principle of participation.

H. Use Participatory Methods of Decision Making

The needs to understand the actor's perspective, attract people's attention and gain their commitment, monitor and adjust programs, and design interventions that are within people's limits of tolerance all point to participatory processes as a solution. The principle of participatory choice, outlined in Chapter 6, is a great contribution of community resource management efforts to the solution of environmental problems. Meaningful participation by the actors makes interventions more tolerable because people are unlikely to design programs for themselves that they find objectionable. Participation is also likely to build support for whatever intervention is ultimately chosen, promote perceptions of fairness, and increase the likelihood that the participants will internalize the new rules, thus reducing political conflict and the need for expensive enforcement. And as we have already noted, participation provides an efficient way of monitoring a program's progress. If the participation includes real influence over the design of the intervention, it is likely to result in programs that are fair in reality and not only in perception. We should emphasize, however, that government agencies and businesses that initiate programs to affect the environment sometimes request public participation only reluctantly, and more for public relations purposes than for serious power sharing. People quickly sense this, and the resulting resentment can be highly destructive, both for the program at hand and for the potential for future cooperative efforts at resource management.

Participatory choice procedures may work most smoothly in small groups of individuals, such as the farmers or fishers of a village, but we strongly emphasize that they can also be employed in larger communities. The energy conservation programs in Fitchburg and Hood River are good examples.

Participatory approaches can be employed with groups of organizations, as well as groups of individuals or households. One successful example is the set of water management arrangements worked out by regional organizations in Southern California and briefly described in Chapter 6. The members of these groups were not individuals but municipalities and other large, organizational water users. Another example is the recycling program in St. Paul, which involved a consortium of neighborhoods. Whether the participants are individuals or organizations, the participatory approach has great benefits by reducing the costs of achieving compliance with whatever rules or regulations are devised.

AN ADDITIONAL VALUE OF PARTICIPATORY METHODS

Participatory decision making has one additional, very important application that goes beyond the eight principles we have outlined in this chapter. We present these principles as guides to help "organize an effective program to achieve a particular environmental goal." But we have not yet discussed how a community or a society sets environmental goals: which environmental policies to pursue, or which technologies to adopt, or what choice to make when protecting one set of values puts another set at risk. The most difficult environmental issues usually raise just this sort of problem. For example, to supply a country's need for increased electricity in a way that protects the environment, should it build new nuclear power plants, coal-fired power plants, hydroelectric dams, solar power systems, or conservation power plants, such as at Hood River? This question raises dozens of other questions, because each option has its own set of possible environmental consequences and its own financial and other costs, and all of these need to be considered. Similar cascades of questions arise when a community must choose a method of solid waste disposal, or a place to build a highway or a school. Each option has environmental impacts, and it is necessary to accept one set of impacts in order to avoid the others. Of course, communities want to minimize environmental impacts, but they do not always want to do so at all costs, and even if environmental quality is the only consideration, there is often room for argument about which choice is best for the environment because experts do not have perfect knowledge, and they often disagree. It is for making such complex and difficult choices that participatory methods have their longest and most well-known record.

We are talking, of course, about democratic decision making. But making choices democratically about the environment can be more difficult than making other public policy choices because of the high level of technical information that must be understood to make wise decisions. For example, decisions on electricity supply require knowledge about each power plant technology, and particularly the safety and risk issues associated with each; decisions on solid waste disposal require knowledge about the chemistry of incineration and the movement of chemicals from landfills into soil and groundwater; and decisions about construction projects may require knowledge about soil structure and the habitats of local animals. Technical knowledge is essential even for making fairly simple community decisions, such as about what to include in a city recycling program. Such decisions can turn on the technical capability of the recycling industry to recover usable materials from trash, which determines both the price of recyclables and the economic feasibility of including particular kinds of materials in the program.

In making such choices, technical knowledge is important, and sometimes is absolutely critical. The most obvious case is choices about developing and managing nuclear power, where a technically incorrect choice can contaminate a region with radioactive material for thousands of years.

Because of the critical importance of technical information in many environmental decisions, public officials rely heavily on experts in making environmental policy choices. Sometimes, the resulting decision procedures are far different from those that citizens in democratic countries have come to expect. A few powerful groups in government and industry, based on consultation with their chosen scientific and technical experts, negotiate sets of decisions that are then presented to the citizenry. A good example of such technocratic decision making has been the national decision in France to build an energy system heavily reliant on nuclear power, a decision that is said to have widespread acceptance because of a relatively high level of public trust in government officials (Slovic, 1993). In countries where the government enjoys less trust, however—and the United States is one of these—technocratic decision making has serious problems, as evidenced by widespread popular opposition to nuclear power and hazardous waste disposal technologies in these countries (Kasperson et al., 1992).

Participatory methods can provide an alternative to the technocratic procedures that have been so controversial. But for environmental policy choices that have a strong technical component, it is necessary to incorporate expert knowledge appropriately into the decisions in order to avoid the possibility of a disaster resulting from a participatory decision made in ignorance of available scientific or technical knowledge. The decisions must not only be *fair*, in the sense of the ability of citizens to participate fully, but also technically *competent* (Webler, 1994; see also Pateman, 1970; Habermas, 1984, 1987).

There are several promising approaches to participatory environmental decision making. As an example, we present one, still experimental, that we believe can help in some situations in making technically difficult environmental choices in a way that is both fair and competent. This approach, which has been developed by Peter Dienel, Ortwin Renn, and their associates in Germany, and has now been tried in Switzerland and the United States with varied results, can be illustrated with an example from the energy policy debate in West Germany in the early 1980s. At that time, national energy policy debates were going on across Western Europe and North America in response to major shocks to the world oil market in 1973 and again in 1979. West Germany was not an oil-producing country, so was particularly shaken by rapid increases in oil prices on the world market, where it purchased all its motor fuel. In 1979, a parliamentary commission of the West German government developed a set of four energy policy options to structure political debate in that country, ranging from a high energy-supply scenario that relied on increased development of nuclear power and other energy supply sources to one that emphasized reducing demand by conservation and developing renewable sources of energy.

In August 1982, the German Ministry of Research and Technology asked Renn and his colleagues to conduct a three-year research study of public reactions to the energy options. They did this by using a three-step participatory process (the most complete English-language accounts of its use in Germany can be found in Renn, Webler, Rakel, Dienel, and Johnson, 1993, and Renn and Webler, 1992). First, the researchers held discussions with thirteen major "stakeholder groups," including industrial, labor, and church groups, to generate lists of criteria they wanted used to judge the energy options (for example, effects on air pollution, releases of radioactivity, etc.). These groups were asked to consider their own values and interests, and to identify the possible effects of the policy scenarios that would matter most to them. The stakeholder groups did not need to agree with each other on which criteria to use. If, for example, an environmental group did not think corporate profitability should be a criterion for judging energy policy, it could give that criterion a weight of zero at the same time an industrial group was giving it a strong weighting; similarly, a corporate group might give possible danger to an obscure species of fish a weight of zero at the same time an environmental group could give it a much stronger weight. Across the thirteen groups, 141 different criteria were generated. The purpose of this step was to link public concerns to expert knowledge by identifying the issues that experts needed to analyze: In what ways would each energy scenario affect the 141 outcomes of concern to stakeholders?

In the second step, a collection of about thirty energy experts gave their best scientific estimates of the performance of each energy scenario on each of the criteria. Some of the more technical outcomes were evaluated by consultant groups; the social, political, and psychological impacts, which are harder to estimate and somewhat more likely to be controversial, were evaluated in a series of discussions among subgroups of experts chosen for their varying opinions on the issue. The reports of these discussions made clear where the experts agreed, where they disagreed, and the extent and nature of the disagreements. Thus, expert knowledge was brought into the process in the form of assessments of the effects of the energy policy scenarios on each outcome that concerned interested groups. The expert assessments did not try to provide a single answer where there was none: The process made clear where scientific agreement existed and where it did not.

It was at the third step that the participatory process, now informed by technical knowledge, actually took place. Twenty-four citizens' panels, each consisting of twenty-five citizens, engaged in extensive, four-day meetings to discuss the criteria and the expert judgments and to arrive at policy recommendations. The panels were chosen at random from the German population, and participants were paid for their time. Although most of the individuals selected did not participate for one reason or another, those who did appeared to represent a good cross-section of the West German population. The panels had the opportunity to consider the criteria put forward by the various stakeholder groups and the experts' scientific evaluations of how each policy choice would affect each criterion, and were encouraged to make and discuss their own value judgments about which criteria were most important and which principles should be used for making difficult public choices. The four-day format allowed the groups time to consider what the experts said, including scientific uncertainty or conflict on some of the important issues, and to evaluate the technical information (and lack of it) in light of various points of view present among the general public, as represented by members of the citizens' panel. In this sense, the process was very much like that of an American jury, which relies on a deliberative process among a cross-section of ordinary citizens to produce wise decisions.

Remarkably, the recommendations from the two dozen independent panels converged: "The panels unanimously rejected a high energy supply scenario and opted for an energy policy that emphasized energy conservation and efficient use of energy. Nuclear energy was perceived as non-desirable, but—at least for an intermediate period—as a necessary energy source. The panelists recommended stricter environmental regulation for fossil fuels even if this meant higher energy prices" (Renn et al., 1993, p. 203).

The process was experimental, and not part of the official procedure by which the West German government made national energy policy decisions. The West German parliament was not compelled to do what the citizens' groups recommended, or even to take the recommendations into consideration. Because of this, it is difficult to determine what effect, if any, the citizens' panels had on the national debate. But it is possible to imagine ways to use a participatory deliberation process like the German one to make policy choices. One is to publicize the panels' recommendations widely, so that elected officials who chose to ignore them would probably to be called to explain their positions by their constituents. This approach was used in national debates on energy policy in other European countries (Midden, 1994). Another is for a policy body or policy maker to agree to abide by the decisions of a citizens' panel. This method is actually being tried in Switzerland, where citizens' panels representing all the communities in the eastern part of the Canton of Aargau (similar to a state in the United States) arrived at recommendations on where to locate a solid waste landfill within the region (Webler, 1994). The canton's building department, which has legal responsibility for deciding on a site, invited Renn's group to set up a citizens' panel process, and a member of the state cabinet gave a personal guarantee that the canton would not put the landfill in any community where it was not wanted. As of this writing, the citizens' panels had arrived at their recommendations, but final decisions had not yet been made, so it is too soon to tell whether the promises from the public officials will result in real influence for the participatory process.

We describe these experiments in participatory environmental policy making not because we know that they work—there is not enough experience with them to reach such conclusions—but to show the potential value of the lessons of experience in resource management for dealing with some of the most difficult of environmental decisions. The experiments with citizens' panels draw on several of the principles described in this chapter. They are designed to address conditions beyond the individual (principle D), specifically the conditions set by public policy choices. They do this in ways that take very seriously the perspectives of those who will be affected by the policy choices (principle B) and the need to stay within their tolerance for intervention (principle G). They also apply principles of community management by using small-group processes to help arrive at decisions and build consensus (principle C).

There remain many questions, however, about the feasibility and usefulness of participatory methods in environmental policy making. Some critics have pointed out possible shortcomings in Renn's procedure—for example, that citizens' panels may be misled by a biased selection of experts. We believe that as with other applications of the principles outlined in this chapter, it will take a certain amount of experience to find the most effective procedures. As with other applications, it will be necessary to monitor the effectiveness of experiments in participatory decision making in order to learn from them.

And we expect that the citizens' panel approach will probably provide only part of the answer. But it illustrates that the principles outlined in this chapter have the potential to be applied—and, no doubt, further refined—for solving some of the most difficult problems in environmental management.

CONCLUSION

Based on the successes and failures of past interventions and basic knowledge in behavioral science, we have outlined some principles, or useful general guidelines, for those who would devise policies or programs to promote proenvironmental behavior. Although the principles are not specific enough to apply to particular behaviors in each particular context where change is desired, they include advice on ways to build the more specific knowledge that is needed.

Many of these principles may seem like no more than common sense. But unfortunately, they are commonly ignored by those who devise environmental policies and programs. What is common is that information programs are oversold and inform too poorly, and have disappointing results (Ester and Winett, 1982; Condelli et al., 1984); that new regulations, taxes, and incentives are devised without paying enough attention to securing public support and solving the problems of implementation; that debates about incentives consider only their size, and do not pay attention to the specifics of implementation that determine whether the policy is cost-effective (e.g., Stern, 1986, 1993). The lessons of community management and public participation seem never to be learned (Kasperson et al., 1992), and the value of informal communication is too often ignored (e.g., Coltrane, Archer, and Aronson, 1986; Stern, 1986). Commonly, those who devise and judge programs assume that the initial design is optimal and permanent, so they do not leave room for a learning process, and create the conditions under which programs are likely to be judged prematurely and against unrealistically high expectations, and concluded to be failures. When programs are evaluated, it is most often for final judgment, rather than for learning.

Too often, environmental programs have derived from the "common sense" of people who have not studied which conditions are conducive to which types of policies or other interventions, how interventions interact, what makes them succeed and fail, and how to build in the possibilities for learning and adaptation. This sort of common sense generally leads to error. As the previous chapters show, systematic study of environmental programs and their interactions with

people, groups, and organizations can improve on untutored intuition. As one of us has remarked elsewhere, this kind of research and analysis can "separate common sense from common nonsense and make uncommon sense more common" (Stern, 1993, p. 1899). The principles we have drawn from experience should enable people to design more effective policies and programs in the future and to continue the learning process.

NOTE

1. *Source:* Personal communications from Mary T'Kach and Martha McDonell, St. Paul Neighborhood Energy Consortium, and Consortium publications, 1994.

PART III

HUMAN BEHAVIORAL PREDISPOSITIONS AS AIDS OR BARRIERS TO SOLUTIONS

CHAPTER 8

STONE AGE GENETIC BEHAVIORAL PREDISPOSITIONS IN THE SPACE AGE

I. Chapter Prologue

II. Introduction
   A. Human Genetic Predispositions and Environmental Problems
   B. A Primer on Anthropology: Biological Evolution and Cultural Evolution
   Box 8-1: Darwinian Natural Selection

III. Natural Stimuli and Human Well-Being: The Biophilia Hypothesis
   A. Informal and Indirect Evidence for the Biophilia Hypothesis
   Box 8-2: Phenotypes versus Genotypes
   B. Limited Grounds and Criteria for Inferring the Existence of Genetic Behavioral Predispositions
   C. Direct and Formal Evidence for Biophilia as a Human Genetic Predisposition
   D. Brief Overview of the Biophilia Research

IV. Genetically Based Sex and Reproductive Urges
   A. Implications of Ehrlich's Argument
   B. Global Population Growth—Other Causes and Solutions

V. Are Human Beings Short-Term Egoists by Nature?
   A. The Argument that Natural Selection Favors Egoism
   B. Arguments that Natural Selection Favors Altruism
   C. Explaining Altruism as Egoism: Sociobiology and Reciprocity
   D. A Cultural Evolutionary Account of Altruism

VI. Stone Age Genetic Perceptual and Cognitive Predispositions
A. Hardin's Hypothesis: Genetically Based Denial
B. Ornstein and Ehrlich's Theory: The "Old" Human Mind in the New World

CHAPTER PROLOGUE

In Chapter 7, the last of Part II (Chapters 2–7), we completed our in-depth discussion of four methods for encouraging proenvironmental individual behaviors. In Part III of the book—this chapter and Chapter 9—we take a more bird's-eye view and examine some very broad and general human behavioral predispositions that might aid or impede the four behavior-change approaches we reviewed in Part II. We begin this chapter with a brief story.

You are in a college classroom listening to an introductory anthropology lecture. The professor is talking about the predecessors of modern humans and the process of biological evolution. She first describes four predecessor species: *Australopithecus,* first appearing more than 4 million years ago; Homo habilis, about 2.4 million years ago; *Homo erectus,* 1.6 million years ago; and the Neanderthals or *Homo sapiens neanderthalis,* 150,000 years ago. The professor goes on to describe the Cro-Magnons—named after the cave in France in which archaeologists first found their remains—who first appeared 30,000 to 40,000 years ago (some archaeologists say 90,000 years ago). She outlines what Cro-Magnon life was like for thousands of years: They lived in small, nomadic groups and hunted and gathered their food, as did their ancestors. They had no agriculture and no permanent villages or settlements, no pottery and no writing. They had no bows and arrows and no domesticated animals. The professor then stops and asks the class a question: "Does anyone know the full Latin genus-species name of the Cro-Magnons?" The correct answer is: "*Homo sapiens sapiens.*" In other words, Cro-Magnons are "us"; we are the same species. Furthermore, the professor points out, due to the extremely long time periods required for major changes through biological evolution, the genetic makeup of our species has not significantly changed since the species first appeared. The genetic makeup of humans alive today is thus almost identical to that of Stone Age or "cave person" ancestors who lived 30,000 to 40,000 or more years ago under conditions of life very different from our own.

Biologist Rene Dubos expresses this important truth in his book, *So Human an Animal* (1968), in the following way:

Every trace of prehistoric [hu]man in the world provides further evidence for the view that [our] . . . fundamental characteristics . . . have not changed since the Stone Age. . . . A Cro-Magnon man [or woman], if he [or she] were born and educated among us, could work in an IBM plant and might even become president of the company. A modern human [conversely] could readily return to primitive life. . . . [pp. 39, 42]

INTRODUCTION

Humans alive today have essentially the same genes as our Stone Age ancestors who lived tens of thousands of years ago. We may now be driving aerodynamic cars, living in glass and steel metropolises, and using computers capable of millions of operations per second, but we are still, from a genetic point of view, "archaic" creatures. What are the implications of this basic truth? Certainly, the genetic makeup of our species does not, by itself, dictate the way humans now perceive, think, and behave; our learning experiences, upbringing, and culture also have an enormous impact on these things. However, the genetic makeup of our species *does* push us in certain general directions and predisposes us to perceive, think, and behave in certain ways. In addition, our genetic makeup sets limits on our plasticity, that is, on what learning, upbringing, and so forth can make of us.

What are the implications for environmental problems? According to some biologists, psychologists, and other scholars, our Stone Age genetic behavioral predispositions and limits actually promote behaviors that damage the environment. Others argue that our Stone Age genes make some methods to encourage proenvironmental behavior effective, while making other methods ineffective. Yet other scholars argue that our archaic genes require us to maintain contact with natural environments like the ones in which our species evolved, or suffer psychological and physical harm.

Human Genetic Predispositions and Environmental Problems

Table 8-1 lists several claims scholars have made that specific archaic genetic predispositions are environmentally relevant. For example, B. F. Skinner's theory suggests that humans are, in essence, genetically programmed to destroy the environment. Skinner argues that humans are short-term egoists by nature (as we discussed in Chapter 5), in other words, that people's behavior is determined mainly by its immediate personal consequences, rather than its long-term consequences or its consequences to others. Skinner claims that modern humans have inherited this predisposition because short-term egoism was a requisite to survival under the conditions of life in which our species evolved. However, in the twentieth century, human short-term egoism is maladaptive given the large size of the human population, the nature of human technology, and the intensity and scale with which we use it. Indeed, short-term egoism, Skinner argues, is exactly the psychological trait that makes modern humans prone to environmental tragedies of the commons. Skinner further argues that intrinsic human short-term egoism makes *incentives* the only effective strategy for solving environmental problems (because they make proenvironmental behavior in each individual's personal short-term interest), while all other strategies (moral/religious appeals, educational programs, and community management) ineffective.

Consider, briefly, the last entry in Table 8-1: Jay Forrester (formerly at the Sloan School of Management at MIT) claims that many large social, political, and environmental systems are now too complex for the human mind to properly comprehend. Also, these systems—which include large corporations, cities, states, countries, and regional and global ecosystems—tend to behave in ways opposite to what human intuition would predict. Forrester (1971, p. 1) argues: "Evolutionary processes have not given us the mental skill needed to properly interpret the . . . behavior of [such] systems. . . ." Further, these systems foil the social mechanisms and institutions that humans have relied on throughout history to deal with complexity. As a result, Forrester claims, human interactions with these systems, including, in some cases, direct attempts to manage them, are likely to produce serious negative results. Forrester goes on to propose a method that may help humans understand complex systems and interact with them more successfully.

Before we discuss Forrester's work further, or the work of Skinner and the other scholars listed in Table 8-1, however, we need to review some important background material. We do this in the section below titled A Primer on Anthropology. We then devote the rest of the chapter to an in-depth look at the hypothesized genetic behavioral predispositions and limits listed in Table 8-1 and their environmental implications.

A Primer on Anthropology: Biological Evolution and Cultural Evolution

To fully understand how our archaic genetic makeup may be playing a role in today's environmental problems, the reader must have a basic understanding of *biological evolution,* the process that shaped the genetic makeup of all of the Earth's plant and animal species, including Homo sapiens; *natural selection,* the driving force behind biological evolution; the long time periods necessary for major changes in a species' genetic makeup via biological evolution; *culture,* highly adaptive nongenetic information passed from generation to generation; and *cultural evolution,* the process that gave rise to our advanced technology and other features of our space-age lives.

TABLE 8-1 Claims That Stone Age Genetic Behavioral Predispositions Are Relevant to Contemporary Environmental Problems

SCHOLAR	PREDISPOSITION OR LIMITATION
	NEED FOR NATURAL STIMULI
Dubos; Iltis; Wilson	Humans have a genetically based need to be near natural environmental stimuli (foliage, sounds and motions of animals, and so on); the absence of natural stimuli in urban environments may be harmful to people's health and well-being.
	DRIVE TO REPRODUCE
Ehrlich	Genetically based sex and reproductive urges are major causes of the global population explosion.
	EGOISM AND ALTRUISM
	(THE THREE BRACKETED ITEMS ARE MUTUALLY EXCLUSIVE)
Skinner	Genetically based human short-term egoism is a major cause of environmental tragedies of the commons.
Wilson; others	Genetically based egoism is tempered by a genetic tendency to live in groups, to cooperate, and to behave altruistically toward one's kin, and also toward nonkin in expectation of altruism in return.
Simon	Genetically based "docility"—a strong tendency to engage in social learning—together with "bounded rationality" may work to counteract any inherent human short-term egoism and to create genuinely altruistic behavior.
Fox and others	Humans inherently function best in small, cohesive communities, like the ones in which our evolutionary ancestors lived. Such communities are, in addition, more effective in encouraging and maintaining proenvironmental individual behavior than are contemporary urban living arrangements. (See Fox, 1985)
	PERCEPTUAL AND COGNITIVE LIMITATIONS
Hardin	Genetically based "denial" is a major cause of human underestimation of the probability and severity of environmental threats. (Chap. 9)
Ornstein and Ehrlich	Our "old," or archaic, mind tends not to perceive and respond to gradual environmental deterioration. (Chap. 9)
Forrester	The human mind can't properly comprehend complex social, political, and environmental systems that humans now interact with or attempt to manage; further, these systems often behave the reverse of the way human intuition would predict. (Chap. 11)

Biological Evolution. The concept of biological evolution predated Charles Darwin's work in the 1800s. Elementary descriptions of the process appeared in ancient Greek writings and detailed accounts of it in Western scientific writings by the 1700s. The essence of biological evolution is that all life on Earth developed from a single source. This first living material was probably a simple protein formed by natural processes perhaps 3 to 4 billion years ago. In the millions of ensuing years, the many different species of plants and animals developed from the original source. In general, more complex species appeared

TABLE 8-2 Fossil Remains and the Sequence of Biological Evolution

MILLIONS OF YEARS AGO	FORM OF LIFE
.04	Homo sapiens sapiens
3–4	Australopithecus and other precursors of modern humans
25	Anthropoid apes
50	Modern mammals
65	Dinosaurs become extinct
200	Height of dinosaur evolution
300	Insects (air dwelling) and amphibians
400	Early land plants (first life on land)
600	Many water invertebrates (and water plant life also)

(Before 600, fossil records are scarce)

later in the sequence, though all species share some basic genetic and biological mechanisms. Homo sapiens thus developed from earlier nonhuman forms of life.

Evidence that life evolved in this way comes from examining the fossil remains of plants and animals (i.e., leaf outlines, skeletons, and other traces embedded in rock formations). Based mainly on the geological layer of the Earth's crust in which a fossil is found, scientists can determine the age of the fossil with reasonable accuracy. A rough timetable of biological evolution based on fossil evidence appears in Table 8-2 (the table reads bottom to top).

Natural Selection. Darwin did not discover biological evolution. His unique contribution was the discovery of natural selection as the main force that drives biological evolution. The natural selection principle first appeared in Darwin's book *The Origin of Species* (1859). The basic idea is that the genes of a species change, and new species evolve, through an interplay between: 1) genetic variability that occurs within all species, and 2) the characteristics of the species' environment. More specifically, those members of a species having genetic traits that are adaptive in the species' environment survive (they are "selected by the environment"), and their traits eventually become predominant in the species' gene pool. Darwin's prin-

ciple of natural selection is widely accepted by scientists today, though it has been elaborated and somewhat altered. There is now, for example, uncertainty concerning the smoothness versus discontinuity of the overall process (i.e., gradual versus sudden appearances of new species). We outline Darwin's principle of natural selection in its simplest terms in Box 8-1.

The Time Course of Biological Evolution. Biological evolution, driven by natural selection, is a very slow process (though not necessarily a gradual or incremental one). Many generations must pass before significant genetic changes occur. Evolution is especially slow for species like Homo sapiens in which more than a decade elapses between the birth of one generation and that of the next. Many people find it difficult to appreciate the slow time course of biological evolution, and they fall prey to misconceptions.

For example, some people mistakenly assume that human genetic makeup has shifted in the direction of increased height in the several hundred years since the Middle Ages. These people have seen suits of armor in museums, noted that medieval knights were quite short by today's standards, and inferred that this change is due to biological evolution. However, such a change is not possible. With the exception of resistance to certain fatal infectious diseases, genetic changes in humans of this magnitude through biological evolution could not

Box 8-1

Darwinian Natural Selection

There are four components of the natural selection process:

1. Genetic variability within a species. All species exhibit genetic variability at the individual level. As a crude, imperfect example, notice the variability in the physical appearance of students in a college classroom or library. Though physical appearance ("phenotype," as we discuss later) is determined by both genetic ("genotypical") and environmental influences, some aspects of physical appearance do strongly reflect genetic influences, for example, eye color, complexion, and hair color (assuming no hair dye or makeup is used).

Genetic variability within a species is the product of two processes: Genetic mutations—changes in genetic material due to the passage of X rays and cosmic radiation (mainly from outer space) through plants and animal bodies—and sexual reproduction—in which portions of a mother's and father's genetic material are randomly shuffled and combined.

2. Organisms produce more offspring than can survive. A classic example is that a female Atlantic codfish lays 85,000,000 eggs at once, only a few of which make it to adulthood. Young hatchlings must try to survive in a hostile environment. Food is limited, and there are predators and diseases. Only recently (the last few millennia) in the eons-long history of biological evolution have humans begun to radically alter this situation by means of technology. Human life was significantly harsher and shorter before the develop-

ment of agriculture, modern medicine, and public health measures.

3. Individuals with traits that are adaptive are most likely to survive. Individuals with traits best suited to the species' environment (e.g., resistance to heat stress in a tropical environment, or teeth that can grind tough vegetation in an environment in which such vegetation is plentiful) are most likely to survive. This step is specifically called "natural selection" in that the characteristics of the natural environment determine, or select, which individuals will survive and which will be eliminated.[10] (Note that Skinnerians emphasize the strong parallels between operant conditioning and Darwinian natural selection: Both processes involve the environment's selecting certain characteristics and making them predominant; in natural selection the environment shapes the genetic makeup of a species; in operant conditioning the environment shapes the behaviors emitted by an individual [i.e., behaviors followed by desirable consequences become predominant, ones followed by undesirable consequences are eliminated].)

Finally:

4. Only individuals that survive contribute to the species' gene pool. Over time their genetic traits become predominant. Obviously, only individuals that survive can have offspring and raise them to maturity. Thus, only those individuals that are naturally selected (step 3) pass their genes on to subsequent generations, and their genetic traits eventually become predominant in the species.

---

have occurred in 500 years. The increased height is due better nutrition and other environmental influences (Bogin, 1988).

The following two simple examples may help convey the long time periods involved in biological evolution.

First, if you had a time machine and could travel backward in time, you would have to travel back about 500,000 years before you would find an ances-

tor of the horse that was perceptibly different from modern ones.

Consider, as the second example, the precursors to modern humans and the timetable of their appearance over a 3 to 4 million year period, as shown in Table 8-3. Though paleontologists are not sure of the exact shape of the human family tree (e.g., whether *Homo sapiens sapiens* descended in part from the Neanderthals or evolved independent of them), they are sure that major

TABLE 8-3 A Chronology: Homo Sapiens and Their Predecessors

40,000–90,000 years ago	*Cro-Magnon*—(Homo sapiens sapiens) Biologically identical to modern humans.
150,000 years ago	*Neanderthal*—(Homo sapiens neanderthalis) Large-boned, heavy eyebrow ridges, receding forehead. Cave living, with hunting, fire, sophisticated flint tools, clothing, artistic achievements, religion.
1,500,000 years ago	*Homo erectus*—Looked more human than Homo habilis. Had fire and tool use (chipped stone).
2.4 million years ago	*Homo habilis*—First genus Homo. Humanlike skeleton; up to 5 ft. tall.
4+ million years ago	*Australopithecus*—Looked like a cross between a human and an ape. The size of a chimp, but erect, or bipedal, most of the time, and used crude tools.

genetic changes and the evolution of new species take tens of thousands of years.

Thus, as we noted in the Prologue, our species, first appearing between 40,000 to 90,000 years ago, has undergone little or no significant genetic change.[1] Modern humans are biologically almost identical to "primitive" humans. This means that our genetic characteristics evolved under conditions of nomadic hunting and gathering life, low population densities, and other features of Stone Age existence. Human genetic traits are ones that enhanced survival under these conditions.

Though we are genetically "archaic," our species now finds itself living in space-age conditions radically different from the conditions under which it evolved. The process that produced these space-age conditions is cultural evolution. But before we can discuss cultural evolution we must first discuss culture and the relationship between culture and biological evolution.

Culture. The relationship between culture and biological evolution is summarized in Figure 8-1. As we discussed earlier, biological evolution proceeded for millennia, producing various plant and animal species, and eventually producing the predecessors of *Homo sapiens sapiens* hundreds of thousands of years ago. But in these new species, biological evolution had created some *unique physical and psychological capabilities and characteristics:*[2]

— high-capacity long-term memory
— advanced cognitive abilities (or, loosely speaking, "intelligence")
— physical and psychological capacity for speech and language

These new physical and psychological characteristics made an important phenomenon possible for the first time in evolutionary history: Now an organism could give its offspring a great deal more than just a genetic heritage, that is, more than just the genes for

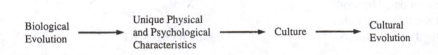

FIGURE 8-1 The Relationship Between Biological Evolution and Culture and Cultural Evolution

traits well-suited for survival in the species' environment. It could give its offspring a wealth of *nongenetic* information known as *culture.*

Cultural information is extremely powerful, adaptive, and even lifesaving. Parents can teach their children about the experiences of many prior generations on many vital topics, for example: How to start and manage fire, how to hunt large animals, how to find shelter in a forest, how to tell harmless from poisonous mushrooms, and so on. This information allows offspring to avoid the time-consuming, and potentially deadly, process of personal trial-and-error learning. Culture also includes values, and specific pre- and proscribed individual behaviors (e.g., prohibitions of murder and incest) which are transmitted from one generation to the next, and that aid both individual and group survival.

Cultural Evolution. Not only is cultural information highly useful and adaptive, it tends to change or evolve over time. For one thing, the teachings of prior generations are continually honed by the experiences of each new generation. The long-term soundness of cultural values, behaviors, and technologies is given the test of time. Any elements of culture that do not contribute to the long-term survival of individuals and of the group tend to be weeded out (either the group doesn't survive, or the group alters its culture); those elements that do contribute tend to be retained. Note that this process is somewhat like biological evolution driven by natural selection, though the underlying issue is the environment's selection of a group's cultural practices rather than of a species' genetic makeup. B. F. Skinner (1971) wrote about this aspect of cultural evolution at length. To quote (p. 137): "A given culture evolves as new practices arise . . . and are selected by their contribution to the strength of the culture as it 'competes' with the physical environment and with other cultures." In a similar manner, each generation may be thought of as *inheriting* the culture of the past generations. The mode of transmission is not biological or genetic but rather via language and learning (see Boyd and Richerson, 1985, for a more complete discussion).

There is another general reason that culture evolves over time. First, each generation can add *totally new items* to the store of knowledge. This new knowledge is passed on to succeeding generations, and thus the total fund of knowledge accumulates. But in addition to this simple accumulation process, other forces fuel an accelerating, snowballing rate of accumulation. For one thing, any new discovery, information, or invention often has more uses than the discoverer anticipated and also tends to spawn further discoveries, information, and inventions, each of which, in turn, may spawn more. Consider, for example, the uses of fire: Fire may have first provided warmth to cold shelters, but the technology subsequently made possible enhanced food preparation (cooking to improve flavor and safety of food), food preservation (smoking), primitive medical measures (cauterization), light for nocturnal activity, torches for weaponry, heat for firing of pottery, and later, for glassmaking, metallurgy, and so on (see Figure 8-2).[3]

Finally, in a closely related manner, the more advances in knowledge and technology exist, the more new ones can be created by combining old ones. For example, the development of the TV picture tube required the technologies of glassmaking, glassblowing, vacuum tube design, vacuum pumps, electron beams, magnetic fields, phosphorescent chemicals, and so on. Similarly, space flight depended on advances in computers, in metallurgy, ceramics,

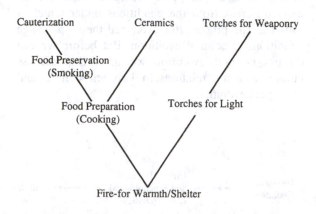

FIGURE 8-2 Example of How One Discovery or Invention Can Spawn Many More (Read from bottom to top.)

meteorology, atmospheric science, medicine, and several other areas.

So, again, there is an intrinsically *accelerating* accumulation of knowledge and technology, one that in a way is similar to exponential population growth. Figure 8-3 illustrates this accelerating accumulation over the 40,000 year period that our species has been around. Note that there were no significant permanent human settlements during the first 30,000 years of human existence before the development of agriculture 10,000 years ago.[4] Writing appeared only 5,000 years ago, printing 500 years ago, steam and other engines 100 years ago, and in the last half-century a technological explosion, with lasers, nuclear energy, computers, heart transplants, space travel, and many others.

It is this accelerating process of technological and cultural change, then, that has thrust our species into the space age. Note how the accelerative pace of technological change and cultural evolution stands in contrast to the glacial pace of biological evolution,[5] as shown in Figure 8-4. Human genetic makeup is still essentially Stone Age while human culture and technology have catapulted our species into the space age and beyond. And, several scholars argue, this discrepancy between human Stone Age genetic predispositions and limits and space-age conditions of life may be playing a major role in the environmental problems that now threaten human survival.

The following quotations further convey this sense of discrepancy between human Stone Age genes and the conditions of space-age life:

. . . *[A]s we consider the human predicament today, we should keep a basic point in mind: while humanity has become the most successful animal on the planet, almost all its biological evolution occurred long before it achieved such exalted status, and it occurred in a very different environment. (Ornstein and Ehrlich, 1989 [p. 28]).*

. . . *[W]e carry the remains of our long history inside our own heads, but we have completely changed the environment in which those heads must function. (Ornstein and Ehrlich, 1989 [p. 24]).*

The common factor that dominated human evolution and produced Homo sapiens *were preagricultural. Agricultural ways of life have dominated less than 1 per cent of human history, and there is no evidence of major biological changes during that period of time. (S. Washburn and C. Lancaster, 1968 [in Lee and DeVore, p. 293]).*

We can be fairly certain that most of the genetic evolution of human social behavior occurred over the five million years prior to civilization, when the species consisted of sparse . . . populations of hunter-gatherers. On the other hand, by far the greater part of cultural evolution has occurred since the origin of agriculture and cities approximately 10,000 years ago. Although genetic evolution of some kind continued during this latter, historical, sprint, it cannot have fashioned more than a tiny fraction of the traits of human nature. . . . (E. Wilson, 1978 [p. 34]).

Having covered the anthropology material above, we're now in a position to take a detailed look at the proposed genetic behavioral predispositions and limits in Table 8-1. We begin with the first item in the table—the claim that humans have a genetically based need for natural stimuli.

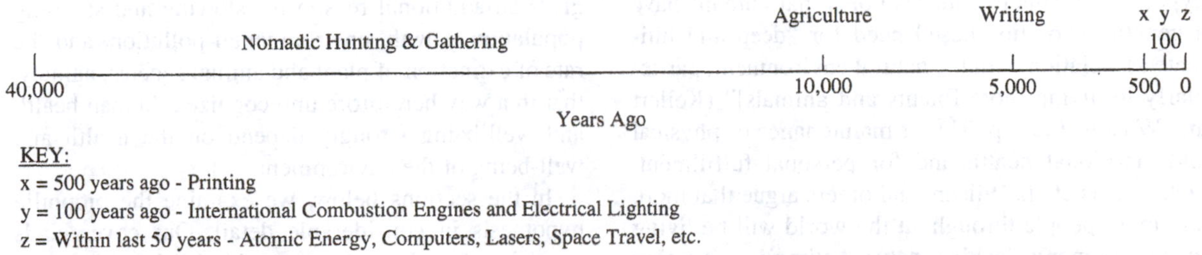

KEY:
x = 500 years ago - Printing
y = 100 years ago - International Combustion Engines and Electrical Lighting
z = Within last 50 years - Atomic Energy, Computers, Lasers, Space Travel, etc.

FIGURE 8-3 The Accelerating Pace of Cultural Evolution

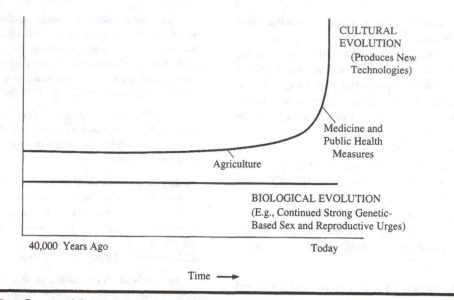

FIGURE 8-4 Time Course of Cultural Evolution versus Biological Evolution

NATURAL STIMULI AND HUMAN WELL-BEING: THE BIOPHILIA HYPOTHESIS

Biologists Rene Dubos (1968), Hugh Iltis, Orie Loucks, and Peter Andrews (1970) have argued that humans have a innate need to be near plants, animals, and other natural stimuli. These scholars claim that if humans are deprived of these stimuli, human emotional health may be impaired. Natural stimuli, more specifically, include the shapes of foliage and vegetation, the sounds and motions of animals and of bodies of water, annual seasonal changes, and so on (see photos on p. 185).

Recently, Edward Wilson (1984) and others (Kellert and Wilson, 1993) have extended Dubos/Iltis et al.'s idea into what Wilson calls the "biophilia hypothesis." This hypothesis holds that humans have a genetic, evolution-based need for "deep and intimate association with the natural environment, particularly its living biota [plants and animals]" (Kellert and Wilson, 1993, p. 21) for maintenance of physical and emotional health and for personal fulfillment. Dubos, Iltis et al., Wilson, and others argue that more and more people throughout the world will be living in environments lacking natural stimuli in coming years. As global population grows, more people will

live in urban areas (Ross, 1994), which are filled with concrete, glass, and steel structures, but generally lacking in natural stimuli. Expanding human settlements will also continue to consume farmlands, woods, and wilderness areas. And more people are likely to be exposed to deteriorated environmental conditions such as air pollution. As a result, these scholars predict, growing numbers of people will suffer impaired emotional and physical health. Indeed, some claim, the higher per capita rates of mental illness and other pathology now found in urban areas versus rural areas may be caused in part by the absence of natural stimuli in urban areas today.

Going a step further, Wilson (1993a, 1993b), Kellert (1993), and others argue that human biophilia gives an additional reason for slowing and stopping population growth, environmental pollution, and the rate of extinction of plant and animal species, namely, that in a way heretofore unrecognized, human health and well-being strongly depend on the health and well-being of the environment.

In the sections below, we examine the biophilia hypothesis in considerable detail. Our coverage is extensive for three reasons: If the hypothesis is correct, it is a cause for considerable alarm to people

Mt. Rundle, Western Canada
(Copyright 1993 Corel Corporation.)

Desert Fox Kits
(Copyright 1993 Corel Corporation.)

Coast Off Australia
(Copyright 1993 Corel Corporation.)

Woods, Snow, and Light
(Copyright 1994 Corel Corporation.)

concerned about the environment and human welfare. Similarly, knowledge of biophilia may further motivate policy makers and the public to take proenvironmental actions. Finally, our examination of biophilia illustrates the methodological difficulties involved in arguing that a particular behavior reflects a genetic predisposition produced by biological evolution. We can then apply our understanding of these difficulties to the other claims in Table 8-1.

Here's how our biophilia coverage is organized: We start by reviewing and critiquing the *informal and indirect* research evidence that proponents of the biophilia hypothesis offer in its support. We then discuss the limited number of basic grounds or criteria that permit an inference, based on research data, of an evolution-based genetic influence on a human behavior. Lastly, we review *formal and direct* biophilia research involving those grounds/criteria. We conclude that there is relatively little direct and compelling evidence at this point in favor of biophilia as a human genetic predisposition. However, we find the large quantity and variety of research results that are consistent with the hypothesis striking.[6] We thus find it impossible to ignore the biophilia hypothesis, and await the results of additional research that tests its validity.

Informal and Indirect Evidence for the Biophilia Hypothesis

A General, Evolution-Based Argument. Biophilia supporters offer the following argument as a justification: For millions of years, our ancestors lived in natural environments, surrounded by natural stimuli; their genetic makeup, which we have inherited, must have been shaped by natural selection to function best in the presence of such stimuli. Quoting Iltis et al. (1970): "Unique as we [humans] may think we are, it seems likely that we are genetically programmed to a natural habitat of clean air and a varied green landscape, like any other mammal [p. 3]." Note that this argument is part of a broader one that holds that humans function best under conditions that mimic the ones in which we evolved, conditions that include a hunting-and-gathering lifestyle, and membership in small, cohesive social groups (Fox, 1985).

While the above line of reasoning is intuitively appealing, exposure over millennia to natural environmental conditions doesn't necessarily mean that humans function best under these conditions and do less well under other conditions. To use a nonenvironmental example, if our human and prehuman male ancestors trimmed their facial hair with crude stone tools for millions of years, this doesn't prove that contemporary men have a specific genetic predisposition to prefer such tools and would be better off using them rather than modern razors.

Informal Observations of People's Reactions to Natural Environmental Stimuli. As a second source of support, Dubos, Iltis, Wilson, and others offer a number of informal, quasi-empirical observations (see Knopf, 1987). For example, Iltis et al. (1970) note that many humans spend money and effort to put green plants in their homes and offices and to keep animals as pets. Also, large numbers of city dwellers choose to travel to rural or wilderness areas for their vacations. Similarly, urbanites use city parks and zoos very heavily. Wilson (1993) points out, ". . . more children and adults . . . visit zoos than attend all major professional sports combined (at least . . . in the United States and Canada) [p. 32]."

However, we can't conclude that a genetically based need for natural stimuli per se is what causes urbanites to use parks and zoos heavily. If more people visit zoos than attend professional spectator sports, we do not know if it is the presence of animals as natural stimuli that draws people to them, rather than, for example, the opportunity to closely view rare and exotic things that can't be viewed elsewhere.

Critics of the Dubos/Iltis position also point out that large numbers of urbanites apparently do not seek out natural stimuli for amusement, vacations, and so on. Indeed, people raised in cities are sometimes frightened by their first encounters with woods and other natural settings (Knopf, 1987). Also, a not insignificant number of people have no plants in their offices and/or homes and keep no pets.

Stronger evidence for the biophilia hypothesis comes from research we review in the next sections.

Quasi-Experimental Studies on the Effects of Natural Stimuli on Stress, and on the Effects of Hospital-Room Window View on Recovery from Surgery. Over 100 research studies have found that people who hike and camp in wilderness areas often report a reduction in stress as a result of their experience. People who use urban parks and arboretums report similar reductions in stress (Ulrich, 1993). Many people also report experiencing emotional solace and comfort after spending time in a park or other natural setting. However, there are several possible explanations for the solace, comfort, and reduced stress that people apparently experience. Is it the natural stimuli that produce the benefit or is it merely a change of scene, the opportunity to explore and master a new environment, or the learning of new skills (e.g., mountain climbing)? Would a vacation centered around a well-designed but natureless urban amusement park do as well? Similarly, if people feel emotionally comforted after spending an hour in a park, we do not know if it is the natural setting per se that produces the comfort, or the passage of time, the quiet of the park, or the physical exercise involved in walking to and in the park. Would an hour spent in an equally distant, quiet library have the same effect?

The design of some research studies rules out some of the alternative explanations above. For example,

Hartig et al. (1991) randomly assigned 102 University of California students to one of three conditions following a period of a taxing mental activity: One-third spent forty minutes walking in a regional park, near a large metropolitan area, which contained various natural stimuli including a body of water; one-third spent forty minutes walking in a well-maintained urban residential and commercial area; and one-third spent forty minutes relaxing in a laboratory room containing comfortable chairs, magazines, and a radio. In a before/after design, the students in the first group reported more positive emotions, greater overall happiness, and showed greater accuracy in performing a mental task (manuscript proofreading) than did the students in the other two groups.

A few other studies have found that dental patients, surgery patients in hospitals, and patients in psychiatric hospitals experience less anxiety, less pain, and/or recover more quickly if they have a window view of natural stimuli or they are exposed to murals or pictures containing natural stimuli, compared to patients not exposed to such stimuli (Ulrich, 1993). For example, Ulrich (1984) did a retrospective study on the postoperative recovery of forty-six hospital patients who had undergone gall bladder surgery. One-half of the patients were assigned (in an essentially random manner) to hospital rooms with views of a grove of trees, the other half to otherwise identical rooms that had views of brick walls. The patients in the two groups were matched with respect to gender, age, obesity, smoker/nonsmoker status, and other variables. Those patients in the rooms with tree views required less pain medication, were described less negatively in nurses' notes (e.g., "patient is upset," "needs encouragement"), and were discharged earlier than the patients with the brick wall views.

Laboratory Studies of People's Aesthetic Preferences for Different Environments in Photos. In scores of controlled studies, hundreds of people have shown a strong preference for photographs of outdoor natural scenes over photos of urban scenes devoid of foliage and other natural features (e.g., Ulrich, 1981, and Herzog, Kaplan, and Kaplan, 1982; see Knopf, 1987; Kaplan and Kaplan, 1989; and Ulrich, 1993, for reviews). In other words, people consistently rate scenes with natural stimuli more attractive than scenes of human-produced settings devoid of plants, and so on. This phenomenon is so strong that even mundane natural scenes are judged more beautiful than many spectacular urban scenes (i.e., scenes with striking architecture). The presence of a lake, river, or other body of water in an outdoor scene is an especially desirable feature. In much of this research, the experimenters ensured that photos of natural and nonnatural scenes did not differ with respect to confounding variables such as weather, season of the year, overall lighting, sun angle, amount of sunlight reaching the ground, and overall complexity or information content.

A few studies of people's aesthetic preferences have taken the biophilia hypothesis a striking and provocative step further: Balling and Falk's (1982) research suggests that people prefer natural scenes that are most like the biome (geographic region and its characteristic vegetation and climate) in which the ancestors of modern humans evolved for millions of years, according to paleontologists: African savannas, which are flat or gently rolling expanses of grassland with scattered clumps of trees and shrubs and with a tropically warm but dry climate (Miller, 1992). The Balling and Falk study involved over 500 subjects of varying ages (from third graders to senior citizens) living in the Washington, D.C., area who rated their preference for living in five biomes depicted in color photographs: savanna, tropical rain forest, desert, temperate deciduous forest, and coniferous (evergreen) forest. Balling and Falk found that children in third and ninth grades preferred the savanna photos, but all other subjects showed no significant preference. Their interpretation of these results: The young children's preference reflects a human genetic predisposition for savanna environments, but with age and repeated exposure to nonsavanna environments (e.g., deciduous forests), people come to equally prefer the latter. There are other interpretations of Balling and Falk's results, and few other relevant studies. Thus the Balling and Falk results and interpretation remain highly speculative. We are on much firmer ground with the controlled studies on photo preferences we reviewed earlier as evidence for the general biophilia hypothesis.

However, we face a fundamental problem in inferring even that the strong, consistent results of the photo-preference research support the general biophilia hypothesis. The hypothesis claims a genetic predisposition to favor and flourish in natural environments, but the research results could be due to *nongenetic* forces. There are many nongenetic influences that may affect how people respond to environmental stimuli. These influences include: People's formal and informal education concerning values, beliefs, and tastes (from family members, friends, schools, and religious institutions); the environments and scenery they have seen through travel; perhaps even their diet and any environmental pollutants to which they may have been exposed. As a result, we simply can't infer that a given behavior, even one commonly observed in all Americans or even all citizens of Western countries, reflects a genetic predisposition or limit. Thus, even if all these people strongly preferred photos of natural scenes, all kept plants and pets, sought out nature and nature alone for vacations and for emotional solace, and so on, this still would not prove that humans have a *genetically* based need for nature. We discuss these issues further in Box 8-2 on Human Phenotypes versus Genotypes.

Limited Grounds and Criteria for Inferring the Existence of Genetic Behavioral Predispositions

Given that there are so many nongenetic influences on human behavior, how, the reader may ask, can we infer that a behavioral trait truly reflects a genetic influence? The answer is that there is no definitive or foolproof method. However, there are three grounds or criteria for inferring the existence of a genetic predisposition that are more compelling than the arguments or evidence we have so far discussed in this chapter.[7] We list the three grounds or criteria in Table 8-4.

One ground or criterion is the appearance of a specific behavioral trait in all human cultures studied by anthropologists and by others. This is sometimes called *cultural universality*. If a certain behavioral trait is found in a wide range of cultures—for example, American, Fiji Island, Japanese, Kalahari Bush, Norwegian, Inuit, and so on—despite wide differ-

TABLE 8-4 Three Grounds or Criteria for Inferring the Existence of a Human Genetic Behavioral Predisposition

1. *Cultural Universality*—A behavior or trait is found in all the human cultures studied by anthropologists and others.

2. *Behavior-Genetic Research*—Research on identical twins, fraternal twins, and other family members suggests that a predisposition for the behavior or trait is heritable.

3. *A Logical Argument Concerning the Properties of Biological Evolution and Darwinian Natural Selection*—A compelling argument that biological evolution and natural selection must have favored a genetic predisposition for a behavior or trait.

ences between these cultures in such things as religion, family structure, geographic location, climate, types of technology, economic system, and so on, then we have greater confidence that the behavioral trait reflects a genetic predisposition. To use a nonenvironmental example, sociobiologist Edward Wilson (1978) argues that a taboo against sexual relations between biological brothers and sisters, and, more significantly, a sexual aversion that brothers and sisters usually feel toward each other, appear in all cultures studied by anthropologists, and are therefore genetically based. (Wilson also argues that the taboo and the aversion are adaptive, and favored by natural selection, in that they discourage genetically similar people from having children, and, thereby, the serious physical and mental pathology often seen in such children. This is an argument based on the third ground, natural selection pressures.)

In the several sections that follow, we look at evidence for biophilia as a genetic predisposition using grounds/criteria 1 and 2 in Table 8-4.

Direct and Formal Evidence for Biophilia as a Human Genetic Predisposition

Limited Evidence of Cultural Universality in Studies of People's Preference for Environments in Photos. As we have noted, a great many studies show that

Box 8-2 _____

Genotypes versus Phenotypes

Human behavior is determined *both* by genes and by the environment, and it is important to understand how these two influences operate and interact. For this purpose, we need to understand the concepts of *genotype* and *phenotype*.

An individual's *genotype* is the genetic material the individual inherits from his/her parents and thereby his/her earlier ancestors. This material is determined and fixed at the moment of conception—when sperm and egg unite to form a one-celled embryo. Except for somatic mutations, which are infrequent, a person's genotype does not change from conception to birth to death.

In contrast, an individual's *phenotype* consists of all the characteristics or traits—both physical and psychological—that the person now possesses, that is, what the person is today: The person's height, weight, personality, intellectual skills, creativity, and so on. Likewise, a person's phenotype includes whether the person behaves as a short-term egoist, whether the person aesthetically prefers natural stimuli, and so on.

Going a step further, a person's phenotype is determined by an interaction between his/her genotype and various environmental influences. This principle is expressed in Figure 8B2-1. The environmental influences are quite diverse and change over a person's lifetime. These influences include: The hormonal and chemical characteristics of the embryo's intra-uterine environment; the kinds of foods a person consumes in infancy and childhood; the presence or absence of injurious substances (like lead in paint chips) in the environment in which the person grows up; the learning from parents and from experience of language, motor, and social skills, as well as values, morals, and religious precepts; informal social influences of peers; formal learning experiences in public school and college; and so on.

Because phenotype is determined by an interaction between genotype and environmental influences,

the behavior of individuals in any research study (i.e., their phenotypes—for example the individuals' aesthetic preferences) can *never* be taken to prove a genotypical cause of that behavior. Even if all subjects in a study strongly preferred natural stimuli, we *cannot* conclude that genetic makeup was a significant cause of that preference, because environmental influences like culture, values, and other components of upbringing might be primarily responsible. Thus, a person from another culture—an Eskimo, Ethiopian, Gururumban, and so on—might behave differently.

This book is devoted to an in-depth exploration of aspects of human behavior that contribute to environmental problems. These aspects may include genotypical ones (e.g., genetic perceptual/cognitive predispositions, as claimed by some theories discussed near the end of this chapter) as well as ones mainly acquired through learning and environment. If we want to claim that a human genetic predisposition influences a certain human behavior, we need stronger lines of evidence or grounds than experimental observation, as we discuss later and list in Table 8-4.

Developmental psychologists and anthropologists conceive of the interaction between genotype and environmental influences that determines phenotype as having *two related manifestations*:

One. Genotype sets limits. A person's genotype sets limits on how far environmental influences can push that person in determining his/her phenotype. For example, Down's syndrome, a condition which includes mental retardation, is caused by a genetic defect (the presence of an extra twenty-first chromosome). A person with this defect, even if given the most supportive and superior environmental influences (diet, medical care, education, etc.), will not likely grow up to become a research physicist, and may not even become a self-sufficient adult who is able to hold a regular job.

While the Down's syndrome example centers on the role of genotype at the level of individuals, genotype plays a similar role at the *species level*. Humans share certain genetic characteristics that set limits on our physical and psychological development. Consider the following prosaic examples: No human

Genotype x Environmental Influences ⟶ Phenotype

_____

FIGURE 8B2-1 The Interaction Between Genotype and Environment

(continued)

BOX 8-2 continued

can train or exercise enough to live underwater for long periods without artificial breathing equipment, lift a large building, or broad jump across the Mississippi River at its widest point. Two examples of species-level *psychological* limits: No human, regardless of training, can fully comprehend a half-dozen simultaneously presented, novel, and unrelated lectures or speeches, or solve in several seconds a novel set of a thousand simultaneous equations without assistance.

To repeat, the human genotype sets general limits on what environmental factors can do in shaping human phenotypes. While humans are incredibly intelligent, adaptive, and creative creatures, humans are *not infinitely plastic*. Our environments cannot push us into virtually any shape or form.

Two. *Genotype provides "pushes" or predispositions* in certain directions: Genotype provides "readinesses," that is, makes certain phenotypical traits easier to acquire than others. As a nonhuman example, consider classic research on chaffinches, a type of bird that sings a distinctive song. Research in the 1940s centered on the influences that "cause" the chaffinch song to develop. In an experiment, chaffinch eggs were incubated, hatched, and the chicks reared in isolation, unexposed to the normal chaffinch song from other chaffinches. Weeks later, the birds began to sing a song, one recognizable as chaffinchlike, but not completely "correct." This outcome clearly shows the operation of a genotypical influence or predisposition for the song-type, one sufficient for the birds reared in isolation to approximate the correct song (the outcome also shows that environmental exposure or lack thereof determines the degree to which the genotypical potential is realized).

A simple example of a genotypical push in humans: Behavioral scientists agree that humans as a species have a genetic propensity for language acquisition, that is, the psychological and physical machinery for quick acquisition of vocabulary and grammatical rules and a strong tendency to use the machinery for that purpose (e.g., Wade and Tavris, 1990, pp. 474–5).

---

people have a strong aesthetic preference for natural outdoor scenes over urban scenes devoid of natural stimuli. While most of these studies involved North American subjects and subjects from European countries, as we noted above, a few studies involved Asian subjects. The results suggest that Asians share with Europeans and North Americans similar patterns of preference for outdoor scenes. One recent study involving Asian subjects was done by Y. Yi (1992) in an unpublished doctoral dissertation at Texas A & M University. The study is described by Ulrich (1993) as follows:

[The study by] . . . Yi (1992) investigated the roles of cultural and occupational differences in influencing the natural landscape preferences of diverse groups of South Koreans and Texans, including farmers, ranchers, and nonfarmer urban groups. Individuals were shown a collection of color photographs depicting diverse natural settings in Korea and Texas. The collection included several scenes from Korea and Texas that contained features having strongly positive associations for one of the cultures but not the other—for example, a Korean landscape with a distinctive mountain known to Koreans but not Texans as the site of a famous Buddhist temple. Despite stacking the deck in this manner in favor of cultural influences, Yi's results reveal high agreement among all groups in their aesthetic preferences. Differences attributable to culture and occupation were statistically significant but comparatively minor, accounting for little of the variance. It should be mentioned that the groups were similar in according especially high preferences to landscapes having water features or savanna-like characteristics [p. 93].

However, there are few other studies of the scenic preferences of non-Westerners (Ulrich, 1993). Further, there are two possible objections to the Yi study: First, Korean and other Asian cultures have been significantly influenced by Western cultures; the scene preferences of the Korean research subjects may reflect this Western influence. Second, the main biome found in Korea as well as in much of the United States and Western Europe is temperate deciduous forest (Miller, 1992); therefore similarities in the scenic preferences of the Koreans and Western subjects may trace from their predominant exposure over their lifetimes to this biome type. Though the results of the Yi study are in the right direction, clearly additional cross-cultural research is needed.

Behavior-Genetic Research, and "Biophobias" as Genetically Influenced Behaviors. The results of some behavior-genetic research on the phenomenon of bio*phobia* are consistent with the biophilia hypothesis. Before we describe this research, we first describe biophobias and evidence of their existence. Ulrich (1993) and others argue that just as humans may be genetically predisposed to favor certain natural stimuli, we may also be predisposed to fear and/or avoid other natural stimuli. The latter stimuli are elements of natural environments—like snakes and spiders—that are dangerous to humans. Several laboratory studies have found that people more easily acquire classically conditioned phobias to these "primal" stimuli than to other natural stimuli, and that phobias to these stimuli are especially resistant to extinction.

Much of this conditioning research has been done by Arne Öhman and his colleagues (e.g., Öhman et al. 1985, cited by Ulrich, 1993). Öhman et al. used a painful, but not harmful, electric shock to create classically conditioned phobias in human subjects toward certain stimuli. Classical conditioning is the form of learning first studied by the Russian physiologist Ivan Pavlov in the early 1900s. In classical conditioning, a human or animal subject is repeatedly presented with a neutral stimulus (e.g., a bell), at the same time it is presented a stimulus that reliably elicits an unconditioned reflex (e.g., food, which produces a reflexive salivation response). The result is that eventually the neutral stimulus (bell) comes to evoke the response (salivation) by itself, without the use of the reliable stimulus (presence of food). When a neutral stimulus (such as a bell, tone, or light) is paired with a *painful or threatening* stimulus (e.g., an extremely loud noise or an electric shock), the result is that the research

subject will acquire a strong phobia or visceral fear response to the once neutral stimulus.

Öhman et al. found that subjects acquired fear responses more rapidly, that is, in fewer trials, to photos of snakes and spiders (primal stimuli) than to geometric figures. Furthermore, conditioned fears of snakes and spiders extinguished (were forgotten) far less readily than conditioned fears of geometric figures. That is, following conditioning, when the feared object was repeatedly presented without the simultaneous electric shock, the fears of the geometric figures weakened more quickly over repeated presentations than did the fears of snakes or spiders. Further, Ulrich (1993) reports research evidence that people readily learn fears of certain natural stimuli, including snakes, spiders, and rats, *vicariously,* that is, by watching *other people* react fearfully to these stimuli. In contrast, the vicarious learning process does not operate for other natural stimuli, for example, berries.

One might object that the subjects in Öhman et al.'s research were predisposed to fear spiders and snakes but not meaningless geometric figures because of the subjects' Western upbringing (which portrays these animals negatively) or prior negative personal experience with these animals. This objection is countered, Ulrich (1993) argues, by other research in which fears classically conditioned to handguns and frayed electric wires—which are greater threats to most urban and suburban people than are spiders and snakes—extinguished more rapidly than fears conditioned to spiders and snakes.

Still, one could argue that the findings of the conditioning research might, in some other subtle way, reflect subjects' prior learning and experience, and that the findings don't forcefully demonstrate the existence of genetically-based tendency toward biophobias.[8] This is where behavior genetic research, as we noted in Table 8-4, provides more compelling evidence for a genetic influence. Behavioral-genetic research closely studies the behaviors and personalities of identical twins, fraternal twins, and other family members. In general, these studies have shown that identical twins (who share the same genetic makeup) are more alike on a variety of behaviors and traits, including neuroticism, social smiling, bodily responses to stress, and others, than are fraternal twins (who have different genetic makeups). The greater similarity of identical twin pairs than fraternal twin pairs indicates that the relevant behaviors and traits have a genetic component. (We remind the reader that as we pointed out at the beginning of this chapter and in the Phenotype versus Genotype box, a genetic predisposition [genotype] *pushes* a person in a certain direction only, but actual behavior [phenotype] is determined by the interaction of genetic makeup and a variety of environmental influences.)

Ulrich (1993) reviews behavior-genetic research by Kendler et al. (1992) that is specifically relevant to the biophobias concept. The Kendler research involved a large number of subjects—over 2,000 female twins—and sophisticated statistical analyses. Kendler et al. found that about one-third of these twins had lifelong phobias of one type or another. Focusing just on phobias toward snakes, spiders, other insects, and bats, further analyses showed that identical twins were significantly more alike (either both having these animal phobias or both not having the phobias) than fraternal twins. These findings provide evidence that genetic differences play a significant role in individual differences in phobias of primal stimuli. Going a step further, if people vary in genetically based predispositions toward *phobias* of certain animals, it is not unreasonable to conclude that people could also vary in genetic predispositions toward *philias,* or attractions, to other animals or environmental stimuli. Thus the behavioral-genetic research lend support to the biophilia hypothesis in that it shows that genetics do play a role in people's reactions to natural stimuli. However, there is one limitation of behavior-genetic findings such as those above. The findings do not prove that humans *as a species* share a genetic predisposition for either biophilia or biophobias (Segal, 1993). In other words, the findings show individual genetic variability, but don't prove an average, or species-level, predisposition. (There are several other specific behavior-genetic research designs that can lend support to a claim of an evolutionarily based genetic behavioral predisposition. A full discussion of these designs is beyond the scope of this book, but Segal [1993] reviews many of these designs and related issues.)

Brief Overview of the Biophilia Research

The tone of our review, in the sections above, of the biophilia research is somewhat critical. When held up to the strictest standards of empirical proof for a genetic predisposition, the evidence for biophilia does not appear at this point compelling. Also, little of the research addresses the claim that humans can suffer psychological or physical harm from a lack of natural stimuli. However, as we noted above, we are impressed by the number and variety of research results that are consistent with the biophilia hypothesis. The claims of Wilson and others are provocative and fascinating, and are of great significance if they are valid. The biophilia hypothesis is relatively new, and there is much research on it yet to be done. We eagerly await the results of this research.

GENETICALLY BASED SEX AND REPRODUCTIVE URGES

The second item in Table 8-1 is biologist Paul Ehrlich's claim that past-occurring superexponential global population growth and current exponential growth (Chapter 1) derive, in part, from a human genetic predisposition to reproduce. Ehrlich uses the third ground or criterion we listed in Table 8-4 for inferring the existence of a genetic behavioral predisposition. This ground or criterion involves a logical argument showing that natural selection and biological evolution would have to favor a genetic predisposition toward the behavior in question. To put it another way, the argument is that, given that Darwinian forces and biological evolution operate in the way they do, a certain genetic behavioral predisposition would have to have become predominant.

Ehrlich's argument is simple and straightforward (Ehrlich, 1978; Ornstein and Ehrlich, 1989). He first notes that the essence of natural selection (as we discussed above) is that individuals with traits beneficial or adaptive in the environment they occupy tend to survive and pass their genes on to subsequent generations. Eventually, the genes of such individuals predominate in the species' gene pool. Ehrlich points out, however, that mere survival is *not enough* to guarantee the passage of an individual's genes to subsequent generations. The behavior of surviving individuals must *also* guarantee there will be subsequent generations. Consider the fate of individual animals that are well adapted to a specific environment and thus survive, but that either lack sex drives or don't want offspring and won't help raise them. Clearly, such individuals wouldn't pass on their genes despite the fact that they survived. The key issue in natural selection is differences in rates of reproduction. Genetic traits that make for greater reproductive success become predominant in the species' gene pool. Thus, Ehrlich argues, strong genetic predispositions toward sex and reproductive urges must be selected for by Darwinian forces and become predominant.

But Ehrlich's argument continues: The genetic predispositions for sex and reproduction were appropriate and adaptive at earlier times in human and prehuman history when food was scarce, disease rates high, and life expectancy short. However, *cultural evolution* radically altered these conditions (Figure 8-4). With the advent of agriculture (10,000 years ago) and of modern medicine and public health measures (within the last 150 years), the population death rate decreased dramatically—an unprecedented event. Never before had a species been able to dramatically change the "game plan" of evolution in this way. Further, given the long time periods needed for major changes in genotype via biological evolution, modern humans retain Stone Age predispositions for sex and reproduction. Ehrlich argues that this discrepancy between Stone Age human genes and the conditions of modern life created by cultural evolution is a major cause of the global population explosion—the superexponential growth (exponential growth but at successively higher rates) that began many hundreds of years ago and continued until 1970 or so, and exponential growth since 1970, which we discussed in Chapter 1. To repeat, the basic argument: Genetic predispositions adaptive and selected-for in the Stone Age became no longer adaptive and became a major cause of a contemporary environmental problem.

Implications of Ehrlich's Argument

Ehrlich's (1978) argument has some simple and direct implications concerning methods to lower birth rates and slow population growth. His analysis underscores the weakness of two birth-control methods—*sexual abstinence* and the "*rhythm method*" (selective abstinence, i.e., a couple having sex only at "safe" times in a woman's menstrual cycle), the only birth-control methods approved by some religions leaders (including the Pope of the Roman Catholic Church). Ehrlich and many others argue that these methods are misguided and ineffective because they push against strong genetically predisposed sex drives. In this view, recommending sexual abstinence is like asking people to stop eating for long periods—it can be done up to a point, but it goes against basic human drives. Note, finally, that the rhythm method, even if practiced with scrupulous care, is not a very effective contraceptive.

Ehrlich argues that a better method of birth control is to *decouple sexual behavior from conception and birth*. Most people want to have sex more frequently than they want to have children. To allow for sex without unwanted pregnancies, Ehrlich advocates (as do many others) making safe and effective contraceptive devices and methods widely available, and educating people in the use of these devices and methods. The need for these services is still great: Approximately 400 million women in developing countries want these services but don't have access to them (Miller, 1992).

Global Population Growth— Other Causes and Solutions

Ehrlich's (1978) analysis, above, reveals only one cause of exponential global population growth. There are several others, as Ehrlich himself pointed out. An additional major cause in developing countries is that the social structure in those countries makes people want many children. Public opinion surveys show that couples in many of these nations want three to six children (Miller, 1992), well above the 2.1 children per family "replacement level" that would eventually produce zero population growth (after a delay of sixty years—recall our discussion of "population momentum" in Box 1-2 of Chapter 1). Aspects of social structure that makes people want many children include: The lack of government social security programs (one's children are the only source of support in old age); high infant and child mortality (a couple must have many children to ensure that one or more survive into the couple's old age); and other aspects we discussed briefly in Chapter 2. Therefore, family planning or planned parenthood programs, which enable people to avoid unwanted pregnancies, are not enough to slow and stop population growth. They need to be supplemented by programs that decrease the number of children many people want.

Programs and measures likely to be effective in doing so, according to Repetto (1986) and others who have studied the issue, include: Providing good medical care for pregnant women and postnatal care for infants and children, increasing educational and occupational opportunities for women, and creating reliable government social security programs to support people too old to work (see Chapters 1 and 12).

Note, finally, that the number of children per couple in many Western nations and even some developing countries is close to 2.1, and it is even lower in some others. This indicates that any genetic predispositions for reproduction humans have can be countered by the right social structures, incentives, contraceptive programs, and so on. In this sense, cultural evolution, which greatly lowered population death rates, especially in the last hundred years, has in many nations succeeded in lowering birth rates to achieve a stable population. (Further discussion of these changes and the "demographic transition" are beyond the scope of this book). Whether developing countries that have high birth rates can lower birth rates rapidly enough to avoid negative self-reinforcing spirals of population growth and environmental deterioration remains to be seen.

ARE HUMAN BEINGS SHORT-TERM EGOISTS BY NATURE?

B. F. Skinner was a prominent proponent of a view of human nature that makes global environmental problems appear as almost inevitable (see the Egoism/Altruism cluster in Table 8-1). In his view, human beings are by nature acquisitive, egoistic, and shortsighted. It is quite predictable that such a species, as soon as it developed technology capable of increasing its material well-being, would use that technology without limit and without regard for tragedies of the commons it might cause, or damage that might result for future generations of the species itself. As we noted in Chapters 2 and 5, Garrett Hardin is also a proponent of this view, which has historical roots in Western political philosophy going back at least as far as Thomas Hobbes. Hardin's characterization of environmental disasters as tragedy implies that as with the Greek tragedies, the root causes lie in human nature.

Because it is so influential, the argument that environmental problems are an inevitable outgrowth of human nature is worth careful examination. The argument's basic premise is that predispositions toward egoism and shortsightedness are genetically inherited and endemic to all human populations. If this is true, proponents reason that egoism and shortsightedness are universal features of human behavior, and that efforts to restrain egoism and promote farsightedness will always be uphill battles, and losing battles in the long run. This argument, if true, holds out a poor prognosis for most of the strategies of behavior change set out in the previous chapters. The only promising strategy for behavior change would be the use of incentives to individuals, and even incentives can be undermined eventually by creative individuals who can get benefits for themselves by evading incentives that are successfully restraining the behavior of others. This tendency to put oneself first is, after all, the dynamic Hardin described so powerfully in his essay on the tragedy of the commons.

The Argument That Natural Selection Favors Egoism

Let us examine the premise of the Skinner-Hardin argument: that egoism and shortsightedness are part of human nature. What is the basis for this twin claim? Essentially, it is that natural selection has, for millions of years, favored egoism and shortsightedness, with the result that over that period of time, only egoistic and shortsighted species survived—Homo sapiens being one of the most successful.

The argument goes like this: Consider the fate of any creature that was not a short-term egoist, that is, one that behaved in violation of the basic principles of operant conditioning. Imagine an individual animal, living long ago, that did not repeat behaviors that led to favorable immediate consequences to itself, or that repeated behaviors that led to unfavorable immediate consequences for itself. For example, imagine that the animal was hungry, ate berries of a certain type for the first time, found that they tasted good and quelled its hunger, and then, when hungry in the future, never again looked for and ate this kind of berry. Such a creature would probably not survive long, nor pass its genes on to subsequent generations. Similarly, an animal that put its paw into a fire caused by a lightning strike, having injured its other paw in a earlier fire, would not survive for long nor pass its genes on to subsequent generations. Thus, natural selection favors animals genetically predisposed to be short-term egoists, that is, to learn from their experiences, and consequently, the argument goes, that predisposition would come to predominate over time in any population of animals. In this way, natural selection ensures that a tendency to pursue immediate individual self-interest is a basic part of the genetic design of human beings and other animals. Skinnerians would further argue that the success of operant conditioning as a theory of behavior can be attributed to the fact that natural selection favors animals whose behavior follows its laws.

According to this argument, the short-term egoism that natural selection favors is very short-term, indeed. If there is a long delay between a behavior and its consequence, it is almost impossible for an animal to discover the relationship between the two (Skinner, 1978, pp. 19–20). For example, if an animal ate a certain root and after the passage of several months became more attractive to potential mates as a result, this consequence, advantageous as it is for fitness, would not induce the animal to eat the root more often. There are too many intervening behaviors that could potentially have been responsible for this result, and the animal would not make the correct connection. Propensities to learning, according to Skinner and his animal studies, are such that only relatively immediate consequences will control or modify behavior. The animal unexpectedly rewarded with access to mates would be more likely to repeat some unrelated behavior that immediately preceded its reward than to return to find the root. (In fact, this is how Skinner explains superstitious behavior—animals, including humans, often repeat behaviors that immediately precede rewards, even if there is in fact no cause-and-effect relationship.)

Skinner recognizes that animals, including human ones, do sometimes work for long-term rewards. He argues that this tendency is learned through processes that link delayed rewards to immediate ones. For example, most people work at their jobs to receive money, a secondary reinforcer (so called because it has value only because of the things one can get with it), even though there is usually a long delay between doing the work and getting anything of immediate value, such as food or shelter. But this qualification merely underlines Skinner's main point that the only way to motivate any behavior is to link it, either directly or indirectly, to immediate consequences to the individual.

In Chapters 2 and 5, we noted the implications of this view for environmental problems. If people are by nature short-term egoists, the exploitative behavior that causes environmental destruction is simply an enactment of human nature. Consequently, the only reliable way to change it must be to devise incentives that harness short-term egoism by providing relatively immediate personal consequences, normally lacking in nature, that capture the delayed and diffuse environmental consequences of people's behavior and bring them into the very near future for the individual whose behavior is to be changed.

But is short-term egoism in fact embedded in human nature to the degree that Skinner and Hardin presume, or are there countervailing tendencies in human nature that predispose people, at least under certain conditions, to act in the long-term interest of larger groups of which they are a part? If there is more to human nature than short-term egoism, appeals to egoism may not be the only way to prevent environmental destruction.

These questions are central to a current debate among researchers in genetic and cultural evolution. In this debate, the question is framed in a slightly different manner: Can altruism be consistent with the principles of evolution? (Among evolutionary theorists, an altruistic act is usually defined as one that helps another individual survive and/or reproduce and that is performed at some cost to the helper's ability to survive and/or reproduce.) If the forces of natural selection are always weighted against altruistic individuals, Skinner and Hardin are essentially right. But if natural selection can favor altruism, at least under certain conditions, human nature is more complex than Skinner and Hardin allow, and the argument is severely weakened that the tragedy of the commons is preordained by human nature. Two basic lines of argument support an evolutionary basis for altruism and thus tend to undermine the claims of Hardin, Skinner, and the others who share their views.

Arguments that Natural Selection Favors Altruism

A good case can be made for altruism as part of human nature on the basis of cultural universality. There is little doubt, for instance, that human beings in all cultures are intensely social animals. Group living, cooperation with other group members, and division of labor appear to be culturally universal features of human life, not only across cultures (Gray, 1991), but also across history, dating back to the Stone Age. Moreover, apparently altruistic behavior, such as the willingness of parents to sacrifice for their children, also seems to be culturally universal.

There is also a strong argument that a genetic propensity toward group living, division of labor, and intragroup cooperation were favored in Homo sapiens by natural selection. First of all, our species, like other primates, has a long period of childhood dependency during which the young cannot survive without continuing protection and nurturance. During this period, the nurturers are handicapped in terms of food production and defense, so that the survival of the young would be promoted by living in a group larger than just the mother and infant. Further, a group of cooperating Stone Age hunters could stalk and kill animals that were too large for a solitary hunter to stalk and kill, and would be able to provide food for the young and their nurturers. Groups acting

in concert could also repel threatening predators more effectively than could individuals, and thus protect their progeny. Note also that a division of labor (e.g., some individuals hunt, other individuals gather) permits individuals to specialize in a subset of the tasks on which survival depends, gaining further advantages for groups. Finally, group living and cooperation greatly increase the ability of our species to survive, because group living made possible all the benefits of cultural evolution: the development of language, and the transmission from person to person and across generations of survival-enhancing information (culture). In fact, culture and the long period of childhood dependency are mutually reinforcing characteristics. Childhood dependency makes it necessary that human beings live in groups where they can develop culture, and at the same time, culture is more easily transmitted when the young have extended periods of dependence on, and close interaction with, caregivers.

But even if natural selection favors a propensity to live in groups and cooperate, it does not follow that human nature predisposes individuals to sacrifice their own prospects for survival in order to advance the welfare of their groups. How is it possible to square the evidence of apparent altruism with the fact that natural selection favors individuals whose characteristics promote their *own* survival?

Explaining Altruism as Egoism: Sociobiology and Reciprocity

One answer is that the human social patterns of group living, cooperation, division of labor, and even altruism can actually be consistent with a view of humans as essentially egoistic, engaged as individuals in a Darwinian struggle for survival. Some of the most influential arguments that apparent altruism is ultimately egoistic come from the interdisciplinary field of human sociobiology.

Sociobiology begins with the insight that although natural selection works on individuals, it is actually genes, not individuals, that become more or less common as a result of selection. If evolution is viewed from the standpoint of a gene carried by a particular individual, "fitness" can be advanced equally well by the procreative success of anyone who carries the gene (Dawkins, 1976). In particular, if the individual who carries the gene sacrifices his or her life so that others who carry the same gene can survive and reproduce, the gene passes on. When the sacrifice is made for close relatives, a large number of the individual's genes can be passed on to the next generation, even if the altruist has no offspring. This particular sort of altruism should be favored by natural selection because any gene that predisposes individuals to make sacrifices for close kin would help pass that individual's genes on to the next generation. In this sense, and as paradoxical as it sounds, it can be to the individual's biological advantage to sacrifice life: Under the proper circumstances, self-sacrifice is the selfish thing to do.

Sociobiologists have developed the concept of "inclusive fitness" (Hamilton, 1964; Wilson, 1975) to measure an individual's reproductive success in a way "that includes the effect of an individual's behavior on its genetic relatives" (Boyd and Richerson, 1985, p. 13). They analyze behavior on the assumption that it tends to maximize inclusive fitness (i.e., to pass on the individual's genes), and they theorize that specific behaviors that maximize inclusive fitness are the ones that survive. An implication is that whatever behaviors people generally engage in at present must tend to increase inclusive fitness. If the behaviors did not, they would have died out. This conclusion must apply to inherited traits and behaviors shaped by natural selection and biological evolution, but also to nongenetic traits and behaviors transmitted as culture across generations, since culture is also subject to a natural selection-like process over time, as we discussed in Box 8-1. Thus, the sociobiological analysis implies that it is not possible to maintain over long periods any cultural innovation (such as a way of managing com mon-pool resources) that is deleterious to the inclusive fitness of the individuals who would adhere to it.

According to the theory of inclusive fitness, natural selection favors altruism only toward close relatives. Parents, children, and siblings share an average of 50 percent of their genes, so helping two of them survive and reproduce is as good, in terms of fitness, as helping yourself. First cousins share 25 percent of

their genes, so helping four of them is as valuable for fitness as helping oneself. More distant relatives share even less genetic material, so that sacrificing oneself for them has very little advantage in terms of inclusive fitness. So, the theory predicts altruism toward very close relatives, but egoism in dealings with nonrelatives.

This sort of sociobiological explanation can account for some very compelling instances of altruism. Mother animals that jeopardize or sacrifice their lives to save their young, for example, by distracting a predator from the nest, are actually increasing their inclusive fitness. Individual animals that act as "scouts" for a herd or flock, and endanger themselves by sounding warning calls, are also enhancing their inclusive fitness if the herd or flock includes many close relatives, as it often does. So, what appears to be altruistic behavior is in reality selfish.

The theory of inclusive fitness implies that the only altruistic behavior that is likely to persist in the face of selection pressures is that in which sacrifice is made for close blood kin. Individuals who sacrifice themselves for nonkin decrease their inclusive fitness, with the result that all their behavioral predispositions (including the ones responsible for altruism) will tend to disappear over time in the population. So, inclusive fitness theory implies that altruism is a very restricted phenomenon and moreover, is not truly altruistic in motive.

Inclusive fitness theory offers little hope for promoting proenvironmental behavior. It implies that if proenvironmental behavior offers no clear advantages to oneself or one's close blood kin, it is selected against. Thus, even if such a behavior can be elicited by appeals to morality, conscience, or concern for others, it will be a fragile achievement unless there are strong benefits to kin. Appeals to protect the environment in order to benefit nonhuman species should have even less success, because none of the beneficiaries are blood kin. Protecting the environment in the name of abstract principles such as beauty, or the balance of nature, is also likely to reduce inclusive fitness, because the benefit of the sacrifice goes equally to all, regardless of blood relationship. So a personal sacrifice offers no competitive advantages for kin, and is therefore selected against. For these reasons, the inclusive fitness theory of altruism is very little different from the theories of Hardin and Skinner in terms of its implications for environmental problems.

There is an evolutionary argument based on underlying egoism that extends the potential for some human social tendencies, particularly cooperation, beyond close kin, under conditions where reciprocity is possible. This argument is best known from the work of the political scientist Robert Axelrod (1984), who demonstrated it with mathematical proofs and a series of laboratory games (on the "prisoner's dilemma"). Essentially, the argument is this: Individuals are often involved in situations in which the best overall outcome is obtained if every individual cooperates (that is, restrains egoism), but in which individuals can do still better if they act egoistically while others are acting as altruists. Because individuals cannot know in advance whether or not others are cooperating, the logic of these situations seems to compel everyone to act as an egoist even though all suffer as a result. Hardin's description of the tragedy of the commons has just these characteristics.

Axelrod has shown that in such situations, if individuals can expect to have a long series of interactions with the same other individuals, they can benefit by using a strategy called *tit for tat,* in which they act altruistically at first, but then reciprocate every egoistic—or altruistic—act of the other. Individuals who follow this strategy are never exploited more than once by anyone else, and they gain considerable adaptive advantages when interacting with others who follow the same strategy because they always achieve the best collective outcome. Over time, this advantage can cause the proportion of reciprocal altruists to grow within a population. Thus, if a human group is structured to promote continued, long-term interaction, its members (even nonkin) can avoid the tragedy of the commons by following a strategy that appears to be altruistic. In fact, of course, the strategy is egoistic at root. This is because apparent altruism occurs only when there is a reasonable expectation that the cost of cooperation will be repaid by the cooperation of others.

The argument that human beings and other organisms can have a genetic propensity for reciprocal altruism is somewhat more hopeful in terms of environmental problems than the argument from inclusive fitness. It suggests that even among groups composed mostly or entirely of nonkin, restraint of egoism can be accomplished and maintained in the face of selection pressures, and without adding incentives other than those already contained in the social situation. The necessary condition is that individuals live in groups in which they have a high probability of continued, long-term interaction with other group members.

It is of more than passing interest that this condition is met by the small bands in which the human species evolved, and also by the small agricultural villages that were apparently the next major innovation of human social organization. As we found in Chapter 6, the fishing community of Alanya, Turkey, the agricultural community of Törbel, Switzerland, and other success stories of community management have in common a condition of long-term stability of group membership. Thus, one of the key conditions for successful community management is also a key condition for reciprocal altruism. This may not be a coincidence: It suggests that one reason small, stable communities have been able to maintain shared resources over long periods is that they have devised social institutions that take advantage of a part of human nature that predisposes people to be reciprocal altruists.

Reciprocal altruism expands the possibilities for long-term success in environmental management beyond those suggested by pure egoism and inclusive fitness. People can increase their fitness by being altruistic not only toward close kin, but also toward people and communities with which they expect to have long-lasting relationships. Still, if these are the only kinds of altruism consistent with natural selection, the prospects for environmental management in modern societies seem limited, because most individuals do not have many long-term relationships, and they frequently break off their relationships to move to greener pastures. For such individuals, incentives still seem to be the only viable strategy. There is, however, another evolutionary account of altruism that suggests that a wider range of intervention strategies can be consistent with evolved human nature.

A Cultural Evolutionary Account of Altruism

Cultural evolutionary theories of altruism suggest that a propensity to social learning, characteristic of human beings in all kinds of communities, can produce altruistic behavior even in modern societies. Possibly the simplest statement of this theory appears in a paper by Herbert Simon (1990). Simon first points out the adaptive advantages of social learning—it provides knowledge at lower cost than trial-and-error learning, and it teaches behaviors that get approval and support from others, something that promotes the fitness of animals that depend on others for survival. He then proposes that individuals vary in a quality he labels "docility," meaning a tendency to accept instruction from others in society. Simon points out that a tendency to learn from others rather than to rely on personal experience has adaptive advantages because it replaces trial-and-error learning with a procedure that leads to the same results at much less cost in time, effort, and dangerous errors. Social learning is uncritical, however—the benefits can be gained only if the behaviors being taught are accepted, as it were, on trust, and not always checked through experience. It is therefore possible for docile individuals to learn things that have net costs for them as individuals, including altruistic behaviors. Docility will still confer adaptive advantages on the individual as long as the cost of the altruistic behaviors is less than the adaptive value that docility confers because of its other advantages. Needless to say, it can be valuable for a society's survival if among the things it teaches its young are sets of altruistic behaviors that benefit the group as a whole more than they cost individual members. So, a society that uses its members' propensities to social learning to teach them a mixture of skills—some that are valuable to them as individuals and some that benefit the whole group—is one that increases the fitness of both the individuals and the group. Individuals living in such a society experience selection pressures for social learning and for altruism, and the proportion of altruists in the society can increase from one generation to the next because of the net advantages of docility.

What would keep individuals from rejecting particular altruistic behaviors because they are personally costly? Simon's answer is that the very feature that provides the adaptive advantage of social learning—its ability to provide useful skills quickly and at low cost—would be compromised if people undertook the effort necessary to evaluate each behavior they are taught to see if it benefited them as individuals. Further, because of intrinsic limits to human perceptual and cognitive capabilities (what Simon calls human "bounded rationality" and which we discuss in Chapter 9), it would probably be impossible for a person to carefully evaluate all the behaviors taught. Thus, to gain the adaptive advantages of social learning, "docile" individuals must forgo evaluating each behavior they are taught—the altruistic behaviors as well as the ones that benefit them personally. Social learning must be uncritical to gain its full adaptive advantage, and if it is uncritical, individuals will not screen out the altruistic behaviors they are taught.

Simon's formulation and other similar ones (Boyd and Richerson, 1985; Caporael et al., 1989) differ from the theories of inclusive fitness and reciprocal altruism in that they do not explain altruism away as a form of covert egoism. They demonstrate that tendencies to make sacrifices to benefit other people—even with no expected benefit to oneself—may be consistent with evolved human nature under strict criteria of selection. Simon is not making the simplistic claim that it is part of human nature to be altruistic as a general trait. Rather, he is making two more specific and more interesting claims. First, he proposes that it is part of human nature for people to inherit various degrees of a propensity to learn from others uncritically, that is, without first considering whether what they learn is immediately beneficial to themselves. Second, he is claiming that individuals can gain adaptive advantages if they live in groups that teach them mixtures of behaviors that include self-sacrifice, provided that the costs to individuals of self-sacrifice are, on the average, less than the benefits they gain from the rest of their social learning. This formulation implies that the amount of social learning and of altruism that is adaptive for individuals depends on what the society teaches. A society that teaches its members skills of great value can demand a great deal of altruism from them without destroying the adaptive value of social learning. The amount of altruism in "human nature" is therefore changeable—societies can structure themselves so that it will pay people to be altruistic often, or rarely, and so that the proportion of altruists increases or decreases over time.

It is important to emphasize that if altruism follows the docility mechanism, it need not be restricted to blood kin or to groups in which there are many long-term relationships between individuals. Also, the groups within which people are altruistic can potentially be much larger than the small bands in which the human species probably evolved. Indeed, in Simon's analysis there is nothing that prevents an entire society from teaching altruistic behaviors to members without running afoul of selection processes, provided that the costs of that altruism are not so large as to wipe out the other advantages of uncritical social learning.

The theory of docility does not, of course, prove anything about the ability of humanity to solve its environmental problems. It does, however, demonstrate that altruism need not run counter to the evolutionary forces that shape human nature, and it thus raises serious question about Hardin's argument that environmental tragedy is inevitable because altruism is against human nature. The argument from cultural evolution also helps make sense of the findings we reported in Chapter 3 concerning human values. Not only have there been long-lasting societies and religions that place intrinsic value on the environment (a set of values that would not seem to advance inclusive fitness or ensure reciprocity), but adherence to a variety of nonegoistic values appears consistently across a wide range of modern societies. In all societies so far studied, researchers find not only the egoistic values that Shalom Schwartz refers to as self-enhancement and openness to change, but also values of upholding tradition and of self-transcendence, including concerns with justice, fairness, peace, and environmental quality. These unselfish values should have disappeared if selection favored only overt and covert egoism, but they should have survived if selection favors altruistic tendencies that benefit the cultural groups in which individuals are raised. Thus,

the persistence of traditional and self-transcendent values provides some indirect support to arguments that natural selection favors a human propensity toward altruism.

Cultural evolutionary considerations suggest that tendencies toward altruism, including proenvironmental behavior, can be learned without diminishing the adaptive success (fitness) of those who learn them. The specific conditions have not yet been spelled out by theorists, but we will hazard some speculative comments. First, it would seem that societies will be more likely to teach proenvironmental behaviors if sacrifices by individuals are seen to benefit the society as a whole. Under those conditions, what the individuals give up is partly compensated by advantages to the group. Sacrifice in the name of other species or abstractions such as "the balance of nature" may be more difficult to motivate unless those responsible for teaching believe that there are ultimate benefits to be gained for society. If this speculation is accurate, proenvironmental behavior may be more easily justified and taught on the basis of a homocentric ethic than on the basis of an ecocentric one. We should qualify this conclusion, however, by noting the prevalence of ecocentric ethics in various human religions that have survived over centuries or millennia.

Second, the cultural evolutionary argument implies that societies can exact more altruism from individuals if they teach skills that offer great adaptive advantages to those individuals. Another way to put this may be that people will sacrifice more out of loyalty to, or identification with, a society if they gain much from participating in it. Societies that teach proenvironmental behaviors may therefore be more successful in good times than bad, and with more advantaged social subgroups than with less advantaged ones. This line of reasoning suggests that improving the conditions of individuals in society, especially individuals in disadvantaged groups, may increase the likelihood that people will accept uncritically demands for proenvironmental behavior. We hasten to add that we would not expect social improvement to lead in any quick and direct way to increased environmental altruism. Selection pressures work more slowly than that. However, the cultural evolution argument does support the conclusion that over the long run, societies that benefit large numbers of their members should have greater success promoting environmentally responsible behavior by simply teaching the relevant attitudes and behaviors.

STONE AGE GENETIC PERCEPTUAL AND COGNITIVE PREDISPOSITIONS

We devote this final section to a review of two provocative theories listed at the end of Table 8-1. Both center on possible Stone Age genetic perceptual and cognitive predispositions that might contribute to environmental problems or interfere with efforts to solve them.

Hardin's Hypothesis: Genetically Based Denial

Hardin (1968) argues that humans have a genetic predisposition to "deny," that is, to underestimate, the risks to which they are subject. People tend to think of themselves as invulnerable or immortal, to think that bad things happen to others but not to themselves. Hardin presents several examples. He notes that in medieval England, convicted pickpockets were sentenced to death by hanging. When crowds formed to view these hangings, other pickpockets worked through the crowd picking pockets. Evidently, these pickpockets could see one of their "brethren" hanging from the gallows and yet think that the same fate could not befall them.[9] If denial of risk is a strong genetically predisposed human tendency, Hardin argues, then people—both the public and government policy makers—may underestimate the severity and probability of such potentially serious environmental threats as erosion of the Earth's ozone layer and climate change due to the greenhouse effect, and will fail to take preventive actions against these threats.

Hardin bases his claim of a genetic predisposition mainly on the cultural-universality argument in Table 8-4, but he also uses a version of the argument on Darwinian forces operating in biological evolution. Here, briefly, are both parts of his reasoning:

Hardin first documents informally the cultural universality of denial—its appearance in both ancient and modern civilizations, Western and non-Western cultures, and so on. He presents, for example, the case of King Croesus of Lydia (a country in Asia Minor on the Aegean Sea) in the fifth century b.c. Croesus was contemplating invading Persia, and he asked the oracle at Delphi whether such an invasion would be successful. To quote Hardin (1968): "[T]he oracle at Delphi . . . replied, with her characteristic ambiguity: 'If Croesus should send an army against the Persians he would destroy a great empire.' Delighted with the reply, Croesus attacked, and the prophecy was fulfilled: A great empire was indeed destroyed—*his* [p. 46]." Since denial is culturally universal, Hardin concludes, it must reflect a human genetic predisposition.

Hardin continues with an argument involving natural selection: If denial reflects a genetic predisposition, then denial must have been selected-for by Darwinian pressures during the evolution of our species. If it was selected for, then it must have been adaptive for our ancestors. But a very high level of denial couldn't possibly be adaptive; an animal, for example, that throws itself off a cliff knowing it can't fly or into deep water knowing it can't swim, wouldn't survive to pass its genes on. Similarly, very low levels of denial couldn't be adaptive: An animal with little or no denial would constantly cringe in fear and never venture forth from its home. However, a *moderate* degree of personal fearlessness and lack of caution *would be* adaptive, Hardin argues, because it would encourage the animal to explore new territories and frontiers—both geographically and culturally—and thus *promote discovery and invention.* An animal with moderate denial might discover new, timesaving routes through the dark and frightening forest, would experiment with new foods, new tools, and new hunting techniques, might discover new territories (e.g., the New World), and so on. Thus neither very high nor very low levels of denial would be adaptive, but moderate levels of denial would, so modern humans have inherited a genetic predisposition to behave this way.

Further, Hardin argues, moderate denial, though adaptive under the Stone Age conditions in which our species evolved, is less so under conditions of modern life. Since humans now can disrupt global ecosystems and erode key resources (because of the size of the human population, the nature of modern technology, and the intensity of its use), even moderate levels of denial toward environmental threats may be disastrous. Thus, Hardin argues, humans have an innate behavioral predisposition that our species must work against to ensure its survival.

While Hardin's analysis above is intuitively appealing, we believe that his natural-selection-based argument is less compelling than the ones used by Simon and others reviewed earlier in the chapter. Further, it is not clear that a tendency to deny immediate threats or dangers to oneself, even if it has a genetic basis, would necessarily apply to people's responses to global environmental problems, which pose diffuse and delayed threats to all of humanity. We discuss these issues in detail in Chapter 9.

Ornstein and Ehrlich's Theory: The "Old" Human Mind in the New World

The final item in our review of possible Stone Age genetic predispositions that may clash with conditions of the space age is the provocative claim advanced by psychologist Robert Ornstein and biologist Paul Ehrlich (1989). They argue that because of our evolutionary history, modern humans tend not to perceive and react to slow, gradual environmental deterioration. Since the environmental problems that now threaten our survival have developed gradually over a period of years, we do not fully notice them or take action against them.

Ornstein and Ehrlich (1989) justify their claim with a simple and intuitively compelling argument concerning natural selection (Table 8-4), though we view their argument as not as well supported as the ones used by Simon or Skinner. Specifically, Ornstein and Ehrlich claim that since our ancestors mainly faced quick and imminent dangers, the human mind is attuned mainly to perceive and take action against such dangers.

Hundreds of thousands or millions or years ago, our ancestors' survival depended in large part on the ability to respond quickly to threats that were immediate, personal, and palpable: threats like the sudden crack of

a branch as it is about to give way or the roar of a flash flood racing down a narrow valley. Threats like the darkening of the entrance of a cavern as a giant cave bear enters. Threats like lightning, threats like a thrown spear.

Those are not threats generated by complex technological devices accumulated over decades by unknown people half a world away. Those are not threats like the slow atmospheric buildup of carbon dioxide from auto exhausts, power plants and deforestation; not threats like the gradual depletion of the ozone layer. . . . (Ornstein and Ehrlich, 1989, [p. 8].)

Thus the human mind evolved to register short- term changes from moment to moment, day to day, and season to season, and to overlook the 'backdrop' against which those took place. . . . [p. 29].

Ornstein and Ehrlich further point out that the particular environment in which our hominid ancestors evolved—East African savannas—remained stable and unchanging for hundreds of thousands of years, thus giving little evolutionary impetus for humans to notice and react to slow environmental change. There were not even significant annual changes of season. Seasonal changes with winters sufficiently cold to require our ancestors' preparation first occurred when the global climate cooled during the Ice Age that began less than 2 million years ago. The necessity to prepare for and respond to a winter season may even have begun, Ornstein and Ehrlich suggest, only 1 million years ago, when some of our ancestors spread out of Africa to colder regions. Overall global cooling trends were slow enough to be undetectable in an individual's lifetime (though after global cooling had occurred, the annual seasonal changes that our ancestors faced were, of course, sufficiently rapid for them to notice and react to). Ornstein and Ehrlich argue further that not only couldn't our ancestors perceive changes much slower than annual climate changes, there was also little they could do about them; they either migrated or perished. Perhaps, Ornstein and Ehrlich speculate, biological evolution favored those individuals who did not even try to worry about slow changes.

Cultural evolution began to radically alter the situation about 10,000 years ago with the advent of agriculture (not sufficiently long ago for changes to occur in genetic makeup through biological evolution).

Humans began raising plant crops and livestock. The result was a stable and abundant food supply that made permanent human settlements possible for the first time (our ancestors before this were nomadic).

Approximately 5,000 years ago, new civilizations—including the Babylonian, Sumerian, and other Mesopotamian cultures—arose between the Tigris and Euphrates Rivers in the Middle East. These civilizations depended on elaborate systems of irrigation canals to provide water for arid cropland. Agriculture heavily dependent on irrigation, however, is not permanently sustainable. Silt continually fills in irrigation canals, which must be dredged, and, more important, continual evaporation of irrigation water leaves deposits of salt that build up in the soil, lowering its fertility. These problems apparently contributed to the collapse of the Mesopotamian cultures. In other words, with the rise of Mesopotamian cultures humans first became able to significantly alter and damage their physical environments. For the first time in human history, our "old minds," shaped by Stone Age conditions and unresponsive to gradual environmental deterioration, became a serious liability.

Over the next several thousand years, cultural evolution accelerated, and new technologies arose with increasing rapidity. With the advent of public health measures and modern medicine, human death rates plummeted, accelerating human population growth to superexponential levels. In just the last few hundred years, major new technologies appeared, such as steam and internal-combustion engines, electric motors, advanced metallurgy, atomic energy, and so on. Thus *Homo sapiens sapiens* made a dramatic alteration of the evolutionary game plan: Because of our technology, the intensity with which we use it, and our increasing numbers, we became able to damage or destroy global ecological systems and resource bases upon which all life depends and which have operated without human interference for millennia. Though these most recent changes occurred in only a few hundred years (barely an instant of geological time), the resulting rate of damage to the environment is, tragically, too slow for humans—with our "old" or Stone Age minds—to fully notice and properly respond to, according to Ornstein and Ehrlich.

Ornstein and Ehrlich (1989) claim not only that evolutionary history selected out a tendency for humans to perceive and respond to immediate threats but not "slow motion disasters," but also that evolution has biased human perceptual and cognitive processes in several other ways: We are especially responsive, they argue, to immediate and acute scarcities of food, water, and other personal resources; we give more weight to things we see and hear than to things we smell or taste; we make quick perceptual judgments about people, things, and events based on initial impressions, judgments that are hard to change even given subsequent contradictory information; we tend to simplify complex, multidimensional issues and problems into simple one-dimensional conceptions; and we are especially prone to make invidious distinctions between the "ingroup" and "out-group," that is, to consider people who are different or are outsiders as unlikable, inferior, and/or threatening.

Thus, Ornstein and Ehrlich argue, our Stone Age genetic heritage—the product of tens of thousands of years of biological evolution—predisposes us to perceive and think in several biased and distorted ways, ways that may now be harmful to us, or even disastrous. Ornstein and Ehrlich further argue that the process of cultural evolution has not yet taken account of and compensated for the genetic behavioral predispositions that threaten our survival. Ornstein and Ehrlich conclude that *Homo sapiens sapiens* must now deliberately and consciously intervene in the cultural-evolution process and develop institutions and practices that, as much as possible, compensate for our archaic Stone Age behavioral tendencies. This "conscious cultural evolution" would include radically new public school curricula, with much greater coverage of such topics as biological and cultural evolution, basic ecological principles, the explosive characteristic of exponential growth, and the nature of human perceptual and cognitive biases and distortions. Conscious cultural evolution would also require major changes in media programming, such as greatly enhanced TV coverage of environmental hazards that develop slowly, like damage to the Earth's protective ozone layer. Ornstein and Ehrlich further urge the creation of government-funded, multidisciplinary "think tanks" devoted to the study of environmental problems and their solution.

A more detailed discussion and critique of Ornstein and Ehrlich's (1989) book is beyond the scope of this chapter. We believe that their book makes a major contribution by raising the possibility of human perceptual/cognitive genetic predispositions that have been heretofore little discussed in professional journals or in popular literature on environmental problems. On the other hand, and as we discussed at length in Chapter 4, educational campaigns like the ones that Ornstein and Ehrlich emphasize are likely to have little positive impact by

themselves in the short run. In the long run, and in concert with the other strategies we discussed in Chapters 5 to 7, educational campaigns could make a difference.

Despite the limitations of Hardin's hypothesis and of Ornstein and Ehrlich's, the role of human perceptual and cognitive processes in people's reactions to environmental hazards is important and requires a more intensive and systematic look. We devote the next chapter (Chapter 9) to that goal. We return to the issue again in Chapter 11, where we discuss Jay Forrester's claim that environmental problems arise because the natural environment and humanity's effects on it are too complex for humans to understand.

NOTES

1. In 30,000–40,000 years, possible changes include differences in skin color (in response to differences in climate and solar radiation at different latitudes); adaptation to different altitudes; dietary adaptation; and, of course, resistance to some fatal infectious diseases (Bogin, 1988).

2. Note that this description oversimplifies a cyclical process involving our prehuman ancestors that occurred over hundreds of thousands of years: Biological evolution produced increases in prehumans' physical and psychological capacities for language and culture, which, in turn, permitted the beginnings of culture, which enhanced survival, which selected for higher genetic levels of these capacities, which enhanced culture and cultural transmission, which enhanced survival, and so on.

3. As a modern example: The first electronic computer made possible quick calculations of the trajectories of military shells and bombs, the purpose for which it was originally designed, but it later made possible a host of changes and improvements in scientific research, business practices, manufacturing processes, weather forecasting, and so on.

4. Figure 8-3 might appear to suggest that there was little change until 10,000 years ago. In fact, there were advances in stone and other implements between 40,000 and 10,000 years ago, as well as other changes (Jones, Martin, and Pilbeam, 1992).

We should also note that some anthropologists claim that humans were healthier *before* the appearance of agriculture and permanent settlements approximately 10,000

years ago (Armelagos and Cohen, 1984): They argue that the diets provided by early agriculture were inferior to those provided by hunting and gathering.

5. The biological evolution curve in Figure 8-4 is drawn as completely flat. However, we remind the reader that although our species has undergone little genetic change over the last 30,000–40,000 years, biological evolution during that period may have produced small genetic changes, such as adaptations to different altitudes, as we discussed in note 1.

6. This review does not extend beyond the one in the 1st edition. A number of relevant studies have been published subsequently.

7. There are other possible grounds or criteria (e.g., see Wilson, 1978, Chap. 2).

8. Also, Jared Diamond, in a chapter in Wilson's edited volume (1993), provides informal evidence suggesting that fears, specifically of snakes and spiders, are not culturally universal.

9. Another possible explanation: Acute poverty and unemployment may have given these pickpockets no other choice of vocation.

10. Note that natural selection operates on phenotypes, not genotypes (see Box 8-4 on phenotypes versus genotypes). Thus, for example, an individual with the genetic potential to attract a large number of sexual partners and have many offspring will pass on that genetic material only if, by the personal characteristics and behavior patterns he or she develops, that potential is actualized.

CHAPTER 9

HUMAN REACTIONS TO ENVIRONMENTAL HAZARDS: PERCEPTUAL AND COGNITIVE PROCESSES

CHAPTER PROLOGUE

In the final sections of Chapter 8, we discussed two theories, one proposed by Hardin (1969), the other by Ornstein and Ehrlich (1989). Both theories claim that modern humans have Stone Age genetic tendencies that could lead them to underestimate environmental threats and fail to take action against them. The two theories are provocative, but, as we saw in Chapter 8, theories about human genetic nature are difficult to prove or disprove.

There is, however, another body of information that can help us predict and understand people's reactions to environmental threats. For several decades, behavioral scientists have done laboratory and survey research on people's responses to a variety of specific hazards and risks. This research has been concerned with people's perceptions of, thoughts about, and actions toward threats, but generally unconcerned with whether genetic influences are involved. Some of this research centers on people's reactions to *natural hazards* like earthquakes and tornadoes. Some centers on people's reactions to *technological hazards* like the risks of autos and commercial air travel (including, in some cases, the environmental impacts of these technologies). Finally, some of the research centers on people's reactions to *threats to their personal health* (such as the threat of heart disease aggravated by smoking, or the threat of AIDS resulting from unsafe sex).

One general finding that emerges in this research is that people tend to underreact to certain risks under certain conditions. That is, people sometimes underestimate risks and/or fail to take protective action against them—a finding compatible with Hardin's

and Ornstein and Ehrlich's claims. However, a second main finding is that people also tend to *overreact* to certain risks and hazards under certain conditions. In other words, people may overestimate some risks and/or take unnecessary protective action against them—the opposite of denial.

We will argue in this chapter that many of the research findings on people's reactions to natural, technological, and health hazards apply also to people's reactions to *environmental* threats and hazards—the main focus of this book. (Indeed, it is technology, together with the scale and intensity of human activity, that causes environmental damage; also, some of the research specifically studied people's perceptions of the environmental impacts of technologies.) If the same patterns of under- and overreaction do, in fact, apply to people's perceptions of, and actions toward, environmental risks, consider that *both* tendencies could be maladaptive and harmful: Underreaction implies that the public and government policymakers may underestimate and not take action against some potentially serious environmental threats such as global climate change, ozone depletion, or the increasingly rapid extinction of plant and animal species (as Hardin, 1969, argued). Conversely, overreaction to some environmental risks may be as counterproductive as underreaction if it causes the public and policymakers to waste limited time, attention, and funds responding to risks that are relatively inconsequential.

Before we begin the main text of the chapter, we present three new examples—one of people's apparent underreaction, and two of people's apparent overreaction, to natural and technological hazards. Our first example is similar to ones we discussed in Chapter 8 in our coverage of Hardin's and Ornstein and Ehrlich's work:

EXAMPLE A: Underreaction to a Natural Hazard: UCLA Students and Earthquakes

In September 1985 two massive earthquakes struck Mexico City, killing approximately 10,000 people, injuring 30,000, and leaving 150,000 homeless (Ford Foundation Letter, 1986). These quakes reminded people in other quake-prone parts of the world—including the West Coast of the United States—of the dangers they faced. Geologists had warned for years that pressure was building along the San Andreas and other California fault lines, and that large earthquakes were inevitable. The U.S. Geological Survey (Dieterich, 1983, cited in Lehman and Taylor, 1987), for example, estimated that there was a 90 percent chance that Los Angeles would, in the next twenty years, have a quake severe enough to kill 20,000 people and badly injure 90,000 (references quoted by Lehman and Taylor, 1987). Broad television, newspaper, and other media coverage informed the California public of this earthquake risk. The media also outlined how people could prepare in advance for a quake (e.g., having a battery-powered radio and flashlight on hand) and what protective actions they should take during a quake (e.g., stand in a doorway or get under a heavy piece of furniture).

The same year, the University of California at Los Angeles (UCLA), a school with over 20,000 students, made public an internal report on the earthquake vulnerability of the campus and its buildings. The report estimated that a major quake could kill 2,000 students, staff, and faculty, and seriously injure 4,000 (UCLA Earthquake Safety Committee, 1985, cited in Lehman and Taylor, 1987). The report also indicated that different campus buildings were differently vulnerable to damage. Some buildings were likely to survive a major quake, whereas others were poorly designed and subject to serious structural damage or to collapse. The poorly designed buildings included four multistory dormitories housing over 3,000 students.

In February 1986, five months after the Mexico City quakes and after the publication of the UCLA report, two psychologists, Darrin Lehman and Shelley Taylor (1987), interviewed fifty UCLA students living in the seismically poor dormitories. Lehman and Taylor asked the students various questions about their dormitory buildings and about earthquakes. The overwhelming majority of students had heard about the campus earthquake report and knew that it found their dormitories to be poorly designed.[1] Surprisingly, however, these students were more likely than were students in another group who lived in good dormitories to agree with statements implying an underestimation or denial of risk, statements like: "There may be an earthquake, but it won't be so bad"; "The likelihood [of] a major earthquake . . . has been greatly

Multistory Buildings in Mexico City Badly Damaged or Collapsed by the September 1985 Earthquakes
(David Tenenbaum/AP/Wide World Photos)

Multistory Dormitory Buildings (Upper Left) on the University of California at Los Angeles (UCLA) Campus
(Courtesy UCLA Public Information Office)

exaggerated"; "I don't think about the earthquake"; and "I ignore it" [p. 551]. Even more surprising, one-third of the students living in the poor buildings did not know of the right actions to take and actions to avoid during an earthquake—although they were widely publicized in the media. Furthermore, *none* of the students had taken any of the simple earthquake precautionary measures recommended in the media, measures that might help save their lives.

EXAMPLE B: Overreaction to a Natural Hazard: Lightning versus Tornadoes

Lightning killed more people in the United States in recent years than any other natural hazard except for flash floods (Bureau of the Census, 1986, cited by Greening and Dollinger, 1992). Thus, lightning claimed more lives than did tornadoes. Research data, however, suggest that people misperceive the relative fatality risks posed by lightning, tornadoes, and other hazards. Consider, for example, a study by Sarah Lichtenstein, Paul Slovic, Baruch Fischhoff, Mark Layman, and Barbara Combs (1978). These psychologists asked seventy-four University of Oregon undergraduates to estimate the number of Americans who died each year as a result of forty-one different causes of death, including lightning and tornadoes. The subjects estimated, on average, that lightning caused ninety-six deaths each year, quite close to the actual number—107—according to government statistics. However, the subjects estimated that tornadoes caused around 575 deaths, whereas statistics show that tornadoes claimed on average only ninety lives each year. In other words, Lichtenstein et al.'s subjects *overestimated* the number of deaths caused by tornadoes by a factor of 6. In a similar study, Greening and Dollinger (1992) asked 455 high school students how likely a person "like themselves" would die as a result of twenty-four different causes of death, including lightning and tornadoes. On average, Greening and Dollinger's subjects estimated that the fatality risk for tornadoes was *ten times* greater than the risk for lightning, despite the fact that tornadoes actually posed the smaller threat.

EXAMPLE C: Overreaction to the Risks of Nuclear Power and Other Technologies

Example C differs in format from Examples A and B above in that it consists of four brief quotes from articles or books. The quotes accurately reflect the thinking of several writers, scientists, and engineers concerning the inaccuracy of the public's perceptions of technological risks (e.g., see Wilson, 1979). These writers, et cetera argue that the U.S. public overreacts to the risks of certain technologies.

Society's attitudes towards risks such as cancer and nuclear reactors are not readily distinguishable from its earlier fears of the evil eye (Clark, 1980, p. 137).

Public perceptions [of technological risks] and reality dramatically differ . . . (Raiffa, 1980, p. 340).

The risk [of underground repositories of radioactive wastes from nuclear reactors] is as negligible as it is possible to imagine. . . . It is embarrassingly easy to solve the technical problems, yet impossible to solve the political one [of public opposition to these repositories]. (Lewis, 1990, p. 246).

. . . [T]he general public . . . flocks to cities prone to earthquakes while it frets about pesticides in foods. . . . [T]he . . . general public . . . ban[s] an artificial sweetener because of a one-in-a-million chance that it might cause cancer. . . . [T]he general public['s] concern for the safety of nuclear reactors—which have claimed a total of three lives in accidents [in this country] in the last 30 years—[has] brought the industry to a virtual standstill. . . (Allman, 1985, p. 31).

INTRODUCTION—MORE ON PEOPLE'S REACTIONS TO HAZARDS AND RISKS

As the above examples show, people may underreact to some natural, technological, and personal health hazards but also, apparently, overreact to others, depending upon the specific hazard and on the conditions involved. As we also noted, people may respond to environmental threats and hazards in many of the same ways, underreacting *to* some, but overreacting to others. And both under- or overreaction to an environmental threat can be maladaptive and harmful.

In this chapter, therefore, we take an in-depth look at research on people's reactions to natural, technological, and personal health hazards to see how the findings and theories from this research can help us understand people's reactions to environmental hazards. We begin, in this *second part* of the chapter, by presenting more examples of people's apparent under- and overreaction to natural, technological, and health hazards. We look especially closely at people's reactions to the risks of *technologies* (because, as we noted above, technologies are a major cause of damage to the environment). We quickly discover, however, that it's often hard to tell if the public under- or overreacts to the risks of a given technology, including the technology's environmental impacts. For one thing, there is sometimes disagreement or uncertainty among scientists as to what the risks and impacts really are. Further, people's risk estimates and actions must be judged in the context of controversial issues of ethics, morals, and politics that underlie all societal debate about the risks of technologies and their management.

We discuss these moral, ethical, and so forth issues—many of which have relatively little to do with perceptual and cognitive processes per se—in a brief "side trip," but then refocus our attention on human perception and cognition for the remainder of the chapter. In the *third part* of the chapter, we review three psychological theories that can help explain hazard underestimation and inaction and that might apply to environmental threats and hazards. These theories involve human mental health, human emotions, and human responses to stress. In the *fourth part,* we examine a single psychological theory involving cognitive and perceptual biases, errors, and shortcuts that can help explain both risk under- and overreactions. In the *fifth part*, we present a general framework that encompasses and summarizes all the theories and research findings we covered in the third and fourth parts. Finally, the *sixth part* is a brief conclusion.

A. Underreactions to Natural, Technological, and Personal Health Hazards: A Closer Look

As we just noted, one main outcome of the research on people's reactions to natural, technological, and health hazards is that it partly supports Hardin's (1969) claim. Risk underestimation and failure to take preventive action *are* relatively common features of human behavior, but they by no means occur for every person or for every risk. We now present four new examples of people's underreaction to risk to

supplement the examples in Chapter 8 and the Prologue.

The behavior of people who live in areas prone to recurring natural hazards (like floods and earthquakes) provide a further example of risk underreaction. Many residents claim that no reoccurrence of the hazard is possible for several years following a prior occurrence they know of. Some residents even claim that a reoccurrence in their lifetimes is impossible. (Scientific knowledge and historical data indicate that these hazards can reoccur any year with equal probability, and that a reoccurrence is relatively likely during the lifetimes of many residents; we'll say more about this later in the chapter.) Geographers Ian Burton, Robert Kates, and Gilbert White (1978) reviewed studies of twenty cultures or communities in different parts of the world located in places subject to natural hazards. Burton et al. found that, on average, 30 percent of the public interviewed denied the possibility that the hazard could recur for several years following a prior occurrence. In the most extreme cases, 80 percent of the public claimed that the natural hazard could not reoccur in the near future, and 75 percent denied the possibility of any reoccurrence in their lifetimes. In other communities, however, the incidence of underestimation was much lower.

A related example is the failure of people who live in natural-hazard-prone areas to take simple preventive actions that would lessen their risk of death, injury, and/or property damage. This behavior is *more common* than people's denial of hazard reoccurrence that we discussed in the paragraph above. Kunreuther (1978) found that many homeowners in flood- and earthquake-prone parts of the United States fail to buy flood and earthquake insurance (note that standard homeowners' insurance policies, required by banks and other lending institutions, exclude flood or earthquake coverage). This failure to protect happens even when hazard insurance is relatively cheap (e.g., when it is subsidized up to 90 percent by the federal government [Slovic et al., 1978]). As an example, when Rapid City, South Dakota, suffered a flood in 1972 that caused over $150 million in damage, only 29 flood insurance policies were in effect, though the policies had been available for over a year. More generally, Kunreuther claims, banks and government agencies have found that many people will not even buy standard homeowners' insurance or auto insurance unless required to do so by bank regulations or by law.

People apparently fail to take preventive actions toward some technological hazards as well as natural hazards. For example, few people voluntarily wore auto seat belts in the United States before mandatory belt-use laws were passed, though it was well established that thousands of lives would be saved each year if all people wore them (Bick and Hohenemser, 1979; Reppy, 1984). In fact, for some age groups, regularly "buckling up" was one of the most effective actions that people could take to increase their life expectancy. Yet less than 20 percent of the public used belts in states lacking mandatory belt-use laws (Reppy, 1984). Mandatory-use laws now found in many states have boosted the figure to over 60 percent (Associated Press, 1993).

A tendency for people to underestimate the hazards to which they are subject appears to extend to several other common risks to life and health. For example, Neil Weinstein (e.g., 1980, 1983, 1987) had Rutgers University undergraduates rate their chances of: having a heart attack before age forty, developing an alcohol problem, getting lung cancer, being divorced soon after marrying, and other events. Students rated their chances of each event *compared to* other Rutgers students of the same age and sex. In some of the studies (e.g., 1983), students used a 7-point rating scale: −3 = "much below average chance of the event happening to me" up to +3 = "much above average chance"; a 0 meant "the same chance as others."

In analyzing the data, it was impossible for Weinstein to tell whether or not any given student was underestimating his/her vulnerability to any given negative event. For example, a student who rated her chances of getting lung cancer as much lower than average might be correct, given that she was a nonsmoker, had no family history of cancer, and lived where there were no radon or other air pollution problems. However, Weinstein reasoned, if the average rating for a large group of students is well below 0,

TABLE 9-1 Mean Comparative Risk Judgments for Different Health and Other Hazards Made by 296 Randomly-Sampled Members of the General Public (in Central New Jersey)

EVENT	MEAN JUDGMENT**
Drug addiction	−2.17*
Drinking (alcohol) problem	−2.02*
Attempting suicide	−1.94*
Food poisoning	−1.25*
Homicide victim	−1.14*
Pneumonia	−.80*
Lung cancer	−.77*
Ulcer	−.55*
Mugging victim	−.54*
Diabetes	−.53*
Stroke	−.29
Serious auto injury	−.27*
Heart attack	−.24
High blood pressure	−.02
Cancer (nonlung)	.08

Adapted from Weinstein, N. Unrealistic optimism about susceptibility to health problems: Conclusions from a community-wide sample. *Journal of Behavioral Medicine*, Volume 10, 481–500. Copyright 1987. Plenum Publishing Corp. Used with permission.
*Result statistically significant at p < .05 or better
**Entries are mean values on a −3 to +3 scale (see text)

then this is strong evidence that some, if not many, of these students are underestimating their vulnerability (thus showing what he calls "unrealistic optimism"). Indeed, the subjects' average ratings for most (though not all) of the negative events turned out to be significantly below 0. Since college students and other young adults are supposedly famous for their belief in their own immortality, Weinstein (1987) repeated the study with a random sample of the general public and got the same results. Table 9-1 shows some of the data from his 1987 study.

B. Overreactions to Natural, Technological, and Personal Health Hazards: A Closer Look

As we discussed earlier in the chapter, people rather than solely underreacting to the hazards they face also significantly *overestimate* and *overrespond* to certain risks under certain conditions. We gave two examples of apparent risk overestimation in the Prologue—Example B, on tornadoes/lightning, and Example C, on nuclear power and other technologies. Slovic, Kunreuther, and White (1974) provide an additional example. They cite anecdotal evidence of people's overreaction to the rare cases of grizzly bear attack on campers in U.S. national parks. Even though the overall frequency of attack is only 1 per 2 million park visitors, and the frequency of death is still lower (Herrero, 1970, cited in Slovic et al., 1974), a single, widely publicized bear attack apparently can induce hundreds or thousands of people to alter their vacation plans or to be unduly anxious during a park visit.

A stronger case can be made that the public overreacts to terrorist attacks and hijacking attempts on commercial passenger planes. Though there are usually few such attacks/attempts per year, though commercial air travel is, overall, several times safer per mile than other modes of transportation, and though there are literally thousands of safe, uneventful domestic and foreign flights annually, a single hijacking attempt or terrorist attack usually produces thousands of ticket and reservation cancellations (Anon, 1986b; Hudson, 1989).

Even stronger evidence (methodologically speaking) that people may overestimate hazards comes from the well-known Lichtenstein, Slovic, Fischhoff, Layman, and Combs (1978) study we mentioned in Example B of the Prologue. Let's now take a closer look at this research. Recall that Lichtenstein et al. asked their college student subjects to estimate the number of people killed each year in the United States as a result of forty-one different causes. These causes ranged widely from diseases (e.g., smallpox, breast cancer), to natural hazards (e.g., tornadoes and lightning, as we noted above), to technological hazards (e.g., train collisions, auto accidents).

Figure 9-1 shows a more complete picture of the Lichtenstein et al. (1978) data. The X-axis of Figure 9-1 indexes true frequencies of death per year, based on historical data collected by a government agency. The Y-axis indexes subjects' estimates of the frequencies of death per year. If the subjects' estimates were perfectly accurate, their data points should have fallen along the diagonal straight line in the figure.

This was not the case. Instead, the data points generally follow the *curved* line in the figure. The curved line shows that the subjects underestimated the frequencies of many common causes of death (like heart disease, stroke, and stomach cancer) but *overestimated* the frequency of rare causes of death (like botulism [a food poisoning], floods, and complications of pregnancy). Lichtenstein et al. used the term *primary bias* to refer to this overall pattern of data. The pattern has been observed in several subsequent studies, and may reflect a more general (though not universally operating) tendency for people to underestimate large probabilities and overestimate small ones (Crocker, 1981, cited by Lehman and Taylor, 1987).

One might argue that primary bias is not a very good example of people's misestimation of fatality risks since it is apparently a general pattern that shows up in people's frequency estimates for some other events. However, Lichtenstein et al. found that their subjects showed an *additional,* and perhaps more impressive, *secondary bias* in their fatality estimates, one *superimposed over* the primary bias trend shown by the curved line: Subjects tended to *exaggerate* dramatic and sensational causes of death compared to causes that are prosaic or ordinary. For example, note that motor vehicle accidents and stomach cancer kill, on average, about the same number of people per year in the United States. However, the subjects judged that many more people were killed in auto accidents than by stomach cancer. Similarly, diabetes kills more Americans each year than homicides, but subjects judged homicides as significantly more frequent. To repeat, the subjects tended to overestimate dramatic, sensational causes of death, but underestimate common, prosaic causes. Later in the chapter, we will discuss a psychological process that may underlie and explain this secondary bias pattern.

We briefly described yet another apparent case of hazard overreaction in Example C of the Prologue, above. Widespread public opposition in the United States to nuclear power is a major reason that few new nuclear plants are being built, despite the fact that a number of scientists and engineers believe the technology to be a good source of electricity. Public opposition stems, these scientists and engineers

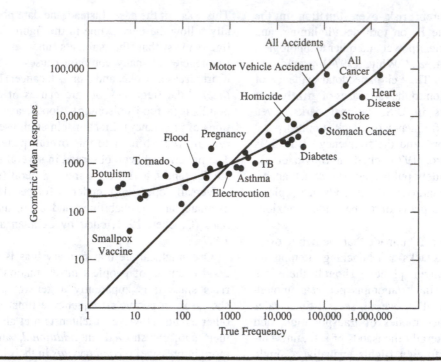

FIGURE 9-1 Results of the Lichtenstein, Slovic, Fischhoff, Layman, and Combs (1978) Study: Subjects' Mean Estimates of the Number of Deaths Per Year in the United States as a Result of 41 Different Causes versus the Actual Frequencies of Those Deaths

Reprinted from Lichtenstein, S., Slovic, P., Fischhoff, B., Layman, M., and Combs, B. Judged frequency of lethal events. *Journal of Experimental Psychology, Human Perception and Performance*, Volume 4, 551–578. Copyright © 1978 by the American Psychological Association.

claim, from the public's overestimation of the risks of nuclear plants, overestimation caused by the public's lack of technical knowledge and/or misinformation about the technology in the media and/or even an irrational public phobia toward nuclear technology (Dupont, 1980). The public seems unable to understand, these engineers and scientists argue, that a technology might have severe and frightening potential risks (like a catastrophic nuclear accident) but that the probability of these risks actually occurring is extremely low.

The public also overreacts, some engineers and scientists claim, to the low-probability risks of other technologies. An apparent example is the Alar scare of the late 1980s. Alar is a chemical sprayed on apples to delay their ripening and to increase their red color.

Consumer purchases of apples plummeted after Ralph Nader (1988) and the Natural Resources Defense Council (an environmental group) reported that the chemical posed a cancer risk. Some scientists argued, however, that the public and public-interest groups had greatly overestimated the cancer risk posed by the chemical. Quoting Martin (1990): An ". . . influential biochemist estimated that the [cancer] risk . . . [from drinking] a glass of apple juice [made] from apples treated with Alar was 18 times less than the cancer risk of [eating] a peanut butter sandwich . . . and 1,000 times less than the [cancer] risk of [drinking] one beer" [p. 434–437]. Concerning the public's perceptions of technological risks more generally, Allman (1985) writes: "[T]he . . . general public [is] irrational, uninformed, superstitious, even stupid. . . . [It doesn't] understand probability, [is] . . . biased by the news media, and . . . [has] a fear of some technologies that borders on the primeval."

Inferring That the Public Overestimates the Risks of Technologies—Some Cautions and Qualifications. Despite the claims discussed above that the public overestimates the risks of Alar and of nuclear power, the process of inferring whether the public, in fact, overestimates a technology's risks can be extremely tricky. For one thing, it is sometimes difficult to arrive at an objective, factual risk level with which to gauge the accuracy of the public's perceptions. In a number of cases, knowledgeable scientists disagree as to what the risk levels of a given technology actually are, for example, the degree to which Alar can cause cancer and the carcinogenicity of other chemicals (Lowrance, 1976; Graham et al., 1988).

There is also scientific disagreement concerning such environmental impacts of technology as erosion of the Earth's ozone layer and possible global climate change. In the case of climate change (traceable mainly to fossil fuel combustion in autos, homes, factories, etc.), knowledgeable scientists disagree on the size of the possible changes, their likely geographic distribution, and their onset in time (Silver and DeFries, 1990). Since global climate change due to human activity and erosion of the ozone layer are unprecedented in human and geologic history, the inevitable result is uncertainty about what the future will bring.

In the case of nuclear power, scientists and engineers must base their risk estimates on complex analyses that require some subjective judgments, rather than solely on historical safety statistics (Lewis, 1980). The reason is that nuclear power is a relatively new technology. Nuclear plants are extremely complicated, and they subject key internal components to unprecedented conditions. Thus, Allman's (1985) claim, quoted in Example C of the Prologue, that ". . . nuclear reactors . . . have claimed a total of three lives in accidents in the last 30 years," is somewhat misleading. Though some nuclear power plants, including ones used in the military (e.g., on submarines), have been around for three decades, the technology is newer than other technolo-

gies used to generate electricity. More significantly, major differences in the designs of different commercial and military nuclear plants limit the ability of analysts to generalize from historical safety data. The accident at the Three-Mile Island power plant in Pennsylvania in 1979—an accident said to have been impossible before it happened—and the more serious accident at the Chernobyl plant in the Ukraine in 1986 (Read, 1993) demonstrate that there is still uncertainty concerning the overall reliability of this technology. (We discuss the safety of nuclear power plants in greater detail in Chapter 11.)

In the absence of extensive historical safety statistics for nuclear power plants, scientists and engineers must use complex formal analyses to estimate the future reliability and safety of such plants. These analyses are known as *fault-* and *event-tree* analyses (Lewis, 1980). They attempt to assess the overall reliability of a plant by determining the reliability levels of its individual component parts, such as pipes and valves (for which there are some historical data), and then combining these levels mathematically. (We'll discuss fault-tree analyses in more detail later in the chapter.) These analyses, however—like other methods that attempt to "crystal-ball" the future—require that the analyst make a number of subjective estimates and educated guesses. The results therefore cannot be considered as the "factual" or "actual" risks of commercial nuclear power with which we can evaluate the accuracy of the public's risk estimates.

But we discuss, in the section that follows (An Important Side Trip) an *even more important* qualification and caution concerning the accuracy of the public's perceptions of technological risks, and the appropriateness of their responses to those risks.

C. An Important Side Trip: A Major Problem When Judging Whether People Over- or Underreact to Risks of a Technology—The Moral, Ethical, and Political Issues that Underlie Societal Debates about Technological Risks

Because technologies (together with the scale and intensity of human activity) are the cause of damage to the environment, we'd like to examine the research on people's reactions to technological risks and the context of that research in greater detail. We focus, in this section, on a critically important fact: It is impossible, as we noted above, to study and assess the accuracy of people's perceptions of the risks of technology and the appropriateness of their actions (as well as the psychological processes that underlie perceptions and actions) without running into controversial issues of values, morals, ethics, and politics. These fundamental issues underlie all societal debates about the safety and regulation of technologies, though they are usually hidden and not readily observable.

The issues, more specifically, center on how to define *risks* and *benefits,* and on what makes a technology, overall, "acceptable" to society. Any technology, such as the automobile or nuclear power, creates benefits and advantages, while at the same time creating risks or hazards. The benefits of autos, for example, include facilitating commerce, creating jobs, increasing individual mobility, convenience, enjoyment, and so on. The risks and hazards include the risks to human life and health via accidents, the production of air pollutants that may impair the health of people and of nonhuman forms of life, the longer-term negative effects of the pollutants on the biosphere (the parts of the Earth and its atmosphere that support plant and animal life), and so on. But how do we evaluate and weigh all the various risks and benefits of the automobile or of any other technology? Which risks are most important, which least important? Should impacts on nonhuman forms of life be considered at all? Should the risks about which there is greater scientific uncertainty be weighted more or less than risks about which there is a scientific consensus? How should the various benefits be weighed against the various risks? Is, overall, the technology—given its risks and benefits—being properly regulated by government agencies and institutions? Are stricter regulations called for, is the technology "acceptable" to society as things stand now, or should regulations be less strict?

These questions are not mainly scientific. They, instead, involve values, ethics, morals, politics, and philosophy. Something is not a "risk" unless it threatens things that people value. For example, for people who think that every new generation should "fend for itself," any long-term environmental damage that autos may cause is inconsequential. There are parallels here with our discussion in Chapter 3 of environmental values, morals, and ethics. Recall our review of "deep" versus "shallow" ecology, and of Merchant's (1992) three ethical bases for judging whether an environmental problem is really a problem—egocentrism, homocentrism, and ecocentrism. For example, people who uphold the egocentric ethic define an environmental problem worth worrying about only as one that threatens them and their families. Thus, people's perceptions of the environmental risks of a technology, its benefits, and its acceptability to society, cannot be judged and studied except in the context of the moral, values, ethical, and so on, issues above.

The psychologists—Slovic, Fischhoff, Lichtenstein, and others (e.g., Slovic et al., 1979)—who did the pioneering research in the late 1970s and early 1980s on the accuracy of people's risk perceptions quickly discovered this truth. They found that members of the general public define the "risks" of a technology and the technology's "acceptability to society" in *systematically different ways* than do scientists and engineers. These differences, again, mainly reflect differences in values, ethics, and philosophy, not differences in education or rationality.

More specifically, the researchers found that many engineers, scientists, and professional risk managers tend to equate the risk of a technology (such as nuclear power) with the probable number of human deaths the technology will cause. They tend to overlook damage caused to ecosystems and to nonhuman forms of life (recall our discussion of "externalities" in Chapter 5). These experts then advocate judging whether the technology is, overall, acceptable to society based on whether the technology's benefits (e.g., the quantity of electricity generated), usually measured in dollars, outweigh its risks (again, human deaths) and costs (e.g., the costs of building and operating a power plant). If the benefits exceed the risks and costs, the technology should be acceptable to society. If not, then stricter government regulations or industry self-regulations may lower risks enough for the technology to be acceptable. Going a step further, these experts have developed a widely used formal, quantitative method known as *risk analysis* for making risk estimates and for judging societal acceptability along the lines we just described.

In contrast, the researchers found that the risk ratings of the general public, while influenced by people's estimates of the number of human deaths a technology will likely cause, are also strongly affected by their judgments of several *qualitative characteristics* of the technology. These characteristics include the degree of apparent disagreement in the scientific community about the risks of the technology, the degree to which the technology can kill many people in a single catastrophic accident (rather than the same number of people one at a time in small accidents), and the degree to which the risks can affect future generations. The general public also appears to consider the damage to ecosystems and nonhuman forms of life the technology may cause (Gardner et al., 1982). Furthermore, the research indicates that these and other qualitative characteristics— for example, whether the benefits of a technology are equitably, and voluntarily, distributed among those that bear its risks—shape not only public risk perceptions but also public judgments of whether a technology, overall, is acceptable to society. In other words, the general public appears to use a *broader and more complex definition of risk and acceptability* than does the technical community. Thus the public and the technical community are "speaking different languages," and therefore speaking past each other. (The above research is surveyed in more detail by Gardner, 1982; Gould et al., 1988; and Gardner and Gould, 1989). Lastly, the public's perceptions of risks and its judgments of acceptability correlate with its *behavior* toward technology and risks, such as joining environmental groups like the Sierra Club, and voting for political candidates based on the candidates' stance on the environment (Gardner et al., 1982; Gould et al., 1988).

It is important to reemphasize that how one defines *risk* and *acceptability* is fundamentally a matter of values, ethics, morals, and philosophy, rather than of

scientific fact or deduction. As we said above, something is not a risk unless it endangers things that people value. To quote psychologist Colin Green (1980):

Any measure of risk involves making some moral decisions: Is it worse to have 1,000 people die in one accident or the same number one at a time, or doesn't it matter? Is one death of a child by leukemia as bad as 100 [adult] lives shortened by chronic bronchitis? Any measure of risk involves making value judgments such as these.

Thus, there is no reason that the "number of human fatalities," which scientists and engineers focus on, is the sole and correct index of a technology's risks to society. (To provide an additional, though crude, example: Imagine a flawed consumer food-preparation product that caused users to lose a finger, or that trapped them in their kitchen for days, or that badly polluted bodies of water into which kitchen drains emptied. Most would agree that the product had serious risks, even though it caused no human deaths.) Similarly, it is a matter of values, morals, and philosophy—not scientific fact—whether the definition of *risk* and *acceptability* should include such qualitative characteristics as the effects of a technology on future generations, and the impact of the technology on nonhuman forms of life.

Let's carry this discussion a step further. Slovic, Fischhoff, and Lichtenstein and other researchers argue that differences in the *risk* and *acceptability* definitions used by the public versus by scientists and engineers can cause *significant mistrust and miscommunication* between the two groups. The differences in definition can make engineers and scientists view the public as "hysterically" overestimating risks, and, at the same time, make the public view scientists, engineers, et cetera as incorrectly, or even dishonestly, underestimating risks. The following interesting example illustrates how this can happen: Several years ago, a number of scientists and engineers (e.g., Wilson, 1979; Sowby, 1965; Cohen and Lee, 1979) attempted to assuage public concerns about the risks of certain technologies, especially nuclear power, by means of "risk compendia"—lists of risks of death due to a variety of different hazards. Table 9-2 shows portions of one of these compendia put together by Harvard University physicist Richard Wilson (1979). Wilson calculated that all the items in the compendium create *equal* increases in the probability of one's death. Thus, smoking 1.4 cigarettes, traveling six minutes in a canoe, traveling thirty miles in an auto, and living fifty years within 5 miles of a commercial nuclear reactor are equally risky to one's life. An intended conclusion is that since the risk of death from nuclear reactors is low relative to conventional and currently accepted technology, commercial nuclear power should be considered as societally acceptable. It is based on this reasoning that a number of scientists and engineers judge the public's opposition to nuclear power unreasonable or pathologic, as shown in Example C in the Chapter Prologue.

However, the results of the psychological research we discussed above would predict that Wilson's analysis would do little to "educate" the public or successfully assuage public concerns about nuclear power. From the point of view of the general public, Wilson's approach is too narrow and simplistic. It ignores many of the qualitative properties of technology that the public deems important in judging risk and acceptability—scientific uncertainty about the risks, the maximum number of deaths per accident, intergenerational equity, and so on. (Note that the risk compendium may be criticized for other reasons as

well, for example, the fact that it "frames," or defines, the problem too restrictively in the first place: The Wilson analysis doesn't consider alternative ways of generating or conserving electricity besides nuclear power—see our discussion of framing effects later in this chapter. The Wilson analysis also relies on risk estimates for nuclear power that are derived from fault-tree analyses [see our discussions above and below].)

Finally, we'd like to briefly review three other values/ethics/political/et cetera components of debates about technologies, their risks (including environmental impacts), and their regulation. One component is what Slovic and his colleagues (e.g., Slovic, 1993a) call the "signal value" of a technological/environmental accident. They argue that a relatively minor environmental accident, especially in a new and complex technology, can cause what may seem like disproportionately great public concern. However, this public concern, on closer inspection, occurs because the accident "signals" that something may be fundamentally wrong with the technology and its regulation, and that future serious accidents may be probable if regulatory changes are not made. An apparent example of this is the strong negative public reaction to the accident in 1979 at the Three Mile Island nuclear power plant in Pennsylvania, an accident which we discuss in considerable detail in

TABLE 9-2 Nine Exposures or Events That Produce Equal Increases (by One in a Million) in the Probability of One's Death, According to R. Wilson (1979)

- Smoking 1.4 cigarettes
- Living 2 days in New York City or Boston (air pollution)
- Traveling 6 minutes by canoe (accident)
- Traveling 30 miles by auto (accident)
- Flying 1,000 miles by jet (accident)
- One chest X-ray in a good hospital
- Eating 40 tablespoons of peanut butter
- Drinking 30 12-ounce cans of diet soda containing saccharin
- Eating 100 charcoal-broiled steaks
- Living 150 years within 20 miles of a nuclear power plant
- Living 50 years within 5 miles of a nuclear power plant

Adapted from Wilson, R. Analyzing the daily risks of life. From *Technology Review*, copyright 1979, Volume 81, 40–46.

Chapter 11. The accident caused no immediate human injuries or deaths and released only a small amount of radioactive material into the environment. Further, several of the safety systems in the plant worked exactly as they should have during the accident and averted a much worse disaster. But because engineers and scientists had claimed that an accident of this type was not possible, because the accident involved previously unknown risk elements (such as the formation and explosion of a large hydrogen bubble in the reactor building), and because the accident caused enough damage to permanently disable the multimillion dollar reactor, public reaction, gauged via public opinion polls, to the accident were strongly negative.

A second, related, component of societal debates about risks involves the low level of public trust in government regulatory agencies, private industry, and professional groups. Slovic (1993b) and others argue that what may appear to be irrational and untutored public opposition to a technology really involves the lack of trust in government and industry management of the technology (see also Sandman, 1987). Thus, a study by Slovic, Flynn, and Layman (1991) concluded that the massive public opposition to the U.S. Department of Energy's attempts to create a national storage site for high-level radioactive wastes from nuclear power plants mainly reflects public distrust of government and industry, not a blind general opposition to nuclear technology. Slovic and his colleagues also point out that public trust is fragile and easily destroyed, and that trust has been breached in the past by unfortunate episodes of cover-ups and untruths.

This brings us, lastly, to work by Kasperson et al. (1988) and others, for example, Renn et al. (1992), on the "social amplification and attenuation" of risk. These researchers argue that the actions of the information media, government institutions, and public and private interest groups cause the overall level of societal concern about a technological risk to grow or shrink in a complex, closed-loop process. First, the public can't, in most cases, be concerned about a technological risk unless it is made aware of the risk by television or the other information media. Further, the degree of public concern is influenced by just how the media portray and frame the risk. Media coverage of the risk, in turn, is affected by the input of various environmental groups and private interest groups. Once the public is concerned about the risk, then public input (and input from environmental and other groups) to government agencies causes these agencies to investigate the risk and alter their regulation of it. Media coverage of these investigations and regulation changes can, in turn, increase or decrease the public's concern about the risk. This change in concern is again communicated to government agencies, et cetera. And so on. Kasperson et al.'s and Renn et al.'s analysis of this complex "social amplification/attenuation of risk" process provides yet another reason the public's risk perceptions and actions cannot be fully evaluated and understood by looking just at psychological processes and at the individual level of analysis.

Section Summary and Plan for the Rest of the Chapter. We have seen that some apparently clear cases of public overestimation and overreaction to the environmental (and other) risks of a technology—as in Example C on nuclear power—turn out, on closer inspection, to be far less clear. More generally, we've seen that it is impossible to judge the accuracy of the public's perceptions of risks or understand its actions without looking at the complex underlying set of moral, ethical, political, and social issues that may not be readily apparent.

Having explored these issues (our "important side trip"), we now refocus attention on the psychology of human perception and cognition. We will argue that *there are several perceptual and cognitive processes and phenomena that **may push** members of the public and policymakers **in the direction of** misestimating some environmental hazards and/or failing to take proper action.* Though, as we just discussed, judging the accuracy of people's risk estimates is a tricky and complex matter (it can even be hard to determine the "true" risks of a natural hazard, like floods, and the "proper" actions to take), we nonetheless believe that the public and policymakers are likely to misestimate some environmental risks and fail to take proper actions against them in cases that could be costly to society, or even disastrous.

Evidence comes from examples earlier in the chapter that despite the complex social, political, moral, and so on issues that underlie technological risk debates, we can still say that people misestimate some risks and/or fail to take proper actions. Recall that Lichtenstein et al.'s (1978) subjects *overestimated* the number of deaths due to auto accidents relative to other causes. Recall also that the subjects seriously overestimated the fatality risks of certain natural and health hazards. While, as we have argued, the number of human deaths is not the sole valid dimension of the overall risk of a technology or other hazard, most would agree that it is one important component, and serious public misperception of it might lead to wastes of societal time, attention, and money. Similarly, there is evidence, which we will discuss later, that the public may be highly concerned about some environmental problems that have dramatic and anxiety-provoking features (problems like oil spills from ocean-going tankers), but which environmental scientists believe are less serious than some less dramatic but more insidious environmental problems (like damage to the ozone layer) about which the public may be less concerned.

Conversely, we've seen that Lichtenstein et al.'s (1978) subjects seriously *underestimated* the fatality risks of several natural, technological, and health hazards. We've seen that many people will not protect themselves against a major technological hazard, that is, will not wear auto seatbelts, unless compelled to by law.[2] We've also seen that people often fail to take simple, though effective, preventive actions against earthquakes, floods, and other natural hazards, even when they know they live in a hazard-prone area (and even, in some cases, after they've been warned about an imminent threat). Similarly, Weinstein's (e.g., 1980, 1987) data suggest that most people significantly underestimate a variety of health risks to which they are subject. Other studies show that many people fail to take simple actions against health risks, (e.g., fail to use condoms or follow other safe sex practices to prevent AIDS, *FDA Consumer,* 1992). These examples suggest that a reasonable, initial working hypothesis is that—just as people are likely to overreact to some environmental hazards

under some condition—people are also likely to underreact to some environmental hazards under some conditions.

We now move on to discuss the psychological processes that may contribute to misestimations of environmental risks and failures to take proper action.

D. An Overview of Psychological Causes of Risk Misestimation and Failure to Take Appropriate Actions—Four Theories

In the next two main parts of the chapter, we review *four theories about human perception and cognition* that can help explain why people might under- or overestimate risks, including environmental risks, and/or fail to take proper actions against them. We will argue—and this will not surprise the reader—that *all* the theories have some validity. We believe that risk misestimation and inappropriate coping behavior have *several* different psychological causes, each cause the topic of a theory, and that more than one cause can operate at the same time. To use a simple and prosaic analogy, consider that several different design features of an automobile simultaneously affect its overall safety.

Before we discuss the four theories, let's briefly recap two theories we introduced in Chapter 8 that also can help explain why people might misestimate environmental risks and/or fail to take action against them. Recall, first, Ehrlich and Ornstein's (1989) "old mind" concept and the claim that, because of human evolutionary history, we tend not to notice or react to "slow motion disasters," that is, environmental deterioration that occurs gradually over long periods. Recall, second, Hardin's (1969) claim that Darwinian natural selection has produced a human genetic tendency toward a moderate degree of denial, because moderate denial promotes discovery and invention. (Of course, we also discussed in Chapter 8 that it is hard to prove or disprove claims of genetic psychological tendencies traceable to natural selection.)

We'd now like to briefly introduce the four other psychological theories that may help explain risk under- and/or overestimation and failure to take appropriate action. The four fall into two main categories. The *first category* includes three theories with *perceptual, cognitive, evolutionary, and emotional* components. These three theories, which are covered in the next main part of the chapter, all predict and explain risk *underreactions.* The theories are: 1) Slovic, Fischhoff, and Lichtenstein's concept of risk underestimation as necessary for "getting on with one's life"; 2) Taylor and Brown's theory of "illusions," perceived control, and mental health; and 3) psychological stress theory and emotion-focused coping as a response to threats perceived as uncontrollable. The *second category* includes a single general theory involving *cognitive (and perceptual)* shortcuts, biases, and errors that can contribute to misestimation and inappropriate action. This theory, which is covered in the main part of the chapter following the one directly below, predicts and explains *both* people's *underreaction* as well as their *overreaction* to risks, depending on the risks and the circumstances. Let's now examine each of the three theories in the first category.

PERCEPTUAL/COGNITIVE/EMOTIONAL/ EVOLUTIONARY CAUSES OF RISK UNDERREACTION—THREE THEORIES

As we noted above, all three of the theories in this category—1) Slovic et al.'s, 2) Taylor and Brown's, and 3) general stress theory—can explain why people might tend to *underestimate* certain risks and/or *fail to take protective actions* against them under certain conditions.

A. Slovic, Fischhoff, and Lichtenstein's (1978) Concept of Risk Underestimation/Inaction as Necessary for "Getting on with One's Life"

Though the focus of Slovic et al.'s (1978) research is people's failure to use auto seat belts, the theory they advance has wider applicability, quite possibly to environmental hazards. But first, going back to seat belts: Slovic et al. basically argue that the effort and inconvenience of buckling one's seat belt on every auto trip outweighs the edge in safety that the seat belt provides. Here is their argument in detail:

Seat belt use involves continuous, repetitive behavior, that is, "buckling up" before every auto trip. However, for any single trip (such as your trip today to work or school), a person's chances of being killed in an accident are minuscule—about 1 in 3.5 million—and people rightly perceive this. Also, the chances of being seriously, though nonfatally, injured are very small—approximately 1 in 100,000. Given these low probabilities, people's incentive on any given trip to use their seat belts is just too small to outweigh the effort, attention, and inconvenience (belts can be uncomfortable and restrict one's reach to the radio controls on the dashboard, etc.) of using them.

Now, you might object: What is more important to a person than his or her own life? Shouldn't it be worth a little bit of trouble and inconvenience to lower his/her probability of death or disfigurement by even a *small* amount? Slovic et al. argue that the answer to this question is no. People face a myriad of hazards: Elevators fall, dams burst, TV sets explode," hotel walkways collapse, planes crash into buildings, lightning strikes, defective ladders give way, et cetera. (A friend of the authors' was almost killed by chicken soup. The soup had been made with a chicken packaged as a "roasting chicken." The breast was implanted with a small plastic thermometer anchored with sharp hooks and designed to pop up when the meat reached a certain temperature. The person drinking the soup got the thermometer lodged in his throat and almost choked. Fortunately, his wife reached down his throat and pulled the thermometer out.) However, there are limits to the time, energy, attention a person can devote to hazards and threats. Unless a person ignores many hazards, his or her life would be spent in an "obsessive preoccupation with risk" that would prevent any normal, productive existence.

It is therefore necessary for a person to ignore many of the threats and hazards he or she faces. The threats most logical to ignore are those with relatively *low* probabilities. Since the probabilities of death or serious injury in any given auto trip are minuscule (and presumably perceived this way by most people), most people won't voluntarily wear seat belts. Slovic et al. (1978) argue, further, that the same concept explains why many people who live in flood-prone or other hazard-prone areas do not buy hazard insurance unless required to do so: These people perceive the likelihood of a severe flood or other hazard occurring in any given year as quite low and therefore not worth preparing for. The theory might generalize to regional and global environmental problems; it would predict a failure of people to take actions against environmental threats unless they perceive the probability of the threats as above some minimal level. (Note that an alternate form of this theory would hold that people systematically underestimate the perceived probabilities of *all* the hazards they face, and take action only against risks that remain fairly probable after this underestimation has operated.)

To go a step further, one could easily propose a biological evolution/natural selection version of Slovic et al.'s (1978) theory (though Slovic et al. don't do so). We could add to Hardin's idea that some denial is adaptive and has been selected for by biological evolution because it promotes discovery and invention, the idea that denial has also been selected for because it enables an animal to "get out of its cave" and stop cowering from the many threats it faces; in other words, denial enables the animal to get on with the necessary activities of everyday life.[3]

While the "biological evolution/natural selection" version of Slovic et al.'s (1978) theory is merely spec-

ulation on our part, there is a lot of research evidence suggesting that people do tend to ignore low-probability hazards—regardless of the severity of these hazards—and tend to take protective action only against high probability ones. As an example, consider an experiment by Slovic, Fischhoff, Lichtenstein, Corrigan, and Combs (1977). These researchers had college undergraduates play an elaborate "farm game" or "simulation." Each subject "managed" a 240-acre farm worth $200,000 to $300,000 for a "15-year" period; (each trial of the game constituted a year). The manager had to make several yearly decisions, including whether to buy insurance protection against five (unnamed) natural hazards. Table 9-3 shows the properties of the five hazards, labeled A, B, C, D, and E.

The second and third columns from the left of Table 9-3 show, respectively, the probability of each hazard occurring in any given year, and the severity of the hazard (i.e., the dollar loss to the farmer caused by the hazard). Note that there is an *inverse* relationship between the entries in these two columns. Thus, Hazard A has a very small chance of occurring, but if it does occur, the financial losses are devastating; conversely, Hazard E has a relatively high probability of occurring, but the resulting financial loss is small and easily absorbed. (Notice that these five hazards

TABLE 9-3 Design of Slovic et al.'s (1977) "Farm Management" Experiment, and the Results: Probability of Loss as the Main Determinant of Decisions of Whether to Buy Insurance

HAZARD	PROBABILITY OF LOSS (*PROBABILITY*)	MAGNITUDE OF LOSS (*SEVERITY*)	ANNUAL INSURANCE COST	PERCENT OF SUBJECTS BUYING
A	.002	$247,500*	$500	33
B	.01	49,500	500	45
C	.05	9,900	500	52
D	.10	4,950	500	49
E	.25	1,980	500	73

Reprinted from Slovic, P., Fischhoff, B., Lichtenstein, S., Corrigan, B., and Combs, B. Preference for insuring against probable small losses: Implications for the theory and practice of insurance. *Journal of Risk and Insurance*, Volume 44, 237–258. Copyright © 1977, American Risk and Insurance Association. All rights reserved.
*A loss this large would bankrupt the individual and result in his/her loss of the farm.

are, in a certain sense, identical. The overall, long-term expected loss for each of the five is the same. That is, the probability of the hazard, multiplied by the dollar loss caused by the hazard, is $495 for all of the hazards. As a result, over an infinite number of years, the total expected loss would be the same for all of them.)

Slovic et al.'s (1977) key results—the average percentage of subjects who bought insurance for each of the five hazards—appear in the right-most column of Table 9-3. Notice that subjects tended to protect themselves against the high-probability/low-loss hazards, rather than the low-probability/high-loss hazards. To put it another way: The *probability of the hazard occurring*—and not the severity of the damage it causes—appeared to be the main factor in subjects' insurance purchase decisions. Again, the subjects tended not to worry about small-probability hazards, even if these hazards would be catastrophic to them. These results have been replicated in several other studies involving different subject populations and subject matters, including studies of the actual insurance-buying behaviors of the U.S. general public (Kunreuther, 1978).

Note that people's tendency to ignore low-probability but catastrophic hazards contradicts the logic that economists, government officials, and insurance company executives say should guide insurance purchase decisions. Insurance is intended to protect people from hazards that, though unlikely, are unbearable, rather than hazards that, though likely, are easy to bear! The data, instead, show that people are relatively unconcerned about low probability hazards, regardless of the severity of these hazards.[4]

Slovic et al.'s (1978) theory has important implications for campaigns to boost people's seat belt use or to increase other preventive behaviors. Slovic et al. argue that many previously tried "buckle up for safety" media campaigns have failed because they have not boosted people's perceptions of the likelihood of death or serious injury without a seat belt. Thus, campaigns that assert that seat belts are effective ("Seat belts *do* save lives."), or that demonstrate graphically that seat belts are effective (e.g., a photo of a wrecked car that driver and passengers walked away from unharmed because they wore their belts),

or that remind people to wear seat belts because they are wanted and needed by their families or employers (e.g., signs for employees leaving a company parking lot that say "Buckle up—we need you.") all focus on aspects of auto safety and belt use that are *irrelevant* to people who see the probability of death and harm without a belt as extremely low. Research has, in fact, shown the above approaches to be ineffective (Slovic et al., 1978).

Going a step further, Slovic et al. (1978) propose a technique for boosting people's perceived probability of auto injuries and deaths, in an attempt to boost their seat belt use. The technique might be applicable to other hazards and involves "elongating people's time frame" for the hazard. Slovic et al. found, in a pilot study, that when they told college students that the probability of death in a *lifetime* (fifty years) of driving is 1 in 100 and the probability of a serious injury is 1 in 3, 39 percent of the students who were belt nonusers reported they would use their belts regularly from then on. However, as we saw in Chapter 4, what people say and what they actually do are often not the same. Indeed, a subsequent study of *actual* belt use (surreptitiously observed) by Slovic, Schwalm, and colleagues (reported by Allman, 1985) did not replicate the results of the pilot study, though Slovic (1994) views this subsequent study as not definitive and believes that the elongation-of-time-frame approach may work better in other contexts.

We, however, have more to say later in the chapter about people's perceptions of hazard probability and about the Slovic et al. method of elongating people's time frame. For now, we remind you that Slovic et al.'s basic idea—that people tend to ignore hazards they perceive as being of low probability, even if the hazards are potentially catastrophic—might generalize to regional and global environmental hazards. If this is true, then people's perception of the probability of an environmental risk may be a more important psychological variable than their perception of the risk's severity in determining their response to the risk.

However, we must point out a possible limit to applying Slovic et al.'s (1978) explanation of denial (risk underreaction as necessary for getting on with one's life) to environmental risks and problems. The Slovic et al. work centers on people's reactions to immediate, direct threats to their lives, health, and property (from natural hazards and from auto accidents). In contrast, the environmental impacts of modern technologies are broader and more diverse. They do sometimes include threats to an individual's life, health, and property, but also include threats to the lives and health of others, the lives and health of non-human forms of life, and the "health" of the biosphere itself. Further, environmental hazards are usually diffuse and develop slowly over time, unlike the immediate personal impacts of auto accidents and some natural hazards. Therefore, a psychological tendency à la Slovic et al. to deny direct, immediate threats to health, life, and property in order to get on with one's life may not generalize to the nondirect, diffuse, and slowly developing threats posed by environmental problems. In other words, many people *may* take seriously some diffuse, nonpersonal, long-term environmental threats (like damage to the ozone layer) whereas they are not very concerned with the myriad immediate threats to their personal lives (bridges collapsing, the need for seat belts, etc.), including the immediate personal impacts of environmental pollution.

Our discussion earlier in the chapter and in Chapter 3 of Merchant's three ethical orientations toward the environment—egocentrism, homocentrism, and ecocentrism—is relevant here. People who have the egocentric orientation will probably care little about environmental risks other than those that—like autos and natural hazards—directly impact them (or their family); for these people, a Slovic et al.-like tendency to deny risks to "get on with one's life" *might, indeed,* apply to environmental risks that pose immediate and personal threats. However, people who are homocentric or ecocentric will care about a broader spectrum of environmental hazards and are less likely to deny them.

Only a few studies address the specific issue of whether people's perceptions and actions toward environmental impacts of technologies depend on whether they think of them in personal terms or as social risks. One such study, by Gardner, Tiemann, Gould, DeLuca, Doob, and Stolwijk (1982), examined the risk perceptions, benefit perceptions, acceptability judgments, and self-reported actions toward *nuclear power* of 367 subjects (a nonrandom sample of the U.S. population). For half of the subjects, the questions asked about risks and benefits were worded to refer only to "oneself and members of one's immediate family"; for the other half, question wordings referred to "people in this country." The results revealed no difference in the relationships between risk perceptions, benefit perceptions, acceptability judgments, and self-reported actions for the two different question wordings. These results suggest that our subjects thought of the environmental and other risks of nuclear power in social—not just personal—terms. Another interpretation is that the subjects in our "societal"-question-wording group were actually just thinking of themselves despite question wordings about risks, and so on to "people in this country"; this possibility is not supported by the results of an unpublished study by Sjoberg (1993). The results of a marginally relevant study by Tyler and Cook (1984) are at least not in contradiction to our general conclusion. Thus, Gardner et al.'s results suggest that Slovic et al.'s explanation of denial of auto risks may not apply to people's reactions to the nondirect, diffuse, slowly developing aspects of environmental hazards.

B. Taylor and Brown's (1988) Theory About "Illusions," Perceived Control, and Mental Health

Of the four psychological theories we present in this chapter, the one proposed by Shelley Taylor and Jonathan Brown (1988) is probably least promising as an explanation of people's misestimation of environmental risks, for reasons similar to those that may limit the applicability of Slovic et al.'s (1978) theory to these risks. However, Taylor and Brown's theory is so provocative and counterintuitive that we describe it briefly here (the theory may help explain why people misestimate other, nonenvironmental, hazards). Note that the main focus of Taylor and Brown's work is not risk perception but mental health. Taylor and Brown argue that most people overestimate their own skills and positive traits, overestimate the degree to which they control events in the outside world, and are overly optimistic about their futures. These misperceptions, or "illusions," Taylor and Brown claim, are necessary for people's mental health. However, Taylor and Brown point out, these illusions might also—as side effect—cause people to significantly underestimate the severity and probability of some hazards, and to fail to take preventive actions toward them.[5] Let's look more closely at Taylor and Brown's reasoning.

Taylor and Brown (1988), first, provide empirical evidence that mentally healthy people are subject to the three "illusions" just described. Concerning the first illusion, they cite studies showing a tendency for (normal) people to see themselves as better than others on a variety of personal traits. For example, Svenson (1981) and Svenson et al. (1985) found that most of their subjects judged themselves as significantly safer, more skillful drivers than average. Svenson and Svenson et al. used "Weinstein-type (1980)" logic to infer that their subjects' ratings were unrealistically positive: Most of the subjects in a large group could not possibly be "better than average," but that is, in fact, how most saw themselves. Conversely, Taylor and Brown cite studies showing that moderately depressed people actually see their own abilities and traits in a less biased, more accurate way!

Concerning the second illusion (perception of control), Taylor and Brown, cite evidence that people often believe they can influence events that are, in fact, determined entirely by chance. For example, subjects in studies on gambling often believed that they had more control of the outcome when they, rather than someone else, rolled the dice. Concerning the third illusion, Taylor and Brown cite Weinstein's (1980, etc.) findings, which we reviewed earlier in the chapter, that most people were "unrealistically optimistic" about the future state of their health (see Table 9-1).[6]

Showing that most normal people are subject to the three illusions, however, does not prove that the illusions, in fact, enhance mental health. Taylor and Brown therefore go on to review research suggesting that compared to individuals who do not have the illusions, individuals who do are more likely to: make friends and be popular, do creative and productive work, and be generally happy—all important components of good mental health. For example, people who view their own skills, their degree of control, and their future quite positively are more likely to persevere at their work, despite setbacks, and achieve their goals than are people who view their own skills, and so forth, negatively.

While the three illusions in Taylor and Brown's theory may contribute to mental health, each could possibly contribute to risk underestimation and a failure to take preventive actions. Thus, a person who believes that he is a superior driver, that he has great control of his car and events on the highway, and that he has a rosier future than others, is more likely to deny or underestimate the hazards of auto travel than someone who does not think in these ways. Several research results, however, suggest that, of the three illusions, *exaggerated perception of personal control* plays the biggest role in causing hazard underestimation and unrealistic optimism. For example, Weinstein found a strong relationship between perceived controllability and degree of unrealistic optimism toward health and other hazards in several of his studies (e.g. 1980, 1983, 1984, 1987). If you look back at data from his 1989 study shown in Table 9-1, you'll notice that the hazards that evoked the most unrealistic optimism (the hazards high in the list) tend to be under an individual's personal control (e.g., drug addiction), while those that evoked the least unrealistic optimism (those low on the list) tend not to be under an individual's control (e.g., high blood pressure and non-lung cancer).

Consider, as another example, David DeJoy's (1989) findings on perceptions of auto accident risk. He had 106 University of Georgia undergraduates read brief descriptions of ten different accidents and rate their chances of having each accident *relative* to other students (a Weinstein-type measure of realistic versus unrealistic perceptions of personal risk). The ten accidents covered a broad spectrum and included "scraping the side of your car at a drive-in window," "losing control of your car at high speed and hitting another vehicle," and "having your vehicle struck by a speeding hit-and-run driver." (As it turned out, DeJoy's subjects showed a great deal of unrealistic optimism about their chances of having some accidents, but less unrealistic optimism, or none at all, for other accidents.) Subjects then rated each of the ten accidents on several other characteristics, including: the accident's seriousness, the degree that one would worry about it, and the degree of driver control in causing/avoiding it (i.e., controllability). Using statistical regression analysis, DeJoy found that controllability was, by far, the main predictor of subjects' unrealistic optimism. In other words, his subjects showed the most unrealistic optimism for accidents they rated as most personally controllable. No other characteristic (seriousness, worry, etc.) was a statistically significant predictor of unrealistic optimism.

At least for personal health and auto safety risks, then, *exaggerated perception of personal control* appears to be a major factor underlying people's underreaction to these risks. The key role of perceived personal control, further, can explain how a person might deny his/her personal risks of auto travel and fail to wear seat belts, while at the same time might overestimate the overall national death rate from auto accidents, as we noted earlier in the chapter (we say more on the overestimation part later in the chapter). It is this key role of perceived control that also leads us to conclude that Taylor and Brown–type mechanisms are unlikely to be a major cause of people's underreactions to environmental hazards: What relevant data we have on people's perceptions of environmental risks (e.g., Slovic et al., 1979; Sjoberg, 1993) suggest that people perceive little personal control over global and regional environmental problems; people are therefore unlikely to underreact to these problems due to exaggerated perceptions of control.

More generally, we must point out the possibility, as we noted earlier, that theories—like Taylor and Brown's—that explain people's underreaction to some direct threats to their health and safety may not apply to people's reactions to longer-term, more diffuse, and less personal, environmental threats like the erosion of the ozone layer. It is, for example, easy to imagine a scientist who dedicates her research to helping solve global environmental problems but who fails to fasten her auto seat belt, doesn't get regular exercise, and/or eats foods high in saturated fat.

Though, as we just discussed, exaggerated perceptions of control may cause denial and inaction in some cases, paradoxically, the reverse also can occur: Under some conditions, unfounded perceptions of noncontrol can cause people to deny risks and fail to take protective actions. We discuss this phenomenon, which may generalize to environmental hazards, in the next section.

C. Psychological Stress Theory: Denial as a Response to Threats Perceived as Uncontrollable

The *theory of psychological stress* is a third theoretical framework that can shed light on people's underreactions to hazards. This theory holds that denial and/or a failure to act can occur when a person perceives a threat in his or her environment as uncontrollable, even though the person might, in fact, have some control over it. The denial and inaction serve to lessen the person's fear and anxiety—unpleasant emotions caused by a hazard perceived as uncontrollable—and also to protect the person from experiencing the uncomfortable feelings that noncontrol itself produces.

The theory of psychological stress was first proposed by Richard Lazarus in the 1960s (e.g., 1966) and later refined by Lazarus, his colleagues (e.g., Lazarus and Folkman, 1984), and others (e.g., Glass and Singer, 1971; Evans and Cohen, 1987). Stress theory, a key theory in health psychology and in environmental psychology, provides a framework for understanding how people react to a wide variety of threats and challenges. The theory defines an *environmental stressor* quite broadly—as an element of a person's environment that is unpleasant and/or that threatens the person's well-being in some way (Bell et al., 1990). Lazarus and others identify four main types of stressors, and, again, consider all of them within their single theoretical framework: cataclysmic events (such as floods and earthquakes), stressful life events (such as divorce, or death of a close relative), daily hassles (such as pressure at work or a long commute to work or school), and ambient stressors (such as chronic air pollution, noise, and crowding) (Cohen and Evans, 1987). Psychological stress theory may be broad enough to also shed light on how people respond to global and regional environmental hazards. Let's examine the theory carefully.

Several of the environmental stressors mentioned above (e.g., noise) can have direct negative impact on a person's body (e.g., permanent damage to hearing) (Selye, e.g., 1976). However, according to psychological stress theory, negative *psychological effects* can occur only after a person *perceptually and cognitively appraises* an environmental element as a stressor. Two perceptual/cognitive processes are involved: *Primary appraisal* assesses the nature and magnitude of the environmental element; *secondary appraisal* evaluates the degree to which the person's coping skills and resources are sufficient to meet the challenge and threat of the potential stressor.

If the primary appraisal process determines that the environmental element poses a threat but the secondary appraisal process determines that the person has immediate and direct control over the environmental element, then the person will suffer few if any negative psychological effects. For example, if a person hears loud, raucous music from the next apartment while he/she is trying to study, but also knows that the neighbor will graciously turn off the stereo if asked, the person will experience little distraction or stress, or any of the negative psychological effects of non-control we will outline shortly. Note that perceived personal control significantly or totally negates the negative effects of many stressors, *even if the person does not actually take action*. In other words, the *mere knowledge or perception of control* is sufficient to reduce or remove the effects of many stressors (e.g., Gardner, 1978; Evans and Cohen, 1987; Bell et al., 1990).

Now, what if a person is confronted with an environmental element that he or she perceives as unpleasant or harmful and over which he or she perceives no immediate and direct control? What if, for example, the neighbor is perpetually noisy and consistently refuses to turn down his or her stereo? According to Lazarus and his colleagues (e.g., Lazarus 1980), the person uses one of two types of coping strategy: *problem-focused coping and emotion-focused coping*. In problem-focused coping,

the person exerts control over the stressor, but not in the simple and immediate way we described above: The person thinks up and executes new behaviors that diminish the stressor (such as, in the case of the noisy neighbor, wearing earplugs, installing a soundproof wall, or moving to a new apartment). When people don't have the option of problem-focused coping, they often resort to emotion-focused coping. Emotion-focused coping involves cognitive and behavioral efforts to lessen the unpleasant emotions caused by the stressor or to better tolerate them (in the case of the stereo example, efforts such as clobbering a punching bag to vent the person's anger towards the neighbor, trying to think mainly about the many compensatingly positive features of the apartment, or deciding that studying is not so important after all).

There are several other types of emotion-focused coping strategies that people sometimes use, especially when confronted with serious natural and other hazards they perceive as being uncontrollable. Rippetoe and Rogers (1987) identify: *avoidance* (actually, "denial," i.e., an attempt to repudiate or refuse to admit the existence or the size of the threat), *wishful thinking* (the use of simplistic and unrealistic solution efforts or "panaceas"), *religious faith* (". . . the use of one's spiritual beliefs and faith in God's will to cope with the possibility of [a serious threat] [p. 598])," and *fatalism* (acceptance that the threat is unavoidable and a cessation of any attempts to avoid it). Note that these emotion-focused coping efforts do help allay a person's fear and anxiety, but they also stop a person from trying to cope with a threat in situations in which further efforts to gain information or exert control might be effective.

The behavior of the UCLA students studied by Lehman and Taylor (1987) and described in the Prologue of this chapter appears to provide a good example of emotion-focused coping, specifically, "avoidance" or denial.[7] Recall that students who lived in seismically poor dormitories were more likely than students who lived in seismically good buildings to endorse statements suggesting denial or minimization of the earthquake risk (statements like "There may be an earthquake, but it won't be so bad," "The likelihood [of] a major earthquake . . . has been greatly exaggerated," and so on). Lehman and Taylor, incidentally, cite similar data from an unpublished study on residents of the Santa Ynez Valley of California: People who were at risk because they lived near a dam that might collapse in an earthquake were more prone to deny the earthquake threat than residents living farther away and less at risk. (Other supportive data were reported by Vaughan, 1993, based on her correlational field study of 282 California farm workers exposed to agricultural pesticides: Those workers who perceived they had few other job possibilities were less likely to report taking preventive actions to minimize their risks from pesticide exposure than were workers who perceived they could find other jobs.)

As Lehman and Taylor (1987) argue, the students in the seismically poor dormitories (and the residents of the Santa Ynez Valley living near the dam) apparently used an emotion-focused coping strategy— avoidance or denial—because they perceived themselves as powerless against the earthquake threat. The students knew that their dormitories would likely suffer major structural damage or would collapse in a large earthquake. Furthermore, these buildings were *multistory* buildings. Taking earthquake preparation measures would not make much of a difference, nor would doing the right thing during an actual quake (i.e., getting under a heavy table or standing in a doorway). Further, the students living in the poor buildings had not put themselves in this situation knowingly; they either had been randomly assigned to their building or had requested a certain building before they knew about the safety report. In short, these students were faced with a potentially deadly environmental threat but had almost no control over what would happen to them. Deprived of the ability to use problem-focused coping, their only alternative was emotion-focused coping, specifically, denial or minimization of the hazard. By denying, they lessened their unpleasant feelings of fear and anxiety. They also lessened the unpleasant feelings of helplessness itself.

Before we further discuss emotion-focused coping, let's examine *helplessness* in more detail. The

state of helplessness—a person's perceived inability to control key parts of his/her environment—is not only unpleasant but can also be psychologically harmful. Prolonged noncontrol can cause a pathological condition called "learned helplessness" (Seligman, 1975; Evans and Cohen, 1987). The main symptom is that a person, having been seriously thwarted in efforts to control his or her environment, no longer tries, even though efforts to regain control might be effective at a later time. Learned helplessness, further, is often accompanied by depression.

One of the earliest research studies on learned helplessness in animals provides a dramatic illustration of the phenomenon. Seligman (1975) physically restrained dogs from escaping from a series of painful, though not physically harmful, electric shocks. The next day, the dogs were shocked again but could avoid the shocks by jumping from one side of a box to the other after a warning light came on. Rather than try to escape from the shocks, however, the dogs lay down and suffered. Thus the dogs stopped trying to control their environment though control was possible. (Dogs not shocked and restrained the day before quickly learned to jump and avoid the shock.) Some similar findings emerged in research on humans (e.g., Hiroto, 1974).

The phenomenon of learned helplessness underscores the psychological importance of the perception of personal control. Several psychologists and biologists argue that humans have a strong innate tendency to exert and maintain control over their environments (Cohen et al., 1986; Evans and Cohen, 1987). A creature that strives for mastery and control is more likely to survive than one that doesn't. Thus, these psychologists and biologists claim, humans (and some other animals) are strongly predisposed to learn, understand, predict, and control their environments and are constantly trying to do so. It is because severe noncontrol opposes this strong innate tendency that noncontrol can be so harmful, that is, can cause depression and the pathological state of learned helplessness.

Returning to our discussion of environmental threats and emotion- versus problem-focused coping, recall that the UCLA students had no control over the natural hazard they faced and could take no effective actions against it. However, in many other cases, people living in hazard-prone areas *do* have some control, that is, do have effective protective actions they can take. But to exert control, people must perceive the control as available and be motivated to use it. It appears that this is not always the case. For example, Sims and Baumann (1972) found that residents of tornado-prone regions in Alabama tended to view "God" or "luck" rather than their own behavior as responsible for their fates, and therefore took few preventive actions toward tornadoes—an apparent example of emotion-focused coping. In contrast, residents of Illinois who faced a similar threat were more inclined to use problem-focused coping: While they understood they could not alter the path or power of a tornado, they knew there were actions that lessen the risk of injury or death (such as moving from a wood-frame home to a stronger structure or to the right part of one's basement), and they took those actions. As a result, Sims and Baumann argue, tornado fatality rates (deaths per 100,000 inhabitants) were significantly lower in Illinois than in Alabama (even controlling for possible differences in the strength, size, and number of tornadoes per year between the two areas).

The cause of the difference in coping styles used by the Alabama versus the Illinois residents is not clear. Sims and Baumann speculate that the Illinois residents were higher in a personality trait known as "internal locus of control." The prevalence of this characteristic differs somewhat across ethnic groups and socioeconomic levels. A simpler alternative is that social norms that govern behavior were different in the two regions. (We say more about the influence of social norms later in the chapter.)

Though we are unsure of the factors that influenced people's choice of coping strategies in the Sims and Baumann study, we are convinced that psychological stress theory, in general, applies to people's reactions to environmental, as well as natural and other, hazards. We'll have much more to say about psychological stress theory and about problem-versus emotion-focused coping strategies in the main part of the chapter following the one directly below.

COGNITIVE/PERCEPTUAL BIASES, ERRORS, AND SHORTCUTS AS CAUSES OF BOTH HAZARD OVERREACTION AND UNDERREACTION—ONE THEORY

The three psychological theories discussed in the previous part of the chapter can help explain why people might underreact to certain environmental threats under certain conditions. However, there is little doubt that both hazard underreaction *and hazard overreaction* can stem from a source not covered in the previous part, namely, certain flaws, shortcuts, misunderstandings, and biases that humans often exhibit in their perception, cognition, and decision making. Thus, hazard misestimation and a lack of appropriate action may sometimes arise from perceptual/cognitive/decision processes, not out of a need to "get on with one's life," a need to reduce unpleasant feelings like fear, or other causes we've discussed so far in the chapter.

For example, there are data showing that some people who live in areas prone to hurricanes, floods, tornadoes, and certain other natural hazards underestimate the likelihood that these hazards will occur in the future, because they fall prey to a cognitive error known as the *gambler's fallacy* (Burton, Kates, and White, 1978). The gambler's fallacy is illustrated in its simplest form by the following example: Imagine a fair coin tossed eight times and coming up heads every time. The question then is, what is the probability that a tail will come up on the ninth toss? The correct answer is 50 percent. Individual tosses of a fair coin are independent and random. The coin has no "memory" of its past behavior, and thus the chance of a tail on the ninth toss is the same as it was on the first toss. Yet some people believe that the probability of a tail on the ninth toss is considerably higher than 50 percent. They assume that since eight heads in a row is such a rare, unusual sequence, the sequence can't continue, and a tail is now much more probable.

Belief in the gambler's fallacy can lead people who have experienced a severe flood or tornado—both relatively rare events—to assume that another bad flood or tornado cannot happen in their area in the next few months or years, that is, that there is a temporal "window of safety." But this is a misconception of the random natural processes that cause these weather events. Individual tornadoes and floods are like coin tosses. They are independent of each other. If a severe flood occurs in some area this summer, the chances of another flood later in the same summer or in the next few summers is the same as it was before this flood occurred; there is no window of safety. For example, Burton, Kates, and White (1978) report that two severe floods ". . . estimated to have a probability of once a century took place on the Housatonic River [in New England] during the *same* summer of 1955. Another flood might come the following year—or not again for two centuries [p. 97]."[8]

A. Bounded Rationality

Research on human cognition shows that the gambler's fallacy is only one of several cognitive errors or misconceptions that people are subject to. It's not that humans are basically stupid or think poorly. It's just that people sometimes have trouble understanding complex natural processes and random low-probability events. People therefore rely on overly simple mental models of these processes and events (the gambler's fallacy is an example). Further, people often use a variety of mental shortcuts, "tricks," and biases when they process information and make decisions. People's use of overly simple mental models, their cognitive shortcuts, biases, and so on, usually work reasonably well; any errors are usually not very serious. However, sometimes the simplifications, shortcuts, and biases do cause problems. Of course, the type of problem we are interested here is people's over- or underestimation of hazard severities and probabilities.

Until the early 1950s, most psychologists, economists, and other behavioral/social scientists assumed that humans processed information, understood events in the outside world, and made decisions perfectly, that is, the way that mathematical and other theories indicated were optimal and correct. For example, people supposedly made decisions according to the "rational person" model, that is, they

always chose which action to take only after exhaustively evaluating and comparing the desirability of all possible alternative actions. However, Herbert Simon (e.g., 1956, 1959) and others pointed out that humans could not follow the tenets of the rational-person model because of the large number of decisions humans must make in a limited time, the large quantity of information from the outside world they must process daily, and the limited (though impressive) capabilities of the human mind. Instead, Simon argued, humans, exhibit *bounded rationality* in their thinking and information processing. Bounded rationality approximates the ideal "rational person" formula, but necessarily takes shortcuts, simplifies complex problems, uses certain tricks, and shows certain biases. As we noted above, these shortcuts, tricks, and so on, usually work reasonably well. You, the reader, wouldn't be sitting there with most of your limbs intact, nor would we, the middle-aged authors of this book, be writing it, if human cognitive shortcuts, and so forth, worked poorly or if the errors made were very severe. In other words, none of us would have survived this long!

Consider a specific example of a generally effective cognitive "trick" (or "heuristic," as cognitive psychologists call them) that people frequently use. People rely on this heuristic to estimate brief intervals of elapsed time, something that the human mind is not very good at; if we asked you to signal when thirty-five seconds had elapsed (without looking at your watch), you would do a poor job of it. However, long ago, people learned to meter elapsed seconds by silently saying to themselves: "One one-thousand, two one-thousand, three one-thousand," and so on. And the technique works pretty well.

B. The Availability Heuristic

The availability heuristic is one that people apparently use frequently and with success, but that can sometimes lead them to seriously under- and overestimate hazard probabilities. The heuristic was first identified and studied by two prominent cognitive psychologists, Amos Tversky and Daniel Kahneman (e.g., 1973). People use the availability heuristic when attempting to judge the future likelihood or frequency of an event (e.g., "the chances of snow in November"). Here's how the heuristic works: You judge the probability or frequency of a future event based on the ease with which you can imagine or recall similar events from the past. If you find it difficult to imagine or recall a past instance, you judge the future frequency as very low, (e.g., "It's very unlikely to snow in November"); if you find it is very easy to imagine or recall, you judge the frequency as very high; and so on. Implicit in Tversky and Kahneman's argument is that most people (or at least most Westerners) regularly and consistently use the availability heuristic to predict future frequencies or probabilities.

Now, generally, the availability heuristic works pretty well. It is, indeed, true that events that occur frequently are easier for a person to recall and/or imagine than are events that occur rarely. However, because of certain properties of human memory and imagination, the availability heuristic can lead a person astray. Consider some classic examples of this. If we asked you which was more common in English, words beginning with k or words with k as the third letter, you (if you are like most subjects studied in research) would probably answer, words beginning with k. In reality, however, words with k as the third letter are twice as frequent as words beginning with k. Evidently, words in memory are organized by first letters rather than third letters, and it is therefore easier to retrieve words based on their first, rather than their third, letters. As a result, we incorrectly judge words beginning with k as more frequent. As another example, imagine the following experiment: Two groups of research subjects (group A and group B) are given a list of 100 people's names to try to learn (e.g., "Sam Johnson," "Jane Doe," and so on); fifty are male names and fifty are female names, randomly intermixed. For group A, many of the males named are famous and prominent (e.g., movie stars or politicians), while the females named are less famous and prominent; the reverse is true for group B. The next day, subjects from both groups are asked which were more frequent in the list of names they saw the day before: males or females. Subjects in group A tend to say there were more males than females;

subjects in group B tend to say the reverse. Evidently, when asked the next day which names were more numerous, group A subjects found that male names came easier to mind (were more "available") and group B subjects found that female names came easier to mind. Thus, the availability heuristic led both groups astray, since the actual number of male and female names in the lists was the same.

C. Availability and Risk Underreaction

How could people's use of the availability heuristic lead them to misestimate the future likelihood of a threat or hazard and/or fail to take proper actions? The answer is simple and straightforward. Let's begin with risk *under*estimation/inaction. If, for example, a person lives in a flood-prone area but has not personally experienced a flood there, the person will find it difficult to imagine or recall a flood. Therefore, using the availability heuristic, the person will judge the probability of a future flood as very low, maybe even zero, and will tend not to take protective action, such as buying flood insurance. Geographer Robert Kates (1962) describes the underlying principle in somewhat more intuitive terms: He says that people are "prisoners of their experience." Humans have trouble conceptualizing, much less taking protective actions against, hazards that they have not personally experienced. Kates goes on to point out that people who do actually experience a flood or other hazard are more likely to show concern about future floods and take preventive action, such as buying insurance; however the people seem to have trouble conceptualizing or taking action against a flood *worse* than the one they experienced.

The prisoner-of-experience principle could apply to environmental hazards and risks just as well as it does to floods and other natural hazards. For example, humans cannot directly perceive certain harmful chemical components of air pollution (such as radon in indoor air). We therefore may have trouble being concerned about these pollutants or taking action to decrease them (e.g., see Weinstein et al., 1990). (By "we," we mean members of the general public and also government policymakers.) Similarly, most

Americans have never experienced the extreme air pollution common in many large cities in the developing world or in London in the 1950s. Therefore, air pollution this bad may seem to be almost out of the realm of possibilities. More ominously, no person has experienced some of the possibly catastrophic environmental threats that loom before us, including wholesale destruction of the ozone layer or a significant rise of sea level due to global warming via the greenhouse effect. Thus, these problems may be difficult for us to take sufficiently seriously; we may tend to underrate, ignore, or deny them. Of course, the usual qualifications apply: Some people and groups are quite concerned, for example, about the possible destruction of the ozone layer despite the fact that none of these people have personally witnessed it, as we discussed in Chapter 3 (e.g., Dunlap et al., 1993). A tendency to deny doesn't imply that denial is total or universal, but that there is a "push" or predisposition in that direction under some conditions.

Several research studies support the validity of the prisoner of experience (POE) concept, in that they show that a person's denial decreases and protective actions increase after the person actually experiences a hazard that can recur. An example is Dooley et al.'s (1992) study of 800 residents of earthquake-prone Orange County, California. Dooley et al. found that residents who had personally experienced an earthquake that "had scared them" were three times more likely to express concern or worry toward future quakes than residents who had no personal experience with quakes. In turn, residents who expressed concern or fear were 50 percent more likely to have taken protective measures than those who did not express concern or fear. Dooley et al.'s data, however, suggested that the effects of personal experience were short-lived and wore off after a period of several months. It may be that after a person experiences a major hazard, the person's subsequent experience of day-after-day calm and safety slowly erodes the elevated concern caused by the earlier hazard experience (Slovic et al., 1978).

Results of other research studies on the effects of personal experience are mixed or inconsistent. Sims

and Baumann (1983) cited four prior studies on floods and hurricanes that showed that personal experience was positively related to preparedness and/or evacuation under warning. On the other hand, two studies yielded negative results and one was ambiguous. Weinstein (1989) reviewed twenty-four studies of the effects of personal experience on subsequent protective behavior. Personal experience increased posthazard protective action for some types of hazards,[9] including natural hazards and heart attacks, but didn't for other hazard types, such as being a victim of a crime. Weinstein's review also suggests that the effects of personal experience wear off with time, just as Dooley et al. found. Finally, Weinstein argues that personal experience of a flood, tornado, or other hazard can actually *increase* a person's denial and decrease preventive action; this could happen if the experience makes the person perceive the hazard as *less controllable* than he or she had thought in the past. Weinstein's analysis, if correct, may help explain the somewhat inconsistent findings concerning the effects of personal experience on denial and protective behavior.

D. Availability and Risk Overreaction

Up to this point we have shown how people's use of the availability heuristic might explain why they underestimate risks and/or fail to take protective actions. However, availability can also help explain the opposite pattern—people's overestimation of, and/or overresponse to, certain hazards. Recall, first, our earlier discussion of the Lichtenstein et al. (1978) study and its results, shown in Figure 9-1 (also see Example B in the Prologue). Recall the *primary bias* shown in Figure 9-1 (the curved line), that is, a tendency for subjects to underestimate the frequencies of common causes of death (e.g., heart disease) but to *over*estimate the frequencies of rare causes (e.g., botulism). As we pointed out earlier, this pattern is apparently a general trend found for some other, including *non*risk, subject matters (Lichtenstein et al., p. 574; Crocker, 1981, cited in Lehman and Taylor, 1988). For this reason, primary bias is of less interest to us than the *secondary bias* that Lichtenstein et al. found superimposed on the primary bias pattern.

Note the secondary bias in Figure 9-1, shown as deviations from the curved line: Subjects overestimated the frequencies of *dramatic* and *sensational* causes of death (e.g., auto accidents and homicides) and underestimated those of more ordinary causes (e.g., stomach cancer and diabetes). Lichtenstein et al. suggest that one probable cause of secondary bias is—yes!—people's use of the availability heuristic. Instances of dramatic causes of death—like fiery auto accidents and vicious bear attacks—are so striking and graphic that these instances are more easily recalled and more easily imagined in vivid imagery than are instances of mundane and prosaic causes of death. If people do, in fact, estimate the future frequency of events based on the availability of these events, then they will overestimate dramatic and sensational hazards.

Public overestimation of dramatic hazards can be unfortunate and costly to society, as we discussed earlier in the chapter. If the public overestimates the probability and severity of environmental hazards that are dramatic but are actually unlikely and negligible, the public will want government officials to focus on lessening these hazards, diverting limited resources away from larger and more serious threats. A former administrator of the U.S. Environmental Protection Agency, William Reilly, argued along these lines several years ago (Stevens, 1991). Reilly claimed that too much government attention and funding went toward environmental problems that are momentarily dramatic and/or anxiety-provoking to the public but that scientists and engineers do not consider as very serious; these problems include oil spills from ocean-going tankers, seepage from toxic chemical dumps (as at Love Canal in New York), and radiation from nuclear power plants (Stevens, 1991). Conversely, Reilly claimed that too few resources were directed to environmental problems that are less dramatic and less visible to the public but that scientists and engineers considered more serious (in terms of the magnitude, long-term consequences, and/or irreversibility of the problems); these problems include

species extinction, damage to the ozone layer, and global warming (Stevens, 1991).

Some scientists, engineers, and government policymakers argue, also, that distorted public fears traceable to the availability heuristic make difficult well-reasoned public discussion of the threats of technologies such as nuclear power. When scientists, engineers, and government officials attempt to assuage public concern about the safety of nuclear power plants by identifying various different failures or accidents and showing how improbable they actually are, the public response is increased anxiety: People will think: "I didn't realize there were so many things that could go wrong" (Slovic, Fischhoff, and Lichtenstein, 1980). Apparently, some frightening events can be highly imaginable, even if they have never occurred and people have never experienced them.

There is, on the other hand, an alternate nonavailability-based explanation for the secondary bias shown by Lichtenstein et al.'s (1978) subjects, (a bias shown also by subjects in similar studies): People's "distorted" hazard estimates may reflect the overreporting of sensational hazards by newspapers, TV, and other news media, which are the public's main source of hazard information. (Newspaper editors may overreport sensational stories in the belief that this practice will boost readership.) Some research findings support this explanation. Combs and Slovic (1979) surveyed the coverage of the forty-one different causes of death used in the Lichtenstein et al. study in two newspapers, one on the East Coast and one on the West Coast of the United States. (The West Coast paper was published in the city in which Lichtenstein et al.'s subjects lived.) Combs and Slovic found that coverage (the number and length of articles) in both newspapers did not reflect government statistics on the frequencies of the forty-one causes of death (shown on the x-axis of Figure 9-1); instead the papers overreported dramatic causes of death (like homicide) and underreported prosaic but much more frequent causes of death (like diseases). Further, the estimates of death made by the subjects in the Lichtenstein et al. (1978) study correlated highly with newspaper coverage of the forty-one different causes of death. Thus Combs and Slovic's data suggest that imbalances in newspaper coverage account for inaccuracies in people's perceptions of the probabilities of death from different causes. Of course it is also possible that newspaper editors, in their choice of articles to print, may just be showing the same availability-based distortions of the true frequencies shown by the reading public; a more intuitively compelling explanation, as we suggested above, is that editors choose to cover dramatic and sensational causes of death because they believe these causes make for more interesting reading and therefore higher readership.

In a similar vein, a two-year study of environmental hazard coverage on TV network evening news programs by Greenberg and colleagues (1989a, 1989b) found significantly more coverage of sensational and dramatic environmental incidents, like toxic chemical releases (e.g., the Bhopal [India] disaster) and nuclear power plant accidents, than of more chronic, insidious, and serious environmental problems, like global warming, ozone layer erosion, and so on. Greenberg et al. conclude that this biased media coverage is the cause of the public's apparently distorted environmental concerns claimed by former EPA administrator William Riley, as we discussed above.

So, what is the source of people's misestimation of causes of death and their differential concern about different environmental problems? Is it their use of the availability heuristic, or is it differential reporting in the media? We don't think there is a definitive answer. We say this, in part, because of the small quantity of relevant data now available, in part, because both sources may be involved, and, in part, because of the complexities, which we discussed earlier in the chapter, of defining what a "risk" is and of estimating the true risks of environmental and other threats.

E. Other Heuristics, Cognitive Errors, and Biases

In addition to the availability heuristic, there are other heuristics, cognitive biases, and errors studied by Tversky and Kahneman and colleagues (e.g., 1973)

that could lead people both to *under-* and *over*react to risks, depending on the circumstances. These include: People's overconfidence about the accuracy of their knowledge, the strong effects on perception and decision making of issue "framing," and the "pathological" effects of people's need for certainty.

Work by Slovic, Fischhoff, and Lichtenstein and by Tversky and Kahneman shows that people tend to be *overconfident* about the accuracy of their knowledge and information about the outside world. For example, subjects in the Lichtenstein et al. (1979) research, which we have discussed in detail (Figure 9-1), were asked (in addition to estimating the number of deaths from different causes) which of two given causes of death (e.g., homicides and suicides) was more frequent. Subjects also estimated the probability that their answers were correct. The results showed subjects to be quite overconfident about their knowledge. For example, 30 percent of the subjects who judged *incorrectly* that homicides were more frequent than suicides gave 50:1 odds that their ranking was correct!

In another study, Tversky and Kahneman (1973b) asked subjects "almanac-like" questions such as: "How many foreign cars were imported into the U.S. in [a given year]?" Once a subject gave an estimate, the researchers then asked the subject to give upper and lower estimates, or bounds, so that they were *98 percent sure* that the true value lay between these bounds. The results: For about half the subjects, the true value lay *outside* their 98 percent bounds. Tversky and Kahneman argued that people's use of the *anchoring and adjustment heuristic* was responsible for this overconfidence. In anchoring and adjustment, a person first estimates the number (of cars imported) based on several possible mental strategies; this estimate is the person's "anchor." The person then adds and subtracts—"adjusts"—only a very small amount from the anchor (e.g., plus and minus 10 percent) to produce the "98 percent limits" of the true number. In other words, people don't realize just how inaccurate their original estimates are and that they must allow much greater room for error in setting the 98 percent upper and lower bounds. Given people's strong tendency to both under- and overestimate risks due to their use of the availability heuristic, as we discussed, people's strong tendency to be *overconfident* about their (flawed) perceptions and information about the outside world creates a double whammy!

Further, Slovic, Fischhoff, and Lichtenstein (e.g., 1979) have argued that scientists, engineers, and other technical experts are as prone to overconfidence as are members of the general public, especially when they use "educated intuition" and must go beyond firm historical data. Slovic et al. cite, for example, a study by Hynes and Vanmarcke (1976) in which seven "'internationally known' geotechnical engineers" were asked to give upper and lower bounds concerning how high an embankment of earth and/or stone could be before it caused the failure of a clay foundation. As it turned out, *none* of these experts' estimates (upper and lower bounds) included the true value. Similarly, Henrion and Fischhoff (1986) found that research physicists had significantly overestimated the accuracy of their estimates of important physical constants, such as the speed of light, in decades of work attempting to precisely determine the values of these constants.

Other well-known work of Tversky and Kahneman (e.g., 1981) shows that people's decisions about which of two actions to take can be radically influenced—even totally reversed—by the way the choice is *framed*, that is, the wording of the choice and the way the problem presented to them is structured. Here is a frequently cited example: College student subjects were asked to decide which of two public health programs to enact in an effort to combat a new disease that would kill 600 Americans in the next year if no action was taken. The subjects were divided randomly into two groups. Group I was asked to decide between Programs A and B; they were told: "If Program A is adopted, 200 people will be saved. If Program B is adopted, there is a one-third probability that nobody will die, and a two-thirds probability that no people would be saved." Group II was asked to decide between Programs C and D; they were told that: "If Program C is adopted, 400 people will die. If Program D is adopted, there is a one-third probability that nobody will die, and a two-thirds probability that 600

people will die." Notice that Programs A and C are actually identical, but they are put in different words with respect to lives saved and lost; similarly, Programs B and D are identical. The results: 72 percent chose Program A over B, and 78 percent chose Program D over C. In other words, the different "framings" of the decision produced diametrically opposite decisions. As it turns out, Tversky and Kahneman also did the experiment on medical doctors and got the same results.

Going a step further, Kahneman and Tversky (1979) developed a psychological theory, called "prospect theory," to explain the above results. One main tenet of prospect theory is that people are more upset by losing something than they are happy about gaining the same thing; thus people are bothered more by the loss of $100 than they are happy about winning $100. A second main tenet is that people are disproportionately impacted by an outcome that is certain versus one that is uncertain; thus, for example, people find the possibility of losing $50 with certainty more unpleasant than a 50 percent chance of losing $100; conversely, people prefer winning $50 with certainty than a 50 percent chance of winning $100. (We'll talk more about people's desires for certainty shortly.)

Let's now see how prospect theory can explain the results of the study on people's choices of public health programs that we described above: Concerning the results for subjects in Group I: The subjects tended to prefer Program A because it offered *certainty* that 200 (out of 600) lives will be saved, over Program B which only offered a one-third chance of saving all 600 people. In the case of Group II: First, subjects perceived 400 deaths out of 600 (Program C) as more negative than 200 lives saved out of 600 (i.e., Program A), even though these two programs are actually identical; furthermore, subjects viewed the certain death of 400 people (Program C) as more horrible than a two-thirds chance that 600 will die (Program D).

An important real-world implication of prospect theory and of the research results above is that people's perceptions and decisions can be manipulated by the way a problem is framed, where framing is determined, for example, by how the news media portrays the problem, or how public- or private-interest groups describe the choice at stake, or how government officials define the problem or the decision to be made when discussing it with the general public. Thus, individuals, groups, or institutions that are in a position to frame an issue can manipulate the public's perceptions and preferences toward the issue. We further discuss the possibility of manipulating people's perceptions and/or decisions later in the chapter.

Before we move on, we'd like to cover four additional important points about framing. First, note that different people, groups, and institutions may inherently frame a problem in different ways, and not because they are trying to manipulate others. For example, Vaughan and Seifert (1992) argue that public negativity towards the chemical Alar in the late 1980s (which we described earlier in the chapter) was made stronger because industry groups and the EPA presented statistics on the *overall* health impact on the public of Alar, and did not at first mention that infants and small children were disproportionately at risk; that is, the public thought that industry/government were trying to deceive via the way they framed the problem, whereas industry/government may have merely seen the problem from a different perspective (the overall impact on the public). Second, note that we actually discussed framing earlier in this chapter: Recall that the risk-compendium approach, shown in Table 9-2, was framed around nuclear power and failed to consider alternatives (for example, energy conservation programs). Similarly, Slovic et al.'s method of "elongating the time frame" is an effort to reframe information about a hazard (auto accidents) so as to increase people's perceptions of the risks and thereby increase their protective actions (using seat belts).

Third, we should briefly mention an important connection between Kahneman and Tversky's prospect theory and our discussion in Chapter 4 on providing information and changing attitudes: The Kahneman and Tversky theory predicts (and empirical data confirm) that more people would be willing to install insulation in their homes or buy a high-efficiency furnace if they were informed of how much money they are now *losing* or *wasting* (because of their current lack of insulation, etc.) rather than how much they would save with the new equipment!

Fourth, in a related vein, Gregory et al. (1993) found that college student subjects rated as more desirable proenvironmental policy options framed as improvements back to prior levels (e.g., "back to the lower levels of air pollution 15 years ago") than options framed as a gain from the status quo ("less air pollution than is now prevalent").

We now discuss a final topic under the heading Other Heuristics, Cognitive Errors, and Biases. Kahneman and Tversky, Slovic et al. (1979), Fischhoff et al. (1980), Kaplan (1991), and others argue that an overarching feature of human cognition and decision making is that people have very strong *desires for certainty*. These desires are stronger and more basic than we have so far implied in our discussion of Kahneman and Tversky's prospect theory, above. Based on their research findings, these cognitive psychologists argue that many people find it difficult and discomforting to take gambles (rather than deal in sure bets), to understand the meaning of small probabilities (for example, the meaning of "the probability of a flood this year is .06"), and to make decisions about future events in the face of uncertain information and knowledge. Because people often find that gambling and uncertainty make them uncomfortable, many often deny the uncertainty to reduce their anxiety. For example, some people who live in areas prone to floods, hurricanes, or tornadoes are unable to grasp that these events are low-probability events generated by a random natural process, and that the hazards are equally likely to happen any year. Instead, these people, as we noted earlier in the chapter, may fall prey to a simpler and reassuring, but incorrect, mental model of the natural processes, that is, the gambler's fallacy, which insists that at least for several years after an occurrence of a hazard, people are perfectly safe. Some people may claim, more simply, that a future reoccurrence of the hazard is totally impossible. Note further that people's desire for certainty may also aggravate their tendencies to be overconfident, which we discussed above in the context of the anchoring and adjustment heuristic.

Fischhoff et al. (1980) cite a striking example (Feller, 1968) of people trying to find a graspable structure underlying the occurrence of some extremely hazardous random events: During bombing attacks on London by Germany in World War II, Londoners developed various theories in an effort to explain how the bomb targets were being chosen by German aviators and what places would likely be targets next. Subsequent statistical analyses, however, revealed that the bombing pattern was totally random.

Note finally that the claim by cognitive psychologists that people have a very strong desire for certainty closely parallels Taylor and Brown's (1988) claim, discussed earlier in the chapter, that people need "illusions of control" in order to maintain their mental health. The claim also parallels the phenomenon of defensive denial in general stress theory, that is, a tendency for people to deny serious threats over which they have no control in order to lessen their feelings of anxiety and of helplessness.

F. More on Availability: People's Insensitivity to Missing Items in Fault-Tree Analyses and Diagrams

Let's return to our discussion of the availability heuristic. There is evidence that the heuristic can have an additional important impact on people's perceptions of technological and environmental risks. The impact centers on the fault-tree analyses that scientists and engineers use to estimate the overall safety of new technologies like commercial nuclear power, as we discussed earlier in the chapter. Slovic et al. (1979, 1984), Fischhoff et al. (1978, 1981), and others have argued that the availability phenomenon may make the scientists and others who conduct fault-tree analyses insensitive to the fact that items are missing from the analyses. The role of availability in these situations is that things that are "out of sight" tend to be "out of mind."

Consider Figure 9-2, which shows a fault-tree analysis for one kind of nuclear reactor accident involving a loss of coolant water and resulting damage to the reactor core. As the Figure shows, the fault-tree analysis begins with the particular type of accident and attempts to anticipate all the different component failures and breakdowns that could possibly lead to this accident. Based on estimates of the probability of each component failure/breakdown, the analysts estimate the overall probability of the accident. A major characteristic of these analyses is that, to be valid, the analysts must have thought of and included *all* important possible routes to the failure. If the analysts have missed some routes, their analysis may seriously overestimate the overall safety of the technology. Fischhoff et al. (1978, 1981) outline several reasons fault-tree analysts may, indeed, overlook significant routes to failure. These reasons include: The inability of analysts to anticipate the size and nature of errors made by the human operators of technical equipment like nuclear reactors, and the inability of analysts to comprehend the overall behavior of highly complex technological systems. But in addition, Fischhoff et al. argue that the availability heuristic makes unprecedented or unexperienced events somewhat harder for fault-tree analysts to think of and take seriously. Further, when risk analysts use fault-tree diagrams as a visual device to communicate fault-tree analyses and their conclusions to other experts or to the public, availability may seriously interfere with these audiences' ability to think of possible routes to failure not shown in the diagrams.

Consider one research study that supports this claim. Fischhoff et al. (1978) did an experiment on people's insensitivity to missing items in a fault-tree analysis concerning a prosaic (and nonenvironmental) problem: An auto's inability to start. The fault-tree they used is shown in Figure 9-3. Note that each branch represents a different subsystem of the auto, for example, the fuel system or the ignition system. A defect or failure in any of these subsystems could prevent the car from starting. Fischhoff et al. showed their subjects the fault tree and asked them to estimate, for each subsystem (or branch), the probability that it would fail. The subjects were also asked to indicate the completeness of the tree diagram by estimating the probability of nonstarting caused by the "all other problems" branch. Actually, one group of subjects saw the complete tree shown in Figure 9-3, while the other group saw a severely "pruned" version of it, that is, with half the branches missing.

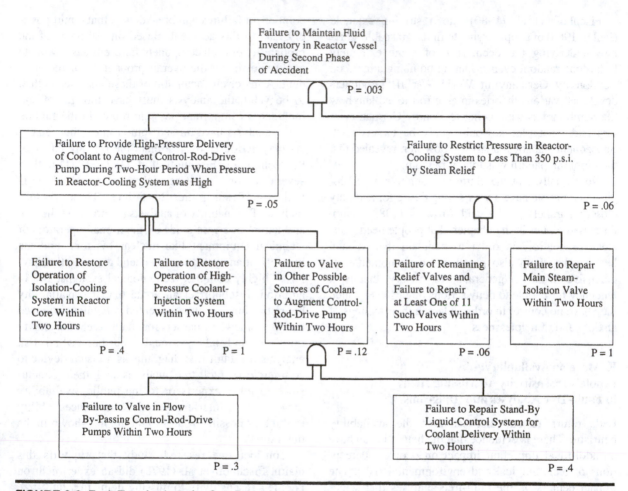

FIGURE 9-2 Fault-Tree Analysis for One Possible Kind of Nuclear Reactor Accident Involving a Loss of Coolant Water and Resulting Damage to the Reactor Core
This analysis comes from a report on an accidental fire in a nuclear power plant at Brown's Ferry, Alabama, in 1975.
From Lewis, H., The safety of fission reactors. *Scientific American,* Volume 242, 53–65. Copyright © 1980. Scientific American Inc., George V. Kelvin. Reprinted with permission.

The subjects who saw the pruned version did not miss the missing branches; their ratings of the probability of "all other problems" were about the same as those of subjects who saw the complete tree. Fischhoff et al. also ran the same experiment using professional auto mechanics as subjects. These expert subjects showed as much insensitivity to missing branches as had the student subjects. While autos are hardly nuclear reactors and auto mechanics not nuclear engineers, these results are striking. Slovic et al. (1979)

argue at length that the results may generalize to engineers who design nuclear plants and who perform fault-tree analyses.

G. Yet More on Availability: Some Speculation

There is yet another possible consequence of people's use of the availability heuristic, one that could lead people both to under- and overreact to hazards and risks, depending on the circumstances involved. In

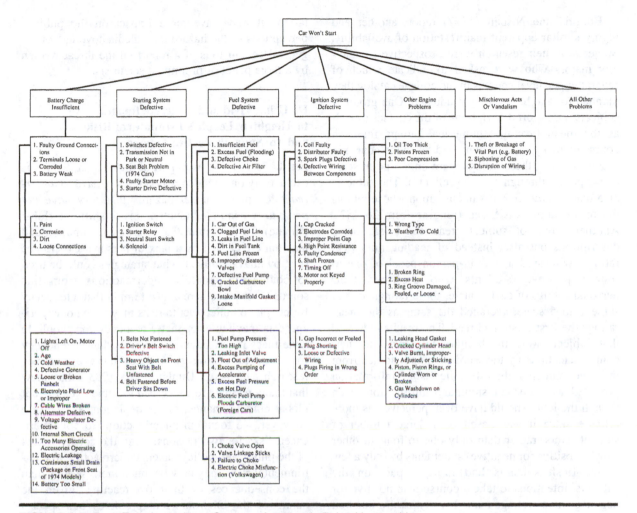

FIGURE 9-3 Fault-Tree Analysis on Causes of an Auto's Being Unstartable
From Fischhoff, B., Slovic, P., and Lichtenstein, S. Fault-trees: Sensitivity of estimated failure probabilities to problem representation. *Journal of Experimental Psychology: Human Perception and Performance,* Volume 4, 330–344. Copyright © 1978 by the American Psychological Association. Reprinted with permission.

essence, this consequence is that people may be unduly swayed by small quantities of word-of-mouth information concerning hazards and risks, and may tend to ignore far more reliable statistical information based on large quantities of data. This bears on our brief discussion in Chapter 4 of the "diffusion of innovation" process and the stronger influence of personal word-of-mouth advice from "opinion leaders" over media information in getting people to adopt a novel technology. It can be argued that the greater influence of word-of-mouth advice than of media information is a manifestation of the availability heuristic similar to the ones we described in previous sections: Personal, word-of-mouth information is more concrete and vivid than media information and is more easily recalled and therefore more influential when a person is making a decision about adopting a new technology.[10]

Borgida and Nisbett (1977) report another and highly similar apparent manifestation of availability. Subjects in their research were prospective psychology majors who were asked to indicate which of several upper-level courses in the college catalog they planned to take before they graduated. One group of subjects read short descriptions of the courses as well as the mean (average) numerical ratings for each course made by students who had taken the course in previous years; these ratings were on a 5-point scale (1 = "poor," through 5 = "excellent"). The subjects also read the *number* of students upon whose ratings the mean rating was based (ranging from 26–142). Another group of subjects read the brief course descriptions but also, instead of reading numerical ratings, witnessed a live, in-person panel of one to four upper-level students who described their personal ratings of each course; these ratings, on the same 1 to 5 scale, averaged the same as the mean ratings the first group had read. The results indicated that subjects were much more influenced in their course selections by the word-of-mouth advice from the live students than by the same information provided as statistical summary information, even though the latter should have been perceived as more reliable since it was based on a large number of student raters, rather than only one to four. In other words, positive (or negative) statements by only a few in–the–flesh students had more impact on the subjects' intentions to take a course than positive (or negative) mean ratings made by, say, over a hundred students. Put the other way, summary statistics based on large numbers of cases were less influential than in-person information from just a handful of raters.

The apparent superiority of word-of-mouth advice and vivid case histories over more-reliable statistical summaries may generalize to environmental subject matters, although there is some dispute in the research literature on the nature and magnitude of the effect that Borgida and Nisbett (1977) reported (see Taylor and Thomson, 1982) and of similar effects reported by some other researchers (see Nisbett et al., 1982). If the effect generalizes, then, for example, the appearance in the news media of two or three telegenic scientists warning about, or, alternatively, assuaging concerns about, an impending environmen-

tal threat may have more impact on the public's perceptions of the hazard than media coverage of the general conclusions of a report on the threat written by a large panel of prominent scientists.

H. Using Vivid and Concrete Images to Heighten People's Estimates of Risk and to Spur Protective Actions

For the rest of this main part of the chapter, we focus exclusively on a tendency people may have to *underreact* to environmental hazards that *they have not personally witnessed* (such as major shifts in global climate). To the extent that risk underestimation and inaction are caused by a lack of personal experience (i.e., by unavailability), what strategies could be used to counteract people's underreactions from this source? Our focus here doesn't imply that a tendency for people to underreact to risks they have not experienced is inevitable, or that a tendency for people to underreact is, on balance, stronger than a tendency for them to overreact. (Indeed, there are survey research data—e.g., Dunlap et al., 1993—that show that many people are concerned to some extent about global climate change, ozone depletion, and so on.) However, we focus on underreactions to nonexperienced risks for two reasons: First, the consequences of the public and policymakers underreacting to environmental risks may now be much more serious than the consequences of their overreacting, given the nature of the environmental problems our species now faces: For the first time in history, human activity has the potential to deplete vital natural resources and to disrupt key global ecological systems. Thus environmental policy errors caused by people's underestimations of risk could be catastrophic and irreversible. The second reason for our focus is that behavioral scientists have conducted scores of potentially relevant studies on ways to lessen people's failure to take preventive actions against natural, technological, and personal health hazards that they have not experienced.

In his classic article on denial, Garrett Hardin (1969) proposed the use of vivid and sensational images of environmental damage to overcome any tendency people might have to deny environmental problems they haven't personally witnessed; in other words these images might unshackle the "prisoner of experience." Specifically, Hardin suggested that the public and policymakers be shown graphic scenes from other places or other times in history that depict the bad things that happen when humans damage the environment. People who view these scenes would find environmental threats more "available" and would be more likely to take the right protective actions. For example, people could be shown, using photos and drawings, that large areas in the Middle East were once thickly forested but were turned into deserts because inhabitants overharvested the trees. As another example, people could be shown the ruins of both Mayan and Mesopotamian civilizations along with graphic depictions of how these civilizations (probably) perished due to rapid population growth and the use of unsustainable agricultural practices. The Tennessee Valley Authority has actually used this method to try to show residents of flood-prone areas their vulnerability and need to take protective action, as Kates (1962) points out. Specifically, the TVA has distributed vivid photos showing well-known buildings in the area in the grips of prior floods.

But what about environmental problems with catastrophic potential, like massive erosion of the Earth's protective ozone layer, that haven't occurred? How could the unavailability of such threats be countered? We can't wait for these problems to actually happen because of their catastrophic and irreversible nature, nor do we have photos or images from other places or times. A possible solution, in the spirit of Hardin (1969), above, would be to show people vivid science fiction–type films and other media, for example, high-quality Hollywood movies on unprecedented environmental hazards. A few of these films have already been made. They include *Soylent Green*, a film about a society racked by pressures of acute overpopulation; *The China Syndrome*, a film about a nuclear power plant accident (an accident similar to the Three Mile Island accident that occurred just months after the film was released); *The Day After*, a made-for-television movie depicting the consequences of global nuclear war (Schofield and Pavelchak, 1985); and *On the Beach*, an earlier antinuclear-war film.

Of course, the concerted, large-scale production and distribution of films of this type in a democratic society raises some philosophical and moral issues. The method presumes that there is scientific consensus about the severity and imminence of the environmental threats involved. It also assumes that there is valid evidence of significant public underreaction. In addition, the method assumes an agreement by the public about the acceptability of educational or "propaganda" campaigns by governments to boost people's concern and actions; (it is one thing for Rachel Carson to try to convince the public that the risks of some pesticides are unacceptable, but another for the federal government to try to do so). But, there are other problems with Hardin-type educational campaigns that we discuss in the next section.

I. Negative Research Evidence on the Use of Vivid, Concrete Images

Hardin's idea of using vivid images to "unshackle the prisoner of experience" has a great deal of intuitive appeal. Unfortunately, considerable research evidence indicates that the method, used by itself, doesn't work. However, we will argue that the method may be effective if it is used as part of a more comprehensive package of methods. With that conclusion in mind, let's first review the negative evidence concerning vivid images used by themselves.

There are literally scores of research studies on the use of vivid imagery to encourage people to take some problem seriously and to take preventive action toward it (e.g., Higbee, 1969), though no studies we know of that have specifically involved environmental subject matter. Thus, public health officials have tried to get smokers to quit by showing them photos of cancerous lung tissue. Educators have tried to encourage high school students to drive safely and use seat belts by showing in driver's education classes vivid photos of terrible auto accidents. Note that attempts like these to encourage preventive behavior are often called *fear* or *threat appeals*. They use vivid imagery to convey as strongly as possible—even by frightening the viewer—the serious consequences of failing to take preventive action. Thus, campaigns that use vivid images almost always have the emotional component of fear (Johnson and Tversky, 1983). Movies like *The Day After* are quite frightening and designed to be so. Hardin's proposed use of images from the past (collapsed civilizations) also has an implicit, if not explicit, fear component.

Unfortunately, as we already noted, research on the above campaigns has consistently found them ineffective in altering people's behavior (e.g., Slovic et al., 1978; Higbee, 1969; Weinstein, Grubb, and Vautier, 1986). Many similar programs have failed, including ones to encourage condom use to stop the spread of AIDS (Aronson et al., 1991), and programs to discourage the sharing of drug hypodermic needles, also to halt AIDS (Sherr, 1990).

Probably the best-known example of a vivid, fearful images campaign in an attempt to change behavior is the "Scared Straight!" program that operated at New Jersey's Rahway Prison in the 1970s. Many Americans know of this program because it was the subject of a TV film (Golden West Productions, 1979) that won an Academy Award. The film, MC'ed by TV/film star Peter Falk, depicted the experiences of seventeen teenage "delinquents" (who had been arrested but not imprisoned) during a day they spent at the prison. The seventeen teenagers were hosted by a group of lifers—inmates with life sentences. The lifers spent several hours telling the teenagers in a vivid and verbally brutal manner how horrible it was to be in prison (e.g., the frequent violence, murders, homosexual rapes, and so on). At the end of the day, the seventeen teenagers appeared on camera, visibly shaken. Each vowed to "go straight"—to stop his or her delinquent behavior—from then on. Finally, Peter Falk said that 8,000 juvenile offenders had gone through the Scared Straight program so far, and that 80 to 90 percent of them had, indeed, gone straight (Presbury and Moore, 1983).

Almost ten years later, a second film was aired on television (Pyramid Film & Video, 1987). The film, MC'ed by TV/film star Whoopi Goldberg, showed what had happened to the seventeen delinquents in the intervening years. All seventeen had been located and were interviewed on camera. Sixteen out of seventeen were law-abiding citizens; only one was in prison. Of the sixteen, fifteen said they had kept out of trouble with the law completely during the years

since their day at Rahway Prison. Finally, this second film mentioned that thousands of teenagers had gone through the Scared Straight program since its inception, with a high rate of success.

Sadly, as we noted above, we must report that well-designed research studies of the Scared Straight program, and the many similar programs that were tried in other parts of the country, have consistently found these programs to be totally ineffective in deterring criminal behavior.[11] There are several possible reasons for the discrepancy between the success of Scared Straight claimed in the two TV programs versus the results of published scientific research. We discuss some of these reasons in Box 9-1, titled The Scared Straight Program: Could Peter Falk and Whoopi Goldberg be Wrong?

We now very briefly review some of the negative research evidence concerning the Scared Straight program. James Finckenauer (1982), a sociologist at Rutgers University, found that a group of young delinquents who had been chosen at random to go through the Scared Straight program actually had *higher* rates of subsequent rearrest than delinquents in a control group who did not go through the program. Lewis (1983) found no difference in the subsequent criminal behavior of a control group (randomly chosen) of fifty-five delinquents versus a group of fifty-three juvenile offenders who went through a program similar to Scared Straight at San Quentin Prison in California (the Squires program). Lewis (1983) also cites the negative results of two similar prison programs: One at Jackson Prison in Michigan (Yarborough, 1979), and one at the Huntsville State Penitentiary in Dallas, Texas (Vreeland, 1981). Kingsnorth (1991) cites two reviews of the research literature that come to the same negative conclusion (Vito, 1984, and Latessa and Vito, 1988). Kingsnorth (1991), incidentally, also reports negative results concerning a Scared Straight–type program aimed at deterring future drunken driving in people arrested for that crime; the program included visits to a morgue to view the bodies of people killed in alcohol-related auto accidents and similar visits to a hospital emergency room.

In short, there is a long, unbroken record of failure of Scared Straight–type programs, as well as of

Box 9-1 _____

The Scared Straight Program:
Could Whoopi Goldberg and Peter Falk Be Wrong?

Though two documentary films shown on TV, one MC'ed by Peter Falk, the other by Whoopi Goldberg, portrayed the Rahway Scared Straight program as highly successful, a number of research studies conducted by behavioral and social scientists found the Program, and similar programs at other prisons, totally ineffective. What can explain this discrepancy?

First, there is some evidence that most of the seventeen teenagers in the Scared Straight films were not the hard-core or lifelong delinquents they were portrayed to be; in other words, the evidence suggests that most of the seventeen were pretty straight to begin with. Jerome Miller (1979, quoted by James Finckenauer, 1982 [pp. 174–178]) claims that thirteen of the seventeen lived in comfortable homes in an upper-middle-class community, a bedroom suburb of New York City; the thirteen, all of whom attended the same 1,200-student high school, were chosen not for their record of delinquency but because they volunteered to be in the film. Further, Miller claims, the school and the community did not have a delinquency problem. Quoting Miller (in Finckenauer, 1982 [p. 178]): ". . . In the last three years [only] 3 or 4 juveniles from the town had been sent away to reform school." (Incidentally, Finckenauer, 1982, provides some evidence that neither the producers of the film nor Mr. Falk had a direct role in selecting the thirteen teenagers.)

Even if we assume, however, that all the seventeen youths *had* been hard-core delinquents, their Scared Straight experience had unique qualities that might have caused them to go straight. None of the other teenagers (thousands of them) who went through the Scared Straight program had the same experience. Specifically, the seventeen shown in the films had been followed by a professional film crew before, during, and after their day at the prison. All seventeen pledged, on camera, to go straight. Further, all viewed themselves on local TV making that pledge and knew that the program was rebroadcast by TV stations across the country. All, presumably, knew that the film had won an Academy Award (and other awards including an Emmy). Thus each had made a national *public commitment* to go straight. Recall from our

discussion in Chapter 4 that even a simple and modest public commitment can cause a significant change in behavior. Given that the seventeen teenagers made about as strong and dramatic a public commitment as it is possible for a person to make, this commitment alone could well have caused a major change in their behavior.

Going a step further, Finckenauer and others argue that the underlying premise of the Scared Straight approach is seriously flawed. It is very unlikely that just a few fearful hours spent inside a prison could cause thousands of hardened delinquents to go straight for the rest of their lives. Delinquent behavior, like any other behavior, is the result of many interacting influences. These can include genetic predispositions (e.g., an extra Y chromosome in males predisposing toward impulsive behaviors [McConnell and Philipchalk, 1992], or a genetic predisposition towards alcoholism), poverty, unemployment, a broken family, an active drug addiction, and so on. We have emphasized throughout this book that there are many psychological, social, economic, political, and physical forces that influence what a person does. One three-hour experience cannot be expected to radically and permanently alter the behavior of large numbers of people. Although many government policymakers and members of the public hope for simple, highly effective, and inexpensive remedies for serious social ills, such remedies, or panaceas, rarely exist in the real world.

Despite the complexities and qualifications we just discussed, a modified Scared Straight approach might be useful as part of a multipart program to encourage prosocial behavior in some young delinquents. Psychologists Jack Presbury and Helen Moore (1983) describe such a program, one focused on children as young as eight. Presbury and Moore argue that delinquent behavior usually has early, childhood roots, and that attempts to intervene in childhood may work, whereas attempts in adolescence come too late. The children in Presbury and Moore's program had committed thefts or vandalism and had been referred to a county mental health center by a juvenile court judge, a social worker, and/or the children's parents.

(continued)

BOX 9-1 continued

In addition to psychological counseling sessions, each child briefly visited a county jail and spent time talking with one (adult) inmate. In contrast to Scared Straight, the inmate didn't try to frighten the child. Instead, the inmate assumed a supportive "big brother" role and urged the child not make the same mistakes he had made as a kid, mistakes that ultimately led to a criminal "career" and to imprisonment. After the jail visit, each child had further counseling sessions at the mental health center. Presbury and Moore report that their program worked successfully for several chil-

dren, though the number of children involved in their research is too small to draw strong conclusions. And, again, it is unlikely that the program will help all delinquent youngsters. Presbury and Moore offer one additional criticism of standard Scared Straight–type programs: They point out that people often react to frightening threats not with changes in behavior but with internal defensive reactions that lessen the unpleasant emotion of fear. We will discuss such emotion-focused coping reactions to environmental threats later in this chapter.

---

programs, described earlier, that used vivid images to discourage smoking or encourage safe driving. Despite this poor record, however, we suggest that the use of vivid imagery to encourage behaviors *not* be abandoned. We believe that the above programs failed for several ascertainable reasons and that if these reasons are addressed by supplementary techniques, vivid imagery may have a major role to play. In the paragraphs immediately below, we discuss one possible reason for the failure of vivid imagery programs, and then go on to discuss other reasons in the next main part of the chapter (titled Putting the Elements Together). We also describe a multidimensional approach we think may be more effective.

J. Conveying Both the Probability and the Severity of the Hazard

Finckenauer (1982) argues that Scared Straight–type programs are ineffective because they focus on the severity of the punishment for criminal behavior but fail to address the probability of being caught and actually suffering the punishment. More specifically, these programs may convince the young offenders who go through them that being in jail is horrible, but the programs do nothing to convince the offenders that the probability they will be caught if they break the law, convicted, and land in jail is high enough to worry about (Finckenauer contends that this probability is actually quite low). If potential offenders view the probability of conviction and sentencing as very

low, they will not be deterred by the prospect of even extremely severe punishments.

Finckenauer's (1982) analysis may seem familiar to you. It is essentially the same as Slovic et al.'s (1978) analysis of why people tend not to use seat belts or to take preventive actions against natural hazards, which we discussed earlier in the chapter. Recall Slovic et al.'s argument that people are forced to ignore many low-probability risks to which they are subject in order to function effectively. Also recall the "farm simulation" experiment in which the probability of a hazard—not its severity—appeared to determine whether subjects purchased insurance against it (see Table 9-3). And recall that campaigns that emphasize that seat belts do save lives or that urge people to wear belts because they are needed by their families focus on aspects of auto safety and belt use that are *irrelevant* to people who see the probability of death and harm without a belt as extremely low.

Incidentally, Higbee (1969) argues that the common use of especially extreme and gory accident photos in high school drivers' education classes to encourage safe driving and seat belt use may actually *backfire*. He notes that there is usually an intrinsic inverse relationship between accident severity and accident probability for autos and other technologies: Extremely severe accidents are very rare, while moderately severe or mild accidents are more likely; Higbee assumes that most people have implicitly learned this relationship. Therefore, showing students photos of unusually severe accidents may lead them to think that auto accidents are "less probable than they had supposed." The students may then become less concerned about seat belt use and safe driving practices. (There is also the possibility that these photos are so emotionally upsetting and anxiety-provoking that something like defensive denial, as in the general stress theory, might occur.)

Finckenauer (1982), further, reviews some research on adult criminal behavior that supports his analysis of the failure of Scared Straight programs (research that also supports Slovic et al.'s 1978 framework). This research shows that an increase in the perceived severity of punishment by itself (e.g., a new law that requires longer prison terms for a certain crime) does little to deter adult criminal behavior. Conversely, increases in the perceived certainty of punishment do have a deterrent effect. A similar pattern emerges in Roth et al.'s (1989) review of research on a specific (adult) criminal behavior: cheating on one's income tax. But Roth et al. also note that a few research studies came to a more complicated conclusion, though caution is called for given the scarcity of relevant data: These studies found that increasing the severity of punishment *does* increase taxpayer compliance, but *only if* the perceived certainty of punishment is *above* some threshold level.

Based on these taxpayer findings and others we've reviewed above, a reasonable conclusion is that programs attempting to deter criminal behavior (or to increase preventive actions against a hazard) can be effective only if they create the perception that the punishment is very severe *and* if they raise the perceived probability of punishment above a certain minimum level. To go a step further, we propose that—just as the ineffectiveness of Scared Straight–type programs can be explained, in part, by their failure to convey a sufficiently high probability of punishment—the failure of Slovic et al.'s (1978, and cited in Allman, 1985) method of "elongating the time frame" (discussed above) may be explained, in part, by its failure to vividly convey the high severity of the punishment for not wearing seat belts. Again, it appears that people must be convinced that there are relatively high levels of *both* risk probability *and* risk severity before they will take appropriate actions.

A few research studies involving environmental subject matter (rather than criminal subject matter) support the idea that both perceived risk probability and perceived risk severity must be addressed in any effective program to encourage preventive action. Modest support, for example, comes from a study by Weinstein, Sandman, and Roberts (1990) on ways to encourage homeowners to test their homes for radon. (Radon is a colorless, odorless gas that can seep into homes through cracks in foundations and can collect in sufficient quantities to cause lung cancer.) Weinstein et al. (1990) mailed questionnaires and different informational brochures about radon to several hundred homeowners living in parts of New Jersey that are radon-prone because of underlying geological formations. Weinstein et al. found that homeowners who received a brochure stressing both the high probability of a radon problem in their homes and the fact that such a problem would pose a severe health threat were significantly more likely to report an intention to test for radon than were homeowners who received a brochure lacking this dual stress. Furthermore, the researchers found that whether a homeowner intended to test was better predicted by the product of the homeowner's perceived probability of having a radon problem in his/her home, times the homeowner's perceived severity of such a problem (i.e., $P \times S$), than by either perceived probability (P) or perceived severity (S) alone.

Keep in mind, however, that increasing people's perceptions of probability and of severity may still not be enough, as we discuss in the next main part of the chapter.

CHAPTER SUMMARY: PUTTING THE DIFFERENT ELEMENTS TOGETHER: PSYCHOLOGICAL CAUSES AND POSSIBLE SOLUTIONS

We've covered a great deal of material in this chapter on the perceptual and cognitive processes that underlie people's underreactions and overreactions to threats and hazards. In this main part of the chapter we summarize much of this material. Our summary involves a broad theoretical framework, one that encompasses the four psychological theories that we've already presented. The framework is a modified version of the "protection-motivation theory" developed by Ronald Rogers and his colleagues (Rogers, 1983; Prentice-Dunn and Rogers, 1985; Rippetoe and Rogers, 1987). Though Rogers originally proposed the theory in the context of personal health threats,[12] the theory appears to have broad applicability, including to natural and technological hazards and to environmental threats.

The protection-motivation framework is especially useful because it shows how the several psychological processes and mechanisms we've already discussed in the chapter can interact, reminds us that all of these processes/mechanisms can contribute to misestimation and inaction at the same time, and suggests multicomponent programs that are likely to be effective in efforts to increase (or decrease, as appropriate) people's estimations of environmental threats and/or their actions toward those threats.

In the paragraphs below, we first briefly outline the protection-motivation framework. We go on to summarize the material we've covered in the chapter in the context of that framework.

A. Overview of Modified Protection-Motivation Theory

Figure 9-4 shows a diagram of the protection-motivation theory as modified by the authors of this book. A main feature of the theory is that it differentiates two major perceptual/cognitive processes that together determine an individual's response to a threat or hazard. The first process—"threat appraisal"—assesses the nature and magnitude of a threat; the second—"coping appraisal"—assesses the type and amount of coping responses the individual has available. The threat- and coping-appraisal processes are basically the same as the two appraisal processes in Lazarus' (e.g., 1966) classic theory of psychological stress we outlined earlier, that is, "primary appraisal" and "secondary appraisal," respectively (Rippetoe and Rogers, 1987). However, the two appraisal processes are conceptualized in greater detail in the protection-motivation model than they are in psychological stress theory.

Figure 9-4 depicts each perceptual/cognitive appraisal process as a large, lightly outlined, labeled box. The *threat appraisal* process (top box) has three subcomponents. A person's appraisal of a potential threat is influenced, first, by the person's relevant *values*. As we discussed earlier, perceived risks don't exist outside a framework of values; for a person to view a feature of his/her environment as a risk, he/she must believe the feature endangers something that he/she values. These values, along with media and other information from the outside world, are the main inputs to the psychological processes that determine a person's perceptions of the severity of, and vulnerability to, the threat. The person's *perception of the severity* of the threat, more specifically, is the person's appraisal of how damaging the outcome of the threat would be to things he/she values if the threat were to actually occur, (e.g.—to use a natural-hazard example—the person's judgment that a powerful earthquake in the area would seriously jeopardize things he/she values highly, such as life, health, or home). The third component of the threat appraisal process, *perceived vulnerability,* is the person's perception of the *probability or likelihood* of the threatening event's actually occurring (e.g., the person's estimate of the likelihood of a severe earthquake in the area in the near future).

The second perceptual/cognitive appraisal process, *coping appraisal* (large bottom box), has three subcomponents. A person's coping appraisal is influenced, first, by his or her *perceived response efficacy,* that is, whether or not he/she knows of any specific actions that he/she believes are effective against the threat (for example, a judgment that certain structural improvements to his/her house would prevent damage from an earthquake). (This assumes that the threat-appraisal process judges the threat as serious to begin with.) The second component, *perceived self-efficacy,* is the person's assessment of whether he or she is personally capable of performing the requisite protective acts (e.g., the degree to which the person feels capable of executing the structural improvements by himself/herself or of successfully contracting this work out to others). The third and last component of the coping appraisal process is *perceived costs and benefits.* This includes many of the barriers that impede people from acting on their proenvironmental attitudes and values, which

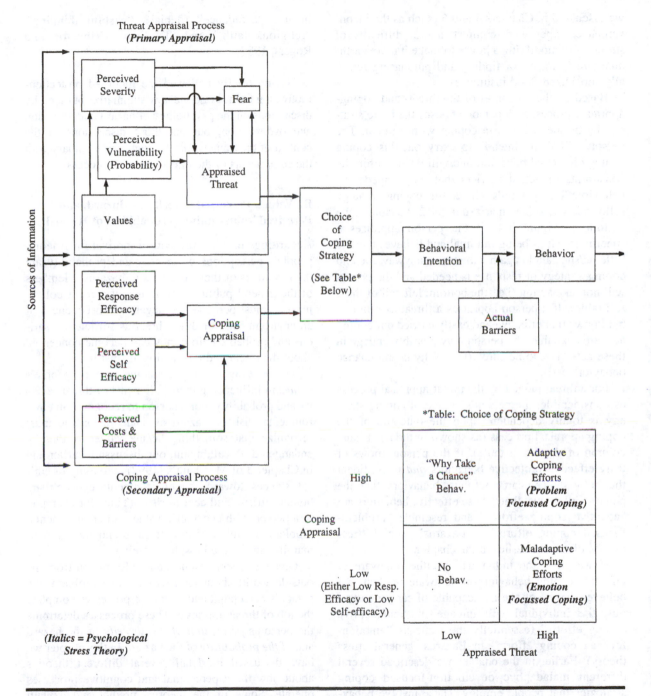

FIGURE 9-4 Protection-Motivation Theoretical Framework

Adapted from Rippetoe, P., and Rogers, R. Effects of components of protection-motivation theory on adaptive and maladaptive coping with a health threat. *Journal of Personality and Social Psychology*, Volume 52, 596–604. Copyright © 1987 by the American Psychological Association. Used with permission.

we discussed in Chapters 4 and 5 (such as the inconvenience, expense, discomfort, and/or difficulty of structurally modifying a house to make it more earthquake-resistant, or of finding and purchasing federally subsidized flood insurance).

Based on the outcomes of the threat- and coping-appraisal processes, a person chooses the basic strategy he or she will use in coping with a threat. The person will then "intend" to carry out this coping strategy in actual behavior, and will do so within the constraints of actual barriers that may impede the behavior. The person's choice of coping strategy follows key rules summarized in the 2 × 2 table at the bottom of Figure 9-4: If the person appraises a specific threat as being minimal and believes there is little he/she can do against the threat anyway, then no coping strategy or response is needed and the person will not take action (i.e., the bottom, left cell of the 2 × 2 table). If a person appraises a threat as minimal but knows that effective and easily carried out coping acts are available, the person may possibly engage in these acts "just to be sure" (the "Why take a chance behavior" cell).

For a threat judged by the threat-appraisal process as considerable, the person's choice of coping strategy is totally dependent upon the outcome of the coping appraisal process (as shown in the right-hand column of the 2 × 2 table). If the person knows of truly effective protective behaviors *and* is confident that he or she can carry out these behaviors, then the person will do just that. These effective behaviors are "adaptive coping efforts," and resemble "problem-focused coping efforts" in Lazarus' general stress theory, discussed earlier in the chapter.

However, if the individual is either unaware of effective coping behaviors *or* is aware of them but believes that he or she is incapable of carrying them out, the individual will engage in "maladaptive coping efforts" (essentially the same as "emotion-focused coping efforts" in Lazarus' general stress theory). Earlier in the chapter we described several different maladaptive or emotion-focused coping strategies that people employ. These involve behavioral and/or cognitive efforts to lessen or better tolerate the feelings of fear, anxiety, and helplessness caused by the threat. Maladaptive coping efforts

include "avoidance" (denial), "wishful thinking," "religious faith," and "fatalism" (Rippetoe and Rogers, 1987).

Having briefly reviewed the modified protection-motivation framework, we now summarize our lengthy discussions on the psychology of hazard underreaction and overreaction, but put those discussions in the context of the framework. We begin our summary with the components of the threat-appraisal process.

B. Values, Perceived Severity of Hazard, and Perceived Vulnerability (Probability) of Hazard

According to the protection-motivation theory, people's perceptions of severity and probability are the keystones of their reactions to hazards. Members of the general public, as well as government policymakers, must perceive the negative consequences of an environmental or other threat as sufficiently serious and probable before they can become concerned about the threat and take action against it.

We once again emphasize the role of people's *values* in influencing their judgments of threat severity and probability. For a person to recognize an environmental risk or hazard as such, he or she must recognize that something he/she values is thereby endangered. Recall, again, our discussion earlier and in Chapter 3 of Merchant's (1992) taxonomy of ethical stances toward the environment: egocentrism, homocentrism, and ecocentrism. (Thus, for example, to a person with ecocentric values, an environmental problem is a threat if it appears to endanger nonhuman life and/or the biosphere itself.)

Given a person's values, and information from the outside world about an environmental problem, the person's perceptual and cognitive processes complete the job of threat appraisal. These processes determine the person's *perception of the severity* of the hazard and *of the probability* of the hazard. In this chapter we have discussed in detail several different theories about how these perceptual and cognitive processes operate. Some of the theories predict and explain people's tendencies to *underestimate* risk severity and/or probability for certain risks and under certain conditions. These theories argue that risk underreaction may occur: because of an innate human inability to perceive "slow motion" disasters; because of a genetic predisposition for moderate risk denial, selected for during biological evolution as it promoted discovery and invention; because some level of risk underreaction is necessary so that humans can go about the business of daily life; because a certain level of risk underestimation is necessary for human mental health; and because of some common human cognitive errors, biases, and heuristics. These cognitive errors, biases, and so forth include people's belief in the gambler's fallacy, and their use of the availability heuristic, which may make it difficult for them to grasp and prepare for some risks they have not personally witnessed; further, overconfidence and desires for certainty may aggravate or increase people's underestimation of risks.

We have also reviewed research on the use of vivid and concrete images as a possible way to make certain risks that are "unavailable" more available, that is, as a way, in some cases, of "unshackling the prisoner of experience." (We remind the reader that the use of this method, and some other methods we have discussed, presumes a public consensus about the appropriateness of such educational/"propaganda" campaigns.) Though the research we reviewed found vivid, concrete images to be ineffective by themselves, there is no research result or theory we're aware of that leads us to conclude that vivid images should be abandoned completely, even as one part of a more comprehensive program, or that some other technique would do a better job of conveying the potential severity of some hazards that people have not experienced.

Slovic et al.'s (1978) work on people's failure to fasten auto seat belts and to purchase natural-hazard insurance adds an important qualification concerning perceived severity and perceived probability: Their work suggests (though the quantity of relevant research is small) that a person will not take action against certain threats—*regardless of the threats' severity*—if the person perceives the probability of the threats' negative consequences as below some threshold level. Slovic et al.'s results suggest that media campaigns attempting, for example, to alert the public to the dangers of household radon (which can cause lung cancer) or of the erosion of the ozone layer (which can lead to increased skin cancer, impaired immune system function, and ecosystem damage) may not work if the campaigns just convey the severity of these outcomes but don't create public perceptions that the probability of these outcomes is above some minimum level. To be effective in boosting public concern and relevant individual action, campaigns—again, there is the presumption of a public consensus on the acceptability of such campaigns—must convince people that the probability of harmful consequences is high enough to warrant action on their part. Slovic et al.'s (1978) method of "elongating the time frame" may help do this for some environmental hazards. Though elongating subjects' time frame apparently failed when used by itself in an attempt to increase people's use of auto seat belts, it is possible that the method may be effective if it is part of a comprehensive program with additional ingredients; more research is needed on the method. Even if the "elongating the time frame" method proves useless, Slovic et al.'s basic point concerning a minimum required perceived probability-level may still be valid.

On the other hand, we have seen that people may seriously *overestimate* hazard severity and probability for risks that are especially dramatic and sensational, because such risks are very imaginable and highly available (we remind the reader of the important cautions and qualifications that should be kept in mind when judging whether people are overreacting or underreacting to a hazard.) Note an implicit further inference: People may overestimate the probability and severity of a risk *even if* they have never themselves witnessed the risk if the risk is especially frightening and imaginable. Further, people's proneness to overconfidence and their desires for certainty may aggravate or increase their overestimation of risks.

C. Perceived Response-Efficacy, Self-Efficacy, and Perceived Costs and Barriers

One important aspect of the protection-motivation theory (and also of general stress theory) is that it argues that people may underreact to a hazard for either of two reasons: One reason is that people may incorrectly perceive the risk as inconsequential and/or unlikely (i.e., an outcome of the threat-appraisal process); alternatively, people may initially perceive a threat as serious, but resort to a maladaptive coping strategy, like denial, because they see no way of controlling the threat (an outcome of the coping-appraisal process). Note that the Slovic et al. seat belt theory does not clearly distinguish between people's initial (mis)perceptions of auto risks as nil, versus people's correct perception of the risk but denial of it in order to "get on with their lives." The same ambiguity appears in Hardin's theory of the adaptivity of "moderate" denial, and possibly Ornstein/Ehrlich's theory (both in Chapter 8). In contrast, both the general stress theory and the protection-motivation theory distinguish between people's initial perceptual reactions and their subsequent coping strategies. For these two theories, underreactions to hazards are a defensive reaction to risks appraised as serious but not controllable. And, again, the stress and protection-motivation theories also allow for the possibility that people's initial perceptions of risks are either under- or overestimates to begin with. The cognitive framework (i.e., errors, biases, and heuristics) focuses on people's initial perceptions or cognitions concerning threats (though the "desire for certainty" has an ambiguous status). Finally, the Taylor and Brown theory (mental health) includes several elements: Rosy perceptions of the future (hence, presumably, initial risk underestimation), and a unique element not found in the other theories: illusions of control that cause a person to not take precautions against a hazard.

A major strength of the protection-motivation framework is that it provides a fully articulated explanation for the consistent failure (and even, on occasion, backfire) of campaigns that solely use vivid images (such as photos of bad auto accidents or frightening descriptions of prison by prison inmates) or other techniques to convey the *severity* of a threat: If such efforts *are not also accompanied* by a description of antithreat actions that people perceive as both effective against the threat and easily performed (and also accompanied by evidence that the probability of the threat is above some minimum level), people will respond with the maladaptive (emotion-focused)

coping efforts we described earlier, such as avoidance or denial (as in the case of UCLA students in seismically poor dormitories), wishful thinking, or fatalism. These efforts allay negative emotions but can prevent people from seeking more effective response strategies. (Rogers and his colleagues provide and cite specific evidence for the *backfire* effect—Van der Velde and Van der Pligt, 1991; Prentice-Dunn and Rogers, 1987.) Thus, protection-motivation theory helps explain the long history of failure of vividness and "scare-tactic" programs, one of the most important research findings we have reviewed in this chapter. To repeat, campaigns that successfully convey the severity and even the likelihood of a threat but don't also present effective and easily performed protective acts will likely fail or backfire.

In the context of regional and global environmental problems (erosion of the ozone layer, and so on) that are the focus of this book, the protection-motivation theory predicts that people *will* take action if they perceive environmental threats as sufficiently serious and likely, and know of actions they can take that will really make a difference and perceive themselves as capable of taking those actions. But there is one other ingredient required in order for people to take action:

We have talked extensively in Chapters 4 to 7 about the costs and barriers that can impede the proenvironmental actions of people who are otherwise inclined to take them. Removing or lessening such barriers is a critical part of any program to encourage proenvironmental behaviors. As an example, consider that, in their study of home radon problems, Weinstein and Sandman's (1992b) concluded that the single most important thing that government policymakers could do to encourage homeowners to test for radon and then take the remediation actions recommended was to lower barriers that interfered with these actions. Thus, the protection-motivation framework explicates the several elements necessary for programs to encourage proenvironmental actions to succeed.

D. Beyond the Protection-Motivation Framework

Protection-motivation theory, above, is a useful framework for understanding the different perceptual and cognitive processes underlying people's reactions to risks and the ways in which these processes interact. However, the theory is not all-encompassing (nor was it intended to be by its authors), because it does not address the social-psychological, interpersonal, and other processes that we discussed in Chapters 3 to 7. More specifically, the protection-motivation framework includes few of the variables affecting people's environmentally relevant behavior that we summarized in Table 4-2 and Table 7-2. For example, while protection-motivation theory stresses that people must learn about effective protective actions before they can act, the theory ignores the material covered in Chapter 4 on the best ways to inform people about those actions.

Weinstein and his colleagues (Weinstein, Sandman, and Roberts, 1990; Weinstein and Sandman, 1992a) and others (e.g., Stasson and Fishbein, 1990) have similarly argued that since protection-motivation theory focuses on individual behavior and on perception and cognition, the theory ignores important social and interpersonal processes that affect protective behavior. For example, Weinstein and Sandman's (1992a) research on home radon abatement led them to conclude that the availability of helpful neighbors and friends strongly affects whether a homeowner tests for radon and/or takes corrective actions if necessary. Weinstein and Sandman point out that friends and neighbors who have already gone through the process can help a homeowner figure out which of several radon testing methods is best, interpret the test results, choose corrective actions if necessary, and hire a reliable contractor to do the work (abatement actions are often difficult for an average homeowner to perform by him- or herself).[13]

Not only are friends and neighbors a source of valuable information and reassurance, they also convey and remind us about prevailing social norms. This, in turn, can strongly affect whether we take preventive action against a hazard. For example, in a study on auto seat belt use, Stasson and Fishbein (1990) concluded that subjects' intentions to use belts were more directly affected by their perceptions of social norms concerning belt use than by their perceptions of the risks of accidents. Weinstein, Sandman, and Roberts (1990) also argue that prevailing social norms have a significant effect on radon testing and remediation behavior.

A related limitation of the protection-motivation framework is that it does not fully represent the strong motivational forces of values and morals. As we pointed out in Chapters 4 and 6, values—especially homocentric and ecocentric ones—can be the basis for moral judgments that carry more emotional force than the protection-motivation concept of "threat appraisal" might suggest. Responses to risk can be motivated by moral judgments based on altruistic concerns for other people (including future generations) or for nonhuman species or the biosphere. In fact, some of the strongest reactions to environmental risks appear to be associated with harm to the innocent—children harmed by chemicals, sea otters killed by oil spills, and so on. Some people tie their environmental concern to a powerful religious ethic: "It is wrong to harm the environment God created" (Kempton et al., 1995). In short, people's judgments of risk are sometimes based more on considerations of morality or justice than on dispassionate threat appraisal.

On the other hand, the perceptual and cognitive processes we have discussed in this chapter, and the protection-motivation framework that summarizes them, are not addressed in prior chapters of the book. Because these processes significantly affect people's reactions to environmental threats and hazards, this chapter complements the others. As we have repeatedly stressed, there are many different psychological components and dimensions of environmental problems, and all of them must be addressed in efforts to lessen these problems if the efforts are to succeed.

CONCLUSION

In an attempt to better understand people's reactions to environmental hazards and risks, we have reviewed a large quantity of research on people's reactions to natural, technological, and personal-health hazards. Early in the chapter we saw that it is difficult to judge the accuracy of people's perceptions of the environmental risks of a technology independent of the moral, ethical, and political issues that underlie societal debates about technologies. We've also seen that scientific disagreement and uncertainty about risks (for example, about the magnitude and onset of global climate change) and about the risks of newer technologies (e.g., commercial nuclear power) also makes it difficult to assess the accuracy of people's risk perceptions and the appropriateness of their actions.

Though it's impossible to fully separate the study of psychological processes that underlie people's reactions to risks from a consideration of moral and ethical issues and of scientific uncertainty, we have argued that there *are* perceptual and cognitive processes and phenomena that may *push people in the direction of* misestimating some environmental hazards and/or failing to take appropriate actions toward them. We have seen that these processes and phenomena are, for the most part, highly adaptive (e.g., people's use of the availability heuristic) but that they can also lead people to significantly under- and/or overreact to risks.

We've reviewed four psychological theories (and two others in Chapter 8) that predict and explain why people might underestimate and fail to act against certain threats and hazards. We've seen, for example, that humans may sometimes have trouble comprehending and taking action against hazards they have not personally witnessed. To the extent that this trouble applies to current global problems, the consequences could prove serious: Government policymakers and the public may fail to fully recognize and take action against the unprecedented environmental threats that now appear before us. Because there are well over 5 billion humans on the Earth, because of the nature of our technology, and because of the intensity with which we use it, our behavior now has the potential to deplete vital resources and to disrupt global ecological systems. Thus, for the first time in human history, environmental policy mistakes caused by people's underestimations of risk could be catastrophic and irreversible.

On the other hand, some of the six psychological theories that predict hazard underreaction are speculative, and some might not fully apply to people's reactions to environmental problems. Further, there is a countervailing psychological tendency: Major components of cognitive theory (including people's use of the availability heuristic, their overconfidence, and desire for certainty) predict and explain a tendency for people to *overreact* to certain threats and hazards.

Further, it is impossible to assess the relative strengths of the two opposing psychological processes, those that push toward underreaction and those that push toward overreaction. In other words, there's no way to determine the overall or net effect, and it is likely that the relative strengths vary in complex ways across individuals, hazards, and situations.

Certainly, there's evidence that the public and policy makers are not massively or universally denying or underreacting to major environmental threats. As we discussed in Chapter 3, public opinion polls show that many Americans are concerned about environmental problems and support environmental causes, and that their concern and support has persisted for many years. Further, many Americans are members of proenvironmental groups such as the Sierra Club, the Audubon Society, the Nature Conservancy, and others. Many candidates for political office have taken proenvironmental stances. We find all of this heartening. However, the ubiquity of human tendencies to deny risk (that it is found in many cultures and for many subject matters), the fact that six relevant psychological theories predict hazard underreaction (however speculative or questionable some of the theories may be), and the fact that underreaction to some environmental threats could have catastrophic and irreversible consequences convinces us it would be irresponsible to ignore the possibility that humans might have a net tendency to deny certain environmental threats under certain conditions. This does not mean that all people will deny all threats all of the time, but that some people may deny some threats, and to a degree that may cause serious problems.

Our review of the cognitive theory (i.e., on errors, biases, and heuristics) has revealed other, striking psychological phenomena. We've seen that people's preferences and decisions are quite changeable and apparently easily influenced by the way alternatives in the decision are framed; that word-of-mouth advice can apparently drown out more reliable statistical summary information based on large samples; that people are apparently insensitive to missing branches of fault-trees used to estimate and communicate the overall reliability of new technologies; and that people are often seriously overconfident about the accuracy of their judgments and perceptions of the outside world.

Further, it appears that scientists, engineers, physicians and others with technical training are, under some conditions, as prone as the public to the biasing effects of framing, to overconfidence, and to insensitivity to missing fault-tree branches. Though we didn't discuss it the text of this chapter, research by Tversky and Kahneman (e.g., 1982) shows that even scientists who have extensive statistical training show biases and make errors in their interpretation and use of their own research data.

The proneness of people to the cognitive errors/biases/heuristics we have reviewed (including availability, the strong effects of framing, insensitivity to missing fault-tree branches, and desires for certainty) suggests another possibility that merits concern: The media and private and public interest groups may inadvertently or deliberately use these principles of cognition to manipulate public opinion. One defense against manipulation is to make the relevant principles of human perception, cognition, and decision making very widely known; however, mere knowledge doesn't guarantee immunity to manipulation (consider the observation about scientists with statistical training, above).

Finally, we'd like to reiterate that hazard underreactions and overreactions can result from several different psychological processes or mechanisms and that more than one can be operating at the same time (as we discussed in our review of the protection-moti-vation framework earlier in the chapter). We remind the reader, more generally, that individual behavior occurs in a complex matrix of physical, social, economic, and political forces. For these two general reasons, it is very unlikely that one TV program—even a superb one—one high-quality, widely distributed educational brochure, one day spent by juvenile delinquents in prison, a high school class period or week of class periods devoted to auto safety and seat belt use, and analogous single-component programs about *environmental hazards* will significantly alter individual behavior toward risks. Instead, attempts to alter people's perceptions of environmental hazards and to encourage actions (assuming the existence of a societal consensus concerning the appropriateness of these attempts) must address the several psychological processes involved, as well as the many social, political, and other forces that affect the relevant individual behavior, as we discussed in Chapters 3 to 7.

In Chapters 10 through 12 we take a broader look at the relationship between individual behavior and the economic, political, technical, and ecological systems in which the behavior occurs. Chapter 10 focuses on ways to ascertain which behaviors of individuals have the greatest impacts on the environment and which changes in behavior will benefit the environment the most. Chapters 11 and 12 focus on complex systems, including global and other ecological systems, and on the characteristics of those systems that may prevent humans from understanding and successfully interacting with them.

NOTES

1. Lehman and Taylor discarded the data of students who had not heard about the campus earthquake report (16 percent of their original sample of sixty). Note also that students were assigned to dormitories before the UCLA report was released. See Lehman and Taylor for the data of other groups of students.

2. Note the apparent contradiction between people's failure to wear seat belts and their overestimation of auto fatalities, as we noted above. The theories we discuss shortly can help explain this apparent contradiction.

3. It may help to keep in mind that the threats to human life and limb in the Stone Age were more numerous and serious than they are today in Western countries. Life expectancies

were significantly lower. Stone Age threats that have been decreased or eliminated by modern technology include animal predators, many diseases, and extremes of weather.
4. As a related, additional point, Slovic et al. (1978) suggest that people tend to view insurance as an investment, like a certificate of deposit in a bank. People apparently tend to think: "I won't take out insurance against a hazard unless I can make a good return on the investment—that is, unless I am likely to collect." In contrast, economists, government officials, and insurance executives argue that people should be thinking: "I'll protect myself against unlikely but unbearable events and I pray that I don't ever collect."

5. While Taylor and Brown (1988) raise this possibility, they do not say that serious risk underestimation and failure to take proper action are inevitable.

6. Weinstein's explanation of this unrealistic optimism is somewhat different from the one Taylor and Brown are proposing: Weinstein argues that unrealistic optimism serves to enhance or maintain a person's self-esteem: If a health hazard (e.g., drug addiction) is viewed as preventable by an individual's actions, then a person must rate his or her chances of having the hazard happen as less than average; otherwise the person is either ignorant about the hazard and how to prevent it, or is a weakling for not taking the necessary precautions.

7. Weinstein (e.g., 1987) and DeJoy (1989) argue that defensive denial does *not* contribute to the unrealistic optimism found in their studies. Specifically, they've found no correlation between their subjects' ratings of the "severity" (and hence the anxiety it engenders) of a health or other hazard (ranging from "not at all serious" to "extremely serious or fatal") and the amount of unrealistic optimism subjects show towards it. However, defensive denial could well play a role in other instances of hazard underestimation. Leyman and Taylor (1988) and Sims and Baumann (1972) make a good case that defensive denial plays a role in the hazard situations they studied. We will discuss the Sims and Baumann study later in the chapter.

8. Earthquakes are important exceptions to this principle. As we noted in the Prologue of this chapter, earthquakes occur when pressure between adjoining tectonic plates, which builds up slowly over time, reaches a certain high level. However, after a major earthquake or series of quakes, there is a window of safety; that is, for a certain period of time, earthquakes are significantly less likely to occur, until pressure between the tectonic plates again builds to the requisite level.

9. Weinstein's (1989) conclusions about the effects of personal experience are to some degree at odds with those of Taylor and Thompson (1982). On the other hand, the research that Taylor and Thompson review generally involves persuasive communications concerning *non*hazard subject matter, for example, political or philosophical issues.

10. Cognitive psychologists generally use the term *availability* in a narrower way than is used here; they mainly use it in the context of people's predicting the probability of future events (which is the context we have mainly in mind later in this section).

11. The only exception we are aware of is an unpublished study of the Rahway Scared Straight program by S. Langer (1979, cited by Finckenauer, 1982 [p. 196ff.]) Langer found no increase in the criminal behavior over twenty-two months of a group of SS participants, versus an increase of the control group. Thus the SS program appeared at least to deter a natural increase in delinquent behavior.

12. The results of several studies, mainly on health behavior, confirm both general and specific aspects of the theory (e.g., Prentice-Dunn and Rogers, 1986; Rippetoe and Rogers, 1987; Tanner, Day, and Crask, 1989; Seydel, Taal, and Weigman, 1990; and Tanner, Hunt, and Eppright, 1991). Note that the theory is one of several similar cognitive or decisional theories of risk-reduction behavior (see Prentice-Dunn and Rogers, 1986; Weinstein, Sandman, and Roberts, 1990). Note also that while several studies confirm the theory, the results of a few studies are consistent with some, but not all, parts of the theory (Wurtele and Maddux, 1987; Mulilis and Lippa, 1990).

13. This discussion relates to our discussion of the *diffusion of innovation* process earlier in this chapter and in Chapters 4 and 6. Recall the key role that word-of-mouth advice plays in the diffusion of innovation process, and that word-of-mouth advice is often more convincing and persuasive than statistical summary (or "base rate") information, as we discussed above. Note, also, that policymakers and others who may want to do so can speed up the diffusion process by intervening at key points via demonstration projects, as we discussed Chapter 4.) Finally note that Taylor and Thompson (1982) argue that case-history information (word-of-mouth advice) is not especially impactful on people, but that, instead, people have trouble inferring the great "diagnosticity" of statistical information. However, the end result still remains: Word of mouth has a stronger effect than impersonal, statistical information.

PART IV

BEHAVIORAL SOLUTIONS IN CONTEXT: ECOLOGICAL AND SOCIETAL SYSTEMS

CHAPTER 10

CHOOSING THE BEHAVIORS TO CHANGE AND THE POINTS OF INTERVENTION

I. Chapter Prologue

II. Introduction

III. A Behaviorally Oriented Analysis of the U.S. Energy System

    A. Major Energy Users

    B. Major Uses of Energy by Individuals and Households

    C. The Conservation Potential of Thirty Different Energy-Conserving Actions

    D. More on "Curtailment" Conservation Actions versus "Increased Efficiency" Conservation Actions

    Box 10-1: Public Conceptions of Household Energy Conservation: Emphasis on Curtailments and a Neglect of Efficiency Increases

IV. A Behaviorally Oriented Analysis of U.S. Litter and Solid Waste Problems

V. The General Superiority of "Upstream" Rather than "Downstream" Solutions, (or of Prevention Rather than Cure)

CHAPTER PROLOGUE

The brief story that illustrates this chapter's main topics and themes takes place in the *mid-1970s*. The protagonists are two psychology professors, Alexandra Mason and Michael Wilson, who have just completed graduate school and have joined the faculty of a large California state university.

The United States is in the throes of an energy crisis, one that began when Middle Eastern oil-producing countries stopped petroleum shipments to the United States. The U.S. government has responded to this embargo by instituting several major energy conservation measures, including a lowered national highway speed limit (55 MPH) and year-round daylight saving time.

Despite these measures, domestic supplies of petroleum begin to run short. Long lines of cars appear at gas stations all across the country (photo on p. 255). A major shortage of natural gas in the winter of 1977 forces some factory closings, school closings, and worker layoffs. President Carter appears on TV and describes the energy crisis as "the moral equivalent of war." He explains the nature and severity of the crisis and urges all citizens to conserve energy.

Like most Americans at the time, our two college professors, Mason and Wilson, are preoccupied and concerned about the national energy crisis. They realize that the crisis is caused as much by the United State's excessive dependence on petroleum—a key nonrenewable natural resource—as it is caused by the actions of the Middle Eastern oil-producing countries. The two professors are concerned also about the

depletion of other vital natural resources, and about air pollution, water pollution, litter/solid wastes, and other environmental problems. Mason and Wilson are young and idealistic and decide to devote all their research to the study of environmental problems, focusing mainly on the energy crisis, and specifically on developing ways to encourage the U.S. public to conserve energy.

Mason and Wilson begin their research by spending several days observing the energy-consuming behaviors of their colleagues, friends, and family. Based on these observations, they construct the following list of wasteful behaviors that Americans now engage in, and that the two researchers will try to find ways to change:

Failing to turn lights off in rooms that are not in use; leaving lights on all night in garages or unused outside areas; setting thermostats above the recommended 68 degree level during the winter heating season or below 78 degrees during the summer air-conditioning season; leaving outside doors or windows partially open when heating or air conditioning is running; making jackrabbit starts in autos and using excessive speed; using large stove burners to heat small cooking pots so that flames or heating elements around the sides of the pots waste energy; and using electric or gas laundry dryers, rather than outdoor line drying, during warm, sunny weather.

Next, the two professors spend several months devising methods to get the general public to lessen the wasteful behaviors outlined above. The two researchers, for example, create highly visible stick-

Long Lines of Cars Waiting to Buy Gas—a Common Sight in the United States in Early 1979
The lines resulted from a petroleum shortage and a national system of gasoline rationing. (UPI/Bettman Newsphotos)

ers to be placed near light switches to remind people to turn off unnecessary lights; they develop behavior-modeling videotapes for use on television that demonstrate the proper use of refrigerators and stoves; and they adopt some of the other techniques (e.g., public commitment) we described in Chapters 4 and 5. Finally, the two professors design and run several research studies over a two-year period to test the effectiveness of the above methods.

As their energy-conservation research nears completion, the two researchers then decide to focus some of their research on one other environmental problem: garbage and litter. Mason and Wilson have been surprised by the large quantity of beverage cans, bottles, snack-food wrappers, and other trash lying around their otherwise attractive college campus and in adjacent neighborhoods and parks.

And they are also aware that litter is a major problem all across the United States, not just in this particular California town. The two researchers, therefore, spend considerable time designing methods to get people to litter less. For example, they produce antilitter modeling tapes for television and design brightly colored trash barrels for use in public areas, each barrel bearing a slogan urging people to "pitch in."

As it turns out, Mason and Wilson's research programs on energy conservation and on littering are both successful: After several years of intense and imaginative work, the two professors have produced several methods that actually get people to save energy in the ways outlined above, and also get people to deposit trash in trash barrels in parks and other public places.

INTRODUCTION

Clearly, Mason and Wilson's research programs, outlined above, are well-intentioned, ambitious, clever, and creative. These researchers, furthermore, devoted large amounts of time and effort to their work. And the changes in public behavior that their methods produce do save energy and reduce litter.

However, despite all this, the authors of this book believe that Mason and Wilson's research efforts suffer from an unfortunate, major flaw: The energy conservation behaviors that the two professors are encouraging in the general public are behaviors that save very little energy. At the same time, the two are ignoring several other conservation behaviors that could save much more energy. Similarly, the decrease in littering that they are encouraging is a superficial goal, one that addresses only the tip of the iceberg of litter and solid waste problems in the United States.

We believe that these shortcomings stem from Mason and Wilson's unfamiliarity with key details: that is, with the way U.S. energy and solid waste systems actually work. The two used intuition and informal personal impressions, rather than quantitative and technical information, as a basis for choosing "target behaviors"—the public behaviors that they attempted to encourage or change. Mason and Wilson could have done a better job of choosing target behaviors if they had first gotten information from engineers, ecologists, and others familiar with the quantitative and technical aspects of U.S. energy and solid waste systems.

But before we go any further, we have a confession to make: Mason and Wilson are not real people; they are fictitious characters. Their story, however, is all too true. Psychologists in the 1970s devoted a great proportion of their research efforts on environmental problems to behaviors such as those described above. Some contemporary psychologists have done the same. Like Mason and Wilson, these real-world psychologists overlooked vitally important quantitative and technical dimensions of the problems they were studying and, as a result, chose less than optimal target behaviors. In fairness, we must point out that these psychologists exhibited what we believe is an almost universal tendency: Researchers who are interested in solving environmental problems—regardless of their discipline—tend to overlook important inputs from disciplines other than their own. For example, ecologists, economists, engineers, and others have often made intuitively appealing, but incorrect, judgments about important behavioral aspects of environmental problems; or they have overlooked these aspects altogether. Indeed, many of the behavioral aspects of environmental problems that we discussed in the prior chapters of this book (especially Chapters 4 and 5) have sometimes been misjudged or overlooked by nonbehavioral scientists.

However, this chapter is mainly addressed to psychologists and behavioral scientists (and their students), not to engineers, ecologists, and so on. Specifically, we devote the chapter to a careful review of key quantitative and technical aspects of environmental problems that psychologists like Mason and Wilson have tended to misjudge or overlook. We believe that psychologists must be familiar with these aspects so that they can choose effective behavior-change targets and contribute most to solving environmental problems.

By the way, the authors of this book don't claim to be any smarter than other behavioral scientists when it comes to these matters! Any wisdom we have can be traced to the rude awakening we experienced while participating for several years in an unusual interdisciplinary program on energy and environmental problems. We constantly rubbed elbows with scientists from other disciplines, and were forced to confront the narrowness of our own psychological perspectives.

In the pages below, we focus first on U.S. energy problems, in the context of the energy crisis of the mid- and late-1970s. We review what we believe are the most important quantitative/technical dimensions of American energy problems, the dimensions most relevant to a choice of effective target behaviors. We go on to perform a similar, but more recent, analysis of U.S. litter and solid waste problems, and, finally, to an analysis of the present-day greenhouse effect and global warming. At the end of the chapter, we apply what we've learned about choosing target behaviors to a somewhat different subject matter: We critique widely publicized programs to encourage proenvironmental public behavior that were designed as part of the U.S. observance of Earth Day 1990.

One of the themes that appears throughout the chapter is that there are different levels of intervention at which behavioral scientists (and others) can work to try to lessen or solve an environmental problem. At one extreme, psychologists can focus on target behaviors related to the negative impacts or manifestations of a problem; at the other extreme, psychologists can focus on target behaviors related to the underlying sources or origins of a problem. Usually, psychologists can be most effective if they work near the latter rather than the former, although it is usually best for psychologists to work at more than one level of intervention. We hope that this chapter provides both specific quantitative and technical information as well as a general framework useful to psychologists in mastering the behaviorally relevant technical dimensions of any environmental problem.

A final introductory note to our readers: There is relatively little psychology in this chapter, but lots of material from engineering, general ecology, and other fields. If you are a psychology student, some of this material might seem tedious, and you may not be convinced that it's relevant. However, we urge you to stick it out. We're confident the material in this chapter will provide you with important insights into the nature and causes of environmental problems and the most effective ways to help solve them.

A BEHAVIORALLY ORIENTED ANALYSIS OF THE U.S. ENERGY SYSTEM

The analysis of the U.S. energy system that we present in this section produces a better choice of conservation target behaviors than does the intuitive and informal approach that Mason and Wilson used and that we described at the beginning of the chapter.

In the analysis below, we *first* examine how energy is being used in the United States. We do this because it's impossible to figure out the best ways to conserve energy without first determining who the key users of energy in this country now are, how much energy each of them uses, and for what purposes. Once we answer these questions with relevant quantitative and technical information, we *then* can go on to determine

which changes of public behavior are likely to conserve the most energy.

Major Energy Users

As we noted above, the analysis begins with a look at who the major users of energy are in this country. Table 10-1 shows the relevant data.[1] The Table serves to remind us that, as we pointed out in Chapter 1, only one-third of the energy consumed in this country is consumed directly by individuals and households, while the other two-thirds is consumed by factories, businesses, and other "large actors." (Recall from Chapter 1 that the direct role of individual/household behavior is similarly limited in U.S. air pollution and solid waste problems.) And since psychology focuses primarily on the behavior of individuals and of small groups, psychology can help us understand only some, but not all, of the causes and possible solutions to energy problems. (As we discussed in Chapters 3–7, individual behaviors can have significant political and economic impact beyond the individual level, as when people join environmental groups, purchase environmentally sound consumer products, and vote for proenvironmental candidates for office; and, as we saw in Chapters 3–7, psychology can help us understand those behaviors.)

However, let's assume that Mason and Wilson already understood this limitation concerning the role of individual/household behavior, and are content, as psychologists, to limit their research mainly to this sector. After all, one-third of U.S. energy use is still a very large contribution to resource depletion, air pollution, global warming, and other problems. Let's now look more closely at individual and household energy use, to see where the greatest energy savings can be achieved through changing behavior.

Major Uses of Energy by Individuals and Households

Table 10-2 lists the major end-uses, or purposes, for which individuals and households use energy in the United States.[2] It also shows the amounts (percentages) of energy consumed for each of the end uses. The figures in Table 10-2 show that a lion's share of

TABLE 10-1 Estimated Percentage of Direct U.S. Energy Use by Economic Sector (Figures Are for 2000)

SECTOR	%
Household/Individual (including transportation)	32.4
Industrial (including feedstocks)	37.7
Commercial/Service	16.6
Transportation- Non-Household/Individual	13.3
Total	100.0

Source: Adapted from Gardner and Stern (1996), with additional data from Brower and Leon (1999) and U.S. Department of Energy (2000; 2002). .

the energy that individuals and households consume in this country is used to run automobiles and heat homes. In contrast, relatively little energy is used to light homes, cook meals, or dry clothes. These big differences in amounts of energy consumed immediately suggest that conservation actions involving auto use and home heating have a greater potential to save energy than conservation actions involving the other end-uses. In the next section, we explore the conservation potential of different conservation actions more systematically and in greater depth.

The Conservation Potential of Thirty Different Energy-Conserving Actions

Table 10-3 presents estimates of the energy-saving potential of thirty different conservation actions that individuals and households could take. These actions cover a broad range: They involve each of the end-uses in Table 10-2. They include many of the actions chosen by Mason and Wilson (at the beginning of the chapter), but also actions mentioned in the popular media, as well as actions mentioned in engineering publications and other technical sources. Note the way to read the entries in Table 10-3: The "4–6" figure near the upper-left corner of the table, for example, indicates that between 4 and 6 percent of all individual/household energy consumption in the

TABLE 10-2 Estimated Percentage of Total Individual/Household Sector Energy Consumed for Different End-Uses in the U.S. (Figures Are for 1999–2000)

END-USE	%
Transportation:	
Auto	40.5
Other	5.7
Subtotal	46.2
In home use:	
Space heat	29.6
Water heat	7.8
Refrigeration and freezing	4.2
Lighting	3.1
Cooking	2.7
Air-conditioning	2.5
Drying	.9
Other	3.0
Subtotal	53.8

Source: Adapted from Gardner and Stern (1996), with additional data from Brower and Leon (1999) and U.S. Department of Energy (2000; 2002).

United States (1-1/2 to 2 percent of the national total) could be saved if all Americans who now drive to work alone car-pooled to work with one or two other people.

Please note that each figure in the table assumes universal, or nationwide, adoption of the corresponding conservation action. Note also that the figures are for the early 1980s. However, they are probably close enough to current figures to correctly indicate the relative energy-saving potential of the thirty different actions.

Table 10-3 is organized top to bottom from the most energy-consuming end-uses to the least, in the same order as are the end-uses in Table 10-2. The data in Table 10-3 confirm the inference we made in discussing Table 10-2 above: Those conservation actions involving auto travel and space heat (which are the most energy-consuming end-uses) have the

TABLE 10-3 Estimated Percentage of Current Total Individual/Household Energy Consumption That Can Be Saved by Thirty Different Conservation Behaviors (in the U.S.) (Figures Are for the Early 1980s)

END-USE:	CURTAILMENT	% ENERGY SAVED	INCREASED EFFICIENCY	% ENERGY SAVED
		Transportation		
Automobile:				
	Car–pool to work with one to two others	4–6	Buy more fuel-efficient auto (27.5 vs. 14 mpg)	20
	Cut shopping trips to one-half of current mileage	2	Get frequent tune-ups	2
	Alter driving habits with mpg or vacuum feedback	2 (or more)	Maintain correct tire inflation	1
		Inside the home		
Space heat:				
	Set back thermo-stat from 72o F. to 68° F. days, 65° F. nights	4	Insulate and weatherize house	10
			Install more efficient heating equipment	8
Water heat:				
	Set back thermo-stat by 20° F.	1	Install more efficient unit	2
Refrigeration/freezing:				
	Decide on items you want in advance and open/close quickly	0.5	Buy more efficient unit	1.6
	Thaw frozen foods in refrigerator before cooking	0.1	Clean refrigerator coils frequently	0.1
Lighting:				
	Do not leave porch light on all night	1.0	Change one-half of all incandescent bulbs to fluorescent	1.0
	Replace all hall and ceiling fixtures with 40-watt bulbs	0.1	Clean bulbs and fixtures regularly	0.3

TABLE 10-3 Continued

END-USE:	CURTAILMENT	% ENERGY SAVED	INCREASED EFFICIENCY	% ENERGY SAVED
Cooking:				
	Do not use self-cleaning feature of oven	0.2	Buy more efficient unit	0.9
	Use right-size pots and do not open oven door to check food	0.2		
Air-conditioning:				
	Set back (up) thermostat from 73° F. to 78° F.	0.6	Buy more efficient unit	0.7
			Insulate and weatherize home (see above under "Space heat")	0.8
Drying:				
	Do not use dryer 6 months of the year	0.5	Buy more efficient unit	0.2
Miscellaneous:				
	Do not use garbage disposal unit	less than 0.1	Fix all dripping hot water faucets	0.1
			Replace leaking refrigerator door seal	0.1

From Stern and Gardner (1981a), see Table 10-1.

greatest potential to decrease total individual/household energy consumption.

A key feature of Table 10-3 is that it divides the thirty conservation actions into two different general categories: Actions in the left column involve *curtailing* the use of existing energy equipment (such as cutting down on auto trips for shopping by 50 percent). Actions in the right column involve *adopting more energy-efficient* equipment (such as buying and properly maintaining a very fuel-efficient auto to begin with). Before we compare the figures in two columns, we need to mention that those in the increased-efficiency (right) column assume that consumers buy autos, refrigerators, furnaces, and so on, that are among the most energy-efficient they can buy. Note also that each figure assumes that consumers buy new equipment when old equipment wears out, that is, when they would normally replace the old equipment; if consumers make purchases before this time, part of the energy they save by using the more efficient equipment is canceled out by the energy used to manufacture the new equipment.[3]

If you compare the left and right columns of Table 10-3, you'll see a consistent and important difference: Behaviors involving adoption of energy-efficient furnaces, autos, and other equipment generally save more energy than do behaviors that curtail the use of existing equipment. So, for example, installing a fuel-efficient furnace and insulating and weatherizing a house saves much more energy than does setting back the thermostat that governs an inefficient furnace operating in a poorly insulated house. Similarly, buying a very fuel-efficient auto and maintaining it properly saves significantly more energy than curtailing the use—even rather severely—of an intrinsically energy-inefficient auto. And buying and properly maintaining an energy-efficient refrigerator saves significantly more energy than trying to efficiently operate an inefficient unit (e.g., by opening and closing its doors quickly). There are similar differences for other end-uses shown in Table 10-3.

More on "Curtailment" Conservation Actions versus "Increased Efficiency" Conservation Actions

Psychological Differences Between Curtailment and Efficiency-Increasing Conservation Actions. In addition to their difference in conservation potential, note that efficiency actions and curtailment actions have different psychological properties. One major difference is that curtailment actions usually involve small, simple behaviors that must be repeated over and over again for long time periods, whereas efficiency-increasing actions are often larger-scale, one-time-only behaviors. Thus, curtailment actions, such as turning off lights when leaving a room, choosing the right size cooking pots, and so on, require that people make continual efforts to monitor and alter their own behavior. It is this type of action that is studied by Skinnerian, or "behavior modification," psychologists.

In contrast, efficiency-increasing actions, such as buying a new high-efficiency furnace, tend to be actions that people perform infrequently or only once, and that don't require continuous attention or effort. For this reason, people's performance of an efficiency action depends on several factors that just don't apply to curtailment behaviors. For example, your purchase of a new, high-efficiency furnace requires that you have several thousand dollars in cash (or access to a loan), that you've done enough

research on furnaces to enable you to make your purchase decision with reasonable confidence, that you are willing to pay somewhat more for a high-efficiency furnace than for a low-efficiency one, that you can find a heating contractor that you can trust, and that you are willing to purchase this technology even if not many of your neighbors or friends yet have such a device. These factors that influence consumer purchase decisions—discussed previously in Chapters 4, 5, and 6—are generally studied by social, cognitive, and other psychologists who study consumer behavior.

A second important psychological difference between curtailment and efficiency actions involves their perceived impact on lifestyle. People may view some energy curtailment actions, such as setting space-heat thermostat back to 68 degrees (or less) in the winter, as decreasing their comfort and quality of life. As a result, they may tend to judge these actions as undesirable. For example, President Reagan was widely quoted as saying that "Energy conservation means being too cold in the winter and too warm in the summer!" In contrast, actions that increase energy efficiency do not interfere with quality of life at all, as such actions permit people to maintain existing lifestyles, but consume less energy in the process. Thus, unless or until there are changes in public values in this country, efficiency actions—the same actions that save the most energy to begin with—may be easier to encourage.

Implications for "Mason and Wilson's" Conservation Program. In the sections above we reviewed major psychological differences between curtailment and efficiency actions as well as differences in their potential to conserve energy. These differences have major implications for Mason and Wilson's research program on energy conservation, which we described in the Chapter Prologue. Most importantly, note that the conservation actions the two psychologists were encouraging the public to take—turning off lights, making thermostat adjustments, avoiding jack-rabbit starts in cars, choosing the right size cooking pots, and so on—were all curtailment actions. The two researchers ignored actions that increase energy efficiency, such as substituting fluorescent lights for incandescent, insulating and weatherizing homes, installing high-efficiency furnaces, and buying high-fuel-efficiency cars. In other words, Mason and Wilson overlooked the category of conservation behaviors with the greatest energy-saving potential, and did so because they chose target behaviors based on intuition and informal impressions.

A second, related implication for Mason and Wilson's research involves the psychological differences between curtailment and efficiency conservation actions, which we discussed above. Recall that curtailment actions are repetitive small behaviors (the kind studied by Skinnerians), whereas efficiency actions are bigger, one-time-only behaviors (the kind sometimes studied by social/cognitive/consumer psychologists). Methods that successfully encourage the two different types of behavior are likely to be quite different. Thus, the stickers, modeling videotapes, and so on, designed by Mason and Wilson to encourage energy curtailment actions may not be of much value in encouraging efficiency-increasing actions.

Reasons That Mason and Wilson's Intuitive and Informal Approach Led Them to Curtailment, Rather Than Efficiency, Actions. Why did Mason and Wilson's intuitive/informal approach lead them specifically to emphasize curtailment actions and overlook efficiency-increasing conservation actions? To try to answer this question we first note that the actions Mason and Wilson selected were the *same* ones that most members of the U.S. general public, lacking relevant quantitative and technical information, tend to select. How can we make such a statement? We do so based on the results of a study by Kempton, Harris, Keith, and Weihl (1985) discussed in Box 10-1. Kempton et al. interviewed 400 randomly selected Michigan residents and asked them to name as many actions as they could think of that would save energy in their own households. The respondents predominately named energy curtailment actions, such as turning off unneeded lights, rather than efficiency-increasing actions, such as upgrading insulation. (We assume that Michigan residents are reasonably representative of the U.S. public in general, at least concerning the subject matter of this study; indeed, Kempton et al. describe

other research on both American and European subjects that yielded similar results.) See Box 10-1 for more detail.

The question then becomes: Why do *Americans and others* intuitively conceive of energy conservation mainly in terms of curtailments rather than in terms of efficiency increases? One possible answer, proposed by Kempton et al. (1985) and others, involves "visibility": People can directly perceive the operation of lights, TVs, stoves, dishwashers, and so on; they know that energy is being consumed by these devices and that energy could be saved if these devices were used less intensively. In contrast, people *cannot* directly perceive the energy consumed by poorly weather-stripped doors, inefficient furnaces, or inefficient water heaters, and also that energy would be saved if these devices were upgraded. There are other possible explanations besides the visibility explanation, which we also discuss in Box 10-1 on Kempton et al. (1885). But whatever the explanation, it appears that the general public is prone to overlook those conservation actions with the greatest energy-saving potential, unless education, feedback, financial incentive, and/or other methods can effectively convince them not to!

Curtailments and Efficiency Actions: Not a Case of Either-Or. We would like to make one final point concerning curtailment versus efficiency actions and Mason and Wilson's research program: We do *not* suggest that psychologists interested in encouraging energy conservation in this country should ignore the curtailment actions that Mason and Wilson were trying to encourage. Some curtailments—especially lowering space-heat and water-heat thermostats—*can* yield reasonably large energy savings. Also, curtailment actions may have important indirect effects by raising people's consciousness about the need to conserve energy. In addition, global energy systems that are permanently sustainable (into the long-run future) may, indeed, require major energy curtailments. However, if psychologists are concerned with saving the most energy in the United States in the immediate future, it is clearly a mistake for them to ignore the efficiency-increasing conservation actions that Mason and Wilson ignored. In short, it's not a case of curtail-

Box 10-1

Public Conceptions of Household Energy Conservation: An Emphasis on Curtailments and a Neglect of Efficiency Increases

Research performed by Kempton, Harris, Keith, and Weihl (1985) at Michigan State University suggests that members of the U.S. general public tend to think of energy conservation mainly in terms of curtailment actions, such as turning off lights, using less hot water, and watching less TV. Conversely, the public tends to overlook such efficiency-increasing actions as weatherizing homes, installing storm windows, and buying more efficient furnaces and appliances.

Kempton and his colleagues interviewed 400 randomly selected Michigan residents by telephone. The researchers asked each respondent: "What things do you know of that a family could do to reduce energy consumption in their house?" After a respondent named several conservation actions and then paused, the researchers asked "Any more?" until the respondent had named six actions, or couldn't name any more. Note that this method of asking questions is completely open-ended in that the researchers did not give respondents any specific conservation actions to choose from.

Kempton et al. also asked half of their respondents to estimate the annual dollar savings that each of the conservation actions would yield. The researchers then compared these dollar estimates with estimates from scientific and engineering journals, government publications, and other sources.

The main results of the study are shown in Table 10B1-1. Notice that respondents mentioned energy curtailment actions much more frequently than they mentioned efficiency-increasing actions (see the left-most column of numbers). Thus 657, or 83 percent, of the total of 793 conservation actions mentioned by respondents were curtailment actions (keep in mind that most of the 400 respondents named several actions, and that different respondents could name the same action). Conversely, only 136, or 17 percent, of the actions mentioned were efficiency actions. Looking at

TABLE 10B1-1 Number of Respondents (Out of 400) Who Mentioned Each of Ten Household Energy Conservation Actions, and Estimates by Respondents and by Technical Experts of Resulting Annual Savings (from Kempton et al., 1985)

ACTION	NUMBER OF RESPONDENTS (OUT OF 400) WHO MENTIONED	ESTIMATED ANNUAL SAVINGS	
		RESPONDENT ESTIMATE	TECHNICAL ESTIMATE
Turn off lights (C)**	235	$49	$10–24
Lower thermostat (C)	234	$75	$28–55
Insulate home (E)	109	$86	$46–138
Use less hot water (C)	50	$30	$40–110
Use less TV (C)	48	$48	$11
Do less cooking (C)	32	$55	$9–21
Install storm windows (E)	27	$50	$69–115
Use clothes washer less (C)	27	$69	$10–36
Use dishwasher less (C)	21	$27	$10–21
Use dryer less (C)	10	$18	$9–22
Total mentions:	793		

Adapted from Kempton, W., Harris, C., Keith, J., and Weihl, J. Do consumers know "what works" in energy conservation? *Marriage and Family Review,* Volume 9, 115–133. Copyright 1985. Haworth Press, Binghamton, NY.

**Legend: (C) = curtailment action, (E) = efficiency action.

(continued)

BOX 10-1 continued

the left-column data in a slightly different way, note that only one efficiency action—installing home insulation—was mentioned by more than 10 percent of the 400 respondents. Several other very effective efficiency-increasing actions were mentioned by less than 10 percent of the respondents (not all such actions are shown in the table) or were not mentioned at all. These actions include: installing storm windows, applying caulking and weather stripping, insulating water heaters, making furnace efficiency upgrades, and purchasing more efficient refrigerators and other appliances.

Note further that respondents generally overestimated the annual energy dollars that the curtailment measures would save, in some cases by very wide margins (e.g., decreasing the use of clothes washer and of TV, and decreasing cooking). Conversely, respondents, if anything, underestimated the annual dollars that the efficiency-increasing measures would save. Kempton et al. also cite two other research studies, one done in Germany and one in the U.S., that obtained similar results.

Why did members of the public mention curtailment conservation actions much more frequently than they mentioned efficiency-increasing actions, and why did they overestimate the savings that curtailment actions, but not efficiency actions, would produce? There's not enough research to answer these questions definitively. Furthermore, it is likely that more than one psychological process or phenomenon is involved. Let's briefly explore several possibilities:

One possible explanation—as we already noted in the main text of the chapter—involves visibility. Since people can directly perceive the operation of lights, TVs, and dishwashers, they know that energy is being consumed by these items and that energy could be saved if these items were used less intensively (Kempton et al., 1985). In contrast, people cannot directly perceive the energy consumed by poorly weather stripped doors, inefficient furnaces, or inefficient water heaters, and that energy would be saved if these devices were upgraded.[4] If this explanation is correct, feedback techniques and some of the other approaches that we discussed in Chapters 4 and 5 may help people perceive what is now almost invisible. As a similar, but alternate, explanation, people may overestimate the *total amount of household energy consumed* by visible items like lights, TVs, stoves, dishwashers,

clothes washers, and dryers *in the first place,* and therefore overestimate the conservation potential of using them less intensively. (If you look back at Table 2 earlier in the main text of this chapter, you'll find that lighting, cooking, drying, and other miscellaneous end-uses actually consume only very modest percentages of the overall household energy budget.)

Another possibility is that people, when asked to think of ways to save energy, find it easier to think of alterations of acts they now frequently perform (e.g., turning lights and TVs on and off), than taking acts they have never previously performed or have performed only rarely (e.g., buying a furnace or storm windows). As a related possibility, such efficiency-increasing devices as furnace upgrade kits and energy efficient kitchen appliances may not have been in widespread use in 1978 when Kempton et al. (1985) interviewed their respondents. As a result, the idea of purchasing these devices may not have come easily to mind when respondents were asked about conservation measures.

Whatever the reason(s), given that members of the general public think of energy conservation mainly in terms of curtailment measures rather than efficiency measures, we would expect that psychologist-researchers—lacking quantitative information to the contrary—would do the same. We're referring, of course, to Mason and Wilson, the two psychologists in the story at the beginning of this chapter. Thus, Mason and Wilson's use of intuition and informal personal impressions led them to stress public target behaviors with little conservation potential, and overlook target behaviors with much greater potential. The result is that even if Mason and Wilson's programs of media ads, reminder signs, and so on, were to be successful—in other words, actually got people to conserve energy via curtailments—relatively little energy would be saved. This failure to save significant amounts of energy, despite public efforts, would be most unfortunate.

But such an outcome might also be unfortunate for a second reason, as Kempton et al. (1985) point out: The failure to save large amounts of energy via curtailment might lead the public to conclude, incorrectly, that energy conservation—besides requiring sacrifices in comfort and convenience—just doesn't work. People would then be unreceptive to any future efforts to encourage energy conservation via more effective efficiency-increasing measures.

ments *versus* efficiency, but a case of *both:* Both curtailment actions and increased-efficiency actions have a significant role to play in any comprehensive program to conserve individual/household energy in the United States (Stern and Gardner, 1981c). But, again, the choice and mix of actions must be based on a technical analysis of the energy system, rather than merely on intuition or informal personal observations.

There's an additional example, one involving a European country, that illustrates our point above. Sweden today faces a great need to decrease energy consumption because it is phasing out its nuclear reactors (due to the outcome of a public referendum in 1980 after the U.S. Three Mile Island reactor accident), and because it has signed an international pact agreeing to freeze emissions levels of carbon dioxide and other greenhouse gases. However an analysis, like the one we outlined for the United States above, of energy use and conservation potential in Sweden done by Ragnar Löfstedt (1993) reveals the reverse of the U.S. picture: Buildings in Sweden are among the most energy-efficient in the world due to strict building codes and use of advanced technology. There is therefore little more that can be done to increase their energy efficiency. Löfstedt's analysis shows that the only efficiency increase that promises significant savings is that of making certain home appliances more energy-conserving. Conversely, there are several curtailment actions Swedes still can take that have some significant conservation potential, including lowering space-heat thermostats and decreasing hot water use. Again, the most effective mix of conservation actions is revealed only by a formal analysis of the energy system, not by intuitions or informal impressions.

A BEHAVIORALLY ORIENTED ANALYSIS OF U.S. LITTER AND SOLID WASTE PROBLEMS

We move, now, from our behaviorally oriented analysis on energy to a similar analysis on litter and solid waste. You will recall that Mason and Wilson, in the Chapter Prologue, devoted some of their research to litter control (in addition to the work they did on energy conservation). Recall that they developed modeling videotapes and other methods that actually get people to litter less, for example, the use of well-designed trash barrels in public places, each barrel bearing a slogan urging people to "pitch in." Finally, recall that the authors of this book believe that such efforts can have only a small impact at best because littering is only the readily visible tip of the iceberg of solid waste problems in the United States. We now outline the analysis that led us to this conclusion:

To begin with, let's assume that Mason and Wilson's litter control efforts are 100 percent effective and that the public places virtually all items that might potentially litter city streets, university campuses, and parks in proper trash barrels. There's now the problem of what to do with this discarded material after the city collects it and trucks it away. Most (80 percent) of the municipal garbage in this country is buried in "sanitary landfills" (Miller, 1990). Sanitary landfills are large, open areas of land onto which garbage is dumped in layers, alternating with layers of soil; when full, sanitary landfills are covered with a final layer of soil.

The problem is, however, a shortage of landfill space in many U.S. cities. Further, few new municipal landfill sites will be available in the future because few remaining sites near existing cities are suitable, and use of these sites is strongly opposed by people who live in adjacent areas—the so-called NIMBY, or "not in my backyard," phenomenon (Miller, 1994). As a result of these problems, some cities are hauling their garbage to distant locations, in a few cases other states, and even other countries. For example, Philadelphia has shipped parts of its municipal waste to Ohio, West Virginia, Kentucky, and also to Panama via boat (Miller, 1990).

But even if the problem of finding landfill space did not exist, the dumping of litter and other garbage in landfills creates yet another serious problem: Dumping in landfills wastes the large amounts of energy and raw materials consumed in producing the items dumped. For example, enough aluminum beverage cans and other aluminum items are littered or discarded in the United States to rebuild this country's fleet of commercial airplanes every three months, and enough ferrous metal is littered or discarded to totally meet the needs of the U.S. auto industry (Miller, 1990)!

These resources and the energy now wasted by dumping garbage in landfills can be recovered via an alternative to dumping known as *recycling*. In recycling, aluminum, glass, paper, and other recoverable materials in solid wastes are separated and then used to manufacture new cans, bottles, newsprint, and so forth. Recycling saves large quantities of energy and raw materials: For example, the manufacture of aluminum beverage cans from melted-down discarded cans consumes 95 percent less energy than mining and processing raw aluminum ore. It also produces 97 percent less water pollution and 95 percent less air pollution (Miller, 1990). Overall, Miller (1990) claims, a comprehensive national recycling program "could save 5% of annual U.S. energy use—more than the energy [now] generated by all U.S. nuclear power plants." Recycling, in turn, requires either voluntary efforts by citizens to collect and sort wastes (and sometimes deliver the sorted wastes to neighborhood recycling centers), or voluntary efforts motivated, in part, by beverage container deposits, or mandatory citizen efforts (ones required by law). In some cases, the sorting and separating of collected wastes is done at a central municipal garbage processing plant.

But desirable as recycling is, there is another method—*reuse*—that can save even *greater* amounts of energy and raw materials. A good example is the use of returnable and refillable glass bottles for soft drinks and other beverages. After each use, consumers bring bottles back to stores or other collection sites. The bottles are then sterilized, refilled, and resold, in a cycle that can repeat as many as fifty times. According to Young (1991), a glass beverage bottle refilled and resold only ten times consumes only one-quarter of the energy needed to produce ten one-use-only glass bottles from recycled materials. (A reusable glass bottle also consumes only one-eighth of the energy used to produce ten one-use-only glass bottles from virgin raw materials.)

But there is an even *better* strategy than reuse, one that saves even greater amounts of energy and raw materials, and that produces less pollution: *waste prevention* (sometimes called "source reduction"). In

waste prevention, fewer materials that could end up as solid waste are produced and distributed in the first place. For example, many items now sold in the United States have packaging that could well be done without. A quick trip into any store will reveal a profusion of large plastic blister packs and cardboard backing sheets, as well as containers that are considerably larger than the volume of material they contain; some of this packaging could be eliminated. Another waste prevention approach is to increase the durability of consumer goods, so that they can be used for longer periods of time before they need to be discarded. For example, tire companies can readily manufacture radial auto tires that last 80,000 miles. (Indeed, at this writing, at least one tire company has already started to sell such tires). Similarly, appliance manufacturers can make stoves and refrigerators that last much longer than do ones that are now available, and/or make them with standardized replaceable parts.

To summarize our discussion, so far, of the four solid-waste strategies, waste prevention is generally preferable to reuse, which is preferable to recycling, which is preferable to discarding. We must, however, make one additional point: It turns out that there are some major restrictions on the ability to choose which of the four solid-waste strategies or methods to use. Many types of waste can be dealt with using *only one or two* of the four strategies. For example, waste prevention efforts can't eliminate *all* packaging of consumer goods, and many cardboard boxes or cans—for example, for food—cannot be readily reused. Conversely, some types of solid waste are reducible via waste prevention, but they cannot easily be recycled. For example, auto tires and the porcelainized sheet metal used in many home appliances cannot easily be melted down and recycled into new tires and appliances, given currently available technologies.

The above restrictions imply that optimal solid waste management requires a complex mix of the four intervention strategies. In other words, different strategies need to be used for each of the basic types of waste: containers and packaging, durable goods (like tires and appliances), nondurable goods (newsprint, office paper, magazines, cloth, and so forth), and food and yard wastes. Clearly, such a complex, multipart management scheme can only be the product of a technically sophisticated analysis of the U.S. solid waste "system"—and this is the punch line of our solid waste discussion!

This punch line brings us back to Mason and Wilson's research program on litter control presented at the beginning of the chapter. Their efforts to prevent littering seem superficial in light of the complex technical analysis, outlined above, required for an informed program on solid wastes. Note especially how Mason and Wilson's antilitter program—based on intuition and informal observation, rather than on a technical analysis—focused only on getting discarded items, which are highly visible, into refuse cans for collection. Conversely, their program ignored less visible but more important issues such as the exhaustion of landfill sites, the waste of the energy and raw materials used to manufacture and distribute the discarded items, the pollution generated, and so on.

The authors of this book are not criticizing the study of litter control by psychologists or the positive aesthetic, public health, and consciousness-raising effects of antilitter efforts. However, we are suggesting that psychologists interested in litter control can, in addition, apply their expertise on human behavior to other important parts of a comprehensive, technically sophisticated solid waste program. So, for example, psychologists can do more research on ways to encourage the public to participate in curbside or other voluntary community recycling programs (Chapters 4 and 6), to buy recyclable and reusable products if there is a choice, and to buy more durable tires, appliances, and other products, even if these items are initially more expensive than products with shorter life spans. Also, because direct proenvironmental individual actions are sometimes blocked by limited market availability, prohibitive costs, and some government regulations (as we have discussed in prior chapters), psychologists can further study the factors that affect people's voting behavior and their joining of environmental groups (Chapters 4 and 5).

THE GENERAL SUPERIORITY OF "UPSTREAM" RATHER THAN "DOWNSTREAM" SOLUTIONS (OR OF PREVENTION RATHER THAN CURE)

Before we move on, we would like to highlight in this section an important general principle, one that emerged in our litter/solid waste analysis above. The principle concerns different levels at which one can intervene in an attempt to solve an environmental problem, and the preferability of some levels over others. The same principle actually emerged in our analysis of the U.S. energy system, though we didn't note it at the time. The principle will emerge again in our analysis of the greenhouse effect in a later section, and in many, many other examples.

Going back to the solid wastes analysis, recall that the four waste strategies discussed in the section above form a progression, or hierarchy, of environmental desirability. This is shown in Figure 10-1. Waste prevention is generally more desirable than reuse; reuse is generally more desirable than recycling; and recycling is generally more desirable than discarding (Stern and Gardner, 1981b; Young, 1991). In discussing progressions of this type (and not just this particular progression on solid wastes), ecologists and other scientists often use the terms *upstream* and *downstream*, as follows: Strategies that attempt to solve an environmental problem by intervening on the left end of a progression like the one in Figure 1 are called "upstream" strategies; those on the right are called "downstream" strategies (Fischhoff et al., 1978; Hohenemser et al., 1983).

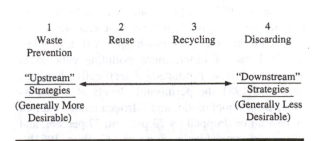

FIGURE 10-1 Order of Preference of Four Solid Waste Treatment Strategies

Technical analyses of many environmental problems reveal that upstream intervention strategies are usually superior to downstream strategies. This superiority repeatedly emerges for a surprisingly diverse set of problems. The superiority of upstream solutions appears to be a manifestation of an even more general prevention versus cure principle, as in the folk expression "An ounce of prevention is worth a pound of cure": It is usually better to stop or prevent *any* problem at its source than it is to deal with the negative consequences of the problem after they have already occurred.

There are many examples of the superiority of upstream solutions in the areas of air and water pollution. Thus, stopping the emission of a harmful chemical air pollutant at its source is usually less expensive and more effective than trying to clean up the pollutant after it is already out of the smokestack and widely dispersed. In some cases it is essentially impossible to cleanse the environment of dispersed nondegradable pollutants—for example, those that have found their way into underground drinking water aquifers or into the world's oceans.

Our discussion earlier in the chapter of the greater effectiveness of energy conservation actions that increase efficiency versus those that curtail the use of existing equipment may be seen as another example of the same general prevention versus cure principle. Thus, trying to save energy by curtailing the use of intrinsically wasteful automobiles, appliances, heating systems, and so on, is less effective than adopting equipment that wastes less energy by design.

Yet another, though similar, example of the principle involves auto exhaust emissions. Federal regulations in the 1970s required automakers to redesign vehicles so as to greatly reduce these emissions. The required reductions were much sharper than people's curtailed use of older, more polluting cars could reasonably have produced. Specifically, between 1976 and 1981 the permissible levels of hydrocarbons, carbon monoxide, and nitrogen oxides emitted by new autos dropped by 73 percent, 77 percent, and 68 percent, respectively (Stern and Gardner, 1981b). Again, it is hard to imagine people's decreasing their driving of older, pollution-emitting cars enough to produce so large an effect.

In Box 10-2, we further discuss the preferability of upstream versus downstream intervention strategies (or of prevention versus cure). First, we illustrate the basic principle and its generality using a nonenvironmental example—fire safety in the home. We then present a new environmental example concerning agricultural pesticides.

BEHAVIORALLY ORIENTED ANALYSES OF GLOBAL ENVIRONMENTAL PROBLEMS: THE GREENHOUSE EFFECT AND GLOBAL CLIMATE CHANGE AS AN EXAMPLE

In previous sections of this chapter, we presented behaviorally oriented analyses of energy and solid waste problems in the United States. Analyses like these help psychologists identify the individual and household behaviors responsible for an environmental problem, and the behavior changes most effective in solving the problem. Such analyses can be similarly helpful in the case of global—rather than regional or national—environmental problems. Global analyses are, however, more difficult to carry out. For one thing, ecologists, meteorologists, and other scientists do not yet completely understand the ecological processes underlying such global problems as the greenhouse effect and global warming (as we noted in Chapter 1). Second, data that can help us identify the best individual/household behavior-change targets are either not available, or are available but require complex analyses yet to be carried out (Stern et al., 1992). With these limitations in mind, consider, below, a very brief behaviorally oriented analysis of the greenhouse effect and global warming.

The Analysis

To begin our greenhouse/global warming analysis, we look at the major types and amounts of greenhouse gases released into the atmosphere by human activity. Recall from Chapter 1 that these gases act like the glass windows of a greenhouse, allowing light from the sun through but trapping the resulting heat and reflecting it toward the ground. The most important of these gases appear as *column headings*

Box 10-2

More on "Upstream" versus "Downstream" Solutions, and Prevention versus Cure

As mentioned in the text, "upstream" solutions to environmental problems are usually superior to "downstream" solutions; in other words, prevention is usually better than cure. (There are, however, exceptions, and cases in which upstream solutions can't be used; for example, we can't use source reduction or reuse for all types of solid wastes, as we pointed out in the main text).

Let's further explore the superiority of upstream solutions by means of the following two examples. The first, a nonenvironmental example concerning home fire safety, illustrates the broad generality of the prevention versus cure principle; that is, for quite a wide variety of subject matters, it is usually better to address a problem at its origin, rather than trying to alleviate (or mitigate) the negative consequences of the problem after they have already occurred. The second example is an environmental one, not discussed in the text, concerning agricultural pesticide use.

The generality of the "prevention versus cure" principle: Fire safety in the home. This simple, prosaic example, adapted from Fischhoff et al. (1978) and Hohenemser et al. (1983), involves the risks posed by a fire burning in a home fireplace. The fire occasionally shoots sparks into the room that could ignite the clothing of people nearby and lead ultimately to injury or even worse. Figure 10B2-1 shows a causal sequence of fire-related events that unfold over time that could lead to a person's death.

Note that the figure shows five different levels at which we can intervene in an effort to prevent or

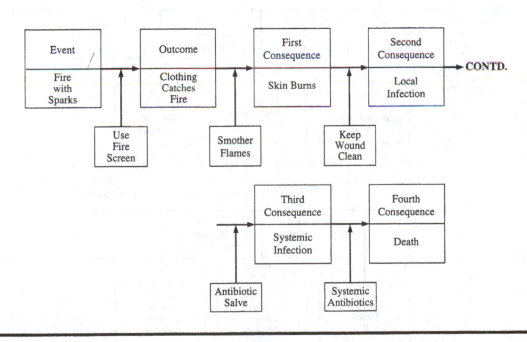

FIGURE 10B2-1 Fire Causal Sequence

Adapted from Fischhoff, B., Hohenemser, C., Kasperson, R., and Kates, R. Handling hazards. *Environment,* Volume 20, 16–20ff, Helen Dwight Reid Educational Foundation. Published by Heldref Publications, 1319 Eighteenth St., N.W., Washington, D.C. 20036. Copyright 1978.

(continued)

BOX 10-2 continued

address the negative outcomes or consequences caused by the flying sparks. The least desirable level of intervention is, clearly, the last, or fifth level—the use of systemic (whole-body) antibiotics to treat a systemic infection caused by the burn from the ignited clothing. At this level, the person faces a direct and immediate threat to his/her life. Obviously, it would be much better to intervene one level earlier—the fourth level, that is, use topical antibiotic ointment to prevent the localized skin infection caused by the burn from becoming a systemic infection. Better still would be to keep the wound from the burn clean so as to prevent even a local infection. Even better would be to use a fire extinguisher or blanket to put out the clothing fire before it burns the skin. The best intervention, however, is to place a wire-mesh screen between the

fireplace and room occupants to prevent sparks from igniting someone's clothes in the first place.

Once again, we see that—as with many other examples from diverse subject matters—upstream solutions are usually more effective and desirable than downstream solutions.

An additional environmental example: The impacts of agricultural pesticide use. The second example, also from Fischhoff et al. (1978) and Hohenemser et al. (1983), concerns a cancer-causing chemical pesticide used in agriculture. As shown in Figure 10B2-2, the pesticide is applied to farm fields to lessen insect damage to food crops. However, after the pesticide is applied to the fields, some of it runs off into adjacent bodies of water. The fish that live in these bodies of water

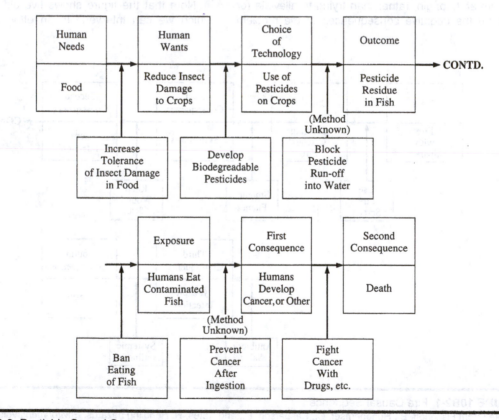

FIGURE 10B2-2 Pesticide Causal Sequence

then ingest the pesticide, which produces pesticide residues in their flesh. These chemical residues, in turn, pose a potential cancer threat to people who eat the fish.

Note that in this example there are six different levels of intervention, as shown in Figure 10B2-2. As is true for the other examples we have discussed, the upstream solution strategies are preferable to the downstream strategies. Indeed, in this case, two of the more downstream strategies are simply not physically possible: There is no currently known way to block the biological action of carcinogens once people ingest them. Similarly, it is not possible to block a chemical widely dispersed in farm fields from running off into adjacent streams, lakes, and oceans.

A more desirable upstream solution would be the development of agricultural pesticides that do their job and then quickly degrade upon exposure to weather. (An alternative to this is the use of agricultural practices that are more organic and that minimize the need for chemical pest control.) An even more upstream intervention would be to change people's tolerance to (presumably minor and nonharmful) insect damage in food.

Causal models and the identification of upstream solutions. Fischhoff, Hohenemser, and their colleagues urge scientists, government policymakers, and others interested in any given environmental problem to develop a complete causal model of the problem, similar to the two models shown in the figures above. Efforts to develop such a model force the individuals involved to consider the full range of possible solutions to the problem, or points of intervention, especially upstream solutions, that might otherwise not come to mind.

Unfortunately, Fischhoff et al. (1978) note, the use of fully developed causal models and of far-upstream solution strategies is, so far, not common in environmental policymaking. Fischhoff et al. discuss several possible reasons for this, which are beyond the scope of our treatment here. We do, however, discuss in Chapter 11 psychological reasons—not addressed in this chapter—that help explain why humans tend to address the highly visible and downstream symptoms of complex environmental problems, rather than less visible underlying and upstream causes.

---

in Table 10-4. The percentage that each gas contributes to the overall global greenhouse effect appears at the bottom of each column. Thus, carbon dioxide emissions are responsible for 55 percent of the overall greenhouse effect, CFCs (which do double duty because they also damage the ozone layer) are responsible for 25 percent, methane for 11 percent, and nitrous oxide for 6 percent.

Each *row* of Table 10-4 refers to a particular *human activity* that causes the release of a greenhouse gas or gases. *Human activity* in this table refers to the activity of *all sectors combined,* that is, the individual/household sector plus the industrial sector plus the commercial sector, and so on. Note that "fossil fuel burning" includes the burning of gasoline, fuel oil, other petroleum products, and also coal, in auto engines, boilers in electric plants, industrial furnaces, home heating equipment, and so on. "Biomass burn" refers to the clearing and burning of tropical forests. "Paddy rice" and "fertilization" refer, respectively, to the cultivation of rice in paddies and the use of agricultural fertilizers.

The percentage that each human activity contributes to the overall global greenhouse effect appears at the end of each row (on the right). Note that *fossil fuel burning* causes almost half (46.5 percent) of the global greenhouse effect; it is by far the single human activity most responsible for the effect.

More on Fossil Fuel Consumption

Because fossil fuel burning plays such an important role in the greenhouse effect and global warming, we give it a closer look. Specifically, we examine how much fossil fuel each sector consumes (individual/household, industrial, and commercial/service sectors), and how much each end-use consumes (transportation, space heat, and so on). The only relevant data we have found so far are for the United States. These data appear in Tables 10-5 and 10-6. Please note that U.S. fossil-fuel use is *not* representative of fossil–fuel use in other parts of the world. Thus, the data in the tables do not necessarily generalize to other countries. On the other hand, the

TABLE 10-4 Estimated Relative Contributions of Human Activities and Greenhouse Gases to Overall Global Warming (Figures Are from the 1980s)

HUMAN ACTIVITY	GREENHOUSE GAS					
	CARBON DIOXIDE	CFCS	METHANE	NITROUS OXIDE	OTHER	TOTAL
Fossil fuel burning	42%		3%	1.5%		46.5%
CFC use		25%				25%
Biomass burn	13%		1%	1%		15%
Paddy rice			3%			3%
Cattle			3%			3%
Fertilization				2%		2%
Landfills			1%			1%
Other				1.5%	4%	5.5%
TOTAL	55%	25%	11%	6%	4%	101%*

Reprinted with permission from Stern, P., Young, O., and Druckman, D. *Global environmental change: Understanding the human dimension.* Copyright 1991 by the National Academy of Sciences. Courtesy of the National Academy Press, Washington, D.C.
*Total is greater than 100% due to rounding of individual entries.

United States is responsible for approximately 20 percent of global carbon dioxide.

Table 10-5 shows that the individual/household sector is responsible for only about 35 percent of all fossil fuel consumed in this country. The industrial and commercial sectors are responsible for the other 65 percent.

Table 10-6 shows that within the U.S. individual/household sector, transportation and space heat are the end-uses that by far consume the most fossil fuel. Therefore, the conservation of fossil fuel used for these two end-uses has the greatest potential to lessen production of carbon dioxide, the main greenhouse gas.

The main findings shown in Tables 10-5 and 10-6 concerning the limited role of the individual/household sector, and the major role of transportation and space heat within that sector, should seem familiar to you. These findings duplicate two main findings from the U.S. energy system analysis we presented earlier in this chapter. Actually, these similarities between the fossil fuel analysis and the energy system analysis are not surprising, given that fossil fuel combustion—rather than solar power or nuclear power—is the main primary energy source in the United States. Specifically, fossil fuels account for approximately 90 percent of all energy consumed in this country each year (Miller, Jr., 1992).

Going a step further, we may infer that those individual/household behaviors identified earlier in the chapter as saving the most *energy* are also the behaviors most effective in reducing *fossil fuel burning* and the release of carbon dioxide. We thus find ourselves, once again, talking about the greater effectiveness of such efficiency-increasing actions as insulating homes and buying fuel-efficient autos, over actions that curtail the use of existing energy equipment. Sound familiar?

The Energy Crisis, Revisited

Though we again find ourselves discussing energy and its conservation, note that we do so in a very different context from before. The energy crisis of

TABLE 10-5 Estimated Percentage of Total Fossil Fuel Use by Economic Sector, for U.S. Only*

SECTOR	%
Household/individual	35
Industrial	43
Commercial/service	22
Total	100

From Congressional Research Service data quoted in the *Washington Post,* 3/10/90, p. A8.
*U.S. data are *not* representative of world fossil fuel use in various ways.

TABLE 10-6 Estimated Percentage of Total Individual/Household Sector Fossil Fuel Used for Different End-Uses, U.S. Only*

END USE	%
Transportation:	
Subtotal	40
In home use:	
Space heat	23
Motors and appliances	14
Water heat	11
Lighting	6
Cooling	6
Subtotal	60

From Congressional Research Service data quoted in the *Washington Post,* 3/10/90, p. A8.
*U.S. data are *not* representative of world fossil fuel use in various ways.

the 1970s discussed at the beginning of the chapter (the Mason and Wilson story) centered on energy supply problems: shortages, embargoes, U.S. dependence on a politically unstable part of the world, and the depletion of nonrenewable fossil fuel reserves. In the 1990s, several of these supply issues are still with us, though they may not be on the public's mind as much.

However, the critical new dimensions of the energy problem in the 1990s involve the gaseous by-products of fossil fuel combustion (Kempton, 1992). As we have just seen, the carbon dioxide produced is the single biggest cause of the greenhouse effect and global warming. Keep in mind that carbon dioxide *is not a "pollutant"* but an inevitable and unavoidable by-product of fossil fuel combustion, as Kempton (1992) points out. It is therefore unlikely that technological fixes, like the antiemission controls in today's autos or the "fluidized bed" combustion methods used in factories, will arise that will eliminate carbon dioxide emissions or keep them from reaching the atmosphere.

Of course, the other by-products of fossil fuel combustion—sulfur oxides, nitrogen oxides, volatile organic compounds, and particulates—cause other serious environmental problems: acid rain, which damages forests and lakes, and urban smog, which poses a health threat to hundreds of thousands of city residents (Miller, Jr., 1990).

To summarize: The use of fossil fuels as a primary energy source is now causing a diverse set of major

problems: The gaseous by-products of fossil fuel combustion contribute to global climate change, as well as to local and regional air and water pollution. The dependence of industrial nations on politically unstable petroleum-producing nations adds to global tensions. And humankind is consuming finite and nonrenewable fossil fuel reserves at a rapid rate. Virtually all of these problems can be lessened and ultimately solved by means of energy conservation measures, especially those that increase energy efficiency, and by a switch from fossil fuels to renewable and nonpolluting sources such as solar energy. This is the inevitable future path that global energy use must take, and a path that psychologists can play a role in helping society follow.

CHOOSING TARGET BEHAVIORS: EARTH DAY 1990, AND THE AS-MANY-AS 750 EVERYDAY THINGS YOU CAN DO TO HELP SAVE THE EARTH

Our major theme in this chapter has been the need for psychologists to choose the most important and effective public target behaviors in their efforts to lessen or solve environmental problems. These choices must be based on data and expertise from fields outside of psychology. There is no other way for psychologists to proceed, given the imperfect nature of human behavior and the real world. Humans have limited information-processing capacity (as we discussed in Chapter 9) and limited time and energy, and can therefore engage in only a limited set of behavior changes or new behaviors. Likewise, government programs, media campaigns, evaluation studies, and so on, are of limited size, scope, and budget, and can only try to encourage a limited set of proenvironmental public behaviors. (Of course, as we discussed in Chapters 4 and 5, taxes that increase the prices of gasoline, electricity, etc., can simultaneously encourage a large number of proenvironmental behaviors; but even for these, media campaigns still need to convey information on the behavior changes that conserve the most energy.) We are therefore forced to carefully choose a limited number of target behaviors to encourage—the behaviors that have the greatest proenvironmental impact, that is, that get "the most bangs for the buck."

Although, so far in this chapter, we have critiqued psychologists for failing to choose the most effective public target behaviors, in this last section of the chapter, we critique *ecologists and experts in related fields* and also *environmental groups*. These individuals and groups have the expertise to perform the quantitative analyses needed to choose the most effective target behaviors, but they have sometimes failed to do so, or have failed to make clear to the public the target-behavior priorities suggested by their analyses.

We specifically have in mind the ecologists and the environmental groups that had a role in organizing Earth Day 1990, an event, which—like its predecessor, the original Earth Day in 1970—was one of the largest educational and media events about the environment that ever took place in this country. In hundreds of American schools, colleges, and universities, thousands of people learned more about air pollution, water pollution, toxic chemical wastes, and the depletion of fossil fuels and other nonrenewable natural resources.

Probably the most publicized, often quoted, and widely distributed book published in connection with Earth Day 1990 was one entitled *50 Simple Things You Can Do to Save the Earth,* by the Earth Works Group; (note that an edition of this book was published by the Natural Resources Defense Council). A similar, though less well publicized and distributed, book was *Save Our Planet: 750 Everyday Ways You Can Help Clean Up the Earth,* by Diane MacEachern.

These and several other books attempted to answer a question that millions of environmentally concerned Americans were asking: "What, exactly, can I and my family do to help solve regional and global environmental problems?" Each book featured a long list of specific recommended proenvironmental individual/household actions—in some cases, hundreds of such actions. The recommended actions ranged from the small (e.g., Don't waste water by letting the faucet run while you brush your teeth) to the large (e.g., Start a comprehensive solid waste recycling program in your community). The environmental problems addressed ranged broadly, and included regional air and water pollution, toxic wastes, destruction of tropical rain forests, depletion of energy resources, and global warming.

Although the numerous individual/household actions recommended in these books were clearly valid and constructive, none of the books ranked the listed actions in order of importance—a significant shortcoming. Again, humans can devote only so much information-processing capacity, time, and effort to solving environmental problems (or any other problems). Few individuals can perform fifty, let alone several hundred, specific actions. Readers of these books, therefore, very much need to know which are the most important and effective actions to take so that they can at least take those actions. The issue is, again, the need for ecologists and environmental groups writing books for the general public to identify the most important proenvironmental individual/household actions, that is, the actions that yield the most proenvironmental bangs for the buck.

Given that a ranking of actions is desirable, the question then becomes: How should the quantitative analysis to produce the ranking be carried out? It is much easier to determine which individual and household behaviors will conserve the most energy than it is to determine which behaviors will have the greatest overall proenvironmental impact. An overall ranking is difficult because there are many different global environmental problems to be considered, including the greenhouse effect, ozone layer destruction, water pollution, depletion of resources, and so on.

However, "difficult" needn't mean "impossible." Alan Durning, a senior researcher at the Worldwatch Institute in Washington, D.C., has noted the lack of rankings in the books discussed above and has actually proposed an overall ranking, at least a very general one (Durning, 1990). Durning is a coauthor of the Worldwatch Institute's widely read annual "State of the World" reports. These describe the results of comprehensive environmental "physical exams" of the Earth. The annual exams look at a broad set of global and regional pollution, population, and resource-depletion problems. Durning and the institute are thus in a good position to look broadly at environmental problems and rank their importance as well as the effectiveness of different possible individual and household corrective actions.

Durning chooses as most serious the multiple global and regional problems caused by fossil fuel combustion—the problems we discussed in the section directly above. The major individual/household action he recommends is, not surprisingly, the conservation of energy. Durning writes:

The most tenacious and threatening environmental challenges facing industrial countries—things like air pollution, acid rain, and the greenhouse effect—are by-products of burning massive quantities of fossil fuels. . . . Consequently, most people's first priority should be to minimize energy consumption both at home . . . and in transportation. . . [p. 40].

Durning also notes that individual/household energy conservation is best achieved by increases in energy efficiency, as we stressed earlier in the chapter.

Durning ranks second in priority the "waste disposal crisis and the enormous energy squandered by a throwaway society." His third priority involves water conservation problems. (He goes on in his article to list problems of lower priority.)

We've discussed above the lack of a rank order of environmental problems or recommended actions in the several books published for Earth Day 1990. We'd like to go a step further and suggest that the *order of appearance* of actions recommended in each book provides an implicit rank ordering for the reader: Those actions discussed early in the book are more likely to be read and attempted by readers than actions that come later in the book. Further, the order of appearance of the actions in the books may not be optimal terms of overall proenvironmental impact.

As an example, let's consider briefly the *50 Simple Things You Can Do* book and focus on Durning's first priority of conserving individual/household energy. Twelve of the fifty "things" in the book concern energy and its conservation; approximately nine of the twelve involve increasing efficiency rather than curtailment, which, as we have argued in this chapter, is good. However, consider the order of some of the energy efficiency things as they appear in the book: Increasing hot-water-heater efficiency is the *sixth* "thing"; proper auto tire purchases and maintenance is *ninth*; buying an energy-efficient car and keeping it in tune is *fourteenth*; and insulating and weatherizing one's house is *thirty-eighth*. A comparison of these implicit rankings with the data in Table 10-3 on the effectiveness of different individual/household conservation actions suggests that the ordering of actions in *50 Things* is not an optimal one.

However, we should also point out that the more effective energy efficiency actions just mentioned, like buying a fuel-efficient car, and insulating and weatherizing your home, are not "simple" actions and are thus not the main focus of a book called *50 Simple Things. . . .*! In fact, the book is divided into three sections: twenty-eight simple things, fourteen "takes some effort" things, and eight things "for the committed." Finally, consider the discussion in Chapter 4 of commitment and of overcoming behavioral inertia. These psychological phenomena justify putting very small and simple actions at the beginning of a book, though such actions may not have great direct proenvironmental impact.

But, on the other hand, and holding all other things equal, a simple but important principle remains: Given that people can take only a limited number of proenvironmental actions, it's vital to identify and emphasize the most important and effective actions.

NOTES

1. Figures are estimates based on Gardner and Stern (1996), and data from Brower and Leon (1999) and U.S. Department of Energy, (2000, 2002).

2. Figures are estimates based on Gardner and Stern (1996), and data from Brower and Leon (1999) and U.S. Department of Energy, (2000, 2002).

3. For a more complete discussion of "net energy" analyses, see, for example, Odum and Odum (1976). Note also that interactive effects among actions are not considered in the table; the total savings from a series of actions is generally less than the sum of the percentages saved by each action alone.

4. *Visibility* may see like the *availability* concept we discussed in Chapter 9. However, cognitive psychologists define *availability* in a specific way and use the concept mainly to explain people's judgments of the *probability of future events*. People use the availability heurisitc when they judge the probability of an event in direct proportion to the ease with which they can recall and/or imagine instances of the event. Psychologists use the availability concept to explain, for example, why people underestimate the likelihood of hazards that they have not personnally experienced: People find it difficult to imagine or recall instances of such hazards.

CHAPTER 11

HUMAN INTERACTIONS WITH COMPLEX SYSTEMS: "NORMAL" ACCIDENTS AND COUNTERINTUITIVE SYSTEM BEHAVIOR

CHAPTER PROLOGUE

We devote this chapter and the next (Chapter 12) to an in-depth analysis of human interactions with environmentally relevant complex systems. The first half of this chapter concerns human management of complex technological systems, such as nuclear power plants, that can pollute or damage the environment. The second half and all of Chapter 12 focus on complex environmental systems, such as regional and global ecosystems, with which humans and human institutions interact. We especially focus on the over-all *global system,* that is, the Earth's environmental systems together with the influences of human activity on them, including the size and affluence of the human population, global levels of industrial activity, of pollution, of resource consumption, and of food production.

The main issue we address in both chapters is whether human beings, given the characteristics of the human mind and of human institutions, can successfully coexist with the natural environment. We ask whether people can understand environmentally relevant complex systems well enough to manage them and, if they cannot, whether they can rely on the same social mechanisms and institutions our species has used throughout its history to live with complexity. Our answer to both parts of this question is negative. We conclude that humans cannot understand these systems well enough to successfully interact with them and that the methods our species has relied on in the past may also fail us. We end Chapter 12 by discussing alternative strategies for living in the Earth's environment that require considerable change in "business as usual."

Our starting point is a discussion of the most serious accident in the history of U.S. commercial nuclear power—the accident in 1979 at the Three Mile Island plant in Pennsylvania. The TMI accident illustrates the role of individuals and institutions in managing complex technological systems that can pollute or damage the environment.

Nuclear Accident at Three Mile Island

It started as a minor malfunction at 4 a.m. on March 28, 1979, but in a few hours, the plant had suffered major and irreparable damage. The danger from the accident persisted for days, and millions of Americans watched on TV as engineers and scientists tried to stabilize the plant and avert a hydrogen explosion that the public was warned might release large amounts of radioactive material.

To properly describe what happened at TMI, we need to cover some important background material.

The Three Mile Island Nuclear Power Plant, Located on the Susquehanna River Near Harrisburg, Pennsylvania
(UPI/Bettman Newsphotos)

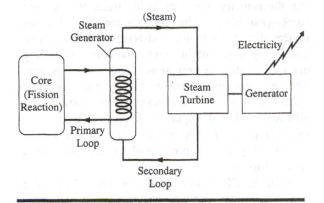

FIGURE 11-1 Basic Design of a Nuclear Power Plant (Pressurized Light-Water Type)

Figure 11-1 shows the basic design of a pressurized light-water nuclear power plant, the type at TMI. Enriched uranium in the reactor's core sustains a controlled nuclear fission reaction that generates large amounts of heat. This heat is absorbed by water under pressure that surrounds the core and circulates through the *primary loop*. The heat of this water turns the water circulating in the physically separate, but adjacent, *secondary loop* into steam. The steam drives a conventional turbine attached to a mechanical generator that produces electricity.

Figure 11-2 shows the design of the plant in more detail. (Though the diagram is elaborate, the key features are simple and straightforward.) Note the steel pressure vessel that surrounds the reactor core, the pumps that circulate the water in the primary and secondary loops, the auxiliary pumps and pipes, and the reinforced-concrete containment building that houses the pressure vessel, steam generator, and related equipment.

Consider the sequence of events at TMI that began at four in the morning on March 28. First, the secondary loop's main feedwater pump stopped

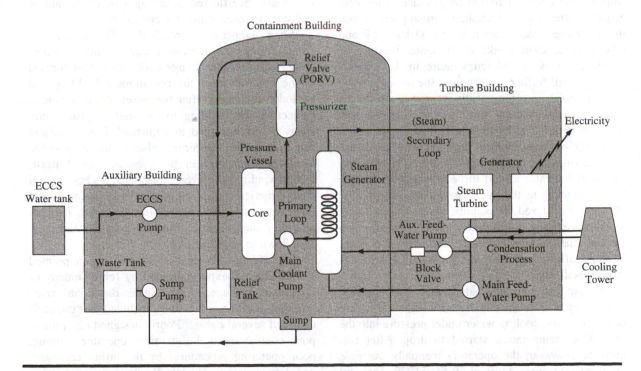

FIGURE 11-2 Detailed Design of a Nuclear Power Plant (Pressurized Light-Water Type)
From Lewis, H., The safety of fission reactions. *Scientific American,* Volume 242, 53–65. Copyright © 1980 Scientific American Inc., George V. Kelvin. Used with permission.

because of a minor mechanical problem. Because water in the secondary loop was no longer circulating and absorbing heat from the primary loop, temperature and pressure in the primary loop began to rise. An automatic safety system sensed this and turned on the secondary loop's auxiliary feedwater pumps to maintain water circulation. Unfortunately, maintenance workers had left two block valves in the auxiliary line closed, a violation of operating rules. Further, the control room operators were not aware the valves were closed because an unrelated cardboard repair tag obscured the indicator light on the main control panel. Thus, though the auxiliary feedwater pumps were on, no water was circulating in the secondary loop, and temperature in the primary loop continued to rise.

At this point an automatic safety system "scrammed" the reactor, that is, stopped the fission reaction by inserting graphite control rods (not shown in the figure) into the reactor's core. Though fission stopped, the decay of fission by-products continued, causing "afterheat." Afterheat, a normal phenomenon in a scrammed reactor, can reach 5,000 degrees F. and take days or even weeks to dissipate. Because of afterheat, pressure and temperature in the primary loop rose still further. In response, the *pressure relief valve* ("PORV") automatically vented some of the water from the primary loop, lowering the pressure, just as it is designed to do in this situation. However, the PORV had a mechanical flaw and failed to reclose after venting the right amount of water. Also, a poorly designed indicator circuit turned on a control-panel light, reporting to the operators incorrectly that the PORV had reclosed. Thus, unbeknownst to the operators, water was pouring out of the primary loop through the open PORV and into the relief tank. Because of this loss of water, the reactor core began to boil itself dry.

In response, an automatic safety system—the *emergency core cooling system* (ECCS)—cut in and began injecting cooling water under pressure into the core. Core temperature started to drop. After two minutes, however, the operators manually overrode the safety system and turned off the ECCS. They did this for several reasons. First, they did not understand

that the primary loop was losing water through the stuck-open PORV: The light on the control panel indicated that the PORV had reseated. Readings of other gauges that might suggest the core was boiling dry were, with one exception, ambiguous or wrong. The operators thus inferred that the primary loop was intact and that the ECCS had somehow come on by mistake. Further, the operators knew from their training that if the ECCS injects water into a closed primary loop, the plumbing could rupture, causing a serious loss of coolant accident.

With the ECCS turned off, and part of the core exposed (not covered with water), core temperatures climbed to 1,600 F. causing serious structural damage: Parts of the uranium fuel rods and metal supports began to swell and crack, releasing radioactive gases into the water pouring from the stuck-open PORV. The radioactive water overfilled the relief tank, leaked into the containment building sump, and was pumped into the auxiliary building's waste tank, which also overflowed, releasing a small amount of radioactive material into the environment.

When temperatures reached 2,000 degrees F., zirconium metal in the core began to oxidize, generating hydrogen gas. A large hydrogen bubble formed in the upper part of the containment building and exploded ten hours after the onset of the accident, subjecting the building to one-half the maximum stress it was designed to withstand. Engineers and scientists were concerned also about a possible hydrogen explosion in the pressure vessel itself, which would cause even more damage. For several days, Americans watched anxiously as engineers and other experts worked to stabilize the plant. Eventually, they succeeded, bringing temperature and pressure down to safe levels.

Soon after the accident, President Carter appointed a committee of experts, headed by John Kemeny of Dartmouth College, to investigate. Based on extensive research, the committee concluded that the accident had several causes: Poorly designed equipment, poor control room design, poor operator training, poor operating procedures by the utility company, and deficiencies in Nuclear Regulatory Commission oversight (NRC is the federal agency that regulates commercial nuclear power). The Kemeny Commission also concluded that the main immediate cause of the accident was human error, specifically the operators stopping the ECCS that had come on automatically. The commission argued that the operators should have deduced from the information they had that a major loss of coolant from the primary loop was occurring and should not have stopped the ECCS. Some observers noted that "if the operators had just stood there with their hands in their pockets and not overridden the automatic safety systems, the accident would have been far less serious." The Kemeny Commission went on to recommend sweeping changes in NRC structure, procedures for licensing new nuclear plants, procedures for operator recruitment and training, and many other changes. (Perrow, 1984; Lewis, 1980; Anon,1979; Sheridan,1980.)

INTRODUCTION

Though the Kemeny Commission concluded that the Three Mile Island accident had many causes—especially operator error—and recommended many changes to boost the safety of nuclear power, some other analysts have come to a radically different conclusion. Prominent among them is Yale University sociologist Charles Perrow (1984). Perrow argues, provocatively, that the control room operators are not mainly to blame for the accident, that the changes recommended by the Kemeny Commission will not greatly reduce the risks of commercial nuclear power, and that other accidents as serious or more so are inevitable in the not-too-distant future. Perrow contends that certain types of highly complex technological systems like nuclear power plants are intrinsically unmanageable and subject to serious accidents, accidents he calls "normal" in the sense that they are unavoidable. Normal accidents, Perrow claims, are due to unforeseeable combinations of multiple small failures, each one of which is innocuous by itself, but that together are "fatal." When these combinations of failures occur, they cannot be understood and corrected by the human operators of the equipment, regardless of their training or intelligence, until it is too late. Perrow's ultimate conclusion is that if a technology is very complex, is prone to normal accidents (as, he says, is nuclear power), and can cause catastrophic damage to people and the environment, then society should not deploy it.

In this chapter and the next (Chapter 12), we examine several lines of argument that, like Perrow's, suggest that human minds and human institutions may be inherently incapable of preventing catastrophe in environmentally relevant complex systems. After reviewing Perrow's work, we turn to the work of Jay Forrester, a professor emeritus at the Sloan School of Management at MIT. The systems that Forrester focuses on are not complex technological systems, like nuclear power plants, but complex social, economic, political, and environmental systems, such as large corporations, large cities, countries, and global ecosystems. Forrester contends that humans can't understand and manage these systems successfully because the systems are too complex for the human mind to grasp. He also claims that these systems often work in the *reverse* of the way human intuition would predict, compounding the problem humans have in interacting with and managing them. Further, the properties of these systems thwart the operation of social mechanisms and institutions that humans have used throughout history to deal with complexity. These mechanisms/institutions include the usual way governments handle complex problems—by making multiple, small, trial-and-error changes in current policies over time (a method sometimes called "disjointed incrementalism")—and the coordination of many decisions via the "invisible hand" of the market. Forrester claims there is a way around these difficulties: The development of computer models of complex systems that can help humans understand, manage, and control the systems.

In the first half of Chapter 12, we go on to discuss *chaos theory,* a major new theoretical framework that has gained increasing acceptance in the natural sciences and mathematics and which social and behavioral scientists are beginning to adopt. Chaos theory claims that many complex natural and social systems can exhibit wild, erratic, and unpredictable swings of behavior under some conditions, and that the conditions that provoke this chaotic behavior are often only very slightly different from those that produce stable and predictable behavior. The magnitude of this instability and unpredictability is sufficient to challenge claims like Forrester's that humans can manage and control complex systems, such as global ecosystems, even with the use of computer models. However, chaos theorists also claim there are forces in many complex systems that tend to oppose chaos and create order and organization from within. These forces are responsible for the coherence and order, for example, in biological evolution. Thus, even though humans can't "manage" complex systems, humans can influence them, and the systems have their internal tendency toward self-organization.

In the last half of Chapter 12, we discuss the work of Donella Meadows, Dennis Meadows, and Jørgen Randers on simulating the behavior of global environmental systems in response to human activity. Meadows et al. claim that these systems and human influences on them have three properties that create instability and can lead to disastrous environmental collapse: Certain long time delays, the exponential nature of human population growth and of the environmental impacts of human activities, and the possibility of irreversible and catastrophic damage to the environment. The combination of these properties, unprecedented in the history of our planet, tends to defeat the self-organization or antichaos that chaos theorists claim reside in natural and social systems. The properties also tend to defeat the processes such as disjointed incrementalism that humans have used in the past to live with complexity. Though Meadows et al. are disciples of Forrester, they do not propose more effort by humans to model and control nature. Instead, they advocate a radical change in human values, morals, institutions, and behaviors, including a cessation of global population and material growth. We discuss these and other strategies at the end of Chapter 12.

Three major themes directly relevant to this book run through the work we discuss in this chapter and Chapter 12. First, the work is concerned with the health of global and regional environmental systems in response to human activity. Second, the work examines possible mismatches between the charac-teristics of environmentally relevant complex systems humans now interact with or attempt to manage and the characteristics of the human mind and of human institutions. Third, the work offers both reasons for concern and reasons for optimism—ways to lessen or solve the serious environmental problems that now confront *Homo sapiens sapiens*.

COMPLEX TECHNOLOGICAL SYSTEMS AND "NORMAL" ACCIDENTS

We now take a closer look at the Three Mile Island accident and Perrow's (1984) claim that accidents like it are inevitable in highly complex technological systems. We begin with several qualifications: First, Perrow claims that only systems that are both "tightly coupled" and "highly interactive"—terms we define later in the chapter—are subject to normal accidents. Further, only some tightly coupled and highly interactive systems can pose threats to the environment, for example, ones that can release large amounts of toxic chemicals or other pollutants, including petrochemical plants and oceangoing tankers. (In Box 11-1 we describe a "normal" marine accident that caused an oil spill: The grounding of the tanker *Transhuron*.)

Second, *normal,* to Perrow, means *unavoidable* and *inevitable* but not *frequent*. Normal accidents are relatively rare. Perrow argues, however, that in systems that pose potentially catastrophic hazards, even rare accidents are unacceptable. (To qualify further: *rare* to Perrow means more frequent than "only once in a hundred years," i.e., sufficiently frequent that it is not worth taking the risks of these technologies.)

Finally, we note that Perrow's work is controversial. Several scholars and writers—for example, McGill (1984), Sills (1984), and Strauss (1985)—have especially criticized what they see as Perrow's extreme and unreasonable bias against commercial nuclear power. Others have argued that Perrow's conclusions are subjective and need to be confirmed by more rigorous, systematic research—for example, Hirschhorn (1985), Rossi (1985), and Simms (1986). We talk about these and other criticisms later in the chapter. Our main concern, however, is with Perrow's

(1984) idea that human failures in managing technological systems that impinge on the environment may be unavoidable, not with his particular conclusions about the acceptability of commercial nuclear power.

Characteristics of Normal Accidents

According to Perrow, normal accidents have *four main characteristics,* all of which are clearly displayed in the Three Mile Island accident: 1) There are multiple small and simple contributing causes; 2) the multiple causes interact or combine to produce a serious accident; 3) the designers and operators of the equipment can't predict or foresee this interaction or combination; and 4) the operators can't comprehend what is happening as the accident occurs until it is too late.[1]

Let's examine these key features and the TMI accident in more detail.

Multiple Simple Causes. Before the TMI accident, most safety experts had assumed that nuclear power plants were most vulnerable to dramatic, major failures (Lewis, 1980). For example, one scenario focused on a massive loss of cooling water due to a rupture in primary loop piping (Figures 11-1 and 11-2) caused by a violent earthquake. In contrast, the TMI accident was caused by several small and ordinary mishaps: Two minor valves (in the auxiliary secondary loop plumbing) left closed by errant maintenance workers, a simple design flaw in an instrument (that erroneously reported to the operators that the PORV had reclosed after its proper opening), even a cardboard repair tag on the control panel that hid a key indicator light.

Similar simple causes play major roles in the other normal accidents Perrow (1984) describes. For example: A mundane mishap involving a deep-fat fryer in a kitchen of a cruise ship starts a fire that, together with other simple causes, envelops the ship and destroys it; the improper installation of a pipe fitting made from iron, rather than bronze, in bronze air-conditioning equipment in the oil tanker *Transhuron* causes corrosion and a water leak that, with other causes, leads to a major oil spill and the loss of the ship (see Box 11-1); and the wrong voltage applied to

an oxygen tank in the *Apollo 13* moon-shot spacecraft in a routine prelaunch test, combined with other causes, mushrooms into major mid-flight failures that almost cause the loss of the spacecraft and the lives of the astronauts aboard it.

"Fatal" Combinations of Simple Causes. The multiple causes of a normal accident, though individually small and simple, interact in "fatal" ways. The operators of the equipment could easily have resolved any one or two or even three of the causes, but not all of the causes together. The TMI control room operators could easily have dealt with the closed auxiliary feedwater valves, or the failure of the PORV to reclose, or the flawed PORV instrument, or the blocking of the indicator light on the control panel. But they couldn't deal with the combination of closed valves, PORV failure, light blockage, flawed PORV instrument, and so on, and the result was a serious accident. In the *Transhuron* tanker accident in Box 11-1, at least *twelve* simple mishaps or failures contributed to the loss of the ship and/or interfered with the rescue of the crew members; the crew could easily have resolved most of the twelve individually or in small numbers.

The "Fatal" Combinations Are Unforeseeable. The third trait of a normal accident is that the designers and operators of the equipment cannot predict or foresee the combination of multiple, interacting small causes of the accident. Perrow argues that the harmful interactions often involve functionally *unrelated* subsystems that just happen to be in close physical proximity. Thus, at TMI, the repair tag obscuring the "block valves closed" indicator light hung down from an unrelated instrument that happened to be above it on the control panel. In the *Transhuron* tanker accident discussed in the Box, a water leak in the air-conditioning system short-circuited and incapacitated a unrelated electric panel that happened to be near it, and that controlled the ship's propulsion. In a nuclear power plant (not TMI), a maintenance crew needed demineralized water to do some routine cleaning. While waiting for a water faucet to be activated by the control room, the crew briefly opened the faucet to see if it was operating yet. A short time later, alarms in the building

Box 11-1

Marine Accidents and the Environment:
The Grounding of the Oil Tanker *Transhuron*

Large quantities of environmentally hazardous substances are routinely shipped in oceangoing vessels. These substances include crude oil (often shipped in supertankers carrying several hundred thousand tons), sulfuric acid, trichloroethylene, and vinylidene chloride (Perrow, 1984, p. 171). Just as with other forms of transportation, marine shipping is not accident-free. Every day, somewhere on the world's oceans a ship either runs aground or sinks and is a total loss. Approximately 17 percent of oceangoing cargo ships are lost before the end of their useful lifespan (Perrow, 1984, p. 171). If a ship carrying an environmentally hazardous substance grounds or sinks, significant quantities of the substance can be released into the environment. (Many readers will recall the grounding of the *Exxon Valdez* off the coast of Alaska in March of 1989, an accident that released about 11 million gallons of crude oil that fouled 1,000 miles of the Alaskan shoreline and the ocean floor nearby.)

Charles Perrow (1984) contends that some marine accidents are *"normal accidents,"* having the properties discussed in the main text (though most are traceable to simple component failures plus the inherent risks of marine travel such as harsh weather extremes). In this box, we closely examine one "normal" marine accident—the grounding in 1974 of the oil tanker *Transhuron*—described at length by Perrow (1984). This accident dramatically illustrates a key feature of normal accidents: the unanticipatable, unpredictable sequence of small—even trivial—difficulties, any one or two or three of which could be handled but that are "fatal" in combination.[2]

The *Transhuron* accident involved a system larger and more complex than just the ship and its immediate environment: In this accident, nearby Indian ship-to-shore radio facilities, Indian government authorities, the U.S. owners of the ship, and tugboats and other ships in the vicinity also played a major role. Thus, rather than just a complex technological system (the ship itself), the accident involved a complex technological-economic-political-social-environmental system.

The *Transhuron*, owned by the U.S. Department of Commerce, was carrying fuel oil for U.S. naval facilities in the Philippines when it went aground and broke open in the Arabian Sea. The first link (#1) in the almost unbelievably long and bizarre chain of mishaps that ultimately caused the grounding was the improper installation of an iron fitting in a cold water line connected to the bronze condenser of the ship's air-conditioning system. The dissimilarity of metals led to corrosion, which produced a major leak between the fitting and the cold water line. (The dissimilar metals problem was somehow missed when the system was serviced a few years earlier, after its original installation—link #2.) The leaking water sprayed upward, passing through a cable channel in the steel ceiling and then hitting an electric control board on the floor above used to control the ship's propulsion. The water caused a short-circuit and a large fire (#3). The crew, attempting to contain the damage and stop the fire, used an improper method of trying to turn off the control board (#4). Further, the carbon dioxide fire extinguisher system failed to operate properly (#5), and the crew finally extinguished the fire with hand extinguishers.

The *Transhuron* was now without power and drifting in bad weather toward an island ringed with reefs. The ship repeatedly and urgently radioed its owners in the United States explaining the trouble and asking for the assistance of a tugboat. They radioed via the closest ship-to-shore operator on the coast of India. The *Transhuron*'s officers found out, however, a day and a half later, that the Indian ship-to-shore operator routinely disregarded and deleted the word urgent from any messages it forwarded and simply sent messages on a first-come, first-served basis (#6). Shortly after this the ship finally received a radio response from the U.S. owners, but the response asked for more details about the fire and damage and said nothing about sending a tug (#7). With no evidence of help coming from its owners and its situation worsening, the *Transhuron* sent out an urgent distress signal to other boats in the area. A Japanese ship, *the Toshima Maru,* steamed to the *Transhuron*'s rescue, but its efforts to shoot a rope via cannon failed, the cannon broke, and a crew member was injured (#8). The *Toshima Maru* radioed that it was leaving immediately to take the injured crew member to a hospital. The *Transhuron* pleaded by radio to the *Toshima Maru* to stay just long enough to tow it and its thirty-five crew members out of danger, and after a delay the *Toshima Maru* agreed. But it was too late. The *Transhuron* had struck a reef, suffered damage, and began to leak its cargo of oil. Shortly after this, the *Transhuron* lowered lifeboats, and all its crew, except officers, abandoned the ship (successfully making it to shore on the island).

The next day, the five remaining crew received, finally (more than three days after the fire), a message of help from the U.S. owners. But fate continued to frown on the *Transhuron*: The owners in the United States radioed that a tug was on its way from Bombay, but that it was forty-eight hours away, and no other tug was available "from Singapore to the Persian Gulf" (#9). The tug finally arrived, but once again, too late. The *Transhuron*'s hull by now was ripped the entire length of the ship. The weather worsened and the remaining crew began to abandon ship. However, the motor in a motorized lifeboat failed (though the *Transhuron* was, by this point, already lost, let's keep counting mishaps—this is #10), and the crew had difficulties lowering the other lifeboats (#11). Perrow notes:

> . . . [l]aunching lifeboats in stormy waters (the time when you most have need to launch them) is quite difficult, as many accounts of marine accidents make clear. Even the ultra-sophisticated, totally enclosed, circular survival boats used on off-shore drilling rigs, with every imaginable safety device, flip over and leave their imprisoned workers to perish (eighteen in one tragic case). Inflatable liferafts are forever failing to fill, or having filled, blow away in gales, or fall on the swimmers awaiting them [pp. 227–228].

Finally the crew managed to launch an oar-powered lifeboat and all but one crew member left the ship. The last crew member, who had helped launch the lifeboat, was stranded because currents drew the boat quickly away from the *Transhuron*. He then deployed an inflatable rubber life raft, but the wind blew it away (#12), ripping a rope that was claimed by the manufacturer to withstand a pull of 3,000 pounds. The crew member finally jumped over the side of the *Transhuron* directly into the water and was, fortunately, rescued safely!

As it turned out, a single piece of good luck did befall the *Transhuron*. After the crew abandoned ship, the Indian government sent another ship, which managed to pump the remaining oil out of the *Transhuron*'s holds, thus limiting the environmental damage caused by the accident. But what is so dramatic about the *Transhuron* story is the incredibly long series of unanticipated and even anticipatable mishaps that ultimately led to the demise of the ship. The disaster resulted from what seems like a highly improbable conjunction of events—but many "normal" accidents have similar improbable stories.

---

signaled a radioactivity leak and the building was evacuated. It turned out that radioactive gas had escaped from the open faucet because of an unexpected interaction in the plant's complex plumbing system between maintenance and steam-generation subsystems, both of which used demineralized water. The gas could escape only if the crew opened the water valve before the control room activated the relevant piping, an interaction that neither the designers nor operators of the plant foresaw or predicted.

According to Perrow, the main reason that designers and operators of systems like nuclear plants can't foresee the interactions of small malfunctions that cause normal accidents is that these systems are so complex. The systems have so many constituent parts, and there are so many potential interactions between them, that the human designers and operators cannot grasp and keep track of them all. Figure 11-3, which is a more detailed and abstract version of Figure 11-2, illustrates the complexity of a nuclear power plant. Keep in mind, however, that even Figure 11-3 oversimplifies the actual complexity of a nuclear power plant.

A related reason is that some malfunctions and combinations are too improbable to even imagine in advance. For example, how could anyone have predicted the combination of human error and design flaw that caused a potentially disastrous fire at a nuclear power plant in Browns Ferry, Alabama? A technician used an open candle flame to check for an air-conditioning duct leak in a crawl space and inadvertently ignited some insulation on wiring, causing a six-hour fire. The crawl space happened to hold the main wiring going to and from the control room, including the wiring for all safety systems. The control room operators thus lost direct control of the

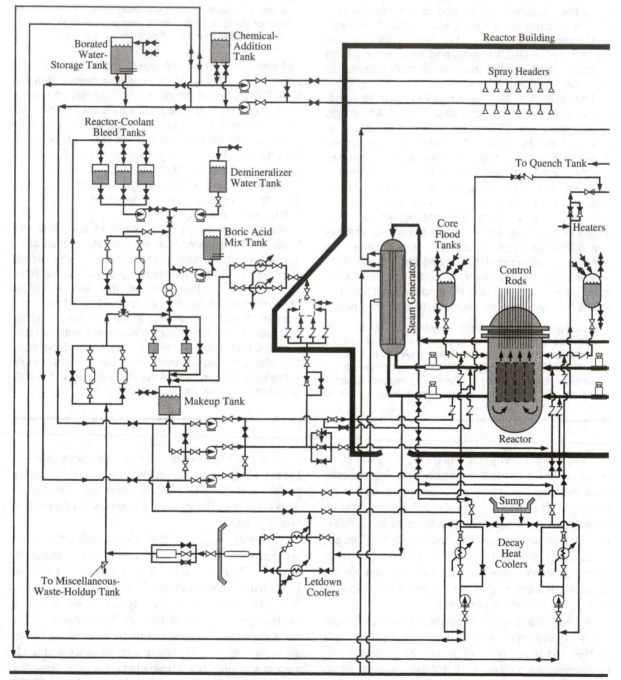

FIGURE 11-3 More Detailed Schematic Diagram of a Nuclear Power Plant (Pressured Light-Water Type—Babcock & Wilcox Co.—Similar to TMI)

From Lewis, H., The safety of fission reactors. *Scientific Americans,* Volume 242, 53–65. Copyright © 1980 Scientific American Inc., George V. Kelvin. Used with permission.

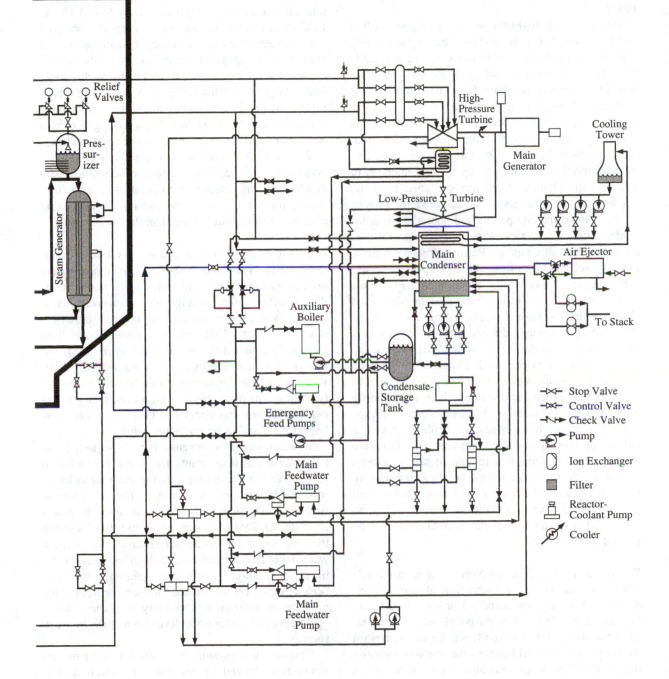

plant, and the reactor core overheated. (Fortunately, clever and creative eleventh-hour efforts by the plant staff averted a major disaster [Slovic and Fischhoff, 1983]).

Designers of highly complex equipment often build in *multiple* (or "redundant") backup and safety systems to minimize the impacts of unforeseeable mishaps and to increase safety. Thus, the TMI plant had three separate auxiliary feedwater pumps in its secondary loop, two electric and one steam-driven (Lewis, 1980). This should increase safety, as it is extremely unlikely that all three pumps would fail at the same time. Perrow (1984) argues, however, that the additional safety and backup systems intended to decrease the chance of an accident often have the opposite effect: These systems increase the overall complexity of the equipment, the potential for unintended interactions with other subsystems, and thus the probability of a mishap. Perrow offers the following relevant examples from the U.S. space program: The Ranger 6 mission to photograph the moon's surface (in advance of the *Apollo* missions discussed above) failed because a *safety* testing device short-circuited and seriously drained the spacecraft's vital power supply. Similarly, in a manned Mercury suborbital mission, a safety hatch (designed to explode open only in an emergency to permit the quick exit of the astronaut aboard) inadvertently blew open shortly after the capsule landed in the ocean. As a result, the capsule filled with water and sank, almost taking a helicopter with it and the astronaut, Gus Grissom. These examples show that redundant safety systems can sometimes cause accidents, but they do not prove that, on balance, these systems increase the overall probability of a serious accident. Perrow offers no other evidence, and it is not clear that his claim can be proved or disproved.

The Incomprehensibility of Normal Accidents While They Are Occurring.

A fourth trait of normal accidents is that they are usually (though not always) incomprehensible at first to the people who are responsible for stopping them. You'll recall that the control room operators at TMI believed that the automatically triggered ECCS system was injecting too much water into the primary loop. The real problem, however, was *too little* water in the loop, due to the stuck-open PORV. Because of their incorrect diagnosis, the operators did exactly the wrong thing: They turned off the ECCS that was cooling the reactor core. But not even the engineers from the company that designed and built the TMI equipment, Babcock and Wilcox, who arrived several hours after the accident began, could initially figure out what had gone wrong. Also, it took several hours, if not several days, for the experts assembled at TMI to realize that a hydrogen bubble had formed and exploded in the containment building.

Perrow (1984) suggests two reasons that normal accidents are initially incomprehensible. One involves the information that operators receive from the equipment, the other the mental "models" that operators form about the accident that is occurring.

Incomprehensibility due to problems with information. Perrow argues that critical information operators need to diagnose what is happening in a normal accident is often absent, ambiguous, or incorrect, or is hidden by excessive amounts of other information. Recall that in the TMI accident a control-panel light incorrectly signaled that the PORV had reclosed, and that another light, indicating a closed auxiliary feedwater valve, was hidden by a cardboard tag. Further (we did not mention this above), a computer printing out key information about the TMI plant was overloaded and printing several *hours* behind schedule.

In many complex systems, Perrow argues, some key gauges and instruments are absent for technical reasons, or are present but provide ambiguous information, forcing operators to rely on *indirect* information about what's going on. For example, there was no gauge at TMI (and similar plants) that indicated the level of water inside the reactor core because water level is technically difficult to measure (water in the core surges under high pressure) and also because designers are reluctant to make orifices in the pressure vessel unless absolutely necessary. Thus, operators had to infer water level from other, indirect, sources.

Conversely, operators of complex systems are sometimes plagued by *too much* information. The control panel of a nuclear power plant is eight feet high and a total of 100 feet long and bristles with thousands of lights, meters, and gauges (Sheridan, 1980). There are as many as 3,000—1,600 at TMI—red and green status lights, a pair for each electric switch and plumbing valve in the plant. Each pair signals whether the corresponding switch (or valve) is on (or open), versus off (or closed). According to Thomas Sheridan, professor of engineering and applied psychology at M.I.T. (1980), in the first sixty seconds of a major loss-of-coolant accident, as many as 500 of these lights change from green to red, or vice versa, and 800 can change in the next sixty seconds. Then there are the annunciators near the top of the control panel: small rectangles of frosted glass which, when illuminated, signal to the operators various malfunctions and abnormalities in the plant. Many of the annunciators were on or blinking during the minutes following the TMI accident, and there were several associated auditory alarms. All in all, there was an enormous amount of information bombarding the control room operators, more than they (or any humans) could properly digest (recall our discussion of *bounded rationality* in Chapter 9).

One potential way around the problem of information overload in the control rooms of complex systems is the use of simplified computer-based displays and controls. Rather than banks of thousands of lights and dials, the operators would view computer-driven color video screens that would summarize critical information about the plant's condition, highlight trouble, and even suggest appropriate actions for the operators to take (Sheridan, 1980). Control systems of this type have already been designed and deployed. However, Perrow's (1984) critique of the unintended effects of multiple backup and safety systems, discussed above, might also apply to computer-based control systems: The added complexity of the computer system itself—both hardware and software—over that of conventional control panels might well increase the probability of unforeseen negative interactions and thus of a normal accident. Major flaws (or bugs) even in the most widely used computer programs for businesses (such as word processors) are very common. Efforts to correct these

flaws often create new problems that are sometimes worse than the old ones. James Gleick (1992) writes:

> *Software bugs defy the industry's best efforts at quality control. Manufacturers may spend far more time and resources on testing and repairing [flaws in] their software than on the original design. . . .*
>
> *Computer programs . . . are machines with far more moving parts than any engine: the parts don't wear out, but they interact and rub up against one another in ways the programmers themselves cannot predict. When a program doubles in size, the potential for unexpected bugs more than doubles—far more. [pp. 40–41]*

Incomprehensibility due to operators' faulty mental models. The second factor Perrow (1984) identifies as causing incomprehensibility involves incorrect mental models operators form about the system they are operating and about the accident that is occurring. These models are often *too simple;* that is, the operators' mental models do not accurately reflect the complexity and nature of the system they are managing and the trouble the system is having. Forced to make sense out of a system they cannot fully comprehend, they form a simplified mental model of the system and the accident. (Recall our discussion in Chapter 9 of *bounded rationality,* and of the *gambler's fallacy* and other simple but incorrect mental models people sometimes have of how the world works.) They then tend to ignore information from their instruments that is inconsistent with their model, and to perceive ambiguous information as supportive of their model. The operators are therefore unsuccessful in halting the accident.

A related phenomenon, Perrow (1984) suggests, is that the mental models of operators of complex equipment undergoing normal accidents often err on the side of *excessive optimism.* As we discussed in Chapter 9, there are several psychological processes that can contribute to people's unreasonable optimism in the face of risk and hazard, including people's use of the *availability heuristic.* Availability appears to have played a role in the TMI accident. Quoting Perrow: "Uncovering the reactor core . . . [was] . . . unheard of; it had never happened in a large . . . commercial . . . reactor in the 380 or so 'reactor years' of large commercial . . . reactor operation [p. 28]." Given the lack of prior experience with such an accident, the operators were insensitive to any information suggesting that this type of accident was occurring. They perceived ambiguous control-panel information as indicating too much water in the primary loop, rather than too little. The operators also assumed that the one instrument unambiguously indicating low pressure in the core (not mentioned above) was wrong and disregarded it. Instead they believed the light on the control panel that indicated that the PORV had reclosed. Hence the operators overrode the ECCS.

Summary; Other Considerations

To summarize Perrow's (1984) argument: Normal accidents in complex technological systems are caused by multiple small failures that interact fatally in ways that designers and engineers can't foresee. Further, these accidents can't be stopped because they are initially incomprehensible to the operators of the systems. Returning to the TMI example, we see a fatal combination of a small equipment failure (the PORV valve's failure to reseal), a small design flaw (the control-panel light that incorrectly indicated that the PORV had reclosed), a minor error by maintenance workers (who left the block valves closed in the auxiliary feedwater line), and other small failures, combining to cause a serious accident. We also see control room operators overwhelmed at the start of the accident by hundreds of rapid changes in instrument readings, who received incorrect or ambiguous information from some of their instruments, and who lacked other important information altogether. We see that the engineers who had designed the plant and who arrived at the scene of the accident took hours to figure out what was happening. On balance, it doesn't seem reasonable, Perrow argues—and the authors of this book agree—to blame the TMI accident mainly on the operators. The accident, instead, appears attributable to the intrinsic properties of complex technological systems, such as nuclear power plants, together with the intrinsic characteristics and limitations of the individuals and institutions responsible for designing and running them.

Types of Complex Technological Systems Prone to Normal Accidents. As we noted earlier, Perrow (1984) argues that not all highly complex technological systems are prone to normal accidents. For example, while there are accidents in large automobile assembly line factories (highly complex systems) and other manufacturing plants, these accidents are not usually normal accidents. Accidents in these factories, instead, typically involve component failures—for example, the electric motor powering a conveyer belt burns out, a steel bracket holding a ceiling-mounted crane breaks— that are foreseeable and comprehensible to the humans operating the plant when they occur; the accidents don't usually involve unforeseeable and incomprehensible combinations of many small failures.

Normal accidents happen mainly, Perrow (1984) claims, in technological systems that are both "tightly coupled" and "highly interactive." Most auto and other manufacturing plants have neither of these properties. *Tightly coupled technological systems* are ones in which delays in operations are usually not possible (e.g., when the fission reaction in a nuclear plant must be stopped, or scrammed, this operation usually can't be postponed); in which operations must be performed in a fixed, invariant order (other orders are dangerous or not permissible); in which it is not possible to make substitutions in personnel, supplies, or equipment; and which have other, related characteristics. *Highly interactive technological systems* are ones in which there is limited information available to operators about some ongoing processes (e.g., about some conditions inside a nuclear reactor core); about which there is incomplete scientific understanding (e.g., uncertainty about the long-term effects of radioactivity, pressure, and temperature on metal components of a nuclear plant); in which pieces of equipment that are not functionally related are spaced close to each other; in which isolation of failed components is not possible; and which have other, related characteristics.

Figure 11-4 shows how several technological systems rate, according to Perrow (1984), on degree of coupling and degree of interactivity. Perrow claims, as we noted earlier, that normal accidents happen mainly in technological systems that have high levels of *both* coupling and interactivity, that is, the ones that lie in the upper right-hand quadrant of the figure. However, note that the ratings in the figure are Perrow's personal judgments. McGill (1984), Kates (1986), and other critics have objected to the use of these subjective ratings, rather than a numerical and empirically based rating scheme.

Lastly, keep in mind that only *some* technologies prone to normal accidents can cause significant damage to people and the environment. These include technologies that can release large quantities of pollutants or toxic chemicals. Note, for example, that space missions like Apollo (upper right corner of Figure 11-4) do not have this property, whereas nuclear and chemical plants do. Perrow (1984) argues that nuclear plants have much greater potential than chemical plants to cause catastrophic and long-lasting environmental damage because plutonium, a nuclear fission by-product, is highly radioactive and has a half-life of tens of thousands of years. Perrow there-

fore concludes, as we noted earlier, that commercial nuclear power generation should be ended.

Criticisms of Perrow. As we noted earlier, several researchers, scholars, and writers have criticized Perrow's (1984) work. They argue that his conclusions, especially his claims about the inevitability of normal accidents in nuclear power plants, require additional empirical support to be taken seriously. They also point to what they perceive as an unwarranted prejudice against commercial nuclear plants on Perrow's part. Hirschhorn (1985), a professor of management and behavioral science at the University of Pennsylvania, argues that Perrow's intuitively appealing analysis of the causes of normal accidents does not prove that such accidents are inevitable *despite* any *future improvements* in the technologies or in the managerial systems that operate them. Hirschhorn further claims that accidents such as at TMI mainly reflect remediable weaknesses in organization and management, and such accidents can be made "non-normal."

In a similar vein, LaPorte and Consolini (1991), and others (see Clarke and Short, 1993) argue that Perrow's analysis is undermined by the existence of complex, tightly coupled, interactive, high-potential-risk technological systems that have proved to be very reliable. These systems include military aircraft carriers, air traffic control systems, and the Diablo Canyon (California) electric power plant—systems these scholars call "high-reliability organizations," or HROs. Clarke and Short (1993) review and critique the HRO literature, arguing that LaPorte and others do not provide specific statistical evidence for their claims of high reliability, and that in several cases the organizations studied are operating under unusually favorable conditions (e.g., aircraft carriers studied only in peacetime, not in combat).

We see no way to resolve this dispute about inevitability and await additional relevant research and analysis. Further, we are not, in a book like this, in a position to make definitive statements about the overall safety or desirability of commercial nuclear power. (As we discuss in Chapter 9, there are political, economic, and philosophical issues as well as scientific and technical ones that underlie debates of

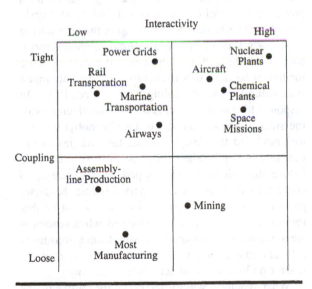

FIGURE 11-4 Degree of Coupling and Interactivity of Different Technological Systems

From Perrow, C. *Normal Accidents: Living with High-Risk Technologies.* Copyright © 1984 by Basic Books, Inc., a division of HarperCollins, Inc.

this type.) We can only say that Perrow's analysis is provocative, and that, since nuclear power plants have the potential to do considerable human and environmental damage, they must be regarded in ways that other technologies need not be. Similarly, Perrow's analysis of normal accidents alerts us to problems humans and human institutions may have in future attempts to manage large technological systems, if these systems continue to increase in complexity, scale, and in their ability to negatively affect the environment.

As a final word on the issue of normal accidents, we look at the nuclear power plant accident that occurred in Chernobyl, Ukraine, in 1986—after the publication of Perrow's book. We ask: Was the Chernobyl accident a normal accident? If so, it would lend some support to Perrow's analysis and point of view. We conclude that the Chernobyl accident had, in fact, many properties of a normal accident.

Does the Chernobyl Accident Fit the Normal Accident Framework?

The accident at Chernobyl, which occurred on April 26, 1986, was much more severe than the one at TMI. A core meltdown and explosion spewed highly radioactive material over hundreds of miles. The accident may have released as much radioactive material as was released by all prior nuclear bomb tests and explosions combined (Anon, 1987). The Chernobyl accident initially killed scores, if not hundreds, of people. In the ensuing forty years, according to one estimate, 30,000 people will die from cancers caused by the accident (Read, 1993). Chernobyl is arguably one of the most serious environmental accidents ever to occur.

To what extent does Perrow's concept of a normal accident apply to the Chernobyl disaster? We must, first, emphasize that Chernobyl was in several major ways a different type of accident from TMI: The Chernobyl plant had no containment building, unlike the TMI plant and all others in the United States, and had there been one, the release of radioactivity into the environment might have been prevented. Further,

there were basic design flaws in the Chernobyl reactor, a type RBMK-1000 (Perney, 1993). Also, the Chernobyl plant appears to have been plagued by defective materials, shoddy workmanship, and failures to inform engineers and operators of flaws and accidents at similar plants in the former Soviet Union (Read, 1993). Despite all this, there are many elements that fit Perrow's normal accident pattern.

First, there were *multiple* causes of the Chernobyl accident, some of them quite *modest*. The most distant cause (cause one) was concern on the part of engineers and administrators at Chernobyl about a possible safety malfunction detected in a similar reactor at Kursk. The RBMK-1000 reactor requires electricity to run the major pumps in the plant (including feedwater and emergency core cooling pumps) as well as to operate the control rods. Electricity is available from three sources: the main generators of the plant itself (unless the plant is scrammed or not operating), electricity from other plants in the power grid, and emergency, on-site diesel generators. In the case of the standby generators, approximately forty seconds are needed to start and engage them after a power failure. Following a power failure at Kursk, considerable afterheat had built up in the core during these forty seconds when all pumps stopped; fortunately, the afterheat was kept below the level of a core meltdown because it was absorbed by the unpumped (i.e., natural) flow of cooling water (Read, 1993). In response to this near-accident, the manufacturers of the steam turbines that drive the Chernobyl generators modified the design so that the turbines would generate enough electricity for safety functions before they slowed and stopped after a scram, to cover the entire forty-second period before the diesel generators could be started. However, a test of this improved design was not performed when reactor 4 went online, due to intense pressure from government officials who hoped to be rewarded for getting the reactor on line ahead of schedule (*cause two*).

When reactor 4 was closed down for routine maintenance two years later, the officials decided to test the change in turbine design. The idea was to simulate a scram and stop the flow of steam to one turbine and see how much electricity its generator produced before the turbine stopped spinning. The test had been scheduled for April 24, but had to be postponed because the Kiev power grid needed the electricity. When testing began, the chief administrator in charge, A. Dyatlov, ordered that the emergency core cooling system be manually turned off (*cause three*). He ordered the override because he wanted the reactor core operating at a constant low level in case the test of turning off the turbine had to be repeated. If he had not overridden it, the ECCS might have been triggered by automatic sensors that picked up the decreased flow of coolant water to and from the turbine and/or the starting of the auxiliary diesel generators; if water was injected by the ECCS it might have stopped the reactor altogether.

Cause four was another flaw in the RBMK-1000's design. The reactor (consisting of a large graphite block penetrated with fuel rods and boron control rods) was unstable when producing low levels of power. To avoid a stall or premature cessation of the fission process, the operators routinely removed more of the control rods than allowed by regulations (*cause five*). An even more serious design flaw (*cause six*) was a tendency for the core to *increase* fission activity for a brief period right after control rods were dropped into the core (e.g., in a scram) and before fission finally ceased. The operators and engineers of the Chernobyl reactor were not informed of this flaw (*cause seven*). Further, all information, operating rules manuals, training provided by government agencies and other sources indicated—in some cases deceitfully—that an explosion and/or serious accident was not possible for the RBMK-1000 design (*cause eight*), unlike the pressurized light water design of the TMI plant (the TMI accident preceded the one at Chernobyl, and the Chernobyl administrators and operators were aware of that accident, its cause, nature, etc.).

After Dyatlov ordered the ECCS and some other safety systems disconnected, he ordered that fission in the reactor be lowered to below one-half power, at which time the reactor began to exhibit its characteristic low-operating-level instability. At first it looked as if the reactor would stall (all activity precipitously stop), and Dyatlov ordered an increase in power

(*cause nine*) by removal of more control rods. The control room operators on duty thought that rather than increasing power, the unstable condition of the reactor dictated an immediate shutdown, but Dyatlov insisted that the test continue and reactor activity be maintained. The intended test of the turbine redesign (i.e., a cessation of steam delivery to one of the turbines) began. Soon, however, gauges indicated that the level of fission activity was climbing dangerously. The operators quickly scrammed the reactor to terminate fission. At this point the reactor's scram or shut-off fission surge (a design flaw—cause six above) together with the climbing reactor activity just before the scram caused enough of a heat buildup to produce a core meltdown and an explosion. By that time, efforts to start the ECCS and cool the reactor were futile. Much of the plant had already been destroyed.

The *second characteristic* of a normal accident— *a fatal combination of simple causes*—was clearly present at Chernobyl. As a matter of fact, the removal of just one of the causes described above (e.g., the disconnection of the ECCS, or the failure to inform the engineers and operators of the plant about the scram-power-surge design flaw) might have totally prevented the accident.

The *third characteristic—the fatal combination is not foreseeable*—appears also to have been present. A government report given at a meeting of the International Atomic Energy Agency in Vienna, August 1986, claimed that the designers of the RBMK-1000 reactor "never envisaged '. . . the deliberate switching off of technical protection systems coupled with violations of . . . operating regulations'; they had 'considered such a conjunction of events to be impossible' (Read, 1993, p. 205)." (Though other parts of this report by the Soviet government were not completely frank, it appears that the above quoted part of the report was accurate [Read, 1993].)

Finally the *fourth characteristic* of a normal accident—*the incomprehensibility of the accident while it was occurring*—was present. Information problems, including poor design of control room gauges and displays, made it difficult for the operators of the reactor to figure out what was going wrong during the

accident. Further, as at TMI, some important information about conditions inside the reactor available
only from a computer was not easily accessible—at
Chernobyl, the printer was located 150 yards away
from the main control panels and reported conditions
only every five minutes (Read, 1993). It also appears
that the operators and officials at Chernobyl had
faulty mental models of the system and the accident.
All people on duty knew that a major explosion had
occurred (there were loud thuds, the walls of the
control room shook, the lights went out, and large
cracks appeared in the ceiling), but a core meltdown
did not seem possible to them. Dyatlov, the head official, then sent a control room operator into the plant
after the explosion to manually open valves to permit
the ECCS to work and to report on conditions there.
The operator returned and reported that the reactor
had exploded and was destroyed. Dyatlov did not
believe him and left the control room to check the
plant himself.

Quoting Read (1993):

*Although Dyatlov knew that there had been an explosion
. . . , it did not occur to him that the explosion could
have been in the core. He had worked with reactors for
twenty years; he had taken numerous courses to bring his
initial training up to date; he had studied the voluminous
documentation that had accompanied the new RBMK
reactors; he had seen the building of the fourth unit and
had supervised its commissioning; and never had it been
suggested that the reactor itself could explode. He . . .
realized that the hermetic containment might have been
ruptured [by the explosion], but he could not envisage
anything worse [p. 69].*

Thus, the four main characteristics of a Perrowian
normal accident were present at Chernobyl. A quote
from a report on the accident by N. Steinberg, a
respected Soviet engineer, further conveys this:

*[The accident began] with an accumulation of small
[design] breaches of the regulations; more than ten were
identified in the design of the safety system of the No. 4
reactor. These produce[d] a set of undesirable properties
and occurrences that, when taken separately, do not
seem to be particularly dangerous, but finally an initiat-
ing event occur[red] that, in this particular case, was the
subjective actions of the personnel that allowed the*

*potentially destructive and dangers of the qualities of the
reactor to be released (Read, 1993, p. 325).*

COMPLEX SOCIAL, ECONOMIC, POLITICAL, AND ENVIRONMENTAL SYSTEMS

While Perrow's work focuses on complex technological systems, such as nuclear power plants, Jay
Forrester's work focuses on complex social,
economic, political, and ecological systems, such as
large cities, countries, and global ecosystems.
However, some of Forrester's main conclusions parallel those of Perrow: Forrester argues that humans
have trouble understanding and managing contemporary social, economic, environmental, et cetera
systems because these systems are too complex for
the human mind to grasp. Forrester further claims that
many of these systems tend to behave in the reverse
of the way human intuition would predict,
compounding the problems humans have in understanding them. Also, the properties of these systems
tend to defeat political and economic mechanisms
humans have traditionally used to manage complex
social and other systems. Most relevant is Forrester's
claim that the environmental problems that now
threaten human survival are caused by a mismatch
between the properties of environmental systems, on
the one hand, and the characteristics of the human
mind and of human institutions that have impact on
the environment, on the other.

Urban Systems and Global Environmental Systems

To understand Forrester's claims about complex
systems, let's begin by exploring the properties of a
complex urban system. The diagram in Figure 11-5
depicts a model Forrester (1969) developed of the
decayed inner-city area found within many large U.S.
cities. The model is designed so that numerical values
can be entered for each element, and the behavior of
the whole system can be simulated by computer
calculations.

The details of Forrester's model are not important
for our purposes right now, nor is the question of

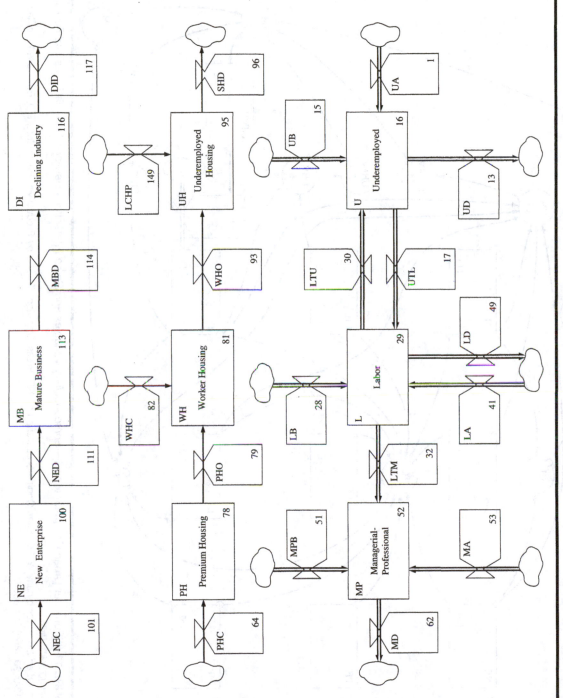

FIGURE 11-5 Forrester Urban Computer Model—*Main Variables and Rates*

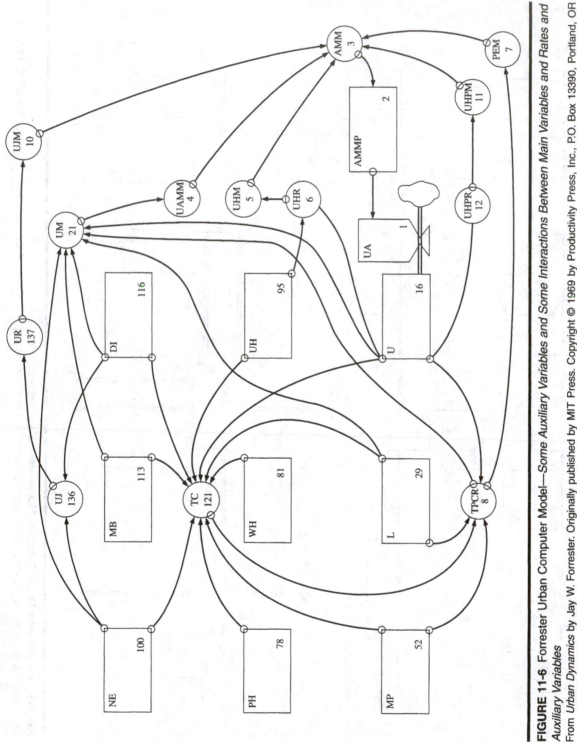

FIGURE 11-6 Forrester Urban Computer Model—*Some Auxiliary Variables and Some Interactions Between Main Variables and Rates and Auxiliary Variables*

From *Urban Dynamics* by Jay W. Forrester. Originally published by MIT Press. Copyright © 1969 by Productivity Press, Inc., P.O. Box 13390, Portland, OR 97213-0390, (800) 394-6868.

whether the model correctly portrays the properties of the real-world system it attempts to portray. The main issue is the complexity of the system the model depicts and whether people have trouble understanding the dynamic properties of such a system, that is, its behavior over time. In the diagram (Figure 11-5), each horizontally oriented rectangle represents a major *variable* of the inner-city area, such as the number of mature businesses in the area (labeled MB), the number of housing units for workers (WH), or the number of un- or under-employed workers (U). Each small rectangle with a triangular end represents an important *rate* that affects a variable, such as the rate at which unemployed people move into this section of the city (UA—unemployed arrivals), or the rate at which new working-class housing is constructed (WHC) built. Variables and rates influence each other via a complex, intricate network of *auxiliary variables* that Figure 11-5 does not show. Figure 11-6 shows several of these auxiliary variables.

Figure 11-7 shows a model of an even larger, more complex system (from Meadows et al., 1972, and Meadows and Meadows, 1973). The model is an ambitious one: It's an attempt to represent the main environmentally relevant aspects of the entire globe. As the figure shows, the model includes major variables involving human population, food production, industrial production, resources, and pollution. The model also includes many related rates and auxiliary variables, as well as the intricate web of interactions between all the rates and variables.

For our purposes here, once again, the details of the models in Figures 11-5 (with 11-6) and 11-7 are not important. What is important is the immense scope and complexity of the real-world systems that the models attempt to describe. It is systems like these that Jay Forrester (1969, 1971, 1980, 1987) claims are too complex for the human mind to comprehend. He argues that the behavior of these systems over time cannot be understood or predicted by humans, and he concludes that humans are likely to interact with these systems with poor results. The complex systems for which Forrester makes these claims include large cities, countries, international economic systems, global climate/atmospheric systems, and some large corporations.

As partial evidence for his argument—at least for his specific claim that people have trouble comprehending the dynamic characteristics of some large corporations and urban systems—Forrester cites managerial failures, widely known in business circles, of certain large corporations, and the failure of many city governments to control and reverse urban decay despite long and costly programs. Concerning urban decay, Forrester (1969) writes:

> *Among political leaders, managers of [urban] redevelopment activity, and political scientists interviewed in connection with [our] . . . study [of urban systems], there was the overwhelming opinion that the problems of the urban area remain as severe in spite of the variety of programs that have been tried over the last three decades [p. 10].*

To help explain why humans can't comprehend complex systems, Forrester (1971) invokes a variant of the Stone Age genes in the space-age idea of Chapter 8. Forrester writes:

> *In the long history of evolution it has not been necessary for [hu]man[s] to understand these systems until very recent historical times. Evolutionary processes have not given us the mental skill needed to properly interpret the dynamic behavior of the systems of which we have now become a part.*

(His claim also parallels our discussion of the limited powers of human cognition, i.e., of *bounded rationality,* in Chapters 8 and 9.)

But remember, Forrester is making an *additional* important claim: that not only are complex systems incomprehensible to humans, but that these systems often behave in the exact reverse of the way human intuition would predict. The systems fool us by behaving the opposite of what we would expect, often with disastrous results. Before taking a closer look at the alleged counter-intuitive behavior of complex systems, however, let's first take a closer look at their complexity per se. What exactly are the characteristics of complex systems that make them incomprehensible to the human mind?

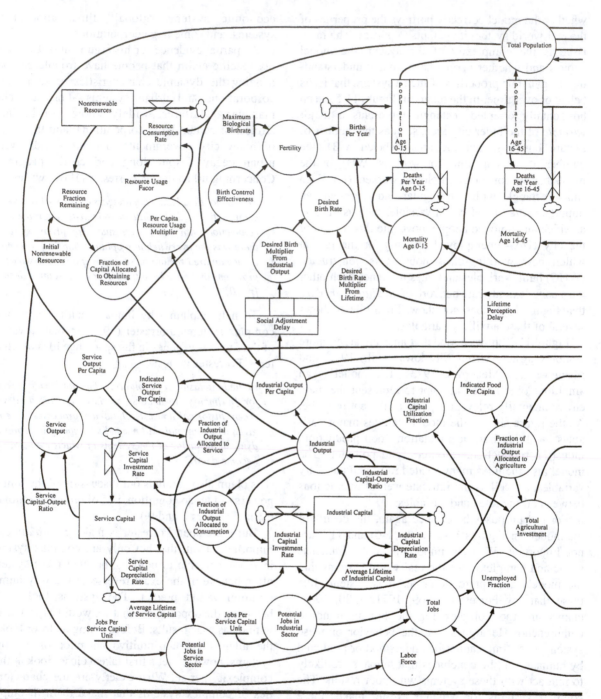

FIGURE 11-7 Meadows et al. (1972, 1973) Global Computer Model

From *Toward a Global Equilibrium: Collected Papers,* edited by Dennis L. Meadows and Donella Meadows. Copyright © 1973 by Wright-Allen Press, Inc. Available only through Productivity Press, Inc., P.O. Box 13390, Portland, OR 97213-0390, (800) 394-6868.

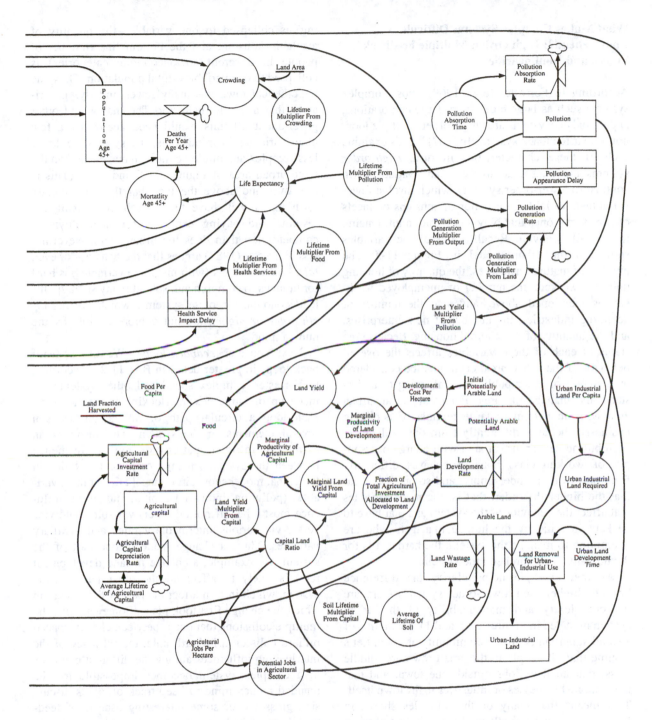

What Makes Complex Systems Difficult to Comprehend?: High Order, Multiple Feedback Loops, and Nonlinearities

According to Forrester (e.g., 1969), most complex systems such as large cities, very large corporations, and global ecosystems are "high-order, multiple loop, non-linear feedback systems [p. 107]." Let's examine each of these characteristics in turn. *High-order* systems are ones that involve many variables, (in contrast to lower-order systems, which involve fewer variables). Variables are major dimensions or facets of a system—ones that play a key role in influencing its overall behavior. Recall some of the variables shown in the urban model in Figure 11-5: The number of mature businesses, the quantity of housing units for workers, the number of unemployed workers, as mentioned above, and also the number of declining industries, the number of new enterprises, and the quantity of "premium" housing. Because the status of each of these variables affects the overall behavior of the urban area, each must be considered by a mayor or other public official trying to understand and manage the area. For example, an urban area with an extremely large number of declining businesses behaves differently from one with fewer such businesses; an urban area with a large oversupply of working-class housing behaves differently from one with an undersupply; and so on. It is clear that the higher the order, that is, the more variables that affect the behavior of the system and that have to be kept in mind by the human managers who are attempting to manage the system, the harder it is for the humans to comprehend the system.

Another example helps clarify the difference between higher- and lower-order systems: Compare the complexity and manageability of the urban system modeled in Figures 11-5 and 11-6 with that of a small suburban bedroom community or town. Let's assume that all people in the small town are middle class, that all have jobs outside the town, and that there are no businesses or industries in the town itself. This means that many of the variables shown in Figure 11-5 simply don't apply to this hypothetical town: The variables on the top line (number of mature businesses, etc.) don't apply; the middle line (hous-

ing) is collapsed to one variable—the quantity of available housing; and the bottom line (number of people in different socioeconomic categories) is collapsed to one variable—total population. Thus, the hypothetical town has many fewer underlying variables that affect its behavior. The mayor and other government officials would need keep only a few main variables in mind; thus, the system would be less complex and might be more manageable than the large urban area of Figures 11-5 and 11-6. (This is true over and above the fact that the middle-class town might well have fewer problems raising tax revenue and paying expenses for unemployment compensation than does the urban area). Nevertheless, Forrester (1987) argues that the behavior of even *third-order* systems (with just three variables) is hard for humans to comprehend, so that many social/political/economic/corporate systems, which are *twentieth-order or higher,* are not comprehensible by the human mind.

Next, consider *feedback loops.* (We discuss feedback loops in greater detail in Box 11-2.) According to Forrester, complex social and other systems are made up of numerous interlocking feedback loops, each loop a circular path of interaction between several elements in the system. The result is an extremely intricate pattern of mutual influence between the system's many components. Notice in the global model shown in Figure 11-7 that most variables (pollution, industrial capital, and so on) influence most other (if not all other) variables, and vice versa, via various intermediary rates and auxiliary variables. Thus, a change made in one part of the system (for example, a change in land development rate) is likely to affect all the other parts of the system, which in turn affect the part of the system in which the change first took place. Somewhat like the multiple collisions that occur between closely spaced billiard balls on a billiard table, the intricacy of the interlocking influences among the different elements in a complex system becomes impossible for the unaided human mind to keep track of. (This discussion glosses over some interesting aspects of feedback loops, which we discuss in Box 11-2.)

Finally, consider the *nonlinear* aspect of complex systems. This involves the "irregular" nature of the

Box 11-2

Feedback Loops

Dynamic systems, that is, systems that exhibit change over time, are found in a variety of settings. There are mechanical, biological, ecological, business, economic, and other dynamic systems. (The urban and global systems modeled in Figures 11-5 to 11-7 are examples.) According to Forrester (1968, 1969, 1980, 1987) and his colleagues (and others), all of these systems—regardless of their specific context—share certain common properties. The systems can thus be understood within a single conceptual framework that Forrester calls "systems dynamics." In all dynamic systems, according to Forrester, positive and negative feedback loops are the basic underlying structural elements, and the complex pattern of interaction between these loops is what determines how the system behaves over time.

Positive and negative feedback loops are circular paths of influence within a system. Positive feedback loops are "goal-diverging," that is, they tend to move some part of the system away from an initial state in an exponential manner. In contrast, negative feedback loops are "goal-seeking," that is, they tend to push some part of the system toward some objective (Forrester, 1969; Meadows et al., 1972, 1992). Positive feedback loops are responsible for all growth in a system over time, be it population growth, biological growth, economic growth, and so on, whereas negative feedback loops tend to oppose growth and to maintain stability in a system.

The accompanying figures show some specific examples of positive and negative feedback loops. Figure 11B2-A illustrates the positive feedback loop that underlies the accumulation of compound interest in

a savings account. Each year, a fixed percentage of the balance is added to the account. As a result, the balance increases exponentially. Figure 11B2-B illustrates a second example of a positive feedback loop. In human or animal populations (assuming fixed birthrates, that is, a fixed number of births per thousand females each year), the larger the population, the more babies born in any given year; when the babies reach maturity, the same birthrate means that even more babies are born, and so on. This positive feedback, if unchecked, leads to exponential growth of population.

But positive feedback loops do not operate in isolation. Their growth-generating action is usually opposed by the action of negative feedback loops and by various limits found in most systems. In the case of human and animal populations, the action of the birth positive feedback loop (Figure 11B2-B), which increases population each year by a certain fixed percentage, is opposed by the action of the death negative feedback loop, which removes a fixed percentage of a population each year. The death negative feedback loop is, thus, continually pushing population levels in the direction of zero.

Figure 11B2-C, adapted from Meadows et al. (1972, 1992), illustrates the opposing influences of the positive (birth) and negative (death) feedback loops. Let's further explore the interaction of the two loops. If no deaths occurred in a given population (that is, if death rate were zero and the action of the positive [birth] feedback loop were unopposed), the population would grow exponentially, as we noted above and as the small inset at the lower left of Figure 11B2-C shows. On the other hand, if birth rate were zero (if the action of the negative feedback

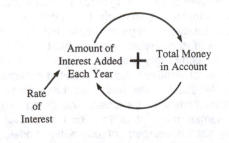

FIGURE 11B2-A Positive Feedback Loop: Compound-Interest Savings Account

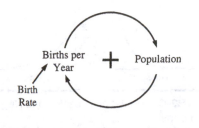

FIGURE 11B2-B Positive Feedback Loop: Births and Population   *(continued)*

BOX 11-2 continued

loop were unopposed), population would shrink to zero as the small inset at the lower right of Figure 11B2-C shows. (For the sake of clarity and simplicity, this example ignores the age structure of a population and assumes only that a fixed percentage of the population dies each year.) Note that the curve shown in the right inset is concave upward—the population shrinks to zero "asymptotically." (Explanatory example: For a fixed annual death rate of 10 percent, a population of 1,000 would decline by 100 to a level of 900 the first year, decline by 90 to a level of 810 the second year, decline by 81 to a level of 729 the third year, . . decline by only 23 to a level of 206 the fifteenth year, and so on.)

What happens, however, in a population that has both births and deaths each year, as do all populations in the real world? If birth rate exceeds death rate, there is a net positive rate of growth, and the population grows exponentially. The slope or steepness of that exponential growth depends on the size of the difference between birth and death rates; the bigger the difference the steeper the slope. On the other hand, if death rate exceeds birth rate, the population declines asymptotically to a level of zero, the slope or speed of the decline depending on the size of the difference between death rate and birth rate (the bigger the difference the faster the decline). In most countries of the

world, birth rate is higher than death rate, and therefore population is growing exponentially, as we discussed in Chapter 1. Again, changes in population level over time are determined by the interplay of opposing positive and negative feedback loops.

In reality, however, the interaction between the positive (birth) and negative (death) feedback loops is more complicated than our discussion and Figure 11B2-C would suggest. For one thing, there are several different time delays involved, such as the delay between the birth of children and the age at which they can bear children of their own. Also, birth rate and death rate are usually changing over time, and are not fixed. Thus, advances in medicine and public health have lowered death rates in most countries. Further, birth rates, for various reasons, have also decreased in many countries in the last decade or two.[3]

But even this description is too simplistic. Birth rates and death rates are themselves influenced by many other variables via feedback loops, such variables as average age at marriage, food supply and quality of diet, availability of other key resources, environmental pollution levels, quality of medical care, public health measures, education, urbanization, and so on. Thus an extremely complicated set of factors and feedback-loop interactions determine population size and growth rate in any given country or part of the world. In particular, death is not the only negative feedback that can oppose population growth.

Some of this complexity is reflected in the Meadows et al. (1972) global model shown in Figure 11-7 of the main text. Many of the variables we just discussed appear as influences on the population variables in the model (shown in the top center of the figure—note that the model breaks the population down into three different age groups). Figure 11-7 also shows the complicated web of interacting feedback loops involving these variables. (Although the loops are not labeled as positive or negative in the figure, a little thought should reveal which are which). It is the kind and level of complexity shown in Figure 11-7—the large number of variables, the complicated network of interacting feedback loops, (and also the nonlinearity of the interactions)—that exceeds the grasp of the unaided human mind. It is this kind and level of complexity that developers of computer models of complex systems must try to capture and represent in their models.

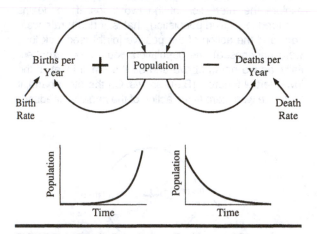

FIGURE 11B2-C Population: Positive and Negative Feedback Loops

Adapted from *Beyond the Limits*. Copyright © 1992 by Meadows, Meadows, and Randers, Chelsea Green Publishing Co., White River Junction, Vermont.

influence that any one variable or component of the system has on another. In a nonlinear interaction, more specifically, the magnitude of the effect of one variable or component on the other cannot be characterized by a simple, straight-line graph. To better understand this, consider, first, a concrete (though prosaic) example of a *linear* interaction—the influence a wall-mounted electric dimmer switch has on the brightness of the light to which it is connected: As you turn the dial of the dimmer switch in a clockwise direction, the brightness of the light it controls increases in a proportional way (though all dimmer switches may not operate in precisely this way, let's assume for the sake of the example that they do). The relationship between knob setting and resulting intensity is, therefore, described by the straight-line function shown in Figure 11-8A.

In contrast, consider an example of a *non*linear interaction, shown in Figure 11-8B, taken from Forrester's (1969) urban model. The example involves the effect of land occupancy in a given part of a city on the amount of new construction in that part over a period of many years. Initially—decades back when the city was just a town—a given part starts off almost completely vacant and undeveloped, that is, there is little or no construction on it. To quote Forrester:

> At [such] very low levels of land occupancy, the area has not yet shown economic importance and the nearly empty land tends to discourage aggressive construction. As land occupancy increases toward [a] middle [level, the pace of construction increases because the area has begun to show] . . . economic excitement and potential. Also all of the common urban activities and services are now present to support further expansion. But as the area approaches the fully occupied condition, the more favorable construction sites have already been used. The less and less attractive locations begin to depress the tendency to build until the [eventual] absolute unavailability of land drives the construction rate to 0 at full land occupancy. [pp. 173–4]

Thus the graph of new construction as a function of percent of land occupied is the curved, inverted U-shape function in Figure 11-8B. Another example of a nonlinear relationship—one unrelated to urban models or complex systems—involves environmental

noises and a threshold effect in human hearing: Noises below 85 decibels (dB) of intensity do not harm the human ear; above 85 dB, however, damage occurs and it increases with the intensity of the noise, as in Figure 11-8C. Figure 11-8D shows graphs of three other nonlinear relationships.

Clearly, linear relationships are simpler and easier to understand than nonlinear relationships. Unfortunately, many of the relationships in complex real-world social, political, economic, and ecological systems are nonlinear. (Nonlinear relationships also play a key role in *chaos theory*; see Chapter 12.)

To summarize: Most complex real-world systems involve many variables that are connected by multiple paths of mutual influence, with many of these paths operating in a nonlinear way. Systems with these characteristics, Forrester claims, cannot be comprehended by the unaided human mind. Just as no human can lift the World Trade Center buildings or pole-vault over the Great Pyramid, no human can comprehend the complex systems in Figures 11-5 to 11-7 well enough to predict their behavior over time. Thus humans are likely to bring about negative consequences when they interact with highly complex systems. Further, any direct, unaided attempts by humans to *manage* such systems, especially global and regional environmental systems, Forrester claims, are likely to fail.[4] We evaluate these claims later in the chapter.

Augmenting the Human Mind: Computer Modeling of Complex Systems

If the human mind can't comprehend the properties of complex social, economic, political, technological, and ecological systems, is there any way to boost human understanding of these systems? According to Forrester one method holds promise: The development and use of computer models, or simulations, of the systems, like those depicted in Figures 11-5 to 11-7. This use of models is more or less analogous to the physical modeling used in engineering (Forrester, 1971). Thus, before a test pilot flies a full-size prototype of an airplane under development, engineers subject a miniature model of the plane to stresses in a wind tunnel. Design flaws often appear in such tests

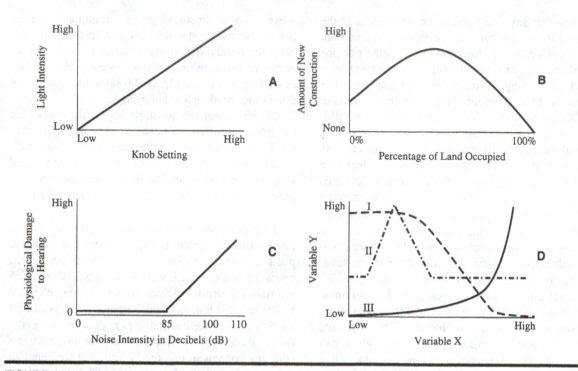

FIGURE 11-8A Light Intensity as a Function of Knob Setting (Linear)

FIGURE 11-8B Amount New Construction as a Function of Land Occupancy (Nonlinear)
From *Urban Dynamics,* by Jay W. Forrester. Originally published by MIT Press. Copyright © 1969 by Productivity Press, Inc., P.O. Box 13390, Portland, OR 97213–0390, (800) 394-6868.

FIGURE 11-8C Damage to Hearing as a Function of Noise Intensity (Nonlinear)

FIGURE 11-8D Variable Y as a Function of Variable X (All Nonlinear)

and can be corrected without risking the life of a pilot or losing a full-size prototype of the plane.

Obviously, it's not possible to make a physical model of a social system such as an urban area. But it is possible, according to Forrester and others, to construct an abstract computer model of a social system, one that correctly depicts the system's structure and predicts its dynamic characteristics, that is, its behavior over time. Mayors and city managers could use an urban computer model to help them understand the properties of a city. They could get a general idea of the causes of current problems, of future conditions in the city if current trends and

government policies continue, and of future conditions if specific government policies are changed.

Some Cautions and Qualifications. Before we go further, it's important that we alert you to several major cautions and qualifications concerning computer modeling. First, no computer model of a complex social system can yield an accurate and detailed crystal ball prediction of the system's future. Real-world social systems are far more complicated than a computer model can adequately portray. Also, there is no way to know of all future events that might affect, say, an urban area, especially events that come from outside the area. At best, good computer models can provide a crude and very general picture of the causes of current problems and the directions a system might go in the future.

Second, the task of building a good computer model of a complex social system such as a city is a daunting one. Even Forrester (1980) acknowledges that this process is more art than science, and that it can take *several years* to build a single model. Note further that Forrester's urban and global computer models and those of his followers have been criticized by several scholars as having specific weaknesses (Legasto and Maciariello, 1980). Even the overall approach of computer modeling a complex social system—as against a complex natural system such as an atmospheric system—is controversial. For the above reasons, some critics of Forrester's global model and that of his students dismiss these models as tentative academic exercises rather than well-developed models that are of real use to policymakers. We discuss these cautions and criticisms in more detail later in Chapter 12.

Regardless of their value for prediction or for making policy, we believe that Forrester's modeling efforts have produced important insights about the properties of complex systems that are valid even in the face of the cautions and criticisms we just reviewed. We now explore in more detail the efforts of Forrester, his students, and his colleagues to use computer models to study complex systems.

How Forrester and Others Develop Computer Models of Complex Social, et cetera Systems. Let's use as an example Forrester's inner-city model shown in Figures 11-5 and 11-6. To develop this model, Forrester first formed a group of city administrators, university professors, and others with expertise on urban problems. This group worked to identify the key *units* of an urban system, that is, the main variables, rates, and auxiliary variables; the group then identified which units *influenced or interacted* with which other units in the system. At this point, Forrester's group had completed diagrams of the urban system like the diagrams in Figures 11-5 and 11-6. Next, the group worked to determine for each interaction between

units, the *form of relationship (often nonlinear),* that best characterized that interaction.

Forrester then took all the information above—the list of key units, the paths of their interactions, and the forms of the interactions—refined the information, and entered it into a computer in the form of a series of mathematical equations. These equations represented in abstract terms the basic structure and properties of the system, as understood by the group of city administrators, and so forth; the set of equations thus constituted the model of the system.

Next, Forrester and his colleagues tested the completed model to see whether it correctly depicted the behavior over time of an actual inner-city area. They did this by seeing if the model could predict the past behavior of an inner-city area, behavior that could be ascertained from historical data. Thus, Forrester and his colleagues set the initial values of variables and rates in the computer model to match those of an actual inner-city urban area several years in the past; they then ran the model to simulate the passage of time and saw if the variable and rate values the model predicted were close to the actual variable and rate values today. The results of these tests were positive.

Their model-development work now complete, Forrester et al. began to apply their model. They ran the model to give them some idea of how the inner-city area might behave in the future—whether conditions would improve or worsen over a period of years—first assuming that present trends and government policies remained the same, and then assuming a variety of other possible government policies and programs. They also identified several factors suggested by the model as possible causes or aggravators of the area's current problems. We present some of the results of their work later in the chapter.

Computer Models versus Human Mental Models. Forrester (1971, 1987) makes an interesting additional claim about the computer-modeling process: He argues that the use of a computer model to help understand and predict the behavior of a complex system is conceptually similar to a person's use of his or her own *mental model* or *cognitive map* to make decisions about the future. As we briefly mentioned in Chapter 9, all people have elaborate mental models of relevant parts the outside world. These models include geographic information (i.e., the layouts of streets and buildings), and also information on the characteristics of various people, things, places, and processes. Further, each of us absorbs information from the outside world during our daily lives to augment our cognitive maps. In turn, we use our maps to guide our decisions about the future. For example, before a person decides whether to ask his or her supervisor for a raise and how to ask for it, the person first considers all he/she knows about the supervisor, the firm, and his or her own past behavior at work. In other words, we use our knowledge of the makeup and properties of the relevant parts of the outside world to try to anticipate that part's likely future behavior and its responses to alternative actions we might take now.

Mayors, city administrators, executives of large corporations, and others who attempt to guide complex social systems use the same approach: They use their mental models of the system to predict its future behavior and to choose policies to enact. However, while the unaided human mind can deal with simple systems (as in the example of asking for a raise), Forrester claims that it can't successfully cope with systems as complex as a large corporation or city because of the many variables involved and the myriad nonlinear interactions between them. Thus, Forrester argues, for complex systems, we must augment the limited powers of the human mind with a formal model of the system linked to the computational power and speed of computers. But Forrester's claim is that computer modeling of complex systems is not a fundamentally novel endeavor, but a formal and quantified version of what humans do all the time.

Forrester (1971, 1980, 1987) makes a fascinating further claim: He argues that—at least in the case of some large corporations, but not necessarily large urban areas or global systems—the mental models that administrators have of the system they are trying to guide are often correct. The executives *already*

know a great deal about what the key components of their corporation are and how these components interact. What the executives, however, cannot figure out is the dynamic properties of the corporation implied by their mental model of it, that is, how the corporation will behave over time, and how it will respond to different policy changes made now. The reason is that although the mental model is reasonably correct and complete, the human mind can't keep track of all the interactions between the components as they reverberate through the system over time. Again, this limitation involves only the executives' use of their mental model to predict the future behavior of the corporation, not the validity of that mental model! To quote Forrester (1971):

> [The development of a computer model of a corporation] . . . starts with the concepts and information on which [the executives] are already acting. Generally these are sufficient. The available perceptions are then assembled in a computer model which can show the consequences of the well-known and properly perceived parts of the system. Generally, the consequences are unexpected [by the executives] [(emphasis added)] [p. 3].

What an intriguing point: The administrators of such systems often understand the structure of the systems but can't comprehend the dynamic behavior that the structure implies.

As a further demonstration of this phenomenon, Forrester (1971, 1987) describes cases in which the executives of large corporations have correct mental models of the corporations, but their misunderstanding of the models' dynamic behavior is actually the cause of policy choices that led the corporations to fail! To quote Forrester (1971):

> Time after time we have gone into a corporation which is having severe and well-known difficulties. The difficulties can be major and obvious such as a falling market share, low profitability, or instability of employment. . . . One can enter such a company and discuss with people in key decision points what they are trying to do to solve the problem. Generally speaking, we find that people perceive correctly . . . [the structure of the corporation]. . . . One can combine these perceptions into a computer model. . . . In many instances it then emerges that . . . the known policies [of the executives] are fully sufficient

to create the difficulty, regardless of what happens outside the company or in the marketplace [p. 3].

Counterintuitive Properties of Complex Systems

The above paragraph embodies one of Forrester's most intriguing claims about complex systems and human interactions with them (Forrester 1969, 1980, 1987): Not only can't humans understand the dynamic behavior of many complex systems, but these systems often behave in the *reverse* of the way human intuition would predict, further compounding the problems humans have interacting with, and, in some cases, managing them. Forrester makes this claim about human interactions with *all* highly complex systems—large urban areas, global environmental systems, and others—not just large business corporations. And his claim applies whether or not peoples's mental models of the systems they are trying to manage are valid.

Forrester's claim is based on his experiences in developing computer models for a number of different complex real-world systems over a period of years. When he compared the unsuccessful policies that human administrators of the systems were using with the computer models' predictions of the systems' response to those policies, he usually found that the systems often worked just the opposite of the way the administrators thought they did. Thus the human administrators were often doing exactly the worst thing, or, alternatively, their policies were the cause of the problem to begin with (as with the corporate executives discussed in the section above). Furthermore, when administrators then adopted the policies that the computer models indicated should work well, the systems usually responded positively, just as the models had indicated. In the paragraphs that follow, we examine in depth some of the specific counterintuitive properties that, according to Forrester, complex social, economic, political, and ecological systems frequently exhibit.[5]

Counterintuitive Property 1: Wide Separation of Symptoms and Causes.
The true causes of a problem and its symptoms or manifestations are often widely separated in space and time; humans attempting to solve a problem in a complex system usually focus their efforts near the symptoms, and, as a result, either fail to solve the problem or make the problem worse. Forrester argues that because of people's extensive everyday experiences with simple systems, they assume that causes and effects lie in close proximity. For example, when a tree branch breaks or a rock moves, the cause usually immediately precedes the effect in both space and time. The same holds true for much of the modern technology with which people are in frequent contact: Thus, the on and off cycling of an air conditioner or heater is an immediate response, via the unit's thermostat, to temperature changes in the room. (The thermostat together with the air conditioner or heater constitute a "first-order negative feedback" system such as we describe in Box 11-2 on feedback loops.) If the heater or air conditioner fails to respond, the cause of the problem is usually in the thermostat, the heater or air conditioner, or the wiring or fuses of the house. Occasionally, the cause is remote, for example, a power outage caused by a problem at the generating plant or in power lines, but even this possibility is quickly discernible by looking at neighboring houses, and it immediately precedes in time the failure of the air conditioner or heater. In contrast, Forrester argues, highly complex social and ecological systems rarely work this way.

Going a step further, Forrester claims that complex systems often present to the humans interacting with them an apparent but false "cause" of a problem that lies near the symptoms. To quote Forrester (1969):

. . . [T]he complex system is far more devious and diabolical than merely being different from the simple systems with which we have experience. Although it is truly different, it appears to be the same. The complex system presents an apparent cause that is close in time and space to the observed symptoms. But the relationship is usually not one of cause and effect. Instead both are coincident symptoms arising from the dynamics of the system structure. . . .

In a situation where coincident symptoms appear to be causes, a person acts to dispel the symptoms. But the underlying causes remain. The treatment is either ineffective or actually detrimental. With a high degree of confidence we can say that the intuitive solutions to the problems of complex social systems will be wrong most of the time [p. 110].

Forrester (1969, 1971, etc.) provides several specific examples of the wide separation between causes and symptoms in complex systems and the mistaken human focus on coincident symptoms. We present two of them. The first example is prosaic and a bit simplistic. The second is richer and more interesting but problematic because it involves Forrester's (1969) urban model, which some scholars and writers have criticized. The two examples together, however, do an excellent job of conveying Forrester's claim.

The first example involves *overcrowded urban facilities.* In many cities of the world (though not all), public facilities such as highways, parking structures, libraries, tennis courts, and so on become more crowded and congested over time, often to the point where they are unpleasant and difficult to use. The natural impulse of city officials is to build more facilities (new highways, new parking structures, etc.). This, however, Forrester claims, just treats a symptom of a not fully recognized though simple underlying cause: exponential growth of the city's population. As a result of building new public facilities, this cause is ignored and continues to operate—see Figure 11-9. The strategy of building new facilities guarantees that

at a later time (T2), the problem of overcrowding will return, but in an even more hard to manage or virulent form: At point T2, the rate of growth (the slope of the curve) is much greater than at T1 and is therefore more difficult to keep up with; the total population is greater than at T1, which means more people are inconvenienced by the poor facilities; and the cheapest and most available resources to address the problem have already been used up at T1. In addition, not only does building more facilities ignore the underlying cause of population growth, it *encourages* or exacerbates this cause. For example, many cities grow because people in the countryside perceive greater opportunities there. The economic activity involved in building more facilities reinforces that image. Also, the construction of more facilities at T1 diverts attention from the more basic causes of the overcrowding; after all, Forrester argues, the overcrowding of facilities provides one of the few tangible reasons for citizens and government officials to take seriously the problem of population growth.

The second example of how complex systems often mislead their managers involves *urban decay.* As we noted earlier, many large U.S. cities have neighborhoods with severely dilapidated housing and high rates of crime, delinquency, and drug addiction. In the 1950s and 1960s, many city officials, university professors, and interested citizens had assumed that the poor housing in these neighborhoods (badly decayed and rat-infested homes), besides being inhuman, was a major *cause* of the crime and pathology. The solution they championed was "urban renewal": Substandard housing units were demolished and replaced with new, government-subsidized units. However, urban renewal was largely a failure. And Forrester, based on his computer model, purports to know why it failed, and what policy approach would work better.

Before we discuss Forrester's claims in detail, however, it is interesting to note why Forrester, a professor of business administration, did work on urban problems in the first place. As it turns out, his faculty office at the Sloan School of Management at M.I.T. happened to be next to that of former Boston mayor John F. Collins, who was a visiting faculty member in the late 1960s. Forrester and Collins

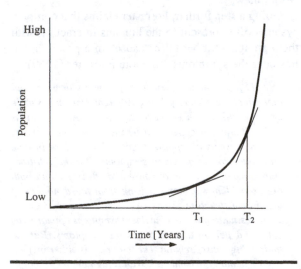

FIGURE 11-9 Population as a Function of Time (Overcrowded Facilities Example)

began to talk about the problems of the city. Collins told of the "pervasive sense of failure and frustration" he and many other officials in city government had concerning their efforts to lessen the problems of decayed inner-city areas. To Forrester, who had discovered the counterintuitive properties of complex systems in his earlier work on large corporations, Collins's description suggested that similar counterintuitive properties were also foiling the attempts of governments to understand and solve the problems of inner-city neighborhoods. Forrester decided to begin work on a major urban computer model. As we described earlier, Forrester and Collins held intensive meetings with government officials, policymakers, university professors, and others to help develop the model. The group provided the basic structure of the model—the major variables, interactions, and nonlinearities of a decayed inner-city area. Forrester refined the computer model and used it to try out several different policy alternatives. He came up with some surprising results:

First, the computer model suggested that decayed housing in inner-city areas was merely a *symptom*, but not the cause, of the crime, delinquency, and so on, found in these areas. The cause, according to the model: unemployment, coupled with an oversupply of low-income housing. High unemployment and excess housing, in turn, were part of a final, "decay" phase of an overall growth cycle that, according to the model, all urban areas tend to go through. This growth cycle is a many-decades-long process that has its own momentum: Initially, a new and largely vacant area of a city undergoes strong economic and physical growth. Then growth slows and finally stops as the area becomes filled with businesses, factories, and homes. In the final phase of the cycle, buildings in the area begin to age and factories become out-of-date. Businesses and industries decline. As they decline, jobs disappear, and more and more people become unemployed and economically disadvantaged. Increasing numbers of people are thus forced to move to smaller, poorer quality, and more crowded housing, producing many empty housing units.

According to Forrester's computer model, the urban renewal approach not only fails to stop inner-city decay but actually worsens it: When substandard housing is torn down and new government-subsidized housing is built, poor and unemployed people from outlying areas are attracted to the area in question. There are thus more people but no additional jobs. In other words, the result is more people with an even higher rate of unemployment, which leads to more crime and drug addiction, and the area declines further. The area becomes even less attractive to the businesses and industries remaining, and some leave, lowering employment still further. Again, the government policy of tearing down substandard housing and building new subsidized housing, which was intuitively appealing to city officials at the time, only served, Forrester claimed, to make matters worse.

Forrester's computer model suggested an altogether different strategy for overcoming urban blight and reversing the final decay phase of the urban growth cycle, a strategy that was very counterintuitive to city officials of the time: City government should demolish *unoccupied* substandard housing but should not build new housing to replace it. City government should then use changes in zoning and tax laws to encourage businesses and industries likely to employ residents of the area to move into the vacant land created by the demolition. Forrester argued that if these efforts were successful and new jobs resulted, the poverty and decay of the area would reverse. At the same time, the lack of excess housing (due to demolition of vacant housing) would discourage unemployed people from surrounding areas from moving into the area, keeping the newly created jobs from being swamped by more unemployed people.[6]

Forrester's urban model may have been right or wrong. It may well be invalid today, because U.S. cities have changed since Forrester's (1969) analysis. In particular, Forrester's model did not include two factors that play a major role in U.S. urban problems in the 1990s: The large number of deinstitutionalized mental patients among our urban homeless, and the large problems we now have with addictive drugs. But the model, even if not entirely correct, serves our present purpose, which is to show how a computer model can reveal the properties of complex social and other systems that are counterintuitive to the people trying to manage them.

Counterintuitive Property 2: Few Leverage Points.
There are few points of policy "leverage" in a complex system; these points are not easily identified via human intuition; in the unusual event that humans do identify a leverage point, they often apply pressure in the wrong direction. Based on his extensive experience with computer models of urban and complex corporate systems, Forrester (1969, 1980, 1989) concludes that these systems are frequently unresponsive to most changes in administrative policy. For example, though city governments can make many policy changes—such as changes in property tax rates, sales tax rates, utility rates, and so on—only a few of these changes will significantly affect the overall health of the city, either for better or for worse. Because of the complexity of urban and other systems, humans who attempt to guide these systems usually cannot intuitively identify the few changes in government policy that can make a difference. Forrester claims that in cases where the administrators do identify an effective leverage point, guided by intuition they often enact the reverse of the policy that will help the system.

An example: According to Forrester's (1969) urban model, the supply of low-income housing is one of the few points of policy leverage in an urban system. In other words, a change in this variable can strongly affect the overall health of the system. However, as we discussed above, the intuitively obvious policy—to increase the supply of low-income housing—is the reverse of what the computer model claimed should be done, namely, to decrease the supply via demolition of unoccupied housing.

Counterintuitive Property 3: Contradiction Between Short-Term and Long-Term Solutions. The short-term and long-term solutions to a problem often contradict each other. Policy changes that solve a problem in the short run (5–10 years) often worsen the problem in the long run (beyond ten years). Conversely, policy changes that permanently and sustainably solve a problem often cause an initial, short-term worsening. The "overcrowded facilities" example we used above to illustrate the separation between the symptoms and causes of a problem (Figure 11-9) also illustrates the contradiction between short-term and long-term solutions. Building new

parking lots and other public facilities to alleviate crowding relieves the effects of population growth in the short run, but this policy only aids and abets continued growth, guaranteeing worse crowding in the long run. Similarly, in the urban-decay example, building government-subsidized low-income housing temporarily relieves physical decay in the short run, but encourages in-migration that produces a worsening of unemployment and decay in the long run.

We discuss another example of the contradiction between short-term and long-term solutions in Box 11-3 on the Tucson (Arizona) Paradox. This example is highly similar to the overcrowded facilities example above, but there's an important and striking difference: The short-term solution in Tucson involved efforts to conserve a scarce resource. While it would initially seem that conservation is good, looked at from a broader perspective, in this case it was not.

Forrester (1969, 1971, 1987) views the contradiction between short- and long-term solutions with concern because he sees it as a potential threat to a democratic system of government. The essence of democracy is that key policymakers are subject to the approval of, or recall by, the public via periodic elections. Terms of office usually run from two to six years. Politicians often win election by promising improvements they can make during their terms, but these changes can have negative long-term effects. Promises by U.S. politicians in the 1980s to lower taxes and increase defense spending clearly worsened the long-term problem of the national budget deficit. It remains to be seen if an educated public will send and return to office political leaders who try to solve long-term environmental problems by means that may worsen economic conditions or people's quality of life in the short-run. (Note, also, that any human genetic tendency toward short-term egoism [Chapter 8] might predispose people to reject such sacrifices.)

Counterintuitive Property 4: Unanticipated Remote Consequences. Policy changes that lessen a problem in one part of a complex system often cause unanticipated problems in some remote part that are as bad as or worse than the original problem. Because humans can't comprehend the many interactions in a complex system, they often take actions to solve a

Box 11-3

Can "Conservation" Ever Be Bad? The Tucson (Arizona) Paradox

James Udall (1985) describes a rather striking example of the contradiction between the short-term and long-term solutions of an environmental problem. The example, which dates back to the late 1970s, involves a water shortage in Tucson, Arizona, a city located in an arid desert valley.

Udall notes that Tucson is one of the largest cities in the world (population three-quarters of a million at the time of the example) that is totally dependent on water from wells rather than surface sources such as rivers or lakes. The wells draw water from an aquifer—a series of underground reservoirs in rock and soil formations filled with rainwater that has slowly filtered down and accumulated over thousands of years. This water is therefore a partially renewable resource. However, beginning in 1940, Tucson began drawing more water from the aquifer each year than is naturally replenished.

In 1976, several elected city council members became concerned about Tucson's water overdraft; they realized that the situation—a growing city on top of a finite and decreasing vital resource—was untenable in the long run. To get Tucsonians to take the problem seriously, the council voted to double the water rates that consumers paid. Their decision proved to be politically unpopular: Critics immediately began a recall drive, and the drive was successful. A few months later several new council members were installed. This set of events is an example of the problem of political resistance to the imposition of incentives that we discussed in Chapters 5 and 7.

When the new council members tried to rescind the rate hike as they had promised to do, they learned that the city's bond rating would suffer if they did so. (All cities are rated for quality of government and financial stability by private agencies; these agencies informed Tucson that its rating would drop if it did not take steps to conserve water, an action the agencies deemed prudent; since cities that receive poor ratings find it more difficult to borrow money to pay for roads, bridges, and other public works, the city council complied.) As a matter of fact, to avoid a bad bond rating, the council was forced not only to keep the high water rates but also to begin a major public water conservation campaign! The campaign and high water rates were effective: Within two years, the average Tucsonian was consuming 25 percent less water than before.

Though the new council members had been forced to maintain and strengthen Tucson's water conservation efforts, they and some other Tucsonians, including several members of the business community, soon came to look upon conservation as a blessing in disguise: The lowered per capita consumption of water meant that Tucson could continue unhampered its rapid growth of people and buildings. Indeed, the business community began to donate as much as $250,000 per year to support the conservation effort. A business-supported conservation public education agency opened an office next to those of the Chamber of Commerce and the Tucson Economic Development Corporation. And Tucson continued to grow: As many as 30,000 new Tucsonians arrived each year. The mayor enthusiastically declared that "[t]here will be as many new homes built in Tucson in the next twenty-five years as have been built since Tucson began [p. 98]."

Thus, the Tucson water conservation effort was diverted from its original intent—to force Tucsonians to address the issue of the limits to growth of a desert city in the face of finite and shrinking water resources—and only succeeded in worsening the basic problem. As author C. Bowden pointed out, ". . . [t]he inevitable achievement of [the Tucson water conservation program] . . . is more people settled over less water (quoted by Udall, p. 99)." Again, a short-term solution to the Tucson water depletion problem worsened the problem in the long run.

The Tucson example is quite similar to the "overcrowded facilities" example (highways, parking lots, libraries, etc.) that we discussed in the body of the text. In both cases, a shortage of a limited resource or public facility caused by exponential growth in their use is temporarily alleviated in a way that encourages that growth to continue, thereby worsening the problem in the long run. The key difference, however, between the Tucson example and the one in the text is that the Tucson example involves conservation—something that most people might assume is automatically good for the environment. Of course, the conservation of a resource is not necessarily bad, but what the Tucson example shows is that it all depends on the big picture. To quote author Helen Ingram:

(continued)

BOX 11-3 Continued

Environmentalists need to recognize that conservation is a complex and contradictory concept with great symbolic appeal that can be employed by various groups for their own benefit. Conservation should not be regarded as an end in itself, but rather as a means to achieve some aim (quoted in Udall, 1985, p. 99).

We'd like to make one final point. The reader might argue that it doesn't require an elaborate computer model to fully appreciate the logic underlying the overcrowded facilities example or the Tucson paradox. Neither example is very counterintuitive when viewed from an overall perspective. Neither greatly taxes the powers of a well-informed human mind. However, let us call your attention to an example of a similar paradox that could not have been appreciated by the unaided human mind. We present it in the main text in our discussion of unanticipated remote consequences of a policy change. The example, from Forrester's (1971) and Meadows et al.'s (1972, 1992) global modeling work, is one in which efforts to stem the

rapid depletion of global natural resources—that is, conservation efforts—cause a more serious unanticipated problem in the long run: A sharp increase in industrial growth and in environmental pollution, the latter so severe that it causes the death of three out of every five people throughout the world (see Figure 11-11). Now, the Forrester/Meadows et al. global model may be seriously flawed, but we think Forrester's main point is sound: Complex ecological, social, and other systems are likely to have paradoxical or counterintuitive properties that cannot be anticipated by unaided human thinking. In the case of conservation, its long-term effects appear to depend on other factors, variables, and interactions, especially the degree of growth of the human population and of the impacts of human activity on resources and the environment. Given that it's hard to tell whether we'll get a Tucson paradox or a more benign outcome, maybe computer models can begin to tell us which to expect, and what it depends on.

---

problem at one location in the system that produce an unanticipated worsening of conditions in some other, far removed, location. There are many relevant environmental examples. Consider the unanticipated negative consequences of the Aswan High Dam on the Nile River in Egypt (Ehrenfeld, 1978). This dam was built to generate badly needed electric power for the region, and it did so as intended. However, the dam also interfered with annual flooding cycles of the Nile that help maintain the fertility of agricultural land in the river's floodplain. Further, the still waters in the lake behind the dam created favorable conditions for the growth of snail-like parasites that cause a serious disease—schistosomiasis—in people who drink and/or swim in the water.

A different, though somewhat less apt, example of unanticipated remote consequences comes from projections of global environmental computer models such as the one we discussed earlier in the chapter (shown in Figure 11-7). Recall that this model attempted to encompass all major environmentally

relevant global variables and their interactions—human population, and so on. The model, based on an earlier global model by Forrester (1971), was developed by Donella Meadows, Dennis Meadows, Jørgen Randers, and William Behrens and presented in their controversial book *Limits to Growth* (1972). An updated version of the model appeared in Meadows et al.'s book *Beyond the Limits* (1992), to which we confine discussion here.

Let's assume for the purposes of this discussion that the Meadows et al. (1992) model is valid. The initial projections of the model indicated that running out of nonrenewable natural resources would be one likely cause of a major decline in global quality of life and of population in the early to mid-twenty-first century. This is shown in Figure 11-10. In more detail: The shrinking supply of natural resources would force down industrial output per capita. This decline in output would mean not only fewer material goods per person, but also a decrease in per capita food production and medical care; this is because agriculture and medical services depend on industrial goods such as tractors, fertilizer, hospital and surgical supplies, and so on.

Meadows et al. (1992) then ran the model a second time, but this time telling the computer to assume twice the quantity of natural resources at the outset, thus overcoming the resource limits that appeared in the initial run; (the greater quantity could be due to larger natural stores to begin with

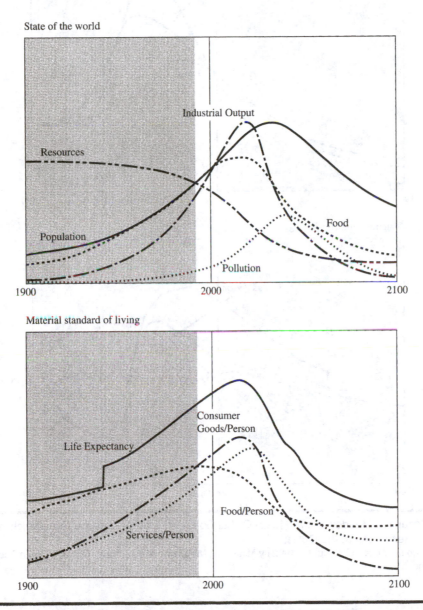

FIGURE 11-10 Results of Meadows et al. (1992) Global Computer Model—Initial Run

Reprinted from *Beyond the Limits*. Copyright © 1992 by Meadows, Meadows, and Randers. Chelsea Green Publishing Co., White River Junction, Vermont.

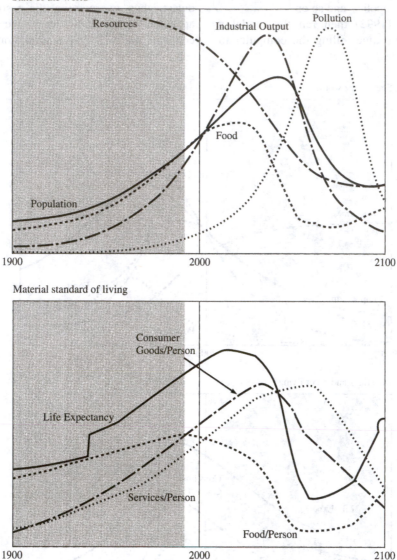

FIGURE 11-11 Results of Meadows et al. (1992) Global Computer Model—Second Run, Which Assumes Double the Natural Resources at the Outset
Reprinted from *Beyond the Limits*. Copyright © 1992 by Meadows, Meadows, and Randers. Chelsea Green Publishing Co., White River Junction, Vermont.

and/or to conservation efforts that stretch small stores). Surprisingly, the model indicated that an even worse global future would result, as shown in Figure 11-11: Because of the ample supply of natural resources, global industrial activity and population levels would both increase sharply through the first third of the twenty-first century; these increases, in turn, would cause an enormous increase in environmental pollution, one that would greatly lower soil fertility and shrink food production. As a result, food per person, life expectancy, and population dropped much more sharply than in the first computer run. Thus, bypassing the resource scarcity problem of the first run actually produced a significantly *worsened* global future.

The example we just reviewed assumes that Meadows et al.'s (1992) global computer model is valid. But the model may not be. If it is not valid, the Meadows et al. results at least should alert us to the possibility that complex systems can exhibit major counterintuitive properties of the type they observed. We will have more to say about this in Chapter 12.

A Caution Before Drawing Conclusions

Before we end the chapter, we must briefly mention one potential objection to Forrester's analysis of complex social and other systems: As we noted in the introduction, there are well-known social mechanisms and institutions that *can* enable large social systems to function well even though the systems are too complex for individual humans to understand. Corporations, industries, cities, and even countries have functioned well for long periods of time without full understanding or the help of computer models. In Chapter 12, we look at how this is done and ask whether these successes give cause for optimism about global environmental problems. We conclude that though these mechanisms worked in the past, they are no longer reliable given the unprecedented properties of today's environmental problems that are the focus of Chapter 12 (second half), for example, the fact that human activity can now do irreversible as well as catastrophic damage to the environment. In addition, these institutions and mechanisms can be foiled by the counterintuitive properties that Forrester alleges characterize complex systems—the properties we reviewed above, including a contradiction between short-term and long-term solutions to problems. If Forrester is correct about these properties, then his arguments for serious concern about the human ability to correctly guide highly complex systems are not contradicted by the fact of successes in the past.

CONCLUSION

In this chapter we have reviewed Charles Perrow's arguments that certain types of complex technological systems that impinge on the environment may be inherently unmanageable by individual humans and human institutions. Perrow's controversial thesis awaits further testing and confirmation. Jay Forrester's argument that humans cannot comprehend many complex social and environmental systems is, we believe, more compelling. In addition, Forrester's claim that many complex systems behave in counterintuitive ways is striking and disturbing, even though much of his argument is based on the characteristics of *computer models* of complex systems. Even if only *some* real-world social and environmental systems have some of the counterintuitive properties that he alleges, there is a much greater possibility of environmental catastrophe. As we noted above, and discuss in detail in Chapter 12, Forrester's argument is more convincing if it is true that the economic and political mechanisms that humans have relied on in the past to live with complexity are thwarted by the counterintuitive properties he claims characterize complex systems. On a more positive note, forewarned about the counterintuitive behavior of complex systems we are forearmed. In addition, the development and use by policymakers of Forrester-type computer models may be a useful way for humans to better understand the behavior of environmentally relevant complex systems in order to predict and control them. But we're getting ahead of the story, which continues in Chapter 12.

NOTES

1. This characteristic is a feature of TMI and most, but not all, normal accidents.

2. Note that the *Transhuron* accident lacks one feature commonly found in normal accidents: incomprehensibility.

3. The decline in death rate (due to advances in medicine and public health measures) during the early 1900s in Western countries and more recently in less developed countries has resulted in *superexponential* global population growth, that is, exponential growth at successively higher rates (see Chapter 1). On the other hand, birthrates in Western countries have declined in the last few decades, and birthrates in many developing countries have declined in the last decade or so, thus lowering the rate of exponential growth.

4. Later in this chapter and in Chapter 12, we discuss the idea that traditional political and economic processes *do* in fact permit humans to successfully manage political and

economic systems too complex to understand. We will conclude, however, that, given the characteristics of the global and regional environmental problems our species now faces, these processes are no longer reliable for preventing environmental catastrophe.

5. Forrester (1980) has written that because of the high degree of nonlinearity found in complex economic, corporate, urban, etc. systems, no set of general properties is likely to apply to virtually all such systems all of the time. He does, however, think that the counterintuitive properties reviewed below are reasonably general and that it is a useful rule-of-thumb to expect most complex systems to exhibit these properties.

6. This description oversimplifies the policy Forrester, 1969, recommends, based on his computer model. The description also appears to ignore the plight of unemployed people living in other areas.

CHAPTER 12

HUMAN INTERACTIONS WITH COMPLEX SYSTEMS: CHAOS, SELF-ORGANIZATION, AND THE GLOBAL ENVIRONMENTAL FUTURE

CHAPTER PROLOGUE

This chapter continues and completes our discussion of human interactions with complex systems, which we began in Chapter 11. The first half of the chapter explores the possibility that environmental systems may exhibit *chaos*—a type of wild, unpredictable, and potentially disruptive behavior—in response to human activity. The second half centers on three properties of the global system that tend to make it unstable and increase the possibility of environmental disaster: Certain long time delays, exponential growth, and the possibility of catastrophic and irreversible damage to the global environment. The chapter ends with a discussion of a broad range of strategies that might help our species avoid catastrophe and live in harmony with our environment into the twenty-first century and beyond.

BEYOND "CONTROL": DETERMINISTIC CHAOS, ANTICHAOS, AND SELF-ORGANIZATION IN COMPLEX SYSTEMS

Introduction; Deterministic Chaos and Noncontrol

We begin with a brief scenario about future advances in computer modeling of complex systems. It is a scenario that many researchers and scholars now believe to be highly improbable.

"Progress in Computer Modeling of Complex Urban and Environmental Systems." It is a decade and a half into the future. Much work by computer experts and policy makers has led to greatly improved computer models of complex social and environmental systems. These models are now in widespread use. Indeed, it would be difficult to find a city anywhere in the world with a population above one million that does not have a computer modeling team and does not rely on models for choosing policies. As a result, most major cities have been able to improve conditions considerably. There are fewer areas of inner-city decay in U.S. cities. The quality of urban life has been edging upward.

Computer modelers also understand global and regional ecosystems significantly better. Policy makers of almost all countries rely on the advice of these modelers to guide their environmentally relevant decisions, and the world has edged away from environmental catastrophe. Human intelligence, creativity, and spirit have proven the equal of the serious environmental problems that threatened human existence.

Implicit in the scenario above and in the work of Forrester on complex urban, corporate, and ecological systems, which we reviewed in Chapter 11, is the idea that humans should interact with these systems through management based on understanding and control—that we should understand the behavior of these systems, with the help of computer models, and use that understanding to control them. Forrester assumes that it is possible to gather enough statistical information and scientific knowledge about complex systems; to build computer models that capture their properties; to use the models to test different policies to correct a problem or reach some goal; and then to enact those policies with success.

This view of management and control of complex urban and other systems, however, has been abandoned by many scholars and researchers in several policy-related disciplines (Dobuzinskis, 1992). A radically different paradigm or way of thinking about the dynamic properties of complex systems is gaining prominence. This paradigm first emerged in physics and mathematics, and then made major inroads into biology, medicine, meteorology, ecology, and other

natural sciences (Gleick, 1987). It has now gained acceptance among some social and behavioral scientists (Dobuzinskis, 1992; Barton, 1994). This paradigm draws on concepts and principles often referred to as *chaos theory, nonlinear systems dynamics,* and *antichaos* or *self-organization* (Kauffman, 1991; Townsend, 1992; Gregersen and Sailer, 1993; Costanza et al., 1993; Barton, 1994; Goerner, 1994).

A central tenet of the new paradigm is that physical, biological, social, and other systems can exhibit radically different behavior depending on the conditions under which they are operating. *Under certain conditions, these systems can behave chaotically.* That is, they exhibit unstable behavior characterized by wild and erratic swings. A system behaving chaotically never exactly repeats itself nor does it ever settle down to some constant and optimal level of activity. Further, its behavior over time is impossible to predict with precision, though the general range of the behavior can be ascertained. What is striking about chaotic behavior is that it can be produced in a *deterministic system,* that is, a system that conforms to basic scientific laws, for example, a system of mechanical parts that conforms to the principles of Newtonian mechanics—or a system described by the kinds of equations in Forrester's (1969) or Meadows et al.'s (1992) computer models. However, under *other* conditions, systems that can exhibit chaos will behave in an orderly, predictable way: They may go rapidly to a stable and high level of functioning; oscillate regularly between, say, a high and medium level of functioning; or quickly go to a constant low level of functioning (e.g., Mosekilde et al., 1990; Gregersen and Sailer, 1993). Importantly, the conditions that produce chaotic behavior in a given system often vary from those that produce stable or periodic but predictable behavior *by only tiny degrees.* Thus, starting conditions that are almost identical can produce radically different forms of system behavior (e.g., Gleick, 1987; Mosekilde et al., 1990; Townsend, 1992; Gregersen and Sailer, 1993).

A well-known hypothetical example of the above phenomenon is the "butterfly effect," so named by meteorologist E. Lorenz (Townsend, 1992). Lorenz suggested that some tiny influence on wind and weather in one part of the world (e.g., butterflies flapping their wings in Brazil) might cause a radically different weather pattern to occur in some other part of the world than would otherwise occur (e.g., a tornado in Texas).

The instability and unpredictability of complex systems is large enough, according to chaos theorists, to seriously interfere with human efforts to predict and control the behavior of most of these systems. Chaos theory thus tends to undercut the belief in human mastery and control, especially control over natural systems, a basic tenet of Western science and the Western traditions of humanism and positivism (e.g., Ehrenfeld, 1978), as we discussed in Chapter 3. And chaos theory contradicts the assumption that humans, *even with* the aid of computer models, can successfully understand and manage many complex systems. This is not to say that humans can't exert a positive influence on complex systems. Analysts who appreciate the possibility of chaos nevertheless believe that actions of policymakers may help guide an unstable system in ". . . situations where a radical break in existing patterns is on the verge of happening in order to avoid the least desirable scenarios (Dobuzinskis, 1992, p. 365)." Further, some chaos theorists believe future advances will improve their ability to understand systems that exhibit chaotic behavior (Townsend, 1992).

To go a step further, scholars and researchers who study chaos emphasize that even *relatively simple* systems can exhibit chaotic behavior (e.g., Gordon, 1992; Townsend, 1992; Gregersen and Sailer, 1993). Thus, first-order systems, that is, systems with a single independent variable, can exhibit the patterns we have described above. The main characteristic responsible for chaotic behavior is *nonlinearity*. Recall from our discussion in Chapter 11 that Forrester emphasized nonlinearity as a main source of the unpredictable and counterintuitive behavior of complex social and other systems. The important role of nonlinearity in both chaos theory and in Forrester's framework shows that there is overlap between the two.

An Example of Chaotic Behavior in a Simple System. Gordon (1992) provides a good hypothetical example of chaotic behavior in a very simple business or management system. The example centers on the use of advertising to boost sales of a product. The essence of the example is shown in the sequence of Figures 12-1 to 12-4. Figure 12-1 shows the nonlinearity in the simple system. Figures 12-2 to 12-4 show how the behavior of the system varies dramatically as a function of the initial conditions (in this case a company's choice of advertising strategy); small changes in policy produce radically different resulting behaviors. Figure

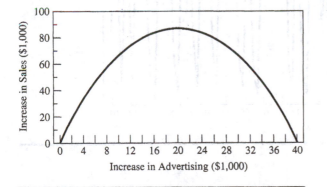

FIGURE 12-1 Increase in Advertising versus Increase in Sales

From Gordon, T. Chaos in social systems. *Technological Forecasting and Social Change*, Volume 42, 1–15. Copyright 1992 by Elsevier Science Inc.

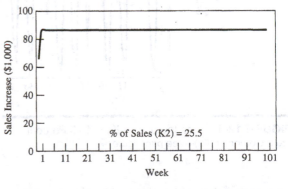

FIGURE 12-2 Sales Increase—K_2 = 25.5 Percent
From Gordon (1992)—see prior figure.

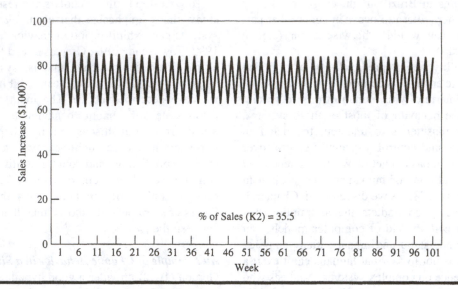

FIGURE 12-3 Sales Increase—K$_2$ = 35.5 Percent
From Gordon (1992)—see prior figure.

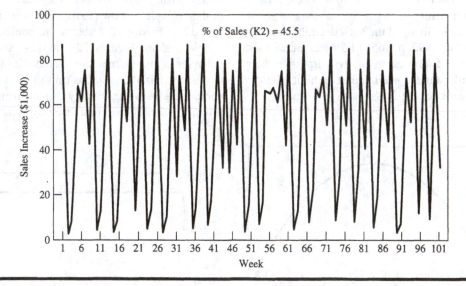

FIGURE 12-4 Sales Increase—K$_2$ = 45.5 Percent
From Gordon (1992)—see prior figure.

12-4 shows the wild and unstable, that is, chaotic, behavior that this simple system exhibits under certain conditions (i.e., for certain advertising strategies). Here is the example in detail:

Company V sells vacuum cleaners and signs a contract with an advertising agency to create and broadcast TV ads designed to boost sales. Rather than just having a fixed budget to buy the same number of thirty-second TV ad spots per week or month, the company and the ad agency agree to *change* the number of TV ad spots *each week*, depending on sales of vacuum cleaners the prior week. More specifically, Company V and the ad agency agree that the company will increase its spending on TV spots in a given week by some fixed percentage—25 percent—of the sales increase the prior week. Thus if, in the past week Company V sold $2,000 more worth of vacuum cleaners than the week before, then the company agrees to spend $500 more on TV ad time this week than the week before. The advertising agency says to the vacuum dealer, "You can't lose [with this arrangement] . . . , you'll only spend more money [for advertising] when our advertising brings you higher sales [p. 5]."[1]

The next detail in the example is the critical one; it is a nonlinearity that is responsible for the chaotic behavior the system can exhibit. This nonlinearity involves the relationship between increases in the number of TV ads and the increases in sales of the product that result. This relationship is shown in Figure 12-1. Note the inverted U-shape curve in the figure: As the number of TV ads for a product increase, so do sales, but only up to some point, at which the market begins to get saturated with the product. This is a point of diminishing returns beyond which additional ads produce much smaller additional sales. (The curve shown in Figure 12-1 is very similar to the one in Figure 11-8B of Chapter 11 that we used as an example of a nonlinear relationship.)[2]

In one possible scenario, the vacuum cleaner company wants only a modest amount of new advertising and proposes that only 25 percent of the sales increases should be put into advertising increases, as in our discussion above. Further, the company agrees to spend $10,000 for TV spots the initial week. Figure 12-2 shows what would happen if this 25 percent policy (with a $10,000 start) were to be maintained unchanged for 102 weeks (actually, Gordon, 1992, happened to use a value of 25.5 percent to calculate the plot shown). Sales increases per week quickly go to around $87,000 dollars and stay there, a very stable pattern of behavior. Figure 12-3 shows the results in another scenario in which 35 percent (actually 35.5 percent) of new sales are devoted to additional adver-

tising for a 102-week period. Sales increases per week fluctuate back and forth over time between around $63,000 and $87,000, a pattern of orderly periodic behavior. Figure 12-4 shows the results of yet another scenario, one in which 45 percent (actually 45.5 percent) of new sales are devoted to additional advertising for a 102-week period. Note how wild and erratic is the resulting pattern of sales increases per week. This is *chaotic behavior.* In summary, depending upon the specific percentage of new sales spent on additional advertising, the resulting pattern of sales increases over time can vary radically and it becomes chaotic at some percentages.

All of the above scenarios assume that the percentage of new sales revenue that goes for increased ads remains stable for a 102-week period. But the vacuum cleaner company might well change the percentage. For example, the company might start with the modest 25 percent figure, as mentioned above. In response to the resulting strong sales increases for the first few weeks (the left-hand part of Figure 12-2), the company might boost the percentage to 35 percent. This would produce the periodic behavior shown in Figure 12-3. Thinking that this behavior is due to some sort of outside influence like some type of national economic fluctuation and still being eager for more increased sales, Company V might try the higher percentage of 45 percent and then encounter the chaotic behavior shown in Figure 12-4. At this point Company V might conclude that there is something seriously wrong with the ad agency and the advertising strategy and decide to take its business elsewhere. Again, the different patterns of behavior—a stable high level of functioning, oscillation, or chaotic behavior—result, not from changing outside influences or from inconsistent ad agency actions, but entirely from inside the system.

We present a somewhat more complicated example of chaotic behavior in a simulated business system in Box 12-1 on The Beer Game.

Processes That Counter Chaos

What does the possibility of chaos in simple social systems imply for the ability of humans to manage the environment? It does *not* necessarily mean that

Box 12-1

Chaotic Behavior: The Beer Game

Here is another example of a simple system that can exhibit chaotic behavior. The system is a network of four businesses that manufacture, distribute, and sell beer. Sterman (1988), Mosekilde et al. (1990), and others have studied the way people manage this system in laboratory and classroom settings; in other words, the people are playing a game or simulation rather than attempting to manage a real-world system. But, though the setting is artificial, and the "beer" subject matter prosaic, the results are quite interesting.

Researchers and instructors at M.I.T. have used the beer game in classes and seminars on business for several years. The game ". . . has been played . . . by thousands of people ranging from high school students to chief executive officers and government officials (Mosekilde et al., 1990, p. 129)." The general structure of the game is depicted in Figure 12B1-1 which shows the four businesses and the flow of products and information between them. Keep in mind that each business is managed by one of the four people who play the game at the same time: The retailer sells beer to retail customers out of store inventory and orders and receives more beer from the wholesaler; the wholesaler sells beer to retailers out of its inventory and orders/receives more beer from the distributor; and so on, until we reach the brewery, which sells beer to the distributor out of its own inventory and which also manufactures new beer. The managerial decision each player makes every "week" (each "week" is round of play that lasts a few minutes) is how much beer to order (or, in the case of the brewery, how much to manufacture). Small tokens are used by the players on the game board to represent cases of

beer. Note that there are delays of one to three "weeks" between the time orders are placed and when the ordered beer is received.

The four-part arrangement in Figure 12B1-1 is actually a common one in the business world. The four levels of inventory help ensure that enough of the product is on hand to meet retail demand, despite fluctuations over time in that demand (Mosekilde et al., 1990, p. 128). Each player has two explicit goals in the simulation: To keep enough beer in inventory to meet customer demand (players are "charged" $1 per case per week for orders they can't fill because of low inventory), and to avoid having too much inventory sitting in the warehouse, not generating revenue (players are charged $0.50 per case per week for idle inventory). One final important detail: The professor running the game arranges it so that total retail customer demand is a constant four cases per week for the first four weeks of play, but total retail demand doubles to eight cases per week on the fifth week and stays at that level for the remainder of the game (thirty-five more weeks of play).

The results we'll describe below come from a specific study by Mosekilde, Larsen, and Sterman (1990) involving forty-eight games, played over a period of four years by 192 participants, including ". . . undergraduate and graduate students at the Sloan School of Management and senior executives from major U.S. firms [p. 138]." One surprising result is that most teams of players had great trouble managing this straightforward system successfully. The average subject suffered ten times the charges (for orders they couldn't fill and/or for unused inventory) than what the

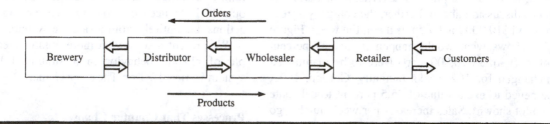

FIGURE 12B1-1   Structure of the "Beer Game"
From Mosekilde, E., Larsen, E., and Sterman, J. Coping with complexity: Deterministic chaos in human decision making behavior. In Bohr, H. (Ed.), *Proceedings of the symposium characterizing complex systems: Interdisciplinary workshop on complexity and chaos.* Copyright © 1990 World Scientific Publishing Co. Pte. Ltd.

mathematically optimal strategy, according to simple equations, would produce. Furthermore, the management (ordering) strategies of almost all subjects produced wide swings in inventory level, from much too little beer on hand to much too much on hand, as Figure 12B1-2 illustrates (note that the game is played for forty weeks but that a computer analyzes and then mimics the subjects' behavior for weeks 41–60, assuming that the players keep using the same management strategies they used before the forty-first week).

Figure 12B1-3 illustrates a second major finding: When the management strategies of the subjects were analyzed and then mimicked by the computer and *run continuously for hundreds of weeks of play*, the strategies used by about one-fourth of the subjects led to *chaotic* system behavior (the wild and erratic patterns shown in Figure 12B1-3). Chaotic behavior, in general, seems to be caused when parts of a system (especially positive feedback loops) cause small perturbations or irregularities to be greatly magnified, producing large changes in the system's overall behavior. This is what was happening in the beer game: When the players used certain management strategies, the one-time doubling in retail demand at week five was amplified into extremely

disordered and unstable swings of inventory levels over many weeks; under these management strategies, used continuously, the four-player beer production/distribution system never became stable. Similar "beer game" results have been reported by other researchers (Sterman, 1988).

The computer simulations also indicated that very small changes in players' management strategies could produce dramatically different results: One type of management strategy produced chaotic system behavior while a slightly different strategy produced behavior that was quite stable and regular—a universal pattern in systems that can exhibit chaotic behavior, according to

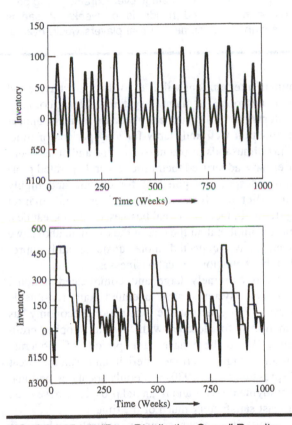

FIGURE 12B1-3 "Beer Distribution Game" Results: *Extended* Computer Simulation of Inventory Levels for Each of Four Players
From Mosekilde et al. (1990)—see prior figure.
Top figure assumes the same management strategies as in Figure 12B1-2. Bottom figure assumes a different specific set of management strategies.

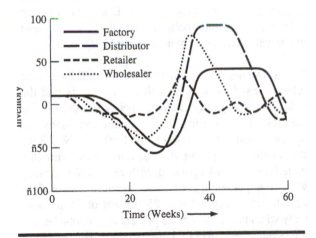

FIGURE 12B1-2 "Beer Distribution Game" Results: Computer Simulation of Inventory Levels for Each of Four Players Using a Specific Set of Management Strategies
From Mosekilde et al. (1990)—see prior figure.

(continued)

BOX 12-1 continued

chaos theorists (as we discuss in the main text). In short, even this apparently simple four-person managerial system has very complicated properties and can exhibit chaotic behavior, and intelligent human players can have a great deal of trouble managing it successfully.

It does not follow from these simulations results, however, that people can't manage real-world business systems. The players were not given much time to learn and master the simulated beer distribution system—only 40 "weeks" or trials. Furthermore, chaotic system behavior appeared only when the computer simulation assumed that players maintained the same, unchanging, poor managerial strategy over extremely long periods—hundreds and hundreds of weeks. It seems reasonable to assume that real players would eventu-

ally learn from experience, modify their suboptimal managerial strategies, and do much better, producing stable system behavior. Indeed, Sterman (1990) notes that in some game/simulations of this type, definite improvements with practice have been found to occur (p. 174). Similarly, it seems obvious that existing real-world beer distribution and other straightforward managerial systems work reasonably well.

Mosekilde et al. (1990), however, point out that many real-world managerial, economic, governmental, and ecological systems that humans influence and/or attempt to manage are far more complicated than the one in the beer game; Mosekilde argue that humans may not be able to learn to manage many of these systems successfully.

---

human interventions will lead to chaotic behavior and to catastrophe in natural systems. Many complex systems, both in nature and made by humans, appear to function well and stably for long periods of time. Thus, thousands of businesses in this and other countries have advertised their products and operated profitably despite the potential for chaotic and highly unpredictable behavior in business and market systems. In fact, since real businesses are a great deal more complex than the one in the example above, we would have expected more unstable and unpredictable behavior for real businesses.

Even extremely large and complicated national economic systems have operated with reasonable success. For example, the U.S. national economy has functioned for decades with few catastrophic problems. While there have been periods of moderate inflation, high government deficit, and even the Great Depression of the 1930s, generally speaking, national employment levels and financial prosperity have been at least satisfactory much of the time.

Similarly, various complex natural mechanisms and processes have operated stably and well for long periods. Billions of complex organisms (plants and animals), with adequate food and other requisites, can maintain their physical integrity for years and produce viable offspring. Biological evolution has produced millions of complex plant and animal

species that are well adapted to their ecological niches. Hugely complicated regional and global ecological systems and processes have operated well for millennia.

So, the observations of chaotic behavior in complex systems leave us with some major unanswered questions: Are local and especially global environmental systems likely to behave chaotically in response to human activities? Does the chaotic behavior have catastrophic potential? If so, how can we prevent environmental catastrophe?

Chaos Occurs Only under Some Conditions. One relevant insight from chaos theory is that despite the wild, dramatic quality of chaotic behavior, systems that manifest it do so only under certain conditions but otherwise can operate stably and effectively. Recall that in the example of the vacuum cleaner firm the underlying system operated with reasonable stability when the percentage of sales increases devoted to advertising increases[3] was 25 percent or 35 percent. Only when the value was 45 percent did chaotic behavior occur. This overall pattern generalizes to other systems, even much larger and complex ones (Gregersen and Sailer, 1993). Thus a complex system may operate for lengthy periods without being exposed to the conditions and circumstances that produce chaotic behavior.

However, the problem with this insight is that we know very little about the conditions that produce chaotic behavior in particular environmental systems, so—at least with present and foreseeable levels of knowledge about the environment—we have no reliable scientific guidance on what to do to prevent chaos and the catastrophes that may result.

Two lines of thinking suggest solutions to this problem—ways that humans and human societies can function effectively within complex systems that are too intricate to understand and that are susceptible to chaotic behavior under certain conditions. We discuss them in the next two sections.

Disjointed Incrementalism: Social Systems Manage Themselves. One line of thinking about how to live with complex and potentially chaotic systems comes from the social sciences. There appear to be political and economic mechanisms and institutions that can stably manage huge social systems for long periods of time. Humans have relied on these mechanisms and institutions throughout history for this purpose. Political scientist Charles Lindblom (1959, 1977) has argued that nations, cities, and other complex systems have been regularly and successfully managed by humans by means of what he calls disjointed incrementalism. Lindblom takes as a given that nations and cities could not possibly be managed directly by elected and appointed officials who tried to comprehensively understand how these systems worked, because the systems are far too complicated. Further, Lindblom claims that no one even *attempts* to run such systems this way. If Lindblom is correct, then Forrester's claim, which we discussed in Chapter 11, that humans cannot manage large, complex systems because the human mind cannot comprehend them is invalid, not to mention unnecessarily alarming, because there is an effective type of management that does not rely on complete understanding.

Disjointed incrementalism (DI), according to Lindblom, works as follows. *Incrementalism* involves the fact that officials trying to manage a complex unit of government or solve a problem in it do so by means of *small trial-and-error steps*. Rather than trying to understand the unit and its problems comprehensively, they make one small change in existing policy at a time (e.g., they increase pollution taxes by 3 percent) and then wait to see the effects. If, following a certain change in policy, things get better, they may try more of the same change, or they may stop if the problem has been alleviated. If not, they try some other small change and wait to see its effects, and so on. (The principle of monitoring and adjustment in Chapter 7 is based on evidence that an incremental approach can work well in small environmental programs.)

The word *disjointed* refers to the fact that rather than there being a single government authority that tries to comprehensively manage a large system, responsibility for the system is divided up among many smaller agencies and departments. Each one is responsible for only a small part of the system. In addition, each of these small agencies and departments receives inputs from various constituents, interest groups, and lobbyists whose own interests are affected by the decisions of that agency or department. To quote Lindblom:

> In the U.S., for example, no [single] part of government attempts a comprehensive overview of policy on income distribution. A policy nevertheless evolves, and one responding to a wide variety of interests. A process of mutual adjustment among farm groups, labor unions, municipalities . . . , tax authorities, and government agencies with responsibilities in [various fields] . . . accomplishes a distribution of income [p. 85].

Thus a multitude of small agencies and departments plus their individual constituent groups, each with a circumscribed mission and view of the system (none with an overall knowledge of the system or responsibility for it) *and* each taking small trial-and-error steps, successfully manages the system. Indeed, the past successes of various large state and national governments provides evidence that DI can work.

Lindblom calls the success of DI in running large units of government "the intelligence of democracy." Other analysts see a magical, deus ex machina quality to DI: There seems no a priori reason that it should succeed, but, claims Lindblom, it does succeed.[4] The successes of DI may make more sense to the reader if we describe what Lindblom considers to be one type of DI: The "invisible hand" of free-market capitalism.

In free-market capitalism, supply and demand in the marketplace determine the prices of resources and commodities as well as their distribution. Each individual actor (buyers and sellers) merely pursues economic self-interest: Sellers try to get as much money for their commodities or services as they possibly can; buyers try to get commodities and services as cheaply as they can. To take a specific example, if fuel oil is in short supply, fuel oil dealers will charge as much for it as the market will bear. If the price of fuel oil goes up, customers, in turn, will buy less of it and compensate for the smaller amount of fuel oil by lowering their thermostats, insulating their homes, and/or using other sources for heat (e.g., a wood stove or an active solar device). Those customers with an extreme need for fuel oil (e.g., elderly people who can not easily lower thermostats or find alternative sources of heat) will be willing to pay more for it. And other individuals, seeing an opportunity to profit, will invest in developing and marketing substitutes for oil. The end result of all this is that society makes do with smaller quantities of the scarce resource, with millions of individual consumers being allocated an amount of the resource that is determined by their own individual situations, and with scarcity encouraging the development of alternative energy forms or conservation methods. All this is accomplished by individual actors, each pursuing narrow self-interest, and without the need for a central government body to try to orchestrate or manage this incredibly complex process. Indeed, it *is* almost as if there *were* an invisible hand that guided all these changes. Some analysts even argue that attempts by large central governments to manage the process above often make matters worse. (Of course, the unrestrained invisible hand has resulted in some elderly people's freezing to death and a number of other negative outcomes.)

Although disjointed incrementalism has apparently worked to keep large social and economic systems functioning in the past, we do not think it can be relied on to prevent environmental catastrophe in the future. The nature of the environmental problems that now threaten human survival violate key requirements that must be fulfilled if DI is to work properly.

One requirement is the possibility of trial-and-error learning on the environment. This requirement is contradicted by the long feedback delays that characterize regional and global environmental problems, and the possibility that errors of policy may cause irreversible and catastrophic damage to the environment. We devote the second half of this chapter to an in-depth discussion of these characteristics of environmental problems.

The second key requirement of DI is the possibility of disjointed management, which requires that officials, agencies, and national governments be able to find solutions to their problems from among the actions they have the power to take. If the symptoms and the true causes of problems in complex societal and ecological systems, however, are widely separated, as Forrester claims and as we discussed in Chapter 11, DI will break down. The essence of disjointed management is that many government agencies and departments, each with responsibility for one small part of the system, manage a large system. But such small agencies/departments cannot control the problems for which they are responsible if the real causes lie far away in some other agency's/department's area of responsibility. Similarly, if negative unanticipated remote consequences often follow the successful solution of a problem in one part of the system, as Forrester claims, DI will fail. Further, if any important environmental system is structured so that there is an intrinsic contradiction between the short-term and long-term solutions of a problem, then DI is in trouble.

Of course, Forrester has not demonstrated to everyone's satisfaction that most or all complex social and environmental systems actually display the characteristics above. But if these characteristics affect the Earth's crucial life-support systems *even some of the time,* there is reason to doubt reliance on DI. We believe there is compelling evidence that these characteristics *are* present some of the time.

To be sure, there are ways to modify traditional economic and political mechanisms to deal more effectively with environmental problems. For instance, free-market capitalism can be made less free, as when a government levies a hefty tax on fuel oil, rather than relying on unfettered market mechanisms to regulate the consumption and distribution of fuel oil; this is indeed one of the basic approaches we discussed in Chapter 5 under the heading of internalizing the externalities.[5] As we suggest in Chapter 7, a combination of changes will be required. The bottom line, however, seems clearly to be that "business as usual," that is, relying on traditional political and economic mechanisms to manage complex social/economic/ecological/et cetera systems, can no longer be trusted to work. We discuss several possible alternative strategies at the end of the chapter.

Antichaos and Self-Organization in Complex Systems. A second way of thinking about how humanity can live in a complex environment comes mainly from the biological sciences. It emphasizes forces and processes within complex systems that tend to oppose any tendencies for the systems to behave chaotically. Several scholars argue that there is a general, order-producing, self-organizing process in many systems (Kauffman, 1991; Dobuzinskis, 1992; Costanza et al., 1993; Ruthen, 1993; Goerner, 1994). The process bears some resemblance to disjointed incrementalism, which we discussed above. These scholars see the process as underlying biological evolution, cultural evolution, the behavior of large economic systems, and the behavior of ecosystems (e.g., Arthur, 1988 [referenced by Costanza]; Costanza et al., 1993; Maxwell and Costanza, in press). The self-organizing, order-producing process creates learning and adaptation over time. It has three key elements: A mechanism that produces diversity and new alternatives; a mechanism that selects and makes predominant those alternatives that are superior on some criterion or criteria; and a mechanism for the storage and transmission of information (Costanza et al. 1993). These three elements are all present in biological evolution as we described it in Chapter 8: Genetic mutations and sexual reproduction create genetic diversity within a species; Darwinian natural selection selects and makes predominant the genetic characteristics best suited to the particular environment the species occupies; and genetic material holds the relevant information in parents' genes and trans-

mits it to offspring. The result is the evolutionary sequence of the Earth's plant and animal species that has continued for eons without apparently resulting in catastrophe.

Some scholars see a spiritual and religious element in the concept of an order-producing, self-organizing, and even creative force underlying the evolutionary process, a concept that also may apply to the creation and regulation of Earth's ecosystems and even of the universe (Dobuzinskis, 1992; Goerner, 1994). Goerner (1994) writes:

> *Classical science painted a picture of a purposeless, passive universe. If one believed this image was fact, then all [religious] traditions that described order/purpose/direction were antithetical to science. This particular irreconcilable difference falls away once order building is seen to have a physical basis. . . . We and all "things" that we see are the net result of this creative drive. A sense of direction, of a process larger than ourselves and of a creative force is restored [p. 152].*

Similarly, Thomas Berry (1988) writes (as we quoted in Chapter 3):

> *[The new proenvironmental religion's] story [of creation will stress] the organic unity and creative power of the planet Earth . . . , the evolutionary process through which every living form achieves its identity and its proper role in the universal drama, the relatedness of things in an omnicentered universe [p. 34]. . . . We will experience an identity with the entire cosmic order within our own beings. This sense of an emergent universe identical with ourselves gives new meaning to the [Taoist] . . . sense of forming one body with all things. . . . That some form of intelligent reflection on itself was implicit in the universe from the beginning is now granted by . . . [a number of] scientists [p. 16].*

The general self-organizing process of learning and adaptation may also be seen as underlying *cultural evolution,* discussed in Chapter 8. Culture evolves because the practices, technologies, and values of prior generations are continually honed by the experiences of each new generation. Any components of a culture that don't contribute to the long-term survival of individuals and of the group get eliminated and components that contribute are retained. The storage and transmission of relevant information is not genetic or biological but via learning and language. But in addition, recall that cultural evolution operates in time periods much shorter than those of biological evolution and that it has, at least for several centuries, produced accelerating social change. Knowledge has accumulated in a snowballing fashion because new discoveries, information, or inventions often have more uses than their discoverers anticipated, and the more the diverse inventions, and so forth that exist, the more new ones that are possible via recombinations. A look back at Chapter 8 and Figure 8-1 shows the curve of cultural evolution over time, a curve resembling that of exponential population growth.

Note that the essence of the order-producing evolutionary process we've been discussing in this section—like that of disjointed incrementalism—is *trial-and-error experimentation and learning.* Its main components are diversity or variety followed by elimination or predominance based on some criteria of usefulness or value. But there's a rub with respect to cultural evolution: Its rapid, accelerative pace tends to subvert a trial-and-error learning process. Because new technologies come so quickly, because many technologies consume resources and generate environmental pollution, because humans now use technologies so intensively, and because there are so many of us on Earth, the "error" potential of cultural trial-and-error is becoming increasingly dangerous. To quote Costanza et al. (1993):

> The costs of . . . rapid cultural evolution . . . are potentially significant. Like a car that has increased speed, humans are in more danger of running off the road or over a cliff. Cultural evolution lacks the built-in long-run bias of . . . [biological] evolution and is susceptible to being led by its hyperefficient short-run adaptability over a cliff into the abyss [p. 551].

More optimistically, Costanza et al. (1993) (citing Arrow, 1962) point out that humans have foresight and may avert the cliff. Further, a closer look at the process of cultural evolution reveals elements that may enhance our foresight and enable us to respond to dangerous developments before they overwhelm

us. Thus, Costanza et al. describe a model of evolution proposed by C. Holling (1987, 1992). Holling argues that all evolving complex systems—be they ecosystems or human societies—go through four basic stages over time in a cyclical, repetitive fashion: *Exploitation, conservation, release,* and *reorganization.* First, in the *exploitation stage,* "pioneers, opportunists, and entrepreneurs" explore, colonize new territory and "capture easily accessible resources." Next, the system creates structures, mechanisms, or institutions that regulate the rate of resource exploitation to establish some stability in the face of potential shortage (the *conservation stage*). Eventually (in the *release stage*) inevitable disruptive events, like "fire, storms, pests, or political upheavals" occur, destroying the stability, and damaging the structures or institutions of the conservation stage. The remnants of this destruction are then reorganized and available for renewed exploitation (the *reorganization stage*), and the cyclic process starts again.

Holling's *release stage*—which is also called the *creative destruction stage*—is an especially important component of his model. Quoting Costanza et al.'s (1993) description:

> The amount of ongoing creative destruction that takes place in the system is critical to its behavior. The conservation phase can often build elaborate and tightly bound structures by severely limiting creative destruction (the former Soviet Union is a good example), but these structures become brittle and susceptible to massive and widespread destruction. If some moderate level of release is allowed to occur on a more routine basis, the destruction is on a smaller scale and leads to a more resilient system. It could be argued that patterns of behavior with moderate levels of ongoing creative destruction evolved in those local communities and human cultures that managed to survive for thousands of years [p. 552].

Costanza et al. offer an example of the desirability of moderate creative destruction. The example involves the large fire in Yellowstone Park in 1988 that burned almost half of the park's area (Peterson, 1993). The fire was so destructive because park officials had previously suppressed the natural fires caused by lightning. If not suppressed, these natural fires occur so often that they are small in area and rarely destroy large quantities of old growth timber; instead, they release nutrients from forest-floor litter and spur new growth. If these fires are suppressed, as they were in Yellowstone, the quantity of flammable material builds up to such high levels that huge and destructive fires, like the ones in 1988, become more probable.

Pursuing the idea of the desirability of moderate "creative destruction" further, Costanza et al. (1993), citing Berkes and Folke (in press), propose that environmental accidents like the chemical spill in 1986 on the Rhine River (a river which provides drinking water to 20 million Europeans) that destroyed all plant and animal life in the upper parts of the river, and the nuclear accident at the Chernobyl power plant in the Ukraine in 1986, which we discussed earlier in the chapter, ". . . may stimulate positive change toward more resilient ecological-economic systems [p. 552]."

However, it is difficult to be confident and optimistic ("it's all part of learning") about an accident like the one at Chernobyl, given its severity and the large number of people whose lives it negatively affected. If cultural evolution is a trial-and-error learning and adaptation process, then some errors, like Chernobyl, begin to appear too costly and unacceptable. Further, there are even larger, more catastrophic, and irreversible environmental accidents now possible, making a trial-and-error learning process unworkable.

The possibility of accidents of this type is the centerpiece of work by Donella Meadows, Dennis Meadows, and Jørgen Randers (1992). We begin the next half of the chapter with an in-depth examination of their work and the related work of others.

A BROAD LOOK AT THE GLOBAL ENVIRONMENTAL FUTURE

We devote this part of the chapter to a broad look at environmental problems at the global level, and at the characteristics of these problems that make trial-and-error, disjointed, "business as usual" increasingly problematic. These include the possibility of catastrophic and irreversible environmental accidents;

long time delays intrinsic to certain environmental systems, including delays between the time that human activity damages the environment and the time at which that damage becomes apparent; and the problem of exponential growth in human population and in the resource- and pollution-intensity of human activity.

We first examine the above three characteristics more closely, drawing on work by Meadows et al. (1992) and others. We next describe the specific "overshoot and collapse" disaster pattern to which the global system, according to Meadows et al.'s (1972, 1992) controversial computer model, is especially prone. We then review criticisms of Meadows et al.'s original computer model and describe subsequent global modeling efforts by several other teams of modelers in various countries. We argue that this subsequent work supports Meadows et al.'s general, qualitative conclusions about the global environmental future. Finally, we consider the changes in human institutions and actions—ranging from the simple and straightforward to the revolutionary—that Meadows et al. and others argue are necessary to prevent a global environmental disaster from occurring.

Three Key Characteristics of the Global Environmental System That Increase the Probability of Environmental Disasters

Long Time Delays. Several types of time delay make it difficult for humanity to learn from its environmental mistakes, and thus contribute to the instability of the global system. These include: Intrinsic delays between the time at which human activity does environmental damage and time at which the damage becomes visible (a delay in feedback to policy makers and others that human activities are causing damage); time delays between the receipt of this feedback and action by policy makers and others to ameliorate the damage; and time delays between such action and its positive environmental effects.

Here are some examples. First, the immense size of global stocks of both renewable resources (e.g., forests) and nonrenewable natural resources (petroleum and minerals) make it possible for human activity to draw down these stocks at rates that are unsustainable in the long run without causing much visible damage in the short run. Similarly, there are important time delays between the release of pollutants into the environment and their actual negative impacts. For example, it takes fifteen to twenty years for CFCs released from air-conditioning and refrigeration equipment on the Earth's surface to actually begin destruction of ozone in the Earth's stratosphere. Likewise, it can take decades for pollutants to seep down into aquifers and contaminate drinking water supplies. For example, environmental scientists now know that a long-lasting chemical contaminant (DCPe) in a soil disinfectant used widely in Dutch agriculture beginning in the 1960s will infiltrate drinking water supplies at concentrations fifty times permissible levels, but this won't occur until after the year 2000 (Meadows et al. 1992). Clearly, time delays of this type temporarily mask the severity of pollution problems. These delays also mean that antipollution measures can come too late to stem damage that is already locked into the system.

Another type of time delay occurs between actions policy makers and others take to lessen global problems and the positive effects of those actions. A good example is the *population momentum* phenomenon: Even if two children per woman were to become the norm all over the world tomorrow, global human population would still continue to grow substantially for the next sixty to seventy years before actual zero population growth (ZPG) would be reached. (We discussed population momentum in Box 1-2 in Chapter 1; the basic idea is that a population that is growing—one with, say, four children per woman—is "bottom heavy" with children who have not yet had their own children, and grandchildren, etc.). A different type of time delay occurs in changing technology. Consider the time it takes to upgrade the energy efficiency of millions of homes and factories in a country, and the longer time it takes to bring in improvements that are only feasible when new buildings are constructed. Or, consider the time needed to replace old infrastructures based on nonrenewable energy sources (fossil fuels) with new ones based on solar or other renewable energy sources.

Exponential Growth. The second key characteristic that contributes to the instability of the global system is exponential growth in the human population and in its use of the environment (see the discussion of exponential population growth in Chapter 1, and the discussion of the positive feedback loop aspect of population growth in Box 11-2 in Chapter 11). The global population is now growing at an exponential rate of 1.4 percent a year (Miller, 2002; Postel, 1994). Although this rate of growth has decreased significantly below the all-time high of 2.06 percent in 1970, it means that global population has a doubling time of fifty-two years (Miller, 2002). This rate of growth means, more concretely, that the global population increases by the equivalent of a new New York City in one month.

Biologists, ecologists, and other environmental scholars and writers go to great lengths to convey to the readers of their books and articles the explosive properties of exponential growth. Indeed, these scholars and writers *themselves* sometimes have trouble appreciating how explosive exponential growth is! Consider the following passage from a speech given by Thomas Lovejoy, a noted tropical ecologist and officer of the Smithsonian Institution, at a meeting of the American Institute for Biological Sciences in 1988:

> *I find to my personal horror that I have not been immune to naivete about exponential functions. . . . While I have been aware that the interlinked problems of loss of biological diversity, tropical deforestation, forest dieback in the Northern Hemisphere and climate change are growing exponentially, it is only this very year that I have truly internalized how rapid their accelerating threat really is (Quoted in Meadows et al., 1992).*

Exponential growth is insidious because although it starts slowly, it becomes extremely rapid in a surprisingly short time. Since exponential growth accelerates constantly, it becomes harder to cope with the longer it has gone on: If you turn back to the graph of exponential growth in Figure 11-9 (Chapter 11) and compare points T1 and T2 you'll see, once again, that society has a much more serious problem at T2 than at T1 because growth is much more rapid (and it is more rapid still at later points in time). Exponential increases in human demands on the Earth's resource base and on its environmental systems, whether caused by increasing human numbers, affluence, technology, or anything else, has this property. Such exponential growth tends to simultaneously stress resources and environmental systems, and also a country's industrial base and infrastructures. It means that all of the following must be provided at increasingly rapid rates: Food, housing, highways and other transportation facilities, electricity, other utilities, medical care, schools, other public facilities, methods of handling pollution, and so on.

The Possibility of Irreversible Damage. The third key characteristic of the global system that contributes to its instability is the fact that human-induced damage to the Earth's environmental and resource systems can last for centuries or even be totally irreversible. Thus, damage to the ozone layer, which shields the Earth from ultraviolet radiation, will take centuries to completely reverse by natural processes (Miller, 1994). Further, certain toxic substances, such as lead and mercury, once dispersed into the environment are permanent because no natural processes can break them down (Miller, 1994). When plant or animal species become extinct as a result of human activity, and when nonrenewable resources such as petroleum and minerals are exhausted, the losses are permanent. Even in the case of potentially renewable resources, damage beyond a certain point is for all practical purposes irreversible. Irreversible damage, for example, occurs when agricultural topsoils blow away due to poor farming methods, when people clear-cut tropical rain forests growing in certain kinds of soil that do not support regrowth, and when salt and pollutants from human activity contaminate underground aquifers.

Note, further, that many of the processes described above that degrade the Earth's environment are especially insidious because they involve vicious circles and also nonlinearities such as thresholds. As an example of a vicious circle, consider the process that can turn productive lands in arid regions of the third world into deserts, according to Brown and Postel (1987), Postel (1994), and others (we discussed this briefly in Chapter 1). As human populations grow in size in these regions, the intensity of human use of the land increases: More and more people cut more and more trees for firewood; farmers, in efforts to feed

hungry mouths, allow livestock to overgraze grasslands; farmers also plant crops in marginal areas subject to erosion and do away with fallow periods. These practices denude the land of vegetation, decrease soil fertility, and cause erosion of topsoil. The result can be the irreversible transformation of productive land into a barren desert. Of course, the more desert area created, the more intensely people must use remaining areas that are productive; this destroys still more land, and so on in a vicious circle.[6] Often threshold effects and other non-linearities (see Figures 11-8B–D) aggravate these vicious circles by masking the damage until it is too late. For example, topsoil on farmlands can be slowly eroded with no apparent decrease in crop yield until the depth of the topsoil is less than that of the root zone of the crop; it is only when this occurs that there is a precipitous drop in yields (Meadows et al., 1992).

The Potential of the Global System for Overshoot and Collapse: the Meadows et al. Computer Model

According to Meadows et al. (1972, 1992), the three above characteristics—long time delays, exponential growth, and the possibility of irreversible damage—make the global system especially prone to what they call an "overshoot and collapse" disaster. They base this claim, in part, on their global computer model, which we briefly described in Chapter 11 (Figure 11-7). The Meadows et al. (1972, 1992) model is based on an earlier one by Forrester (1971). Figure 12-5, from Meadows et al. (1992), illustrates the overshoot and collapse pattern. Over time, both human population and the resource- and pollution-intensity of human activity grow exponentially in the model (solid line). Eventually, however, population, resource use, and environmental pollution reach levels that exceed what Meadows et al. claim to be the sustainable carrying capacity of the Earth (dotted line). (Note that the concept of a *fixed* carrying capacity of the Earth for humans is controversial, as we discuss below.) However, the immense size of global natural resource stocks and the intrinsic time delays between human actions and the actual environmental damage caused by those actions permit a temporary overshoot beyond the

Earth's long-term carrying capacity. During the overshoot period, irreversible damage is done to global resources and the environment, damage that, according to the computer model, dramatically lowers the carrying capacity of the planet. This decrease in carrying capacity, in turn, causes a large and sudden drop in human population and in the quality of life—a disaster of unprecedented proportions.

Meadows et al. argue that if major changes are not made soon, humans and human institutions will not be able to steer the world away from overshoot and collapse. To quote and paraphrase Meadows et al. (1992):

> Any population-economy-environment system that has [the three characteristics above] . . . is . . . [unstable, and headed for disaster]. No matter how brilliant its technologies, no matter how efficient its economy, no matter how wise its decision makers, it simply can't steer itself away from [the] hazards. . . [p. 138].

Keep in mind, however, that Meadows et al. (1992) are also saying that overshoot and collapse can be avoided if we take the proper actions. What are the crucial changes they think are required? We discuss

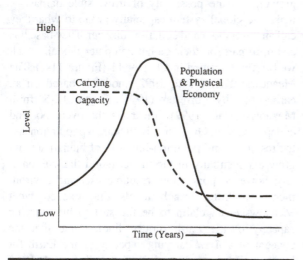

FIGURE 12-5 Overshoot and Collapse
Reprinted from *Beyond the Limits,* copyright © 1992 by Meadows, Meadows, and Randers. With permission from Chelsea Green Publishing Co., White River Junction, Vermont.

them near the end of the chapter. But we should say one thing about the changes now. Meadows et al. are mainly talking about radical moral, economic, political, and social changes. They are *not* primarily talking about changes and advances in technology, management, or computer modeling. Though Meadows et al. (1992) are disciples of Jay Forrester (Meadows et al., 1992, dedicate their book to Forrester), they are not advocates of the "human management and control" paradigm implicit in Forrester's computer modeling work on corporations and urban areas. Rather than trying to better understand, manage, and control the global environmental system, rather than more technocratic business as usual, rather than making a series of minor policy changes, they are arguing that humanity must reverse its course and pursue radical new goals. But we're getting ahead of the story.

Criticisms of Meadows et al.'s Global Modeling Work; General Conclusions That Appear to Be Sustained by This Work

Meadows et al.'s claims about the direction in which the global system is headed are dramatic and frightening. But remember these claims are based on Meadows et al. controversial computer model of the global system (Figure 11-7). In this section, we examine some of the criticisms leveled at Meadows et al.'s initial modeling work, describe later modeling work both by Meadows et al. and others, and consider what general conclusions, if any, we can draw from all this work.

Meadows et al. first presented their global computer model and discussed its implications in a book titled *Limits to Growth* (1972). Their main conclusion, as we have just discussed, was that the world was headed for an overshoot and collapse disaster. Meadows et al. (1972) argued specifically that if then-current trends continued, levels of population and human activity would exceed the Earth's ecological and resource limits within 100 years. (As we discuss below, in their more recently published book, *Beyond the Limits* (1992), Meadows et al. argue that these limits have already been exceeded.)

The book *Limits to Growth* was met by both strong praise and intense criticism. Recall that Forrester's (1969, 1971) similar, earlier attempts to computer-model corporate, urban, and global systems were already controversial: Forrester's overall approach to modeling social, economic, environmental, and so on systems had been criticized by some scholars and writers, as had specific details of his individual models (Legasto and Maciariello, 1980; Machol, 1980; Campbell, 1989). Recall, also, that Forrester himself acknowledged that computer-model building was more of an art than a science, and an arduous and time-consuming art at that.

The controversy set off in 1972 by the publication of Meadows et al.'s *Limits to Growth* was significantly larger than that set off by Forrester's earlier work. Several scholars at the University of Sussex in England, for example, wrote a scathing book-length criticism of *Limits . . .*, titled (somewhat sarcastically) *Models of Doom* (Cole et al., 1973). Here is a partial list of criticisms put forward by Cole et al. and by the other sources cited in the paragraph above: The Meadows et al. global computer model is inaccurate because it is too aggregated, that is, it treats the entire globe as homogeneous and doesn't incorporate important regional differences that could have an effect on future environmental problems. The model ignores the potentially significant effects of wars, conflicts, and other aspects of the relations between countries. The model ignores religion, values, and culture. The model ignores possible future technological inventions and breakthroughs that might solve or lessen global environmental problems. The model does not make allowances for the efforts policy makers are likely to make in the future to solve environmental problems. Furthermore, many specific details of the Meadows et al. model were criticized, such as the structure of the individual feedback loops and the individual nonlinear equations assumed.

Finally, one of the strongest objections to the Meadows et al. global model—especially on the part of economists—was that it appeared to ignore the effects of market mechanisms and technological advances on the supply and demand of scarce resources (Cole, et al., 1973; Nordhaus, 1992; Anon,

The Economist, 1992; Cole, 1993; Byrne, 1994). For example, when petroleum, or some other natural resource, becomes scarce, its price goes up, which results in various adjustments (via the "invisible hand" noted earlier): People buy less of it, and/or people find ways to get more use from smaller amounts of it, and/or the price increase spurs further exploration for more of the resource and technological innovations to conserve it, and/or substitutes are sought and found, and so on. In effect, these mechanisms increase the Earth's carrying capacity for the human species by substituting human ingenuity for natural resources. Indeed, several years ago, Julian Simon bet Paul Ehrlich that five key resources would not be scarcer as measured by market cost in the ensuing ten years. Ehrlich lost on all five. We discuss this Simon versus Ehrlich bet in detail in Box 12-2.

Meadows et al. addressed some of the above criticisms in their more recent book, *Beyond the Limits* (1992). For one thing, they decreased their estimates of the quantity of natural resources consumed by industry per unit of production. Also, they altered their global model to more explicitly include the diversion of industrial capital to the development and use of technology to reduce industrial pollution. Finally, they used widely varying assumptions concerning possible future technological advances that might conserve resources, and also concerning the effects of resource scarcity and market mechanisms. Despite these varying assumptions, their computer model still predicted the overshoot and collapse pattern. Some reviewers of the 1992 book noted with approval the above improvements (Pearce, 1992; Cole, 1993). One reviewer (Common, 1993) pointed to the Foreword of the 1992 book written by Nobel prize–winning economist Jan Tinbergen that says: "We all can learn some lessons from this book, especially we economists. . . ." Other reviewers were unimpressed with the changes and still very critical of the model (Anon, *The Economist,* 1992; Nordhaus, 1992).

Meadows et al. argue that many of the critics of *Limits to Growth* (1972) and *Beyond the Limits* (1992) have misunderstood their work's scope and intent: Though many critics see *Limits . . .* and

Box 12-2

The Julian Simon versus Paul Ehrlich Debate

The economist Julian Simon claims that without any government intervention in economic processes, shortages of materials create incentives that people respond to by curtailing use of those materials, finding substitutes, and inventing technologies that allow us to fulfill our needs while using less of the scarce materials. He is so certain of this argument that in the late 1970s, he challenged all comers to a bet: They could name any natural resource and any date in the future, and he would bet that the resource would be less scarce over time, as indicated by a decreasing price. If the price increased to reflect scarcity, Simon would pay the amount of the price increase; if the price went down, Simon would collect the amount of the decrease (see details below).

Simon's claim and his challenge incensed a number of scientists who believed that human demands on environmental resources were creating scarcities—that human appetites for material things were running the world out of its depletable resources. An exchange of articles in the scientific journals showed that more was at stake here than a simple difference of predictions. The language of the debate was far more heated than what is usual in scientific journals. The biologist Paul Ehrlich criticized Simon's science this way: "One wonders if Simon could not at least find a junior high school science student to review his writings (Ehrlich, 1981b, p. 46), "adding that he would not "eschew words like 'wrong,' 'incompetent,' or even 'moronic'. . . when the shoe fits [p. 48]." Simon, for his part, was not content to reply with data, but described one of Ehrlich's articles as "vintage Ehrlich" and proceeded to focus on Ehrlich's rhetoric and to speculate on the willingness of people to believe "a prophet" who was so often wrong (Simon, 1981, pp. 30, 38). For the full flavor of the Ehrlich-Simon debate, one can turn to their exchange of papers in the journal *Social Science Quarterly* in 1981 and 1982 (Ehrlich, 1981a, 1981b, 1982; Simon, 1981, 1982).

In addition to writing articles, in October 1980, a team of scientists led by Ehrlich bet Simon $1,000—$200 each on copper, chromium, nickel, tin, and tungsten—predicting that scarcity would drive the prices of these metals up over the next ten years. On October 1, 1990, when the bet came due, the prices of each of the metals (in 1980 dollars, as agreed), were $163, $120, $193, $56, and $86. Each price had declined as Simon predicted, and the market basket of metals had lost 38 percent of its value ($382 in 1980 dollars, to be exact). Ehrlich and his colleagues had to write Simon a check for $576.07 (the equivalent in 1990 dollars of the $382 that the prices had declined) (Tierney, 1990).

The bet was an embarrassment to the Ehrlich group. But what does it mean about the future of the environment? It vividly demonstrates the continuation of a trend that has been the basis of Simon's argument: a long-term trend toward decreasing real (that is, inflation-corrected) prices for natural resources over at least the past century. But what about the future? Ehrlich has compared Simon to a man who jumps off the Empire State Building and says, as he passes the tenth floor, "Everything is going fine." We cannot know what the future will bring in advance, and the long history of failed predictions of resource scarcity should teach us that the apparent logic of these predictions can be seriously flawed. Still, for reasons that we make clear in Chapters 1, 11, and 12, we don't have confidence that Simon's trends will continue reliably into the future. There are too many uncertainties in how the environment responds to human activities. We believe that if exponential growth of human population and of affluence continue they will eventually outstrip human ingenuity. Further, the scale of human impacts on the environment is now so large that we cannot rely unquestioningly on past history as a guide to the future.

Beyond . . . as providing detailed, quantitative predictions of future global catastrophe (e.g., a specific type of collapse of a specific magnitude in a specific year), Meadows et al., instead, are looking for general, qualitative behavior patterns of the global system. Their model is only a highly simplified summary of that system, a system that is incredibly complex and imperfectly understood. More specifically, Meadows et al. want to know how, in broad and general terms, the global system will tend to behave over time if current trends continue, and how the system would tend to respond to different types of societal policies and strategies. Meadows et al., for example, don't try to predict what policy decisions future policymakers would make in an effort to lessen environmental problems, because such decisions are impossible to predict. Again, the authors of *Limits* are looking for general patterns of how global population levels, environmental pollution, industrial activity, food production, quality of life, and other major variables tend to interact, and how the global system behaves over time.

Going a step further, it appears that Meadows et al.'s (1972, 1992) general, qualitative conclusions agree with those reached by several other teams of researchers who undertook their own global modeling projects after the publication of *Limits* in 1972. At least twenty groups in various countries, including Japan, the Soviet Union, West Germany, Argentina, and the Netherlands, developed computer models of the global system (Meadows, 1985; Campbell, 1989). Some of these efforts, according to Meadows (1985), were probably motivated by desires to disprove the conclusions of the *Limits* study. The modeling work of these groups differed considerably in methodology, assumptions, and underlying moral concerns (for example, an emphasis on meeting the needs of humanity versus an emphasis on avoiding environmental degradation). As Meadows (1985) points out, "The methods and philosophies of global modelers are so diverse that one would hesitate to call the models a single body of intellectual work [p. 56]." And there were some substantive differences in their results.

However, despite these differences, the results of all twenty modeling projects appear to support two

main qualitative conclusions that we discussed earlier in the chapter and in Chapter 11; at the very least, the results don't contradict these conclusions. The conclusions are, first, that the unaided human mind cannot comprehend the global system and that the system behaves in counterintuitive ways; and, second, that the global system is especially unstable as a result of the key characteristics that we discussed at length—long time delays, exponential growth, and possibly irreversible environmental damage—characteristics that also tend to subvert disjointed incrementalism or any self-organizing process in the global environmental system based on trial-and-error learning.

Here, in greater detail, are some results of the twenty modeling efforts that support the main conclusions in the paragraph above. In support of the first conclusion: The outputs of the twenty models "constantly surprise[d] even their makers [p. 61]," claims Meadows (1985). For example, several models showed that there is greater interdependence between nations than most people imagine: A global model developed by agronomists and economists at two universities in Holland showed that a small change in wheat production in Kansas could have a major negative effect on food policy in Nigeria. Also, several models found that the dramatic changes in oil prices and availability as a result of OPEC actions in 1973 (discussed in Chapter 10) have affected the economies of nations all over the world and for a surprisingly long time period; in fact, "[s]ome models indicate that the economic system still has not settled down from the turbulence caused by the first oil price shock [in 1973], much less the later ones (Meadows, 1985, p. 61)."

In support of the second qualitative conclusion, above, is the fact that according to Meadows (1985), all twenty models conclude that exponential population growth cannot continue forever on a finite planet. This conclusion was reached even by some modelers who objected to what they saw as an "antinatalist" bias in the original *Limits* (1972) model. In addition, many of the models recognize that growth of factories, infrastructure, and housing (i.e., "physical capital") cannot continue forever. Furthermore, all twenty of the global computer models, according to Meadows (1985), when run beyond the year 2000 with current trends and policies in place, show major discontinuities, that is, sharp and significant global changes, rather than merely slow and gradual changes from present conditions. Meadows et al. conclude that "a smooth continuation of present trends can be ruled out as physically impossible [p. 62]."

It is possible that all twenty global models are wrong. For example, in the prior paragraph, though the Dutch global computer model may indicate a large effect on the food economy of Nigeria of a small change in wheat production in Kansas, the real world may not actually operate this way. We would guess that particular predictions like this will very often be wrong. However, the fact that global models developed by researchers in several different countries, using different methodologies, and based on a variety of assumptions, all produce results consistent with the same qualitative conclusions increases our confidence that the real-world global system may actually behave as those conclusions indicate—or at least that there is a plausible chance of global overshoot and collapse. The qualitative conclusions are, for us, the most interesting and significant findings to come out of the global modeling work. That work suggests a significant mismatch between the properties of the global environmental system and the capacities and characteristics of the human mind and of human institutions that interact with that system.

We find it especially difficult to ignore the concerns of Meadows et al. and others about the unprecedented conjunction of the three characteristics of environmental problems today: long time delays, exponential growth, and the possibility of catastrophic and irreversible errors. Our species is, clearly, conducting a huge and risky experiment on the Earth's life support systems. The stakes are enormous. As Canadian scientist-educator David Suzuki put it:

> As a scientist, I know how ignorant we are of the biological and physical world . . . yet we continue to cling to the lie that we know what we're doing. The truth is we have no idea (Quoted by Allen, 1994).

We believe the unavoidable conclusion is that if humanity must err in its degree of concern about, and

efforts to avert, global environmental problems, it should err on the side of protecting the environment.

Solutions to Global Environmental Problems

As we noted earlier, Meadows et al. (1992) argue that our species has already overshot the long-term carrying capacity of the planet. But they also urge optimism. Meadows et al. claim there is still time to avoid catastrophe, but only if we make fundamental changes in human activities and institutions and make them soon. Meadows et al. even argue that we can make these changes and at the same time alleviate poverty and create a good quality of life for the entire global population. While the details of the changes and solutions Meadows et al. and others propose are beyond the scope of this book, we briefly review those changes and solutions below.

"Traditional" Proenvironmental Measures. First, Meadows et al. (1992) advocate the kinds of proenvironmental measures that most people are familiar with and that are already in wide use. These measures involve technologies and methods that lessen the impact of human activity on resources and the environment—for example, improving the energy efficiency of cars, heating systems, and industrial machinery; using agricultural practices that minimize the use of pesticides; recycling and reusing goods and materials; and so on. However, Meadows et al. argue that we must take these measures more aggressively than we do now.

We agree with Meadows et al. about the importance of traditional proenvironmental measures. Table 12-1 lists some strategies for proenvironmental change based on the analyses in this book, and some that are widely recommended by environmental scientists (including Brown, 1991, 1994; Miller, 1994; and Montague, 1994). Here's a brief look at Table 12-1:

Item 1: Many environmentalists now stress the use of upstream rather than downstream interventions. This strategy was central to our discussion in Chapter 10. As a rule, upstream solutions (prevention) can do more to relieve pressure on the environment than solutions farther downstream (cure): More energy-efficient cars save more energy and prevent more pollution than driving less; burning low-sulfur coal

TABLE 12-1 Environmental Actions and Measures—A Partial List of Policy Guidelines

1. *Implement* upstream *proenvironmental measures wherever possible.* This applies to solid waste, air and water pollution, resource depletion, and other problems (see Chapter 10 and this chapter).

2. *Carefully choose proenvironmental policy changes that yield the greatest proenvironmental impact*—for example, energy efficiency changes directed at the sectors and end-uses that consume the most energy (Chapter 10 and this chapter).

3. *Give high priority to promoting energy conservation and speeding the phaseout of fossil fuels and the inevitable transition to renewable energy sources,* because these measures lessen several serious environmental problems simultaneously (Chapter 10 and this chapter).

4. *Implement financial incentives and regulations designed to internalize environmental externalities,* for example, pollution taxes, bottle bills, and regulatory controls on waste disposal (Chapter 5).

5. *Encourage local, grassroots action* to solve local environmental problems, and also grassroots lobbying/political efforts to promote state and national proenvironmental government policies. History shows that these actions can be effective (Chapters 5 and 6).

6. *Address issues of international economic equity as well as issues of gender equity.* Foster new conceptions of national security and sovereignty. Rich-poor equity and gender equity issues are inseparable from environmental issues. (This chapter; Repetto, 1986; Miller, 1994; others).

7. *Use a multiplicity of measures, methods, mechanisms.* The complexity and multiple determinacy of environmental problems require complex, multiple solution strategies, and simultaneous interventions at individual, large-actor, national, and international levels (Chapters 1, 7, 10).

8. *Remain alert for the counterintuitive properties that complex systems may exhibit, and be aware of the paradoxical effects of conservation applied alone* (Chapter 11 and this chapter).

does more for air quality than treating the smoke from high-sulfur coal; reusing bottles is better for the environment than recycling them.

As a relevant example, Montague (1994) argues that the U.S. environmental movement of the 1970s was misguided because it sought the collection, treatment, or neutralization of hazardous substances released into the environment. This policy, he argues, was ineffective, first, because industries have created and used new hazardous substances more rapidly than government regulators can keep up with. Second, he argues that many hazardous substances that are "disposed of" via deep-well injection, incineration, or dumping in landfills (all of which are downstream strategies) never disappear from the environment; instead, they are dispersed and remain a threat. The alternative that Montague urges is the development of closed-cycle industrial processes that avoid or minimize generation of hazardous substances to begin with.

Item 2 is closely related to item 1. As we argued in Chapter 10, comprehensive engineering analyses that identify policy changes likely to have the greatest proenvironmental impact are required in designing effective energy conservation or solid waste programs. Typically, these analyses indicate that upstream interventions are more effective than downstream interventions (item 1 above). But these analyses usually also reveal additional useful strategies. For example the energy analysis in Chapter 10 indicates that energy efficiency measures involving household space heat are likely to save significantly more energy than similar measures involving refrigeration and freezing, because individual/household space heating now consumes seven and one-half times more energy than refrigeration and freezing.

Item 3 in Table 12-1 (related to items 1 and 2) urges that top priority be given to conserving energy, phasing out fossil fuel combustion, and speeding the transition to renewable sources of energy. Fossil fuel consumption contributes simultaneously to many key environmental problems, as we pointed out in Chapter 10. This is true for the "large actors" as well as individuals and households that are the main focus of Chapter 10. The gaseous by-products of fossil fuel combustion contribute to global climate change, as well as to water and air pollution. The dependence of industrial nations on petroleum from politically unstable nations adds to global tensions. Humankind is consuming finite and nonrenewable fossil fuel reserves at a rapid pace. *All* of these problems can be lessened considerably via energy conservation measures, especially those that increase energy efficiency, and by a switch from fossil fuels to renewable and nonpolluting sources such as solar energy. As we emphasized in Chapter 10, this is the inevitable path that global energy use must follow; the sooner we take it, the better for the environment.

Items 4 and 5 of Table 12-1 are self-explanatory.

Item 6 is based on the work of Repetto (1986), Brown (1987, 1995), Postel (1994), Miller (1994), and others. These researchers and writers argue that environmental problems and issues are inextricably linked to equity issues. They point out that though wealthy countries (the United States, Canada, Western European countries, Japan, Australia, and others) contain only 22 percent of the world's population, these countries consume a disproportionately large share of the Earth's resources (approximately 88 percent of the natural resources consumed each year, including 73 percent of the energy resources) and generate most of the pollution and waste (Miller, 1994). Further, the economic gap between wealthy and poor nations has widened in the last decade, and many poor nations are now caught in negative spirals of grinding poverty, rapid population growth, and destruction of their ecological capital (Brown, 1987, 1995). Wealthy nations, Repetto, Brown, Postel, Miller, and others argue, should not ignore the plight of poor nations. In fact, they cannot: People in all countries are dependent on the Earth's environmental systems and its resource bases; all people are harmed by damage to these systems and bases, and all must work to protect them. Further, the situation of poor nations is a global environmental problem because

those nations aspire to follow in the path of the industrialized countries. If they literally do this, the demand on global environmental systems will increase severalfold, as the majority in the world copy the resource-intensive habits of the minority. It is therefore better for the global environment if the rich countries invest in making development in the rest of the world more environmentally benign—for example, by making energy- and resource-efficient technologies available to the developing countries, rather than the obsolete, highly polluting technologies that the rich countries are discarding.

Repetto, Brown, Postel, and Miller (and others) argue, further, that women in many poor countries are especially disadvantaged. Women cannot vote or own property, and they typically have little access to jobs and education. As a result, women marry early and their main role is having and raising children. Demographers, ecologists, and social scientists believe that actions to correct these inequities will have the secondary effect of lowering birthrates. In developing countries that have improved health care, education, and job access for women, birth rates have significantly declined (e.g., Repetto, 1986). The main theme and conclusion of the United Nations Conference on Population and Development held in Cairo in 1994 was that high fertility rates in poor nations are tightly linked to gender inequities, and that urgent action should be taken to correct the inequities and thereby lower birthrates (e.g., Population Action International, 1994). Controlling population growth, it is hoped, will make it easier for developing countries to meet their people's needs without overstressing their environmental systems.

The remaining items in Table 12-1—items 7 and 8—are self-explanatory.

Addressing Long Feedback Delays and the Potential for Catastrophic and Irreversible Environmental Damage. The proenvironmental measures discussed above, while vitally important, are not sufficient. Meadows et al. (1992) and others also urge fundamental changes in government policymaking to address long time delays inherent in global environmental problems and to help avert the irreversible catastrophes to which the global system is now subject. Table 12-2 summarizes some of these proposed changes. Part A of the table suggests an improvement in the monitoring of key environmental conditions and variables so that problems that are developing can be detected earlier than they now can be.

Part B of Table 12-2 proposes government policy rules that would favor technologies if their consequences can be managed by *disjointed incrementalism*. The rules derive from the work of Edward Woodhouse and Joseph Morone, both former graduate students of Charles Lindblom, the Yale political scientist who coined the term disjointed incrementalism and did the most widely cited research on it. Woodhouse (1980) acknowledges that long time delays and the possibility of irreversible environmental damage can make DI ineffective for preventing environmental catastrophe. However, he argues that there are no viable alternatives to DI for making decisions in large political systems and that the only solution is to alter government policymaking so as to avoid the conditions and problems that foil DI. To give a specific example (though not one that Woodhouse, 1980, emphasized), perhaps the deployment of commercial nuclear power should have been avoided in the first place because its characteristics are ones (see list in Table 12-2) that create problems for DI, including long time delays, scientific disagreement about how to address them (e.g., how to store spent wastes that remain radioactive for thousands of years), and the capacity for catastrophic accidents. According to Woodhouse's thinking, the deployment of nuclear power, which was financially supported by the U.S. government, looks especially unwise because energy conservation measures have few of

TABLE 12-2 Policy Strategies to Avoid the Dangers of Long Time Delays and of Possible Catastrophic and Irreversible Environmental Damage

A. *Shorten time delays where possible* (Meadows, et al. 1992), e.g.:
  - Improve the signals: Increase monitoring of, and research on, planetary levels of pollution, resource depletion, and population.
  - Speed up response times: Look actively for signs of trouble. Have coping strategies ready in advance. Anticipate long feedback delays. Give special emphasis and caution to known feedback delays, e.g., population momentum.

B. *Alter policymaking to enable disjointed incrementalism to function properly:*
  - *Woodhouse (1980):* Choose policies and deploy technologies that:
    - expose the fewest people to risks and hazards
    - are least likely to negatively affect future generations
    - are least likely to involve catastrophic consequences
    - are least likely to involve irreversible consequences
    - are least likely to involve long feedback delays
    - involve the least scientific uncertainty and disagreement
    - involve the most tested and conventional technology; (this is related to C, below:)
  - *Morone and Woodhouse (1986):* Use an integrated approach for policymaking concerning technologies with the potential to do catastrophic and irreversible environmental damage:
    - Greatly strengthen devices and methods to contain, limit, or mitigate the damage the technology might cause should a serious mishap occur.
    - Conduct *multiple small-scale pilot tests* of different versions of a risky technology rather than large scale one-at-a-time deployments.
    - Other measures (but see footnote 5 in the main text).

C. "*[Regulate] . . . the deployment of powerful new technologies* that can induce sweeping changes in economic patterns, lifestyles, governance, and social values . . ." and that may therefore create problems faster than democratic governments can handle (Milbrath, 1994, p. 699).

the negative characteristics (see Part B of Table 12-2), and these measures could have made the construction of additional nuclear reactors unnecessary. Pursuing Woodhouse's line of thinking a radical step further, Montague (1994) argues that many new and powerful technologies are being deployed in Western countries at accelerating rates that outstrip the coping abilities of governments. His conclusion is that such new technologies should not be automatically deployed, as is largely the case now in this country, but should be approved in advance and carefully regulated by governments to ensure that their impacts are acceptable and manageable. (Toffler, 1970, proposed a similar strategy.)

The guidelines listed in Part B of Table 12-2 can be interpreted in an absolutist way: that old technology should always be preferred to new, that society should never adopt a technology with any potential for catastrophe, and so on. We do not advocate such a use of these guidelines, but we do accept the argument that society should exercise extreme care before adopting technologies that may turn out to be ungovernable and/or prone to unstoppable catastrophe. For example, a few years ago, a scientist suggested that the buildup of carbon dioxide in the atmosphere might by counteracted by adding iron to the oceans, causing large blooms of algae that would use the excess atmospheric carbon dioxide as a nutrient (Blakesly, 1990). Apart from the question of whether adding iron would have the predicted effect, the ideas in Part B suggest avoiding this sort of "geoengineering" solution to global environmental problems. Not only might the "solution" have long feedback delays, affect future generations, entail great scientific uncertainty, and so on, but it would also be ungovernable in the sense that no single government could control its implementation in the open seas.

Part C of Table 12-2, like Part B, recommends caution about adopting new technologies. The difference is that Part C focuses on technologies that primarily produce social and economic change, and not only environmental change, as in Part B. The difficulty in acting on the strategy in Part C is in identifying in advance the technologies that will produce major social and economic change. Often, new technologies surprise analysts, with some having much larger effects than expected and others having much less.

Controlling Exponential Growth. The two above strategies—traditional proenvironmental measures guided by the principles in Table 12-1, and efforts to change the operation of governments so that they can better manage long time delays and avert irreversible catastrophes (Table 12-2)—are very important. However, in our opinion, that of Meadows et al. (1992), and others, the strategies are not sufficient, regardless of how quickly or zealously they are carried out. Meadows et al. go so far as to assert that even dramatic technological advances—for example, a limitless, risk-free, cost-free, pollution-free source of energy (e.g., perfected nuclear fusion)—will not solve global problems. Meadows et al. base this claim on the results of their computer modeling work. When they programmed into the model possible radical changes in future technology for energy production, resource conservation, or pollution abatement, they got the type of result that we discussed in the previous chapter and that Figure 11-11 illustrated: The new technology permits exponential population and material growth to continue a little longer until the system hits another barrier that limits further growth, a barrier sometimes more problematic than the initial barrier. Also, Meadows et al. got the same results when they programmed in multiple combinations of technological advances. Note, further, that Meadows et al.'s conclusions are consistent with the work of the twenty other global modeling teams that we mentioned above (Meadows, 1985). The conclusions are also ones that other scholars and writers—for example, Ophuls (1977), Daly (1991), and Miller (1994)—have reached based on other considerations, that is, without the specific use of a global computer model.

We do not take the above results from computer models as specific predictions, but they strongly illustrate an important principle: that when growth in human demands on the environment is exponential or nearly so, taking action to control growth in any one aspect of those demands is likely to be inadequate in the long run. What is necessary is to find ways to control the totality of human demands on the environment so that it does not for very long exceed the ability of natural systems to regenerate themselves. It is not safe to rely on the two types of strategy discussed above because of the possibility that people will increase their demands for energy and materials faster than we can find ways to use them more efficiently. Humanity is in a race between the increasing demands of a growing and increasingly affluent population for energy and materials (and the increasing generation of pollutants) on the one hand and, on the other, the ability of technology, economic substitution, and other innovations to reduce the environmental impact of the things people do. Given the problems of time delays and possible environmental catastrophes, we think it unwise to bank on human technical ingenuity.

What further changes do Meadows et al. (1992) and others urge as necessary to avert a global environmental crisis? If technological solutions may fail to keep pace with exponential population and material growth and their impacts on resources and the environment, the only remaining strategy is to slow growth, regulate it carefully, and bring some forms of it to a halt. Meadows et al. (1992) and others argue for strict limits to resource and energy use and pollution generation, the cessation of human population growth, and the slowing and regulation of economic and material growth. Table 12-3 briefly summarizes these recommended changes.

Our interpretation of the set of strategies recommended in Table 12-3 parallels our comments two paragraphs above: What must be accomplished is some combination of the changes listed in points B, C, and D sufficient to achieve the goals listed in point A. Again, the basic conclusion is that total human demands on the environment must not exceed the capacities of environmental systems (i.e., point A). For example, it is unlikely that stopping population growth by itself would end environmental threats because a stable population might continue to increase its demands on the environment (e.g., by increasing affluence) until the demands are unsustainable. Some analysts believe the demands of the present world population are already unsustainable. In the case of economic development, what matters is not total income levels or national products but their environmental impacts. Thus, economies can grow without increasing demands on the environment if at the same time they change so that what people produce and buy is more environmentally benign.

The strategies listed in Table 12-3 challenge some very basic values, beliefs, and institutions. Slowing and stopping population growth (point B) presents such a challenge. Population control has been controversial for decades, partly because of opposition from

TABLE 12-3 Slowing, Controlling, and, in Some Cases, Stopping Growth

A. *Limit resource and energy use and limit pollution generation to preserve the Earth's ecological capital, i.e., to assure long-term sustainability* (Herman Daly, 1991, cited in Meadows et al. 1992, p. 209):
  • Rates of use of renewable resources should not exceed their rates of regeneration.
  • Rates of use of nonrenewable resources should not exceed rates at which sustainable renewable resources are developed.
  • Rates of pollution emission should not exceed the assimilative capacity of the environment.

B. *Slow and stop exponential human population growth* (Meadows, et al., 1992; Miller, 1994; Daly, 1991; Population Statement of Scientific Academies, 1993; Ehrlich, e.g., 1994; others).

C. *Control and limit economic and material growth* (Meadows et al., 1992; Miller, 1994; and others):
  • Economic growth and material growth need not be zero. However, growth should be limited (i.e., within the constraints of A. above) and oriented toward qualitative development rather than physical expansion, and toward "material sufficiency and security for all people [Meadows et al., 1992, p. 211]."

D. *Make concomitant changes in core societal beliefs, morals, and values* concerning population growth, material growth, wealth, and well-being (Chapter 3).

the Catholic Church and its adherents, but also because many developing countries have seen it as a policy rich countries have tried to impose on them instead of controlling their own appetites for resources. In wealthy Western countries that cause the greater per capita demand on resources and the environment, not many leaders have publicly discussed issues of population growth, and the idea of controlling growth goes against prevailing Western values. There are signs, however, that the resistance to population control policies is waning. Many developing countries have adopted such policies because they see them as in their own economic interests; more Western leaders are talking about curtailing growth; and world scientific opinion is on the side of population stabilization. At a Population Summit of the World's Scientific Academies in 1993, fifty-eight academies reached agreement that: ". . . [W]e must achieve zero population growth within the lifetime of our children (Population Summit, 1993, p. 2)." Given the characteristics of *population momentum* we discussed in Chapter 1 (Box 1-2), achieving this goal requires that the world take action now.

The changes proposed in Table 12-3 are radical. Indeed, Meadows et al. (1992) conceive of these changes as constituting a revolution as major and far-reaching as the agricultural revolution, which took place 10,000 years ago (Chapter 8), and the industrial revolution, which began in the late eighteenth century. Ophuls (1977), Brown (1994), Miller (1994), and others also describe the changes they think must occur as revolutionary. Meadows et al. argue that this revolution will take decades to complete but that it has already begun. However, they and others argue that it must be speeded up if we are to avoid environmental disaster. What Meadows et al. call the new "sustainability revolution" will, in their judgment, require profound changes in Western and non-Western institutions, economic processes, values, morals. It will require changes in our basic conceptions of the relationship between humans and the rest of nature. It will require that we acknowledge the enormous complexity of global systems and our inability to manage them and mold them solely to the purposes of humans. And it will require that we more fully accept our responsibilities to future generations.

Note that the values and worldview of Meadows et al.'s sustainability revolution overlap with those of the radical environment movements we discussed in Chapter 3, including *ecotheology, deep ecology,* and *ecofeminism.* As we discussed in Chapter 3, many of these values and beliefs are already widely accepted in the United States. But, as we also argued in Chapter 3, changes in values, beliefs, and worldview will not by themselves produce radical changes in the behavior of individuals. To the extent that the sustainability revolution involves changes in individual behavior, the revolution will also require many of the measures we have discussed in the prior chapters of this book. Educational campaigns designed in accord with the principles in Chapter 4 are needed to inform individuals about which proenvironmental actions to take and how to take them. External barriers to proenvironmental behaviors will need to be removed and incentives for those behaviors provided (Chapter 5). Community processes should be used where possible to encourage proenvironmental behaviors (Chapter 6). Multiple measures should be carefully integrated (Chapter 7). Policymakers should be aware of possible human genetic predispositions and limits that may influence proenvironmental behaviors (Chapter 8). Policymakers should also use knowledge of human perceptual and cognitive processes that bear on whether people take proenvironmental actions (Chapters 9). Technical/engineering analyses should be used to select the individual actions that have the greatest proenvironmental impact (Chapter 10). As we noted in Chapters 1 and 10, the role played by psychology and by individual behaviors, while important, is limited: Changes in the actions of large actors, including political and other institutions, are also critical to the sustainability revolution. But even the actions of these large actors can be significantly affected by the political actions of an informed and concerned public (Chapters 4–6).

We close this chapter, and the book, with a passage written by Donella Meadows (1985). Though the passage is from an article Meadows wrote on global computer models, we think it applies to all aspects of

New York City Skyline Just before Dawn
We include this photo as a symbol of hope. The global environment future can be bright if we humans take the necessary measures now.
(M. Antman/The Image Works)

the transition to an environmentally sustainable global future:

> [Computer models of the global system] . . . have forced us to stand back and look at all the complexity, admit it, be humbled by it, and yet continue to keep confronting it. When we do, we see far too many negative trends to be complacent and far too many positive trends to be hopeless. We mainly see a lot of work to do [p. 62].

NOTES

1. For mathematically inclined readers, the following equation summarizes the agreement between the vacuum cleaner company and the ad agency:

$$A_i = K_2/100 \times S_i$$

where A_i is the change in advertising expenditures in a given week, based on the change in sales in the prior week, S_i, and K_2 is the percentage of the previous sales increase devoted to current advertising, 25 percent in the discussion above.

2. The nonlinear function in Figure 12-1 is specified by the quadratic equation:

$$S_i = (K_1 \times A_i/40) - (K_1 \times A_i/40)^2$$

where A_i is increase in advertising, S_i is increase in sales produced, and K_1 is a constant value determined by the market, in the case of the figure, 3.5.

3. i.e., the value of K_2.

4. Though Lindblom claims that disjointed incrementalism succeeds, other political analysts trace the problems of government in the United States to precisely the disjointed actions of many agencies and interest groups, which they see as causing a sort of gridlock that makes governing increasingly difficult.

5. Two of Lindblom's former students, Joseph Morone and Edward Woodhouse (1986), argue that certain modifications of DI *can* enable it to deal with problems involving long feedback delays and irreversible environmental damage. They further argue that many of these modifications have already appeared over the years in various national and international governments and agencies. They cite as an example of the success of this altered form of DI the rapidly created international agreements to phase out the use of chlorofluorocarbons (CFCs) that damage the ozone layer. Other analysts, however, see the ozone agreement as atypical because only a few manufacturers produced CFCs and because substitutes were available. We are not willing to count on the modifications of DI that Morone and Woodhouse argue have already been informally made in government agencies, et cetera to successfully manage the nations of the world in the face of other global environmental problems that involve long feedback delays and the possibility of irreversible catastrophic damage. Thus, we believe, that additional, more concerted, societal changes are needed to avert these problems.

6. Technically, this vicious circle is a destructive positive feedback loop—see Box 11-2 in the prior chapter.

REFERENCES

Acheson, J. M. (1975). The lobster fiefs: Economic and ecological effects of territoriality in the Maine lobster industry. *Human Ecology*, *3*, 183–207.

Acheson, J. M. (1987). The lobster fiefs, revisited: Economic and ecological effects of territoriality in the Maine lobster industry. In McCay, B., and Acheson, J. (Eds.), *The Question of the Commons*. Tucson, AZ: University of Arizona Press, pp. 37–65.

Agras, W. S., Jacob, R. G., and Lebedeck, M. (1980). The California drought: A quasi–experimental analysis of social policy. *Journal of Applied Behavior Analysis*, *13*, 561–70.

Allman, W. (1985). Staying alive in the 20th century. *Science 85* (Oct.), p. 31–41.

Anderson, V. et al. (1991). Statement by religious leaders at the summit on the environment. New York City, June 3. In: Guide to environmental activities and resources in the North American religious community (pamphlet). Joint Appeal by Religion and Science for the Environment, 1047 Amersterdam Ave., New York, 10025.

Andrews, D. E. (1993). Gore, churches join environment forces. *Chicago Tribune* (Oct. 8).

Anon. (1979). To the brink of the abyss: The first hours of Three Mile Island, *Nucleus* (May), *1*(4).

Anon. (1983). King crab fishing closed in Alaska. *New York Times* (Oct. 3).

Anon. (1986a). Rebuilding Mexico City. *Ford Foundation Letter* (Aug.), p. 4.

Anon. (1986b). *Newsweek* (June 9), p. 47.

Anon. (1987). Eco–update: The big payout. *Acres, U.S.A.* (Feb.), p. 5.

Anon. (1989). Eco–update: A UDMH connection. *Acres, U.S.A.*(Nov.).

Anon. (1992). AIDS Page: High schoolers' practices risky business. *FDA Consumer* (July–Aug.), p. 7.

Anon. (1992). Reading for Rio, *The Economist*, *323*, 89-90.Arcury, T., and Christianson, E. (1990). Environmental worldview in response to environmental problems: Kentucky 1984 and 1988 compared. *Environment & Behavior*, *22*, 387–407.

Arcury, T., and Christianson, E. (1990). Environmental worldview in response to environmental problems: Kentucky 1984 and 1988 compared. *Environment & Behavior*, *22*(3). US: Sage Publications Inc; 1990, 387–407

Armelagos, G., and Cohen, M. (1984). *Paleopathology at the origins of agriculture*. New York: Academic Press.

Arnold, J. E. M., and Campbell, J. G. (1986). Collective management of hill forests in Nepal: The community forestry development project. In National Research Council, *Proceedings of the Conference on Common Property Resource Management*. Washington, DC: National Academy Press.

Aronson, E., Fried, C., and Stone, J. (1991). Overcoming denial and increasing the intention to use condoms through the induction of hypocrisy. *American Journal of Public Health*, *81*, 1636–1638.

Associated Press (1993). Tougher laws get credit for jump in seatbelt use. *Eugene (Oregon) Register–Guard* (Dec. 19).

Axelrod, R. (1984). *The evolution of cooperation*. New York: Basic Books.

Ayres, R. U., Schlesinger, W. H., and Socolow, R. H. (1994). Human impacts on the carbon and nitrogen cycles. In Socolow, R., Andrews, C., Berkhout, F., and Thomas, V. (Eds.), *Industrial ecology and global change*. New York: Cambridge University Press.

Balling, J., and Falk, J. (1982). Development of visual preference for natural environments. *Environment and Behavior*, *14*, 5–28.

Barton, S. (1994). Chaos, self-organization, and psychology. *American Psychologist*, *49*, 5–14.

Becker, L. J. (1978). The joint effect of feedback and goal setting on performance: A field study of residential energy conservation. *Journal of Applied Psychology*, *63*, 428–433.

Bell, P., Fisher, J., Baum, A., and Greene, T. (1990). *Environmental psychology, third edition*. Fort Worth, TX: Holt, Rinehart and Winston, Inc.

Berk, R. A., LaCivita, C. J., Sredl, K., and Cooley, T. F. (1981). *Water shortage: Lessons in conservation from the great California drought, 1976–1977*. Cambridge, MA: Abt Books.

Berkes, F. (1986). Marine inshore fishery management in Turkey. In National Research Council, *Proceedings of the Conference on Common Property Resource Management*. Washington, DC: National Academy Press.

Berkes, F. (2002). Cross-scale institutional linkages: Perspectives from the bottom up. pp. 293–322 in National Research Council, *The Drama of the Commons*. Committee on the Human Dimensions of Global

Change. E. Ostrom, T. Dietz, N Dolsak, P.C. Stern, S. Stonich, and E.U. Weber, eds. Washington: National Academy Press.

Berkes, F., and Folke, C. (1994). Investing in cultural capital for a sustainable use of natural capital. In Jansson, A., Folke, C., Costanza, R., and Hammer, M. (Eds.), *Investing in natural capital: The ecological economic approach to sustainability*. Covelo, CA: Island Press.

Berkes, F., ed. (1988). *Common Property Resources: Ecology and Community-Based Sustainable Development*. London: Belhaven Press.

Berry, L. (1990). *The market penetration of energy–efficiency programs*. Oak Ridge, TN: Oak Ridge National Laboratory, ORNL/CON 299.

Berry, T. (1988). *The dream of the earth*. San Francisco: Sierra Club Books.

Bick, T., and Hohenemser, C. (1979). Target: Highway risks. *Environment, 21*, 16ff.

Black, J. S., Stern, P. C., and Elworth, J. T. (1985). Personal and contextual influences on household energy adaptations. *Journal of Applied Psychology, 70*, 3–21.

Blakesly, S. (1990). Ideas for making ocean trap carbon dioxide arouse hope and fear. *New York Times* (Nov. 20), p. 4.

Bogin, B. (1988). *Patterns of human growth*. Cambridge, England: Cambridge University Press.

Borgida, E., and Nisbett, R. (1977). The differential impact of abstract vs. concrete information on decisions. *Journal of Applied Social Psychology, 7*, 258–271.

Boyd, R., and Richerson, P. (1985). *Culture and the evolutionary process*. Chicago: University of Chicago Press.

Brechin, S. and Kempton, W. (1994). Global environmentalism: A challenge to the postmaterialism thesis? *Social Science Quarterly, 75*, 245–269.

Bromley, D.W., D. Feeny, M.A. McKean, P. Peters, J.L. Gilles, R.J. Oakerson, C.F. Runge, and J.T. Thomson, eds. (1991). *Making the Commons Work: Theory, Practice, and Policy*. San Francisco: Institute for Contemporary Studies Press.

Brower, M., and Leon, W. (1999). *The Consumer's Guide to Effective Environmental Choices: Practial Advice from the Union of Concerned Scientists*. New York: Three Rivers Press.

Brown, L. (1981). *Building a sustainable society*. New York: W. W. Norton & Co.

Brown, L. (1985). Maintaining world fisheries. In Brown, L., et al., (Eds.), *State of the world, 1985*. New York: W. W. Norton and Co.

Brown, L. (1987). Analyzing the demographic trap. In Brown, L., et al., (Eds.), *State of the world, 1987*. New York: W. W. Norton.

Brown, L. (1989). Reexamining the world food prospect. In Brown, L., et al., (Eds.), *State of the world, 1989*. New York: W. W. Norton.

Brown, L. (1990). The illusion of progress. In Brown, L., et al., (Eds.), *State of the world, 1990*. New York: W. W. Norton.

Brown, L. (1991). *Saving the planet: How to shape an environmentally sustainable global economy*. New York: Norton.

Brown, L. (1994). Launching the environmental revolution. In Miller, Jr., G. T., *Living in the environment, eighth edition*. Belmont, CA: Wadsworth. 43-44.

Brown, L. (1995). Nature's limits. In Brown, L., et al., (Eds.), *State of the world, 1995*. New York: W. W. Norton.

Brown, L., and Postel, S. (1987). Thresholds of change. In Brown, L. et al. (Eds.), *State of the world 1987*. New York: Norton.

Bruvold, W. (1979). Residential response to urban drought in Central California. *Water Resources Research, 15*, 1297–1304.

Burton, I., Kates, R., and White, G. (1978). *The environment as hazard*. New York: Oxford University Press.

Byrne, J. (1994). Review of *Beyond the Limits*, *Journal of Urban Affairs, 16*(1), 85-87.

Callicott, J. (1983). Traditional American Indian and traditional Western European attitudes towards nature: An overview. In Eliot, R., and Gare, A. (Eds.), *Environmental philosophy: A collection of readings*. University Park: Pennsylviana State University Press.

Campbell, R. B. (1989). Global modeling and simulation. *Simulation, 52*, 33–35.

Caporael, L. (1989). Mechanisms matter: The difference between sociobiology and evolutionary psychology. *Brain and Behavioral Science, 12*, 17–18.

Carey B. (1992). Vital signs: The AIDS file. *Health* (Jul.–Aug.).

Carson, R. (1962). *Silent spring*. Boston: Houghton–Mifflin.

Cialdini, R., Kallgren, C., and Reno, R. (1991). A focus theory of normative conduct: A theoretical refinement and re–evaluation. *Advances in Experimental Social Psychology, 24*, 201–234.

Cialdini, R., Reno, R., Kallgren, C. (1990). A focus theory of normative conduct: Recycling the concept of norms to reduce littering in public places. *Journal of Personality and Social Psychology, 58*, 1015–26.

Clark, W. (1980). Witches, floods and wonder drugs: Historical perspectives on risk management. In Schwing, C., and Albers, W. (Eds.), *Societal risk assessment: How safe is safe enough?* New York: Plenum Press.

Clarke, L., and Short, Jr., J. (1993). Social organization and risk: Some current controversies. *Annual Review of Sociology, 19,* 375–399.

Cohen, B., and Lee, I. (1979). A catalog of risks. *Health Physics, 36,* 707–722.

Cohen, S., and Spacapan, S. (1984). The social psychology of noise. In Jones, D., and Chapman, A. (Eds.), *Noise and society.* New York: Wiley.

Cohen, S., Evans, G., Krantz, D., and Stokols, D. (1980). Physiological, motivational, and cognitive effects of aircraft noise on children: Moving from the laboratory to the field. *American Psychologist, 35,* 231–243.

Cohen, S., Evans, G., Stokols, D., and Krantz, D. (1986). *Behavior, health, and environmental stress.* New York: Plenum Press.

Cohn, D. (1992). Per–can fees catch on as area's trash mounts. *Washington Post* (Mar. 11), pp. A1, A31.

Cole, S. (1993). Learning to love *Limits. Futures, 25,* 814–818.

Cole, S., and others at Sussex University. (1973). *Models of doom: A critique of The Limits to Growth.* New York: Universe Books.

Coltrane, S., Archer, D., and Aronson, E. (1986). The social–psychological foundations of successful energy conservation programmes. *Energy Policy, 14,* 133–148.

Combs, G., and Slovic, P. (1979). Newspaper coverage of causes of death. *Journalism Quarterly, 56,* 837–843.

Common, M. (1993). Review of *Beyond the Limits. The Economic Journal, 103,* 1084–1086.

Commoner, B. (1970). *Science and survival.* New York: Ballantine.

Commoner, B. (1971). Survival in the environmental–population crisis. In Singer, F. (Ed.), *Is there an optimal level of population.* New York: McGraw–Hill.

Condelli, L., Archer, D., Aronson, E., Curbow, B., McLeod, B., Pettigrew, T., White, L., and Yates, S. (1984). Improving utility conservation programs: Outcomes, interventions, and evaluations. *Energy, 9,* 485–494.

Costanza, R., Wainger, L., Folke, C. and Maler, K. (1993). Modeling complex ecological economic systems. *BioScience, 43,* 545–555.

Couch, J., Garber, T., and Karpus, L. (1979). Response maintenance and paper recycling. *Journal of Environmental Systems, 8,* 127–137.

Cox, S. J. (1985). No tragedy on the commons. *Environmental Ethics, 7,* 49–61.

Craig, C. S., and McCann, J. M. (1978). Assessing communication effects on energy conservation. *Journal of Consumer Research, 5,* 82–88.

Crocker, J. (1981). Judgment of covariation by social perceivers. *Psychological Bulletin, 90,* 272–292.

Cuomo, C. (1992). Unraveling the problems in ecofeminism. *Environmental Ethics, 14,* 351–363.

Daly, H. (1991). Institutions for a steady-state economy. In *Steady state economics.* Washington, DC: Island Press.

Darley, J. M., and Beniger, J. R. (1981). Diffusion of energy–conserving innovations. *Journal of Social Issues, 37*(2), 150–171.

Darwin, C. (1859). *The origin of species.* Reprinted by Penguin Books, 1968.

Davis, S. C., and Strang, S. G. (1993). *Transportation energy data book: Edition 13.* ORNL–6743. Oak Ridge, TN: Oak Ridge National Laboratory.

Dawkins, R. (1976). *The selfish gene.* Oxford, England: Oxford University Press.

De Greene, K. B. (1990). The turbulent–field environment of sociotechnical systems: Beyond metaphor. *Behavioral Science, 35,* 49–59.

DeBell, G. (1970). (Ed.), *The voter's guide to envional politics.* New York: Ballantine.

Dee, M. (1991). Personal communication, Ann Arbor, MI, (seafood wholesaler).

DeJoy, D. (1989). The optimism bias and traffic accident risk perception. *Accident Analysis and Prevention, 21,* 333–340.

Derksen, L., and Gartrell, J. (1993). The social context of recycling. *American Sociological Review, 58,* 434–442.

Deslauriers, B., and Everett, P. (1977). Effects of intermittent and continuous token reinforcement on bus ridership. *Journal of Applied Psychology, 62,* 369–375.

Devall (1985), see Devall and Sessions (1985).

Devall, B. (1988). *Simple in means, rich in ends: Practicing deep ecology.* Salt Lake City: Peregrine Smith Books.

Devall, B., and Sessions, G. (1985). *Deep ecology: Living as if nature mattered.* Salt Lake City: Gibbs Smith.

DeYoung, R. (1989). Exploring the difference between recyclers and non–recyclers: The role of information. *Journal of Environmental Systems, 18,* 341–51.

DeYoung, R. (1993). Changing behavior and making it stick. *Environment and Behavior, 25,* 485–505.

Dhammananda, K. (1982). *What Buddhists believe.* Malaysia: BMS Publications.

Diamond, J. (1993). New Guineans and their natural world. In Kellert, S., and Wilson, E. (Eds.) (1993). *The biophilia hypothesis.* Washington, DC: Island Press.

Dietz, T., and Stern, P. (1995). Toward a theory of choice: Socially embedded preference construction. *Journal of Socio-Economics, 24*, 261-279.

Dietz, T., Kalof, L., and Stern, P.C. (2002). Gender, values, and environmentalism. *Social Science Quarterly, 83(1)*, 353-364.

Dietz, T., Stern, P. C., and Guagnano, G. (1998). Social structural and social psychological determinants of environmental concern. *Environment and Behavior, 30*, 450-471.

Dobuzinskis, L. (1992). Modernist and postmodernist metaphors of the policy process: Control and stability vs. chaos and reflexive understanding. *Policy Sciences, 25*, 355-380.

Dooley, D., Catalano, R., Mishra, S., and Serxner, S. (1992). Earthquake preparedness: Predictors in a community survey. *Journal of Applied Social Psychology, 22*, 451-470.

Dubos, R. (1968). *So human an animal.* New York: Scribner.

Dunlap (1978), see Dunlap and van Liere (1978).

Dunlap, R. (1991). Trends in public opinion toward environmental issues: 1965-1990. *Society and Natural Resources, 4*.

Dunlap, R. and Van Liere, K. (1978). The "new environmental paradigm": A proposed measuring instrument and preliminary results. *Journal of Environmental Education, 9(4)*, 10-19.

Dunlap, R. and Van Liere, K. (1984). Commitment to the dominant social paradigm and concern for environmental quality. *Social Science Quarterly, 65*, 1013-1028.

Dunlap, R., Gallup, Jr., G., and Gallup, A. (1993). Global environmental concern: Results from an international public opinion survey. *Environment, 35(9)*, 7-15ff.

Dunlap, R. E. (2000, April). Americans have positive image of the environmental movement. *The Gallup Poll Monthly*, 19-25.

Dunlap, R. E. (2002). *Long-term trends in American's environmental concerns.* Unpublished manuscript.

Dunlap, R. E., and Saad, L. (2001, April). Only one in four Americans are anxious about the environment. *The Gallup Poll Monthly*, 6-16.

Dunlap, R. E., Van Liere, K.D., Mertig, A., and Jones, R.E. (2000). Measuring endorsement of the New Ecological Paradigm: A revised NEP scale. *Journal of Social Issues, 56(3)*, 425-442.

Dunlap, R. E., Xiao, C., and McCright, A.M. (2001). Politics and environment in America: Partisan and ideological cleavages in public support for environmentalism. *Environmental Politics, 10(4)*, 23-48.

Dunlap. R. E., and Scarce, R. (1991) The polls and poll trends: Environmental problems and protection. *Public Opinion Quarterly, 55*, 651-672.

DuPont, T. (1980). *Nuclear phobia: Phobic thinking about nuclear power.* Washington, DC: The Media Institute.

Durning, A. (1990). Ecology starts at home. *World Watch (magazine)* (March/April), pp. 39-40.

Dwivedi, O. and Tiwari, B. (1987). *Environmental crisis and Hindu religion.* New Delhi, India: Gitanjali Publishing House.

Earth Works Group. (1989). *50 simple things you can do to save the earth.* Berkeley, CA: Earthworks Press.

Eckberg, D., and Blocker, T. (1989). Varieties of religious involvement and environmental concerns: Testing the Lynn White thesis. *Journal for the Scientific Study of Religion, 28(4)*, 509-517.

Ehrenfeld, D. (1978). *The arrogance of humanism.* New York: Oxford University Press.

Ehrlich (1971), see Ehrlich (1978).

Ehrlich, A., and Ehrlich, P. (1994). Simple Simon environmental analysis. In Miller, Jr., G. T., *Living in the environment, eighth edition.* Belmont, CA: Wadsworth. 26-27.

Ehrlich, P. (1978). *The population bomb.* New York: Ballantine.

Ehrlich, P. R. (1981a). Environmental disruption: Implications for the social sciences. *Social Science Quarterly, 62*, 7-22.

Ehrlich, P. R. (1981b). An economist in wonderland. *Social Science Quarterly, 62*, 44-49.

Ehrlich, P. R. (1982). That's right—you should check it for yourself. *Social Science Quarterly, 63*, 385-387.

Ehrlich, P., and Ehrlich, A. (1990). Growing, growing, gone. *Sierra*, March/April, 36-40.

Ehrlich, P., and Holdren, J. (1971). Impact of population growth. *Science, 171*, 1212-1217.

Ester, P., and Winett, R. (1982). Toward more effective antecedent strategies for environmental programs. *Journal of Environmental Systems, 11*, 201-221.

Evans, G., and Cohen, S. (1987). Environmental stress. In D. Stokols and I. Altman (Eds.), *Handbook of environmental psychology.* New York: John Wiley & Sons.

Everett, P., and Watson, B. (1987). Psychological contributions to transportation. In Stokols, D., and Altman, I. (Eds.), *Handbook of environmental psychology.* New York: Wiley.

Everyone's Backyard (1990). McDonalds surrenders!!! (editorial) (Dec.), pp. 2-3.

Federal Highway Administration, U. S. Department of Transportation. (1991). *Highway statistics 1991*. Washington, DC: U.S. Government Printing Office.

Feeny, D., Berkes, F., McCay, B., and Acheson, J. (1990). The tragedy of the commons: Twenty–two years later. *Human Ecology*, *18*, 1–19.

Feller, W. (1968). *An introduction to probability theory and its applications, (3rd. Edition), Volume 1*. New York: Wiley.

Festinger, L. (1957). *A theory of cognitive dissonance*. Stanford, CA: Stanford University Press.

Finckenauer, J. (1982). *Scared Straight! and the panacea phenomenon*. Englewood Cliffs, NJ: Prentice–Hall, Inc.

Fischhoff, B., Hohenemser, C., Kasperson, R., and Kates, R. (1978). Handing hazards. *Environment*, *20*(7), 16–20ff.

Fischhoff, B., Lichtenstein, S., Slovic, P., Derby, S., and Keeney, R. (1981). *Acceptable risk*. Cambridge, England: Cambridge University Press.

Fischhoff, B., Slovic, P., and Lichtenstein, S. (1978). Fault Trees: Sensitivity of estimated failure probabilities to problem representation. *Journal of Experimental Psychlogy: Human Perception and Performance*, *4*(2), 330–344.

Fischhoff, B., Slovic, P., and Lichtenstein, S. (1980). Lay foibles and expert fables in judgments about risk. In O'Riordan, T., and Turner, R. (Eds.), *Progress in resource management and environment planning, volume 3*. Chichester, England: Wiley.

Fisher, A. (1988). Global model: One model to fit all. *Mosaic*, *19*(3/4), 52–59.

Fitchburg Office of the Planning Coordinator (1980). *Fitchburg Action to Conserve Energy: FACE final report*. Fitchburg, MA: Author.

Forrester, J. (1971). Counterintuitive behavior of social systems. *ZPG Newsletter* (June), p. 1ff.

Forrester, J. W. (1961). *Industrial dynamics*. Cambridge, MA: The M.I.T. Press.

Forrester, J. W. (1968). *Principles of systems*. Cambridge, MA: Wright–Allen Press.

Forrester, J. W. (1969). *Urban dynamics*. Cambridge, MA: The M.I.T. Press.

Forrester, J. W. (1980). System dynamics—future opportunities. In Legasto, Jr., A. A., Forrester, J. W., and Lyneis, J. M. (Eds.), *System Dynamics*. Amsterdam: North–Holland Publishing Company.

Forrester, J. W. (1987). Lessons from system dynamics modeling. *System Dynamics Review*, *3*(2), 136–149.

Fox, D. (1985). Psychology, ideology, utopia, and the commons. *American Psychologist*, *40*, 48–58.

Frank, R. H., Gilovich, T., and Regan, D. T. (1993). Does studying economics inhibit cooperation? *Journal of Economic Perspectives*, *7*, 159–171.

French, H. (1995). Forging a new global partnership. In Brown, L., et al., (Eds.), *State of the world 1995*. New York: W. W. Norton.

Gardiner, B., Shanklin, J., and Farman, J. (1985). Large losses of total ozone in Antarctica reveal seasonal C10x/NOx interaction. *Nature*, *315*, 207–210.

Gardner, G. (1978). Effects of federal human subjects regulations on data obtained in environmental stressor research. *Journal of Personality and Social Psychology*, *36*, 628–634.

Gardner, G. and Stern, P. (1996). *Environmental Problems and Human Behavior [First Edition]*. Boston: Allyn and Bacon.

Gardner, G., and Gould, L. (1989). Public perceptions of the risks and benefits of technology. *Risk Analysis*, *9*, 225–242.

Gardner, G., Tiemann, A., Gould, L., DeLuca, D., Doob, L., and Stolwijk, J. (1982). Risk and benefit perceptions, acceptability judgments, and self–reported actions toward nuclear power. *Journal of Social Psychology*, *116*, 179–197.

Gay, J. (1991). Changing times for Bering Sea crabbers. *National Fisherman*, *71*, 16–18.

Gelderloos, O. (1992). *Eco–theology: The Judeo–Christian tradition and the politics of ecological decision making*. Glasgow: Wild Goose Publications.

Geller E. S., Wylie, R. C., and Farris, J. C. (1971). An attempt at applying prompting and reinforcement toward pollution control. *Proceedings of the 79th Annual Convention of the American Psychological Association*, *6*, 701–02.

Geller, E. S. (1981). Evaluating energy conservation programs: Is verbal report enough? *Journal of Consumer Research*, *8*, 331–35.

Geller, E. S., Winett, R. A., and Everett, P. B. (1982). *Preserving the environment: New strategies for behavior change*. New York: Pergamon.

Geller, E., Chaffee, J., and Ingram, R. (1975). Promoting paper recycling on a university campus. *Journal of Environmental Systems*, *5*, 39–57.

Gibson, C., McKean, M., and Ostrom, E., eds. (2000). *People and Forests: Communities, Institutions, and Governance*. Cambridge, MA: MIT Press.

Gill, J. D., Crosby, L. A., and Taylor, J. R. (1986). Ecological concern, attitudes, and social norms in voting behavior. *Public Opinion Quarterly*, *50*, 537–54.

Gimbutas, M. (1974). *The gods and goddesses of Old Europe, 7000 to 3500 B.C.: Myths, legends, and cult images*. London: Thames & Hudson.

Glass, D., and Singer, J. (1972). *Urban stress: Experiments on noise and social stressors*. New York: Academic Press.

Gleick, J. (1987). *Chaos: Making a new science*. New York: Viking Penquin.

Gleick, J. (1992). Chasing bugs in the electronic village. *The New York Times Magazine* (June 14), p. 38ff.

Goerner, S. (1994). *Chaos and the evolving ecological universe*. Amsterdam: Gordon and Breach.

Golden West Productions. (1979). *Scared Straight* (film). Hollywood, CA: KTLA Television.

Goodman, R. (1980). Taoism and ecology. *Environmental Ethics*, 2, Spring, 73–80.

Gordon, T. (1992). Chaos in social systems. *Technological Forecasting and Social Change*, 42, 1–15.

Gould, L., Gardner, G., DeLuca, D., Tiemann, A., Doob, L., & Stolwijk, J. (1988). *Perceptions of technological risks and benefits*. New York: The Russell Sage Foundation.

Graham, J., Green, L., and Roberts, M. (1988). *In search of safety: Chemicals and cancer risk*. Cambridge, MA: Harvard University Press.

Gray, P. (1991). *Psychology, second edition*. New York: Worth Publishers.

Greeley, A. (1993). Religion and attitudes towards the environment. *Journal for the Scientific Study of Religion*, 32(1), 19–28.

Green, C. (1980). Not quite Dr. Strangelove. Paper presented at the Conference on Energy and Planning, Craigie College, Ayr, Scotland, May 27–29.

Green, L. W., Wilson, A. L., and Lovato, C. Y. (1986). What changes can health promotion achieve and how long do these changes last? The trade–offs between expediency and durability. *Preventive Medicine*, 15, 508–21.

Greenberg, M., Sachsman, D., Sandman, P., and Salomone, K. (1989). Network evening news coverage of environmental risk. *Risk Analysis*, 9, 119–126.

Greenberg, M., Sachsman, D., Sandman, P., and Salomone, K. (1989). Risk, drama and geography in coverage of environmental risk by network TV. *Journalism Quarterly*, Summer, 267–276.

Greenberg, M., Sandman, P., Sachsman, D., and Salomone, K. (1989). Network television news coverage of environmental risks. *Environment*, 31, 16ff.

Greening, L., and Dollinger, S. (1992). Adolescents' perceptions of lightning and tornado risks. *Journal of Applied Social Psychology*, 22, 755–762.

Gregersen, H., and Sailer, L. (1993). Chaos theory and its implications for social science research. *Human Relations*, 46, 777–802.

Gregory, R., Lichtenstein, S., and MacGregor, D. (1993). The role of past states in determining reference points for policy decisions. *Organizational Behavior and Human Decision Processes*, 55, 195–206.

Guagnano, G., Stern, P. C., and Dietz, T. (1995). Influences on attitude–behavior relationships: A natural experiment with curbside recycling. *Environment and Behavior*, 27, 699–718.

Habermas, J. (1984). *The theory of communicative action, Volume I. Reason and the rationalzization of society*. Boston: Beacon Press.

Habermas, J. (1987). *The theory of communicative action, Volume II. System and lifeworld*. Boston: Beacon Press.

Hamilton, (1964). The genetic evolution of social behavior, I & II. *Journal of Theoretical Biology*, 7, 1–52.

Hamilton, M. (1991). Big Mac attacks trash problem: McDonald's aims for 80 percent reduction of its solid waste. *Washington Post* (Apr. 17), p. B1.

Hand, C., and van Liere, K. (1984). Religion, mastery-over-nature, and environmental concern. *Social Forces*, 63(2), 555–570.

Hardin (1969), see Hardin (1968).

Hardin, G. (1968). Denial and the gift of history. In Hardin, G. (Ed.), *Population, evolution, and birth control*. San Francisco: W. H. Freeman and Company.

Hardin, G. (1968). The tragedy of the commons. *Science*, 162, 1243–1248.

Hardin, G. (1980). Second thoughts on the "tragedy of the commons." In Daly, H. (Ed.), *Economics, ecology and ethics: Essays towards a steady–state economy*. San Francisco: Freeman.

Hargrove, E. (Ed). (1986). *Religion and environmental crisis*. Athens: University of Georgia Press.

Hartig, T., Mang, M., and Evans, G. (1991). Restorative effects of natural environment experiences. *Environment and Behavior*, 23, 3–26.

Hayes, S., and Cone, J. (1977). Reducing electrical energy use: Payments, information, and feedback. *Journal of Applied Behavior Analysis*, 10, 425–435.

Heberlein, T. A. (1975). Conservation information, the energy crisis and electricity consumption in an apartment complex. *Energy Systems and Policy*, 1, 105–17.

Heberlein, T. A., and Baumgartner, R. M. (1986). Changing attitudes and electricity consumption in a time–of–use experiment. In Monnier, E. et al., (Eds.), *Consumer behavior and energy policy*. New York: Praeger.

Heberlein, T. A., and Warriner, G. K. (1983). The influence of price and attitude on shifting residential electricity consumption from on– to off–peak periods. *Journal of Economic Psychology*, *4*, 107–130.

Henrion, M., and Fischhoff, B. (1986). Assessing uncertainty in physical constants. *American Journal of Physics*, *54*, 791–798.

Herzog, T., Kaplan, S., and Kaplan, R. (1982). The prediction of preference for unfamiliar urban places. *Population and Environment*, *5*, 43–59.

Higbee, K. (1969). Fifteen years of fear arousal: Research on threat appeals: 1953–1968. *Psychological Bulletin*, *72*, 426–444.

Hines, J. M., Hungerford, H. R., and Tomera, A. N. (1987). Analysis and synthesis of research on responsible environmental behavior: A meta–analysis. *Journal of Environmental Education*, *18*, 1–8.

Hiroto, D. (1974). Locus of control and learned helplessness. *Journal of Experimental Psychology*, *102*, 187–193.

Hirschhorn, L. (1985). On technological catastrophe. *Science*, *228*, 846–847.

Hirst, E. (1987). Cooperation and community conservation. Final report, Hood River Conservation Project. DOE/BP–11287–18.

Hirst, E., Berry, L., and Soderstrom, J. (1981). Review of utility home energy audit programs. *Energy*, *6*, 621–630.

Hirst, E., Clinton, J., Geller, H., and Kroner, W. (1986). Energy efficiency in buildings: Progress and promise. Washington, DC: American Council for an Energy–Efficient Economy.

Hobbes, T. (1651). *Leviathan, or the matter, form and power of a commonwealth*. Schneider, H. (Ed.). Indianapolis: Bobbs-Merrill, 1958).

Hohenemser, C., Kates, R., and Slovic, P. (1983). The nature of technological hazard. *Science*, *220*, 378–384.

Holling, C. (1987). Simplifying the complex: the paradigms of ecological function and structure. *European Journal of Operational Research*, *30*, 139–146.

Holling, C. (1992). Cross-scale morphology, geometry and dynamics of ecosystems. *Ecological Monographs*, *62*, 447–502.

Holusha, J. (1990). A setback for polystyrene. *New York Times* (Nov. 18).

Hopper, J. R, and Nielsen, J. M. (1991). Recycling as altruistic behavior: Normative and behavioral strategies to expand participation in a community recycling program. *Environment and Behavior*, *23*, 195–220.

Hudson, R. (1989). Dealing with international hostage–taking: Alternatives to reactive counterrorist assaults. *Terrorism*, 321–378.

Hume, S. (1991). McDonald's. *Advertising Age* (Jan. 29), *62*, 32.

Hutton, R. B. (1982). Advertising and the Department of Energy's campaign for energy conservation. *Journal of Advertising*, *11*, 27–39.

Hynes, M., and Vanmarcke, E. (1976). Reliability of embankment performance predictions. *Proceedings of the ASCE Engineering Mechanics Division Specialty Conference*. Waterloo, Ontario, Canada: University of Waterloo Press.

Iltis, H., Loucks, O, and Andrews, P. (1970). Criteria for an optimum human environment. *Bulletin of the Atomic Scientists*, *25*, 2–6.

Inglehart, R. (1990). *Culture shift in advanced industrial society*. Princeton, NJ: Princeton University Press.

Inglehart, R. (1992). Public support for environmental protection: Objective problems and subjective values. Paper presented at the annual meeting of the American Political Science Association, August, Chicago, IL.

Ingram, R., and Geller, E. S. (1975). A community integrated, behavior modification approach to facilitating paper recycling. *JSAS Catalog of Selected Documents in Psychology*, *5*, 327. Ms. No. 1097.

Ip, P. (1983). Taoism and the foundations of environmental ethics. *Environmental Ethics*, *5*, 335–343.

James, G. (1989). Pupil starts a revolution of her own: Ban plastics. *New York Times* (Apr. 28).

Johnson, E., and Tversky, A. (1983). Affect, generalization and the perception of risk. *Journal of Personality and Social Psychology*, *45*, 20–31.

Jones, M., and Sawhill, R. (1992). Just too good to be true. *Newsweek* (May 4), p. 68.

Jones, R., and Dunlap, R. (1992). The social bases of environmental concern: Have they changed over time? *Rural Sociology*, *57*, 28–47.

Jones, S., Martin, R., and Pilbeam, D. (1992). *The Cambridge encyclopedia of human evolution*. Cambridge, England: Cambridge University Press.

Kahneman, D., and Tversky, A. (1979). Prospect theory: An analysis of decisions under risk. *Econometrica*, *47*, 262–291.

Kaplan, R., and Kaplan, S. (1989). *The experience of nature*. Cambridge, England: Cambridge University Press.

Kaplan, S. (1991). Beyond rationality: Clarity–based decision making. In Garling, T., and Evans, G. (Eds.), *Environment, cognition, and action*. Oxford, England: Oxford University Press.

Kasperson, R. E., Golding, D., and Tuler, S. (1992). Social distrust as a factor in siting hazardous facilities and communicating risks. *Journal of Social Issues, 48,* 161–187.

Kasperson, R., Renn, O., Slovic, P., Brown, H., Emel, J., Goble, R., Kasperson, J., and Ratick, S. (1988). The social amplification of risk: A conceptual framework. *Risk Analysis, 8,* 177–187.

Kates, R. (1962). *Hazard and choice perception in flood plain managment.* Chicago: University of Chicago, Department of Geography, Research Paper No. 78.

Kates, R. (1986). Review: Normal accidents: Living with high–risk systems. *The Professional Geographer, 38,* 121–122.

Katzev, R. D., & Johnson, T. R. (1987). *Promoting energy conservation: An analysis of behavioral research.* Boulder, CO: Westview.

Kauffman, S. (1991). Antichaos and adaptation. *Scientific American,* August, 78–84.

Kellert, S. (1993). The biological basis for human values of nature. In Kellert, S., and Wilson, E. (Eds.), *The biophilia hypothesis.* Washington, DC: Island Press/Shearwater Books.

Kellert, S., and Wilson, E. (Eds.) (1993). *The biophilia hypothesis.* Washington, DC: Island Press.

Kelman, H. C. (1958). Compliance, identification, and internalization: Three processes of attitude change. *Journal of Conflict Resolution, 2,* 51–60.

Kempton (1992), see Kempton, Darley, and Stern (1992).

Kempton, W., Boster, J., and Hartley, J. (1995). *Environmental values in American culture.* Cambridge, MA: MIT Press.

Kempton, W., Darley, J. M., and Stern, P. C. (1992). Psychological research for the new energy problems: Strategies and opportunities. *American Psychologist, 47*(10), 1213–1223.

Kempton, W., Darley, J., and Stern, P. (1992). Psychological research for the new energy problems: Strategies and opportunities. *American Psychologist, 47,* 1213–1223.

Kempton, W., Harris, C., Keith, J., and Weihl, J. (1985). Do consumers know "what works" in energy conservation? *Marriage and Family Review, 9,* 115–133.

Kendler, K., Neale, M., Kessler, R., Heath, A., and Eaves, L. (1992). The genetic epidemiology of phobias in women. *Archives of General Psychiatry, 49,* 273–281.

Kingsnorth, R. (1991). The Gunther Special: Deterrence and the DUI offender. *Criminal Justice and Behavior, 18,* 251–266.

Kinzig, A. P., and Socolow, R. H. (1994). Human impacts on the nitrogen cycle. *Physics Today* (Nov.), 24–31.

Knopf, R. (1987). Human behavior, cognition, and affect in the natural environment. In Stokols, D., and Altman, I. (Eds.), *Handbook of environmental psychology.* New York: John Wiley and Sons.

Kristiansen, C. (1983). Newspaper coverage of diseases and actual mortality statistics. *European Journal of Social Psychology, 13,* 193–194.

Kunreuther, H. (1978). Even Noah built an ark. *The Wharton Magazine* (Summer), p. 28–35.

La Porte, T., and Consolini, P. (1991). Working in practice but not in theory: Theoretical challenges of high–reliability organizations. *Journal of Public Administration Research and Theory, 1,* 19–47.

Langer, S. (1979). The Rahway State Prison Lifers' Group: A critical analysis. Unpublished paper. Union, New Jersey: Kean College Department of Sociology, October.

Layne, L., Kempton, W., Behrens, A., Diamond, R., Fels, M., and Reynolds, C. (1988). Design criteria for a consumer energy report: A pilot field study. PU/CEES Report No. 220. Princeton, NJ: Center for Energy and Environmental Studies, Princeton University.

Lazarus, R. (1966). *Psychological stress and the coping process.* New York: McGraw–Hill.

Lazarus, R., and Folkman, S. (1984). *Stress, appraisal, and coping.* New York: Springer.

Legasto, Jr., A. A., and Maciariello, J. A. (1980). System dynamics: A critical review. In Legasto, Jr., A. A., Forrester, J. W., and Lyneis, J. M. (Eds.), *System dynamics.* Amsterdam: North–Holland Publishing Company.

Lehman, D., and Taylor, S. (1987). Date with an earthquake: Coping with a probable, unpredictable disaster. *Personality and Social Psychology Bulletin, 13,* 546–555.

Leonard–Barton, D. (1980). The role of interpersonal communication networks in the diffusion of energy conserving practices and technologies. Paper presented at the International Conference on Consumer Behavior and Energy Policy, Banff, Alberta, Canada, August 1980.

Leopold, A. (1949). *A Sand County almanac.* New York: Oxford University Press.

Lepper, M. W., and Greene, D. (1978). *The hidden costs of reward: New perspectives on the psychology of human motivation.* Hillsdale, NJ: L. Erlbaum Associates, distributed by New York: Halstead Press.

Levine, B. L. (1986). The tragedy of the commons and the comedy of community: The commons in history. *Journal of Community Psychology, 14,* 81–99.

Lewis, H. (1980). The safety of fission reactors. *Scientific American, 242,* 53–65.

Lewis, H. (1990). *Technological Risk.* New York: W. W. Norton.

Lewis, R. (1983). Scared Straight—California style: Evaluation of the San Quentin Squires program. *Criminal Justice and Behavior*, *10*, 209–226.

Lichtenstein, S., Slovic, P., Fischhoff, B., Layman, M., and Combs, B. (1978). Judged frequency of lethal events. *Journal of Experimental Psychology: Human Learning and Memory*, *4*, 551–578.

Lindblom, C. (1959). The science of muddling through. *Public Administration Review*, *19*, 79-88.

Lindblom, C. (1977). *Politics and markets*. New York: Basic Books.

Lipsett, B. (1990). Plastics industry grasps for straws. *Everyone's Backyard* (Jan.–Feb.), pp. 6–8.

Lloyd, W. F. (1833). *Two lectures on the checks to population*. England: Oxford University Press. Reprinted (in part) in G. Hardin (Ed.). (1984). *Population, evolution, and birth control*. San Francisco: Freeman.

Löfstedt, R. (1993). Hard habits to break: Energy conservation patterns in Sweden. *Environment*, *35*(2), 10–15ff.

Lovins, A. B. (1977). *Soft energy paths: Toward a durable peace*. Cambridge, MA: Ballinger.

Lovrich, N. Tsurutani, T., and Pierce, T. (1987). Culture, politics, and mass publics: Traditional and modern supporters of the New Environmental Paradigm in Japan and the U.S. *Journal of Politics*, *49*, 54-79.

Lowrance, W. (1976). *Of acceptable risk: Sciene and the determination of safety*. Los Altos, CA: William Kaufmann, Inc.

MacEachern, D. (1990). *Save our planet: 750 everyday ways you can help clean up the earth*. New York: Dell.

Machol, R. E. (1980). Foreword to: Legasto, Jr., A. A., Forrester, J. W., and Lyneis, J. M. (Eds.), *System dynamics*. Amsterdam: North–Holland Publishing Company.

Martin, N., (1990). Environmental myths and hoaxes: The evidence of guilt is insufficient. *Vital Speeches of the Day* (May), pp. 434–437.

Marwell, G., and Ames, R. (1981). Economists free ride, does anyone else? *Journal of Public Economics*, *15*, 295–310.

Maslow, A. (1966). *The psychology of science: A reconnaissance*. New York: Harper & Row.

Maxwell, T., and Costanza, R. (in press). An approach to modelling the dynamics of evolutionary self-organization. *Ecological Modeling*.

McCabe, J. T. (1990). Turkana pastoralism: A case against the tragedy of the commons. *Human Ecology*, *18*, 81–103.

McCay, B. J. (1993). Management regimes. Paper presented at the Meeting on Property Rights and the Performance of Natural Resource Syatems, The Beijer Institute, Stockholm, September.

McCay, B.J., and Acheson, J.M., eds. (1987). *The Question of the Commons: The Culture and Ecology of Communal Resources*. Tucson: University of Arizona Press.

McClelland, L. (1980). Encouraging energy conservation in multifamily housing: RUBS and other methods of allocating energy costs to residents. Institute of Behavioral Science, University of Colorado, June.

McConnell, J., and Philipchalk, R. (1992). *Understanding human behavior*. Fort Worth: Harcourt Brace Jovanovich.

McDougall, G., Claxton, J., and Ritchie, J. (1983). Residential home audits: An empirical analysis of the ENER$AVE program. *Journal of Environmental Systems*, *12*, 265–278.

McGill, A. (1984). Book Review: Normal accidents: Living with high–risk technologies. *Human Resource Management*, *23*, 434–436.

McHarg, I. (1971). *Design with nature*. Garden City, NY: Natural History Press.

Meadows, D. H. (1985). Charting the way the world works. *Technology Review*, (Feb./Mar.), 54–63.

Meadows, D. H., Meadows, D. L., and Randers, J. (1992). *Beyond the limits: Confronting global collapse; envisioning a sustainable future*. Post Mills, Vermont: Chelsea Green.

Meadows, D. H., Meadows, D. L., Randers, J., and Behrens, III, W. W. (1972). *The limits to growth*. New York: Universe Books.

Meadows, D., and Meadows, D. (Eds.) (1973). *Toward a global equilibrium: Collected papers*. Portland, OR: Productivity Press.

Merchant, C. (1992). *Radical ecology: The search for a livable world*. NY: Routledge.

Midden, C. (1994). Direct participation in macro–issues: A multiple–group approach. The Dutch national debate on energy policy. In Renn, O., Webler, T., and Wiedemann, P. (Eds.), *Fairness and competence in citizen participation: Evaluating models for environmental discourse*. Boston, MA: Kluwer.

Milbrath, L. (1994). Envisioning a sustainable society. In Miller, Jr., G. T., *Living in the environment, eighth edition*. Belmont, CA: Wadsworth. 698-699.

Miller, G. Tyler (2002). *Living in the environment. Twelfth Edition*. Belmont, California: Wadsworth/Thomson Learning.

Miller, Jr., G. T. (1990, 1992, 1994). *Living in the environment. sixth, seventh, and eighth editions*. Belmont, CA: Wadsworth Publishing Co.

Miller, Jr., G. T. (1992). *Living in the environment: An introduction to environmental science, seventh edition.* Belmont, CA: Wadsworth.

Miller, R. D., and Ford, J. M. (1985). *Shared savings in the residential market: A public/private partnership for energy conservation.* Baltimore, MD: Energy Task Force, Urban Consortium for Technology Initiatives.

Mitchell, R. (1990). Public opinion and the Green lobby: Poised for the 1990s? In Vig, N. and Craft, M. (Eds.), *Environmental policy in the 1990s—towards a new agenda.* Washington, DC: CQ Press.

Mitchell, R. C. (1979). National environmental lobbies and the apparent illogic of collective action. In Russell, C. (Ed.), *Collective decision making: Applications from public choice theory.* Baltimore, MD.: Johns Hopkins University Press.

Moehlmann, J. (1992). The religious community and the environment. *Bioscience, 42,* September, 627.

Montague, P. (1994). A new environmentalism for the 1990s and beyond. In Miller, Jr., G. T., *Living in the environment, eighth edition.* Belmont, CA: Wadsworth. 41–42.

Moore, C. A. (1989). McTruth: Fast food for thought. *Washington Post* (Dec. 10).

Morone, J., and Woodhouse, E. (1986). *Averting catastrophe: Strategies for regulating risky technologies.* Berkeley: University of California Press.

Mosekilde, E., Larsen, E. R., and Sterman, J. D. (1990). Coping with complexity: Deterministic chaos in human decision making behavior. In Bohr, H. (Ed.), *Proceedings of the Symposium Characterising Complex Systems: Interdisciplinary Workshop on Complexity and Chaos.* Singapore: World Scientific.

Mulilis, J., and Lippa, R. (1990). Behavioral change in earthquake preparedness due to negative threat appeals: A test of protection motivation theory. *Journal of Applied Social Psychology, 20,* 619–638.

Nader, R. (1988). Federal insecticide, fungicide, and rodenticide act. Hearing. Washington, DC: GPO.

Naess, A. (1989). *Ecology, community, and lifestyle: An outline of an ecosophy.* Cambridge, England: Cambridge University Press.

Naess, A., and Rothenberg, D. (1989). *Ecology, community and lifestyle: Outline of an ecosophy.* Cambridge, England: Cambridge University Press.

National Research Council (2002). *The Drama of the Commons.* Committee on the Human Dimensions of Global Change. E. Ostrom, T. Dietz, N Dolsak, P.C. Stern, S. Stonich, and E.U. Weber, eds. Washington: National Academy Press.

Netting, R. (1976). What Alpine peasants have in common: Observations on communal tenure in a Swiss village. *Human Ecology, 4,* 135–146.

Netting, R. McC. (1981). *Balancing on an Alp: Ecological change and continuity in a Swiss mountain community.* Cambridge, England: Cambridge University Press.

Nisbett, R., Borgida, E., Crandall, R., and Reed, H. (1982). Popular induction: Information is not necessarily informative. In Kahneman, D., Slovic, P., and Tversky, A. (Eds.), *Judgment under uncertainty: Heuristics and biases.* Cambridge, England: Cambridge University Press.

Nordhaus, W. (1992). Lethal Model 2: *The Limits to Growth* revisited. *Brookings Papers on Economic Activity, 2,* 1–59.

Odum, H., and Odum, E. (1976). *Energy basis for man and nature.* New York: McGraw–Hill.

Öhman, A., Dimberg, U., and Öst, L. (1985). Animal and social phobias: Biological constraints on learned fear responses. In Reiss, S., and Bootzin, R. *Theoretical issues in behavior.* New York: Academic Press.

Ophuls, W. (1973). Leviathan or oblivion? In Daly, H. (Ed.), *Towards a steady–state economy.* San Francisco: Freeman.

Ophuls, W. (1977). *Ecology and the politics of scarcity.* San Francisco: Freeman.

Ornstein, R., and Ehrlich, P. (1989). *New world, new mind: Moving toward conscious evolution.* New York: Touchstone; Simon and Schuster.

Ortner, S. (1974). Is female to male as nature is to culture? In Rosaldo, M., and Lamphere, L. (Eds.), *Woman, culture, and society.* Stanford: Stanford University Press.

Ostrom, E. (1990). *Governing the commons: The evolution of institutions for collective action.* Cambridge, England: Cambridge University Press.

Ostrom, E., Gardner, R., and Walker, J. (1994). *Rules, Games, and Common-Pool Resources.* Ann Arbor: University of Michigan Press.

Pardini, A. U., and Katzev, R. D. (1984). The effect of strength of commitment on newspaper recycling. *Journal of Environmental Systems, 13,* 245–54.

Pateman, C. (1970). *Participation and democratic theory.* Cambridge, England: Cambridge University Press.

Pearce, D. (1992). Review of *Beyond the Limits, International Affairs, 68,* 749–750.

Perakis, S.S., and Hedin, L.O. (2002). Nitrogen loss from unpolluted South American forests mainly via dissolved organic compounds. *Nature, 415,* 416–419.

Perney, L. (1993). Book Review: Ablaze. *Audubon, 95* (March/April), p. 120ff.

Perrow, C. (1984). *Normal accidents: Living with high–risk technologies*. New York: Basic Books.

Peterson, G. (1993). Yellowstone Park fire. *High Country News* (July 26).

Platt, J. (1973). Social traps. *American Psychologist, 28,* 641–651.

Polich, M. D. (1984). Minnesota RCS: The myths and realities. In American Council for an Energy–Efficient Economy (Ed.), *Doing better: Setting an agenda for the second decade, G,* 141–151. Washington: Author.

Population Action International (1994). Washington population update: News and analysis of U.S. and international population assistance. Memorandam, November. 1120 19th Street N.W., Washington, DC 20036.

Population Reference Bureau. (1972). Suburban growth: A case study. *Population Bulletin, 28*(1), whole issue.

Population Statement of Scientific Academies, 1993, see next entry.

Population Summit of the World's Scientific Academies, New Delhi, India, 24-27 October. (1993). Washington DC: National Academy Press.

Postel, S. (1989). Halting land degradation. In Brown, L., et al., (Eds.), *State of the world, 1989.* New York: W. W. Norton.

Postel, S. (1990). Saving water for agriculture. In Brown, L., et al., (Eds.), *State of the world, 1990.* New York: W. W. Norton.

Postel, S. (1994). Carrying capacity: Earth's bottom line. In Brown, L., et al., (Eds.), *State of the world, 1994.* New York: W. W. Norton.

Postel, S., and Flavin, C. (1991). Reshaping the global economy. In Brown, L., et al., (Eds.), *State of the world, 1991.* New York: W. W. Norton.

Prentice–Dunn, S., and Rogers, R. (1986). Protection motivation theory and preventive health: Beyond the health belief model. *Health Education Research, 1,* 153–161.

Presbury, J., and Moore, H. (1983). Taking the 10–year–old offender to jail: An alternative to "Scared Straight," *Personnel and Guidance Journal, 62,* 114–116.

Prester, G., Rohrmann, B., and Schellhammer, E. (1987). Environmental evaluations and participation activities: A social psychological field study. *Journal of Applied Social Psychology, 17,* 751–87.

Pyramid Film & Video. (1987). *Scared Straight! 10 years later* (film for TV). Santa Monica, CA.

Raiffa, H. (1980). "Concluding remarks." p. 340 of Schwing, R. and Albers, Jr., W. (Eds.), *Societal risk assessment: How safe is safe enough?* New York: Plenum Press.

Rappaport, R. A. (1970). Sanctity and adaptation. *Io,* No. 7, 46–71.

Read, P. (1993). *Ablaze: The story of the heroes and victims of Chernobyl.* New York: Random House.

Renn, O., and Webler, T. (1992). Anticipating conflicts: Public participation in managing the solid waste crisis. *GAIA: Ecological Perspectives in Science, Humanities, and Economics, 2,* 84–94.

Renn, O., Burns, J., Kasperson, J., Kasperson, R., and Slovic, P. (1992). The social amplification of risk: Theoretical foundations and empirical applications. *Journal of Social Issues, 48*(4), 137–160.

Renn, O., Webler, T., Rakel, H., Dienel, P., & Johnson, B. (1993). Public participation in decision making: A three-step procedure. *Policy Sciences, 26,* 189–214.

Repetto, R. C. (1986). *World enough and time: Successful strategies for resource management.* New Haven, CT: Yale University Press.

Reppy, J. (1984). The automobile air bag. In Nelkin, D. (Ed.), *Controversy: Politics of technical decisions.* Beverly Hills, CA: Sage Publications.

Rippetoe, P., and Rogers, R. (1987). Effects of components of protection–motivation theory on adaptive and maladaptive coping with a health threat. *Journal of Personality and Social Psychology, 52,* 596–604.

Rogers, R. (1983). Cognitive and physiological processes in fear appeals and attitude change: A revised theory of protection motivation. In Cacioppo, J., and Petty, R. (Eds.), *Social psychology: A sourcebook.* New York: Guilford Press.

Ross, S. (1994). World's populace moving into cities. (An Associated Press article.) *Ann Arbor (MI) News* (Sep. 20), p. A–6.

Rossi, P. (1985). Review of Normal accidents: Living with high–risk technologies. *American Journal of Sociology, 91,* 181.

Roszak, Theodore. (1973). *Where the wasteland ends: Politics and transcendence in postindustrial society.* New York: Doubleday and Company, Inc.

Roth, J., Scholz, J., and Witte, A. (Eds.) (1989). *Taxpayer compliance, Volume 1: An agenda for research.* Philadelphia: University of Pennsylvania Press.

Rousseau, J.-J. (1762). *The social contract.* Frankel, C. (Ed.). New York: Hafner, 1947.

Ruthen, R. (1993). Adapting to complexity. *Scientific American* (Jan.), 130–140.

Sale, K. (1991). *Dwellers in the land: The bioregional vision (2nd ed.).* Philadelphia: New Society Publishers.

Salleh, Ariel. (1992). The ecofeminism/deep ecology debate: A reply to patriarchal reason. *Environmental Ethics, 14*, 195–216.

Sandman, P. (1987). Risk communication: Facing public outrage. *U.S. Environmental Protection Agency Journal, 13*, 21–22.

Schumacher, E. F. (1973). *Small is beautiful: Economics as if people mattered.* New York: Harper and Row.

Schwartz, S. (1992). Universals in the content and structure of values: Theoretical advances and empirical tests in 20 countries. *Advances in Experimental Social Psychology, 25*, 1–65.

Schwartz, S. H. (1977). Normative influences on altruism. In L. Berkowitz (Ed.), *Advances in experimental social psychology: Vol. 10.* New York: Academic Press.

Seed, J., Macy, J., Fleming, P., and Naess, A. (1988). *Thinking like a mountain: Towards a council of all beings.* Philadephia: New Society Publishers.

Segal, N. (1993). Twin, sibling, and adoption methods: Tests of evolutionary hypotheses. *American Psychologist, 48*, 943–956.

Seligman, C., Becker, L., and Darley, J. (1981). Encouraging residential energy conservation through feedback. In Baum, A., and Singer, J. (Eds.), *Advances in environmental psychology: Vol. 3. Energy: Psychological perspectives.* Hillsdale, NJ: Erlbaum.

Seligman, M. (1975). *Helplessness.* San Francisco: Freeman.

Selye, H. (1956). *The stress of life.* New York: McGraw–Hill.

Sessions (1985), see Devall and Sessions (1985).

Seydel, E., Taal, E., and Wiegman, O. (1990). Risk–appraisal, outcome and self–efficacy expectancies: Cognitive factors in preventive behavior related to cancer. *Psychology & Health, 4*, 99–109.

Shaiko, R. (1987). Religion, politics, and environmental concern: A powerful mix of passions. *Social Science Quarterly, 68*, 244–262.

Sheridan, T. B. (1980). Human error in nuclear power plants. *Technology Review*, (Feb.), 22–33.

Sherif, M., Harvey, O. J., White, B. J., Hood, W. R., and Sherif, C. W. (1961). *Intergroup conflict and cooperation: The robbers cave experiment.* Norman, OK: University Book Exchange.

Sherr, L., (1990). Fear arousal and AIDS: Do shock tactics work? *AIDS, 4*, 361–364.

Shine, K., Derwent, R., Wuebbles, D., and Morcrette, J. (1990). Radiative forcing of climate. In Houghton, J., Jenkins, G., and Ephraums, J. (Eds.), *Climate change: The IPCC assessment.* New York: Cambridge Unversity Press.

Shippee, G. (1980). Energy consumption and conservation psychology: A review and conceptual analysis. *Environmental Management, 4*, 297–314.

Shiva, V. (1989). *Staying alive: Women, ecology, and development.* London: Zed Books.

Shofield, J., and Pavelchak, M. (1985). The day after: The impact of a media event. *American Psychologist, 40*(5), 542–548.

Sills, D. (1984). Book Review: The trouble with technology. *Nature, 309*, 185.

Silver, C., and R. DeFries, for the National Academy of Sciences, (1990). *One Earth, one future: Our changing global environment.* National Academy Press, Washington, DC.

Simms, D. (1986). Book Review: Normal accidents: Living with high–risk technologies. *Technology and Culture, 27*, 903–905.

Simon (1991), see Simon (1981).

Simon, H. (1956). Rational choice and the structure of the environment. *Psychological Review, 63*, 129–138.

Simon, H. (1957). *Models of man: Social and rational.* New York: Wiley

Simon, H. (1959). Theories of decision making in economics and behavioral science. *American Economic Review, 49*, 253–283.

Simon, H. (1990). A mechanism for social selection and successful altruism. *Science, 250*, 1665–1668.

Simon, J. L. (1981). Environmental disruption or environmental improvement? *Social Science Quarterly, 62*, 30–43.

Simon, J. L. (1982). Paul Ehrlich saying it is so doesn't make it so. *Social Science Quarterly, 63*, 381–385.

Sims, J., and Baumann, D. (1972). The tornado threat: Coping styles of the North and South. *Science, 176*, 1386–1392.

Sims, J., and Baumann, D. (1983). Educational programs and human response to natural hazards. *Environment and Behavior, 15*, 165–189.

Singleton, S., and Taylor, M. (1992) Common property, collective action, and community. *Journal of Theoretical Politics, 4*, 309–324.

Sjoberg, Lennart (1993). Personal communication.

Skinner, B. (1978). *Reflections on behaviorism and society.* Englewood Cliffs, NJ: Prentice-Hall.

Skinner, B. F. (1938). *The behavior of organisms.* New York: Appleton–Century–Crofts.

Skinner, B. F. (1971). *Beyond freedom and dignity.* New York: Bantam Books.

Slovic, P. (1993a). Perceptions of environmental hazards: Psychological perspectives. In Gärling, T., and Golledge,

R. (Eds.), *Behavior and environment: Psychological and geographical approaches*. Holland: Elsevier Science Publishers.

Slovic, P. (1993b). Perceived risk, trust, and democracy. *Risk Analysis*, *13*(6), 675–682.

Slovic, P. (1994). Personal communication (letter of Jan. 3).

Slovic, P., and Fischhoff, B. (1983). How safe is safe enough?: Determinants of perceived and acceptable risk. In Walker, C. A., Gould, L. C., and Woodhouse, E. J. (Eds.), *Too hot to handle?: Social and policy issues in the management of radioactive wastes*. New Haven, CT: Yale University Press.

Slovic, P., Fischhoff, B., and Lichtenstein, S. (1976). Cognitive processes and societal risk taking. In Carroll, M., and Bayne, J. (Eds.), *Cognition and social behavior*. Potomac, MD: Lawrence Erlbaum Associates.

Slovic, P., Fischhoff, B., and Lichtenstein, S. (1978). Accident probabilities and seat belt usage: A psychological perspective. *Accident Analysis and Prevention*, *10*, 281–285.

Slovic, P., Fischhoff, B., and Lichtenstein, S. (1979). Rating the risks. *Environment*, *21*, 14ff.

Slovic, P., Fischhoff, B., and Lichtenstein, S. (1980). Facts and fears: Understanding perceived risk. In Schwing, R, and Albers, Jr., W. (Eds.), *Societal risk assessment: How safe is safe enough?* New York: Plenum Press.

Slovic, P., Fischhoff, B., and Lichtenstein, S. (1984). Regulation of risk: A psychological perspective. In Noll, R. (Ed.), *Social science and regulatory policy*. Berkeley, CA: University of California Press.

Slovic, P., Fischhoff, B., Lichtenstein, S., Corrigan, B., and Combs, B. (1977). Preference for insuring against probable small losses: Implications for the theory and practice of insurance. *Journal of Risk and Insurance*, *44*, 237–258.

Slovic, P., Flynn, J., and Layman, M. (1991). Perceived risk, trust, and the politics of nuclear waste. *Science*, *254*, 1603–1607.

Slovic, P., Kunreuther, H., White, G. (1974). Decision processes, rationality, and adjustment to natural hazards. In White, G. (Ed.), *Natural hazards, local, national and global*. New York: Oxford University Press.

Smil, V. (1988). *Energy in China's modernization: Advances and limitations*. Armonk, N.Y.: M.E. Sharpe.

Sowby, F. (1965). Radiation and other risks. *Health Physics*, *11*, 879–887.

Stasson, M., and Fishbein, M. (1990). The relation between risk and preventive action: A within–subject analysis of perceived driving risk and intentions to wear seatbelts. *Journal of Applied Social Psychology*, *20*, 1541–1557.

Sterman, J. D. (1988). Deterministic chaos in models of human behavior: Methodological issues and experimental results. *System Dynamics Review*, *4*(1–2), 148–178.

Stern, P. C. (1976). Effect of incentives and education on resource conservation decisions in a simulated commons dilemma. *Journal of Personality and Social Psychology*, *34*, 1285–1292.

Stern, P. C. (1986). Blind spots in policy analysis: What economics doesn't say about energy use. *Journal of Policy Analysis and Management*, *5*, 200–227.

Stern, P. C. (1993). A second environmental science: Human–environment interactions. *Science*, *260*, 1897–1899.

Stern, P. C. and Dietz, T. (1994). The value basis of environmental concern. *Journal of Social Issues*, (forthcoming).

Stern, P. C., and Dietz, T. (1994). The value basis of environmental concern. *Journal of Social Issues*, *50*(3), 65–84.

Stern, P. C., and Oskamp, S. (1987). Managing scarce environmental resources. In Stokols, D., and Altman, I. (Eds.), *Handbook of environmental psychology*. New York: Wiley.

Stern, P. C., Aronson, E., Darley, J. M., Hill, D. H., Hirst, E., Kempton, W., and Wilbanks, T. J. (1986). The effectiveness of incentives for residential energy conservation. *Evaluation Review*, *10*, 147–176.

Stern, P. C., Black, J. S., and Elworth, J. T. (1981). *Home energy conservation: Programs and strategies for the 1980s*. Mount Vernon, NY: Institute for Consumer Policy Research, Consumers Union Foundation.

Stern, P. C., Dietz, T., and Black, J. S. (1986). Support for environmental protection: The role of moral norms. *Population and Environment*, *8*, 204–222.

Stern, P. C., Dietz, T., and Guagnano, G. A. (1995). The New Ecological Paradigm in social-psychological context. *Environment and Behavior*, *27*, 723–743.

Stern, P. C., Dietz, T., and Kalof, L. (1993). Value orientations, gender, and environmental concern. *Environment and Behavior*, *25*, 322–348.

Stern, P. C., Dietz, T., Kalof, L., and Guagnano, G. (1995). Values, beliefs, and proenvironmental action: Attitude formation toward emergent attitude objects. *Journal of Applied Social Psychology*, *25*, 1611-1636.

Stern, P. C., Dietz, T., Kalof, L., and Guagnano, G. (1995). Values, beliefs, and emergent social objects: The social–psychological construction of support for the environmental movement. *Journal of Applied Social Psychology*, in press.

Stern, P., and Gardner, G. (1981a). Psychological research and energy policy. *American Psychologist*, *36*, 329–342.

Stern, P., and Gardner, G. (1981b). The place of behavior change in the management of environmental problems. *Zietschrift fur Umweltpolitik*, Feb., 213–240.

Stern, P., and Gardner, G. (1981c). Habits, hardware, and energy conservation. Comment in *American Psychologist*, 1981, *36*, 426–428.

Stern, P., Dietz, T., and Guagnano, G. (1995). Religion and environmental concern: A reexamination of the White hypothesis. Paper presented to the annual meeting of the Rural Sociological Society, August, Arlington, VA.

Stern, P., Young, O., and Druckman, D. (1992). (Eds.), *Global environmental change: Understanding the human dimensions*. Washington, DC: National Academy Press.

Stern, P.C. (2000). Toward a coherent theory of environmentally significant behavior. *Journal of Social Issues, 56*(3), 407–424.

Stern, P.C., Dietz, T., Abel, T., Guagnano, G.A., and Kalof, L. (1999). A value-belief-norm theory of support for social movements: The case of environmentalism. *Human Ecology Review, 6*, 81–97.

Stevens, W. (1991). What really threatens the environment? *New York Times* (Jan. 29), p. B7.

Strauss, K. (1985). Pulling the plug on nuclear power. *Chemical and Engineering News* (July 29), pp. 30–31.

Suzuki, David. (1994). Quoted by Allen, 1994, "Its a matter of survival." *Amherst, 46*(4), 17–21.

Svenson, O. (1981). Are we all less risky and more skillful than our fellow drivers? *Acta Psychologica, 47*, 143–148.

Svenson, O., Fischhoff, B., and MacGregor, D. (1985). Perceived driving safety and seatbelt usage. *Accident Analysis and Prevention, 17*, 199–133.

Syme, G. J., and Eaton, E. (1989). Public involvement as a negotiation process. *Journal of Social Issues, 45*(1), 87–107.

Tanner, J., Day, E., and Crask, M. (1989). Protection motivation theory: An extension of fear appeals theory in communication. *Journal of Business Research, 19*, 267–276.

Tanner, J., Hunt, J., and Eppright, D. (1991). The protection motivation model: A normative model of fear appeals. *Journal of Marketing, 55*, 26–45.

Taylor, S., and Brown, J. (1988). Illusion and well-being: A social psychological perspective on mental health. *Psychological Bulletin, 103*, 193–210.

Taylor, S., and Thompson, S. (1982). Stalking the elusive "vividness" effect. *Psychological Review, 89*, 155–181.

Thompson, E. L. (1978). Smoking education programs, 1960–1976. *American Journal of Public Health, 68*, 250–57.

Tierney, J. (1990). Betting the planet. *New York Times Magazine* (Dec. 2), pp. 52–53ff.

Toffler, A. (1970). *Future shock*. New York: Bantam Books.

Townsend, J. (1992). Chaos theory: A brief tutorial and discussion. In Healy, A., Kosslyn, S., and Shiffrin, R. (Eds.), *From learning theory to connectionist theory: Essays in honor of William K. Estes*. Hillsdale, NJ: Lawrence Erlbaum.

Toynbee, A. (1973). *The history and culture of China and Japan*. New York: Holt, Rinehart & Winston.

Tuan, Y. (1970). Our treatment of the environment in ideal and actuality. *American Scientist, 58*, 244–249.

Tversky, A., and Kahneman, D. (1973a). Availability: A heuristic for judging frequency and probability. *Cognitive Psychology, 5*, 207–232.

Tversky, A., and Kahneman, D. (1973b). Anchoring and calibration in the assessment of uncertain quantities. Oregon Research Institute Research Bulletin.

Tversky, A., and Kahneman, D. (1981). The framing of decisions and the psychology of choice. *Science, 211*, 1453–1458.

Tversky, A., and Kahneman, D. (1982). Belief in the law of small numbers. In Kahneman, D., Slovic, P., and Tversky, A. (Eds.), *Judgement under uncertainty: Heuristics and biases*. Cambridge, England: Cambridge University Press.

Tyler, T., and Cook, F. (1984). The mass media and judgments of risk: Distinguishing impact on personal and societal level judgments. *Journal of Personality and Social Psychology, 47*, 693–708.

U.S. Department of Energy. (2000). *Regional Energy Profile: Middle Atlantic Data Abstract*. Washington, D.C.: Energy Information Administration.

U.S. Department of Energy. (2002). *Annual Energy Review 2000*. Washington, D.C.: Energy Information Administration.

Udall, J. R. (1985). The Tucson paradox. *Audubon Magazine* (Jan.), pp. 98–99.

Ulrich, R. (1981). Natural versus urban scenes: Some psycho–physiological effects. *Environment and Behavior, 13*, 523–556.

Ulrich, R. (1984). View through a window may influence recovery from surgery. *Science, 224*, 420–421.

Ulrich, R. (1993). Biophilia, biophobia, and natural landscapes. In Kellert, S., and Wilson, E. (Eds.), *The biophilia hypothesis*. Washington, DC: Island Press/Shearwater Books.

Union of Concerned Scientists. (1992). World scientists' warning to humanity. Union of Concerned Scientists, 26 Church St., Cambridge, MA 02238.

United Nations Population Fund (2001). *The state of world population 2001*. New York: The United Nations.

Van der Velde, F., and Van der Pligt, J. (1991). AIDS–related health behavior: Coping, protection motivation, and previous behavior. *Journal of Behavioral Medicine, 14*, 429–451.

Van Liere, K. D., and Dunlap, R. E. (1979). Moral norms and environmental behavior: An application of Schwartz's norm activation model to yard burning. *Journal of Applied Social Psychology, 8*, 174–188.

Vaughan, E., and Seifert, M. (1992). Variability in the framing of risk issues. *Journal of Social Issues, 48*(4), 119–135.

Visaria, P., and Visaria, L. (1981). India's population: Second and growing. *Population Bulletin, 36*(4), whole issue.

Vitousek, P.M., Aber, J., Howarth, R.W., Likens, G.E., Matson, P.A., Schindler, D.W., Schlesinger, W.H., and Tilman, G.D. (1997). Human Alteration of the Global Nitrogen Cycle: Causes and Consequences. *Ecological Applications, 7*(3), 737-750.

Waddington, C. H. (1977). *Tools for thought: How to understand and apply the latest scientific techniques of problem solving*. New York: Basic Books.

Wade, C., and Tavris, C. (1990). *Psychology, 2nd. Edition*. New York: Harper Row.

Wade, R. (1994). *Village Republics: Economic Conditions for Collective Action in South India*. San Francisco, CA: Institute for Contemporary Studies Press.

Walker, J. M. (1979). Energy demand behavior in a master–metered apartment complex: An experimental analysis. *Journal of Applied Psychology, 64*, 190–196.

Washburn, S., and Lancaster, C. (1968). In Lee R., and Devore, I., *Man the hunter*. Chicago: Aldine Publishing Co.

Webler, T. (1994). Experimenting with a new democratic instrument in Switzerland: Siting a landfill in the eastern part of Canton Aargau. Draft final report. Zurich: Polyproject on Risk and Safety of Technological Systems, Swiss Federal Institute of Technology, June.

Weinstein, N. (1978). Cognitive processes and information seeking concerning an environmental health threat. *Journal of Human Stress, 4*, 32–41.

Weinstein, N. (1980). Unrealistic optimism about future life events. *Journal of Personality and Social Psychology, 39*, 806–820.

Weinstein, N. (1983). Reducing unrealistic optimism about illness susceptibility. *Health Psychology, 2*, 11–20.

Weinstein, N. (1984). Why it won't happen to me: Perceptions of risk factors and susceptibility. *Health Psychology, 3*, 431–457.

Weinstein, N. (1987). Unrealistic optimism about susceptibility to health problems: Conclusions from a community–wide sample. *Journal of Behavioral Medicine, 10*, 481–500.

Weinstein, N. (1989). Effects of personal experience on self–protective behavior. *Psychological Bulletin, 105*, 31–50.

Weinstein, N., and Sandman, P. (1992a). A model of the precaution adoption process: Evidence from home radon testing. *Health Psychology, 11*, 170–180.

Weinstein, N., and Sandman, P. (1992b). Predicting homeowners' mitigation responses to radon test data. *Journal of Social Issues, 48*, 63–83.

Weinstein, N., Grubb, P., and Vautier, S. (1986). Increasing automobile seat belt use: An intervention emphasizing risk susceptibility. *Journal of Applied Psychology, 71*, 285–290.

Weinstein, N., Sandman, P., and Roberts, N. (1990). Determinants of self–protective behavior: Home radon testing. *Journal of Applied Social Psychology, 20*, 783–801.

White, Jr., L. (1967). The historical roots of our ecologic crisis. *Science, 155*(3767), 1203–1207.

Whitney, E. (1993). Lynn White, ecotheology, and history. *Environmental Ethics, 15*, Summer, 151–169.

Wilson, E. (1975). *Sociobiology: The new synthesis*. Cambridge, MA: Harvard University Press.

Wilson, E. (1978). *On human nature*. Cambridge, MA: Harvard University Press.

Wilson, E. (1984). *Biophilia: The human bond with other species*. Cambridge, MA: Harvard University Press.

Wilson, E. (1993a). Biophilia and the conservation ethic. In Kellert, S., and Wilson, E. (Eds.), *The biophilia hypothesis*. Washington, DC: Island Press/Shearwater Books.

Wilson, E. (1993b). Is humanity suicidal? *New York Times Magazine* (May 30).

Wilson, R. (1979). Analyzing the daily risks of life. *Technology Review, 81*, 40–46.

Winett, R. A., and Neale, M. S. (1979). Psychological framework for energy conservation in buildings: Strategies, outcomes, directions. *Energy and Buildings, 2*, 101–116.

Winett, R. A., Hatcher, J. W., Fort, T. R., Leckliter, J. N., Love, S. Q., Riley, A. W., and Fishback, J. F. (1982). The effects of videotape modeling and daily feedback on residential electricity conservation, home temperature and humidity, perceived comfort, and clothing worn: Winter and summer. *Journal of Applied Behavior Analysis, 15*, 381–402.

Winett, R., Leckliter, I., Chinn, D., Stahl, B., Love, S. (1985). Effects of television modeling on residential energy conservation. *Journal of Applied Behavior Analysis, 18*, 33–44.

Winkler, R. C., and Winett, R. A. (1982). Behavioral interventions in resource conservation: A systems approach based on behavioral economics. *American Psychologist, 37*, 421–35.

Witmer, J. F., and Geller, E. S. (1976). Facilitating paper recycling: Effects of prompts, raffles, and contests. *Journal of Applied Behavior Analysis, 9*, 315–322.

Wolf, E. (1988). Avoiding a mass extinction of species. In Brown, L., et al., (Eds.), *State of the world 1988*. New York: W. W. Norton.

Woodhouse, E. (1980). The supply and demand of political and technological competence. Unpublished graduate paper, Yale University.

World Resources Institute. (1992). *World Resources 1992–93*. New York: Oxford University Press.

Wright, R. (1994). Population bomb fuse still sizzles. *Los Angeles Times*, reprinted in the *Ann Arbor (MI) News* (Aug. 24), p. A–1.

Wurtele, S., and Maddux, J. (1987). Relative contributions of protection motivation theory components in predicting exercise intentions and behavior. *Health Psychology, 6*, 453–466.

Yates, S. (1982). Using prospect theory to create persuasive communications about solar water heaters and insulation. Unpublished doctoral dissertation, University of California, Santa Cruz, CA.

Yi, Y. (1992). Affect and cognition in aesthetic experiences of landscapes. Unpublished doctoral dissertation, Department of Landscape Architecture and Urban Planning, Texas A & M University.

Young, J. (1991). Reducing waste, saving materials. In Brown, L. et al. (Eds.), *State of the World 1991*. New York: W. W. Norton.